IEE HISTORY OF TECHNOLOGY SERIES 32

Series Editors: Dr B. Bowers
Dr C. Hempstead

Communications: An International History of the Formative Years

Other volumes in this series:

Communications: An International History of the Formative Years

Russell Burns

The Institution of Electrical Engineers

Published by: The Institution of Electrical Engineers, London,
United Kingdom

British Library Cataloguing in Publication Data

Burns, R. W. (Russell W.)
   Communications : an international history of the formative
   years. – (IEE history of technology series ; 32)
   1. Telecommunication – History
   I. Title, II. Institution of Electrical Engineers
   621.3'82'09

ISBN 0 86341 327 7

Typeset in the UK by RefineCatch Ltd, Bungay, Suffolk
Printed in the UK by MPG Books Ltd, Bodmin, Cornwall

Contents

Acknowledgements

Marcus Tullius Cicero (106–43 BC) was in no doubt about the pertinent points that must be followed if a reliable history is to be written:

The first law for the historian is that he shall never dare utter an untruth. The second is that he shall suppress nothing that is true. Moreover, there shall be no suspicion of partiality in his writing, or of malice. (De Oratore II, 62)

The author has endeavoured to follow Cicero's wise counsel and has based, where possible, this book on primary source documents held in archive centres and specialist libraries. (Histories founded on the spoken word should always be treated with some circumspection.) Fortunately a vast amount of documentation on communications exists and since a great deal of my time and effort was spent in archive centres and libraries I should like to express to their staffs my gratitude for the courtesy and support which I received from them. The principal depositories and materials used were the BBC's Written Archives Centre, Caversham; the Post Office Archives and Records Office, London; Marconi Historical Archives, Chelmsford; the Cable and Wireless Archives, London; the Public Record Office, Kew; the archives of the Imperial War Museum; the archives of the Royal Television Society; the archives of Electric and Musical Industries, Hayes; the holdings of the British Newspaper Library, London; the archives of the Institution of Electrical Engineers; the archives of Bell Telephone Laboratories and of AT&T, Murray Hill and New York, USA; the archives of the Radio Corporation of America, Princeton, USA; the archives of BT, London; the National Archives, Washington, USA; the National Record Centre, Suitland, USA; and the personal archives of Mr R.M. Herbert, Sanderstead. In addition much research was undertaken in the libraries of the Institution of Electrical Engineers, the University of Nottingham, Nottingham Trent University, the Science Museum, the British Library, Hastings Public Library and Hastings Museum and Art Gallery. To undertake research work in such comfortable and helpful surroundings was always a pleasure, and the staffs of the above establishments were ever unfailing in their endeavours to be of assistance. I am most indebted to them.

Especially warm thanks are due to all those persons who provided me with photographs and illustrative material, and who gave me – in all cases most willingly – permission on behalf of themselves or their organisations to use the material in this book. To the following I am deeply appreciative of their generosity: the late Mr P. Leggatt, formerly Chief Engineer, external relations, BBC; the late Mr J.A. Lodge, formerly achivist of Thorn EMI's Central Research Laboratories; Mr G.D. Speake, formerly Deputy Director of Research, GEC Research (Marconi Research Centre); Mr S.L. McCrearie, formerly Service Director of Radio Rentals; Mr G. Schindler, formerly of AT&T Bell Telephone Laboratories; Mrs E. Hart, archivist of the Illustrated London News Picture Library; Mrs E. Romano, formerly of the Photo Centre, AT&T; Mr A. Pinsky, formerly of Scientific Information Services, RCA; Miss C. Colvin, archivist of the Royal Television Society; Mr R.M. Herbert; Mr P. Trevitt, librarian of Quinetiq, Malvern; and the Directors of the Imperial War Museum and the Admiralty Library.

For reasons mentioned in the preface, many quotations from both written primary and secondary sources have been included in this book. I am most grateful to the following persons for giving me permission to use extracts from their organisations' publications or private papers: the editors of the *Daily Telegraph*, the *New York Times*, *The Times*, the *Daily Express*, the *Evening Standard*, *Nature, Electronics and Wireless World*, the *Hastings and St Leonards Observer*, the *Sunday Express*, and the journals of the Institution of Electrical Engineers, and the Institute of Physics.

In a few cases, because of the deaths of the copyright holders, and the unknown whereabouts of their dependants, it was not possible to obtain permission to use certain quotations.

Preface

'History is more or less bunk', said Henry Ford. He was wrong: history is fascinating. It is about discovery.

The question may be asked, 'Why study/read communications history?' There are several reasons. First, there is the fascination of discovering the rationale, the decisions, the constraints, the endeavours that led to the implementation of new systems and devices, and new laws and theories which have had a bearing on everyday life: and also of learning about the scientists and engineers whose work has changed the state of civilisation. Second, the study of history enables a perspective to be given to events which occurred serially. When a new country or city is visited, it is usual for a traveller to commit to memory the main features or landmarks of the national or local geography so that a freedom is acquired which allows easy movement from one place to another. However, the members of the human race are travellers in time as well as space; and so a knowledge of events that have unfolded, not only in a person's lifetime but also in the past, is necessary for a full appreciation and understanding of the world. Third, lessons can be learnt from historical studies. Cicero expressed the arguments for studying history elegantly and succinctly when he wrote: 'History is the witness that testifies to the passing of time; it illumines reality, vitalises memory, provides guidance in daily life and brings us tidings of antiquity.'

Today there is much interest in history. The numerous televised programmes dealing with the subject confirm the premise that history is engrossing. Among its various branches that of the history of communications is extensive; it is relevant, it is colourful, it is multifarious, and it is remarkable.

Communications, an international history of the formative years is an attempt to tell how communications have evolved from the time of Aeschylus's Agamemnon (458 BC) to c. 1940 – a convenient breakpoint. The book is not a descriptive chronicle of dates and events; nor is it a descriptive catalogue of devices and systems. Rather it is about discovering the essential factors – technical, political, social, economic and general – that have enabled modern communications to evolve from early primitive stages of development. The history is concerned with narrative, with characters, and with stories. It is also concerned with analysis and synthesis.

A contextual approach has been adopted for this history to stress the influence of one discipline upon another, and some pains have been taken to ensure that the events portrayed have been presented in a coherent, chrono-logical and integrated manner.

Since the author's writings on the history of technology are quite extensive, selected extracts from these have been used in the present book. In particular, the chapters on television are based on material included in his award winning book, *Television, an international history of the formative years* (IEE, London, 1998).

In writing any history it is of course important that the events described are interpreted from a contemporaneous position. Considerable endeavour has been taken to present the views of scientists and engineers, newspaper reporters and editors, cartoonists and others so that the progress of communications is seen from the perspective of the times and not from the standpoint of a later generation. Accordingly, the unfolding saga has been profusely illustrated by contemporary quotations.

Much care and effort has been spent to ensure that an accurate history has been written. Many hundreds of references to primary and secondary sources have been detailed, and the text contains many diagrams, photographs and tables of data.

Finally, the author would like to acknowledge that there is some truth in the statement: 'To understand a science it is necessary to know its history' (Auguste Comte).

Abbreviations

a.c.	alternating current
ATC	Atlantic Telegraph Company
AT&T	American Telephone and Telegraph Company
BBC	British Broadcasting Corporation
BIT	Baird International Television Ltd
BTDC	Baird Television Development Company
BTL	Baird Television Ltd
BTL	Bell Telephone Laboratories
c.r.o.	cathode ray oscillograph
c.r.t.	cathode ray tube
d.c.	direct current
ECSTC	English Channel Submarine Telegraph Company
EMI	Electric and Musical Industries Ltd
FRC	Federal Radio Commission
GE	General Electric (US)
GPO	General Post Office (UK)
HF	high frequency
HMV	His Master's Voice
M–EMI	Marconi–EMI Television Company Ltd
MWT	Marconi (or Marconi's) Wireless Telegraph Company Ltd
PMG	Postmaster General
RCA	Radio Corporation of America
TAC	Television Advisory Committee
TC&M	Telegraph Construction & Maintenance Company
TSC	Technical Subcommittee (of the TAC)
UHF	ultra high frequency
VHF	very high frequency
WEM	Western Electric and Manufacturing Company
WT	Wireless Telegraphy

Chapter 1

Communication among the ancients

Opportunity is of great advantage in all things, but especially in war; and among the several things which have been invented to enable men to seize it, nothing can be more conducive to that end than signals. (Polybius)

Signalling has long played an important role in warfare. It serves to provide the means for transmitting information from reconnaissance and other units in contact with an enemy, and the means for exercising command by transmitting the orders and instructions of commanders to their subordinates. But although good signal communications for military purposes had been considered to be of prime importance for at least 2,500 years it was only as recently as the end of the 18th century that an effective long range system of communications, able to transmit any message without the despatch of runners or riders, was invented. This system, due to Claude Chappe, employed semaphores worked by pulleys and levers to make the signals. Chappe's telegraph system was of crucial importance to the French Government and to Napoleon during the Revolution and the Napoleonic wars. In 1793–1794 the position of France was desperate. The wars of the French Revolution were at their height. France's frontiers were assailed by the Allied Forces of England, Holland, Prussia, Austria, Spain, the Italian States, Hanover and Hesse. Marseilles and Lyons had revolted against the French Government: an English fleet, on 28th August 1793, held Toulon. Between France and perdition there stood several barriers: first the incoherence of the Allies, each having different objects in view; second, the single-mindedness of purpose that Carnot (1796–1832) had succeeded in establishing among all units of the French Army; third, the levee-en-masse which France, the most populous country in Europe, would send into the battlefield; and, fourth, Chappe's telegraph, an apparatus which by a 'few simple motions served to communicate the result of a siege, or of the battle, with the accuracy, if not the minuteness, of a despatch, and with a celerity that in some measure rivalled the progress of sound' [1].

Napoleon was greatly impressed by the speed of signalling possible with the Chappe semaphore and had the system extended so that by 1852 more than

4,800 km of signal lines were in existence. The Emperor knew well the importance of good communications: 'Do not awake me when you have good news to communicate; with that there is no hurry. But when you bring bad news, rouse me instantly, for then there is not a moment to be lost' [2].

The situation that faced the French Government during the last decade of the 18th century was certainly not unique in the history of the world. During the lifetime of Sun Tzu, perhaps the greatest of the ancient Chinese military experts and supposedly a contemporary of Confucius, China was experiencing one of the most dynamic periods of her history. Five of the six major states contested savagely for the empire: 'war was not sporadic, it was endemic' [3]. It consumed the energies and resources of the states implacably, as their rulers attempted in turn to establish hegemony – to 'roll-up All-Under-Heaven like a mat and put the four seas in a bag' [4].

Sun Tzu in his 'Ping Fa', (*The art of war*), written in 550 BC, observed:

The control of large numbers is like that of small numbers – we sub-divide them. If we use the drum, the bell, and the flag, it is possible to control large forces at a distance the same as small forces. According to old books of war, the drum and bell are used as the voice will not reach and flags are used to assist in seeing. The use of bells, drums, banners and flags is to unite and attract the eye and ear [5].

In early historic times, as in modern times, there was a need for long-distance communication as well as for battlefield signalling. Emperors, generals, and the privileged members of society had to maintain control and conduct diplomatic, military, trade and governmental affairs within their, sometimes widespread, empires. Cherry has argued that 'societies can develop and "advance" only so fast as they can develop means of acquiring, recording and disseminating information' and that the progression from the earliest communities to the highly organised industrial states of today is 'one long story of improved means of communication' [6]. The word communication comes from the Latin communico, 'I share'. With its development so the benefits of social sharing – the production of wealth, state protection, democracy and culture – have increased.

Ancient civilisations, particularly those of the Greeks and of the Romans, have left quite extensive accounts of their military and naval engagements, but few accounts of the detailed communication means by which these engagements were strategically and tactically planned and undertaken are extant.

The primary writings on the subject are those of Homer (probably earlier than 700 BC), Herodotus (490–424 BC), Thucydides (471–411 BC), Xenophon (431–355 BC), Aeneas the Tactician (c. 340 BC), Polybius (204–122 BC), Caesar (102–44 BC), Titus Livius or Livy (59 BC–17 AD), Pliny the Elder (23–74 AD), Ammianus Marcellinus (330–400 AD) and Vegetius (383–450 AD) [7].

Ancient Greek literature in style is quite unlike modern English literature.

The Greek writer did not look out on the world around him and write directly of what he saw, and of the emotions with which he saw it. His instinct was to look behind the particular event, and to seek the general tendency or law which it exemplified [8].

Thus in the *Iliad*, which tells of the exploits of Achilles and Agamemnon, Homer does not describe Troy or its surroundings, or the Greek army, or the beginning or the end of the war, or even the way in which Achilles met his death. The Greek poets wrote not for individuals but for public performances. 'It was natural therefore that [the poets] should not write about social life, of the subtle relations of individuals with each other, but of themes of moral or political importance, situations that illustrate some big fact in human existence – the relations (as the Greeks put it) between man and the gods.' [8] The poet would not describe an event in the manner of a modern scientist, but rather, would highlight 'the human courage, skill, glory and their opposites – and the transience of human life'. As a consequence, the extant knowledge of the signalling systems of the ancient Greeks is meagre and not wholly satisfactory.

Among the works of the classical authors who wrote on military subjects, where signalling inventions have been described, the following are of some importance: the *Histories* of Herodotus, of Polybius, and of Caesar; the *Art of war* by Aeneas; the *Duties of a general* by Onosander; the *Stratagems* of Frontinus, and of Polyaenus; the *Tactics* of Aelian, of Arrian, of Leo Imperator, and of Constantinus; the *Tract* ascribed to Cicero; and the four books of Vegetius, *Epitome Rei Militaris* (published between 384 to 395 BC). This last work is especially important since Vegetius collected his materials from the commentaries of Cato, Celsus, Trajan, Hadrian and Frontinus, none of which, except the last, have survived [9].

Most of the authors mention some form of fire signals. Other forms of signalling narrated include the use of flags and pennants [10], trumpets, horns [11] and drums, burnished shields, carrier pigeons [12], runners and pony riders [13,14], smoke signals and lights.

Both Greek and Roman coins display flags and pennants, and the Greek and Roman gods Hermes and Mercury were associated with communications, an indication of the subject's considerable importance. Hermes was the messenger of the Greek gods. Following the conquest of the Greeks by the Romans, and the adoption of the culture and legends of Greece, Hermes became Mercurius, or Mercury. He was depicted with a petasus (the felt hat worn by travellers) on his head and with a caduceus, or herald's staff, which is the symbol of a message, in his hand. The staff was entwined with white ribbons, which indicated peace, and serpents which conferred immunity. On his feet he wore golden sandals with wings for swiftness. Mercury is the emblem of the British army's Royal Corps of Signals, and the caduceus used to be incorporated in the logo of the Institution of Electrical Engineers (UK).

Homer's *Iliad* contains the earliest reference to two important methods of signalling – by beacon fires and by beacon smoke – employed by the ancients in communicating to distant places [15]. The methods are simple, easily implemented, and obvious and have been utilised from the age of Homer to the present times. Their use in Greece and the neighbouring lands controlled by the Greeks was especially effective because of the geography of Greece itself and of the surrounding seas with their numerous islands. The many mountain peaks and the clear atmosphere of the region afforded ideal conditions for the operation of beacon fires and beacon smoke signals (Fig. 1.1).

Herodotus, who wrote a history of the Persian wars in nine books [16], cites several instances of the use of signals during the conflict between Greece and Persia. Following the revolt of the Ionian cities against Persia, Mardonius, son-in-law of Darius the Great, the Persian king, invaded Thrace in 492 BC but his fleet was destroyed in a storm near Mount Athos. Two years later (490 BC) an armada of 600 ships conveyed the forces of Datis and Artaphernes from Samos to Attica, only to meet with disaster in the battle of Marathon. Tidings of the Greek victory were sent, by the first well-known messenger, Pheidippides, who reportedly ran the 40 mile (64.3 km) distance from Marathon to Athens in two days. Darius died in 485 BC and a third campaign was then

Figure 1.1 A fire beacon site in Greece.
Source: A. Belloc, *La telegraphie historique* (Paris, 1894).

undertaken by his son Xerxes in 480 BC. Landing on the Thracian coast Xerxes and his army reached Thessaly without encountering any opposition but at Thermopylae, Leonidas, the Spartan king, attempted to halt the advance while the Athenian fleet maintained station at Artemisium. After the fall of Thermopylae, the Greek ships retreated to the Piraeus followed by the Persian fleet. Thermistocles by a stratagem induced the Persians to enter the narrow straits between the island of Salamis and Attica where they were unable to manoeuvre. The Persian fleet was overwhelmed and Xerxes, who had occupied Athens, then withdrew to the Hellespont leaving Mardonius (Xerxes' field commander) in control in Greece. However, in 479 BC the Greeks, under Pansanias, were victorious at Plataea and the remnants of the Persian fleet were destroyed at Mycale.

Herodotus cites two instances of the use of beacon signals during this conflict. The first was when the vanguard of the fleet of Xerxes had set out from Therma, and joined three Greek ships on picket duty far up the Magnesian coast. News of the event was transmitted by beacon fire from Sciathus to the Greek fleet lying at Artemisium, a distance of less than 20 miles (32 km). The second instance occurred one year later after Xerxes had fled to Asia and had taken up his headquarters at Sardis. Mardonius informed the king by means of a line of beacons placed at intervals throughout the islands that he had captured Athens [17].

Herodotus also wrote that the Persians preferred the reliability of messengers and relays of horses and riders to beacon fires. He ascribed the invention of this mode of news transmission to Xerxes:

Whilst Xerxes was thus employed, he sent a messenger to Persia, with intelligence of his defeat. Nothing human can exceed the velocity of these messengers. It is thus arranged by the Persians that as many days as are required to go from one place to another, so [that] many men and horses are regularly stationed along the road, allowing a man and a horse for each day. Neither snow, rain, heat nor darkness are permitted to obstruct their speed: the first messenger delivers his business to the second, the second to the third, as the truth is handed about amongst the Greeks at the defeat of Vulcan. This mode of conveying intelligence the Persians call 'angareion' [18].

The first organised system of voice relaying using loud-voiced soldiers stationed on hilltops (Fig. 1.2) seems to be that introduced by Darius. Cleomedes de Mundo in his description of the world relates that:

The Persian when making war in Greece, is reported to have had stationed [men] from Athens to Susa, in such manner, that he was able by the voice, to send intelligence into Persia, the intermediate persons repeating the words from one to another. A report, after passing this successive repetition, is known to have been received from Greece into Persia [c. 720 km] in two days and nights [19].

Caesar in the 7th book of his commentaries mentions that the same expedient was used in Gaul:

Figure 1.2 *The megaphone was one of the earliest devices for transmitting intelligible speech over a distance.*
Source: AT&T.

The report quickly spread over the Gallic states (as when any event of great importance happens they send the intelligence through the country by loud acclamations, which are repeated by those inhabitants who chance to hear them); for what was done in Orleans at sun-rise was known on the borders of Auvergne before the close of the first watch (about three hours), which is a distance of 160 miles (257 km) [20].

Aristotle also refers to signal transmission:

There were men stationed to run a day's journey; sentinels on the look out; messengers, guards, and watchmen at the signal towers: in particular, such was the great regularity of signals passed in succession, by fire or torches on the beacon towers, which were established from the extremities of the empire to Susa and Ecbatana, that the king received intelligence of anything new that happened in Asia the very next day it took place [9].

Herodotus's disclosure of a line of beacons placed at intervals through the islands has been questioned because after the battle of Salamis and the return of the Persian fleet to Asia in the preceding year, the Greeks acquired mastery of the sea. Rawlinson [16] has argued that the line must have passed along the European coast to Athos, and thence by Lemnos to Asia.

Probably the most well-known use of fires for signalling occurs in Aeschylus's epic poem *Agamemnon*. This was written in 458 BC and relates how the fall of Troy was reported to Clytemnestra, wife of Agamemnon, at Argos by means of a chain of fire beacons. If Aeschylus is to be believed the chain, in 1185 BC, stretched from Mount Ida behind Troy to the island of Lemnos, across the Aegean Sea to Mount Athos and thence down through Greece to Peloponnesus, a distance of over 298 miles (480 km) to Argos (Fig. 1.3). In Murray's translation [21] the poet says:

line 1   A Fire-god, from Mount Ida scattering flame.
          Whence starting, beacon after beacon burst
          in flaming message hitherward. Ida first
          told Hermes' Lemnian Rock, whose answering sign
          was caught by towering Athos, the divine,

line 13   That word hath reached Messapion's sentinels

line 32   Torch-bearer after torch-bearer, behold
          The tale thereof in stations manifold,
          Each one by each made perfect ere it passed,
          And victory in the first as in the last.
          These be my proof and tokens that my lord
          From Troy hath spoke to me a burning word.

Though the narrative is obscure in numerous details and presents certain difficulties it represents the first extended description of a system of signal communication that has features present in modern communication systems: first, the use of electromagnetic radiation to achieve the transmission of a message over a distance very much greater than could be achieved using sound signals from, for example, bells, drums, trumpets and horns; second, the use of relay stations to overcome the limitations of the line of sight propagation characteristic of electromagnetic radiations; and third, the positioning of the signal stations at elevated situations to achieve a good transmission link range.

Aeschylus's poem and the method of signalling narrated have been subjected to considerable analysis and comment. Verrall (1904) [22] opined that if the notion (that the beacons were watched night after night for ten years to enable the news

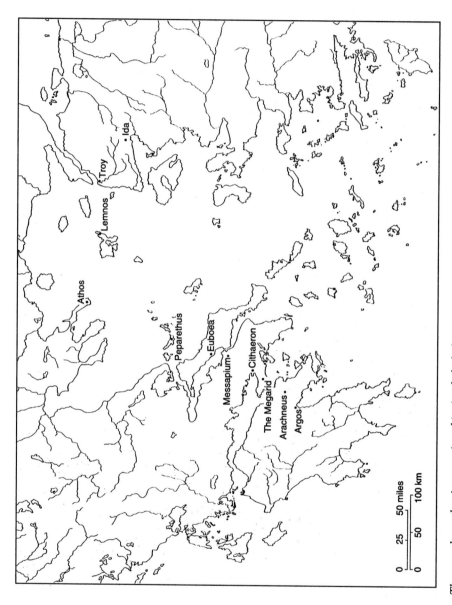

Figure 1.3   The map shows the sites mentioned in Aeschylus's Agamemnon.
Source: L. Solymar, *Getting the message* (OUP, 1999).

to reach Argos some weeks or some days sooner than it would do in any case) 'were not puerile enough, the natural facts are distorted so as to exaggerate the absurdity to the utmost'. His general objections, though serious, were 'nothing to the grotesque and wilful violations of nature which appear in the details'.

Merriman (1890) dealt at length with the physical realisability of Aeschylus's beacon system. He considered the distances between the stations and the heights of the various beacons and seems to have concluded without saying so explicitly that such a chain of beacons could have sent news of the fall of Troy to Argos. As he says: 'Aeschylus knew his land well' [23].

Darmstaedter [24], whose paper contains the heights and distances presented in Table 1.1, concluded that a system of fire signals was theoretically possible. The determining factors were the sizes and luminosities of the beacon fires and the state of the weather.

Even if the authenticity of Aeschylus's account of the fall of Troy is uncertain, there is no doubt that the method he described was in use for a great number of years, indeed until comparatively modern times. The earliest (588 BC) reliable record of messages being transmitted by the 'sign of fire' is to be found in the book of Jeremiah (vi, 1): 'O Yea children of Benjamin, gather yourselves to flee out of the midst of Jerusalem, and blow the trumpet in Tekoa [12 miles/19.3 km south of Jerusalem], and set up a sign of fire in Beth-haccerem; for evil appeareth out of the north, and great destruction' [26]. The Lachish Letters [27], dated 597 and 588 BC, which belong to Jeremiah's age, are particularly important as they make reference to signal stations. Thus from Letter IV:

And Semakhyahu, him has taken Sema'yahu and brought him up to the city [Jerusalem], and thy slave, my lord, shall write thither, (asking), where he is; because if in his turning [rounds] he had inspected, he would know, that for the signal stations of Lachish we are watching, according to all the signs which my lord gives, because we do not see (the signals of) Azeqah.

Table 1.1 The beacon sites mentioned by Aeschylus (according to Darmstaedter [24], but note Solymar's list [25]

Place	Height	Distance between the beacons
Ida	1770 m (5807 ft)	60 km (37 miles)
Teneclos	190 m (623 ft)	75 km (47 miles)
Lemnos	430 m (1411 ft)	69 km (43 miles)
Athos	1935 m (6348 ft)	200 km (124 miles)
Makistos	1209 m (3966 ft)	27 km (17 miles)
Messapin		36 km (22 miles)
Kitharon	1410 m (4626 ft)	27 km (17 miles)
Aigiplankton		45 km (28 miles)
Arachnaion		
Total length c. 540 km		

Such signals are also mentioned in Judges (20, 38) [28]: 'Now there was an appointed sign between the men of Israel and the liers in wait, that they should make a great flame with smoke rise up out of the city'.

One type of beacon fire is mentioned in the Mishnah Rosh Hashanah (II, 2–3) [29] which contains a full description of the massn'oth used to inform the Jews in exile of the appearance of the new moon in Jerusalem:

Before time they used to kindle fires, but after the evil doings of the Samaritans they enacted that messages should go forth. After what fashion did they kindle the flares? They used to take long cedarwood sticks and rushes and oleander wood and flax tow; and a man bound these up with a rope and went up to the top of the hill and set light to them and he waved them to and fro and up and down until he could see his fellow doing the like on the top of the next hill. And so, too, on the top of the third hill.

Although there are many references to the use of the art of signalling by fire in ancient books and treatises the method had a severe limitation in that it was necessary, for both the dispatcher of the signal and the receiver of it, to employ certain definite signals which had previously been agreed upon. As Polybius [30] pointed out:

it was possible by means of the signals agreed upon to send the information that a fleet had arrived at Oreus or Peparethos or Chalcis; but it was impossible to express that 'certain citizens had gone over to the enemy', or 'were betraying the town', or that 'a massacre had taken place', or any of those things which often occur, but which cannot be anticipated.

But it was precisely the unexpected occurrences which demanded immediate consideration and action.

The ancient Greeks were adept in the invention of signal systems and in 341 BC, during the time of Aristotle, Aeneas Tactitus developed a method of signalling that made some progress towards the elimination of the above limitation. Aeneas's system used two clepsydras (Fig. 1.4a) which were well known devices for measuring a certain length of time as a whole rather than its sub-divisions and the gradual passage of it. The water clock [31] itself was invented in Egypt in the second millennium but Aeneas seems to have been the first to adapt it for signalling purposes. In a preparatory treatise on strategy he says:

Let those who wish to communicate any matter of pressing importance to each other by fire signals prepare two earthenware vessels of exactly equal size both as to diameter and depth. Let the depth be three cubits, the diameter one. Then prepare corks of a little shorter diameter than those of the vessels: and in the middle of these corks fix rods divided into equal portions of three fingers breadth, and let each of these portions be marked with a clearly distinguishable line: and in each let there be written one of the most obvious and universal of those events which occur in war; for instance in the first 'Cavalry have entered the country', in the second 'hoplites', in the third, 'lightly-armed', in the next 'infantry and cavalry', in another 'ships', in another 'corn', and so on, until all the portions have written on them the events which may reasonably be expected to occur in the particular war [30].

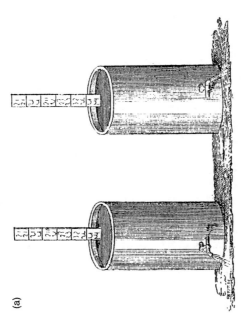

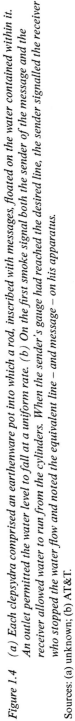

Figure 1.4 (a) Each clepsydra comprised an earthenware pot into which a rod, inscribed with messages, floated on the water contained within it. An outlet permitted the water level to fall at a uniform rate. (b) On the first smoke signal both the sender of the message and the receiver allowed water to run from the cylinders. When the sender's gauge had reached the desired line, the sender signalled the receiver who stopped the water flow and noted the equivalent line – and message – on his apparatus.

Sources: (a) unknown; (b) AT&T.

Each clepsydra was pierced near its base and fitted with a tap, and both clepsydras were filled with water: the rate of discharge of water was the same from the two earthenware jars. In this way both the corks, and the associated rods, would descend at the same rate when the taps were opened.

After this condition had been tested the two vessels were taken respectively to the two places from which the two parties intended watching for fire signals (Fig. 1.4b). Then:

As soon as any one of those eventualities which are inscribed upon the rods takes place, raise a lighted torch, and wait until the signal is answered by a torch from the others: this being raised, both parties are to set the taps running together. When the cork and rod on the signalling side has sunk low enough to bring the ring containing the words which give the desired information on a level with the rim of the vessel, a torch is to be raised again. Those on the receiving side are then at once to stop the tap, and to look at the words in the ring of the rod which is on a level with the rim of their vessel full. This will be the same as that on the signalling side, assuming everything to be done at the same speed on both sides [30].

Aeneas's invention is interesting from the point of view of the history of picture transmission and television as it represented the first attempt ever to use a synchronised system of transmitter and receiver. It was, of course, necessary that the rods in the two clepsydras should descend, not simply isochronously but synchronously. In all television and facsimile transmission systems it is essential that the receiver is synchronised to the transmitter. During the early days of the evolution of facsimile and television communication much effort was spent by inventors in developing methods by which this objective could be achieved. Both of Aeneas's clepsydras were 'free running', or to use modern jargon employed 'open-loop control', but with care in ensuring that the two discharge rates were the same the system could possibly have given reproducible results.

The only known reference in the classics to the operational application of clepsydras is given in the *Stratagems* of Polyaenus [32]. He recorded that when the Carthaginians were ravaging the island of Sicily and required urgent supplies from Lybia, they constructed two identical clepsydras, and fixed on them 'two dials having similar instructions, in those divisions where the figures are usually placed. In one division there it was written, "More transports are wanted". In a second, "More ships of war"; in another "Money" ' and so on. One clepsydra was retained in Sicily and the other was sent to Carthage. By working the system in a way similar to that described above Polyaenus wrote that: 'By this means the Carthaginians obtained a speedy supply of every store requisite for carrying on the war in Sicily'.

The invention was a slight advance on the use of beacons with a pre-conceived code but it was still quite indefinite. Clearly it was not possible to foresee all the contingencies that might arise or, even if one could, to write them on the rods:

and therefore, when anything unexpected in the chapter of accidents does occur, it is plainly impossible to communicate it by this method. Besides, even such statements as are written on the rods are quite indefinite; for the number of cavalry or infantry that have come, or the particular point in the territory which they have entered, the number of ships, or the amount of corn, cannot be expressed [30].

Surprisingly Polybius did not comment on the range of signalling that could be achieved by clepsydras. Unlike a chain of fire beacons where the beacon-to-beacon distance would be of the order of tens of miles in each suitable location, the use of Aeneas's method was essentially limited to short range working, for the indication of the onset and termination of the transmission was by a hand torch [33].

The objection of the indefiniteness of the message was overcome by a method (Fig. 1.5) originated by Cleoxenus and Democlitus and elaborated by Polybius.

It is as follows: divide the alphabet into five groups of five letters each (of course the last group will be one letter short, but this will not interfere with the working of the system). The parties about to signal to each other must then prepare five tablets each, on which the several groups of letters must be written. They must then agree that the party signalling shall first raise two torches, and wait until the other raises two also. The object of this is to let each other know that they are attending. These torches having been lowered, the signalling party raises first torches on the left to indicate which of the tablets he means: for instance, one if he means the first, two if he means the second, and so on. He next raises the torches on the right showing in a similar manner by their number which of the letters in the tablet he wishes to indicate to the recipient [34].

Figure 1.5   An alphabetical fire system was developed by the Greeks in c. 150 BC. It displayed combinations of torches representing various letters of the alphabet.
Source: AT&T.

The grouping of the letters would have been:

	1	2	3	4	5
1	α	ζ	λ	π	φ
2	β	η	μ	ρ	χ
3	γ	θ	ν	σ	ψ
4	δ	ι	ξ	τ	ω
5	ε	κ	ο	υ	

Thus to send the letter κ the signaller would raise two torches on the left to show the receiver that he must look at the second tablet, then he would raise five torches on the right, because κ is the fifth letter in the group.

Essentially Polybius's method involved scanning a two-dimensional array of letters at the transmitter, transmitting information about the position of a given element in the array, receiving the information, and scanning a two-dimensional array of letters at the receiver. The scheme he elaborated did not involve a synchronous scanning action of the arrays at the transmitter and receiver, unlike Aeneas's system

The scanning of two-dimensional rasters is a feature of all facsimile and television broadcasting systems but whereas in Polybius's scheme the scanning of the arrays of letters could be made in any fashion whatsoever, in modern picture transmission links synchronously operated regular scanning is used of the sequential or interlaced type.

In his account Polybius mentioned that the receiver of the message should have a stenoscope, (or sighting tube), with two funnels to enable him to distinguish between the right hand position and the left hand position of the signaller opposite him. 'Near this stenoscope the tablets must be fixed, and both points, to the right and to the left, must be defended by a fence 10 ft long and about the height of a man, in order to make it clear on which side the torches are raised and to hide them entirely when they are lowered.' (The telescope was not available during this period: its use would have greatly increased the range of working of the system.) The method was consequently limited to distances which allowed an observer using the naked eye to resolve the hand held torches, at the transmitter, into discrete objects. Forbes states that groups of torches at say three feet, (0.91m), distance of each other are no longer distinct groups of torches at distances over 0.6 miles (1 km) [35].

A significant point, given by Polybius, which no doubt limited the use of his scheme to emergencies only was the large number of torches, and presumably men employed. Forbes [35] considers that Polybius had to use up to ten men at each station but this figure seems rather high. However, there is no evidence that the system was ever practically employed in antiquity.

Apart from the disadvantages of the many men and stations required to transmit a message over a reasonable distance there was also the added objection, and danger, that an enemy could read the signals. Nevertheless the system put forward by Polybius was an improvement on that suggested by Aeneas. Failing an electric telegraph system the only progress antiquity could make was the development of schemes such as the Greek historian narrated.

Further improvement was suggested in the first half of the third century AD by Sextus Julius Africanus [36] (232–290 AD) an eminent Christian historical writer.

The Romans use a system, according to my opinion a most remarkable one, to tell each other all kinds of things by means of fire signals. They divide the places for signalling in such a way that they have fields in the middle, to the right and to the left; then they divide the letters in such a way that those from α to θ have their places to the left, those from ι to π in the middle, and those from ρ to ω the right. If they want to transmit the sign ρ, they raise the fire signal once to the right, for σ twice, for τ thrice, etc. If they wish to transmit, they raise the fire-brand once in the middle, etc. This they do to avoid transmitting the letters by numbers of fire signals. Those who receive the signals, write down the letters received in the form of fire signals and transmit them to the next station which then transmits them to the next station which then transmits them to the following one and so on until the last of the fire signalling stations.

In diagrammatic form the letters of the Greek alphabet, using Africanus's description of the Roman system may be written:

α	1	ι	1	ρ	1
β	11	κ	11	σ	11
γ	111	λ	111	τ	111
δ	1111	μ	1111	υ	1111
ε	11111	ν	11111	φ	11111
ζ	111111	ξ	111111	χ	111111
η	1111111	ο	1111111	ψ	1111111
θ	11111111	π	11111111	ω	11111111

where the number of short vertical lines denotes the number of times a torch is raised in the given position. Africanus's method was thus the first step on the way to a Morse code but about 2,000 years were to elapse before a person with one signalling apparatus could transmit information in this way.

This simple 'pyrseutic method' was an improvement on that of Polybius for by increasing the distance between the torches to say 10 m the message visibility could be increased to about 6.2 miles (10 km).

There seems to be no evidence that the Romans ever added an extensive system of fire signals to their very efficient cursus publicus [37] (Fig. 1.6). The only known example of a message being transmitted by fire signals was upon the assassination of Sejanus, when the news was transmitted from Rome to the Emperor Tiberius in Capri. Fire signals in wartime do occur in Roman war records and additional information comes from the 157 pictures of the marble scroll of Trajan's column in Rome, and the actual remains of signal stations.

Lino Rossi [38] in his book *Trajan's Column and the Dacian Wars* gives a description of the signal station illustrated on the column. (See Fig. 1.7.)

The scroll begins with a sequence of pictures which are taken by an observer travelling up the Danube on a barge coming from the harbours of the Black Sea. He is facing, in the right bank of the river, the plains of the Roman province of Moesia Inferior. It is March in the year 101 AD.

The flat bank, beyond the wavy water, is protected by a regular series of block-houses and watch towers, which are a local example of how the Roman lines throughout the Empire were defended. The block-houses are ashlar, one storeyed covered by a gabled roof, and surrounded by a circular palisade of pointed stakes. They have a single window over a rectangular door, which faces the gap in the palisade that gives entrance to each. The watch towers are also ashlar built and hip-roofed with two storeys. All round the upper floor runs a balcony with an upright railing and criss-cross slats, from the balcony's door there sticks out a long torch for signalling. These signal towers (burgi) are also surrounded by a square palisade. In the spaces between the buildings there are flares of straw, piled up in tiers on a pole, for signalling by smoke, and tall beacons made of logs, for signalling by fire. Auxiliary soldiers stand in the foreground to represent the garrisoning and patrolling of the lines itself.

Possibly the long torch was used for sending certain premeditated signals while the flares and beacons were used for basic signals in an emergency. (The torch systems of Polybius and Africanus could transmit any message but without the use of a telescope the systems would have been excessively expensive to have been used generally.)

The utilisation of beacons for giving advance warning of the approach of aggressors persisted for many hundreds of years after the Romans had left Britain. At Stirling, on the 13th October 1455, the Scottish Parliament enacted a law [39] stating which beacons were to be lit on the approach of invaders – the English. 'One beacon is a warning of the coming of what powers that ever they be of. Two beacons at once, they are coming indeed; four beacons, each one beside the other and all at once as four candles, shall be sooth fast knowledge that they are of great power'.

For the most part beacons were employed as an alarm signal in the United Kingdom and it is only in recent times that they have been used for the celebration of victories or public holidays. Many hundreds of beacon sites have been traced in Britain but it was not until medieval times that a system of beacons

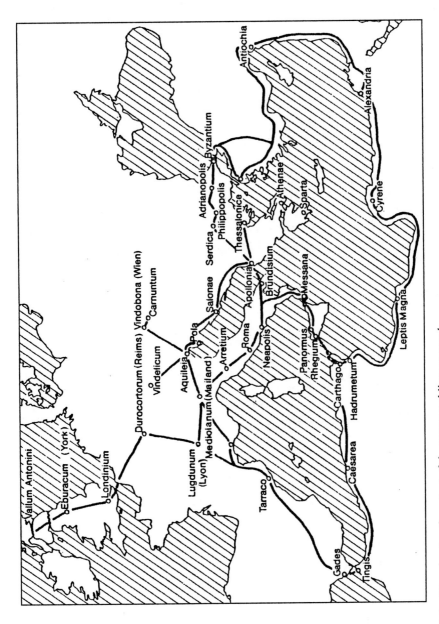

Figure 1.6   Map showing the extent of the cursus publicus network.
Source: R. Oberliesen, *Information, Daten und Signale* (Deutsches Museum, Rowohlt, 1982).

Figure 1.7   Roman watch towers as shown on Trajan's column.
Source: France Telecom, Archives et Documentation Historique.

was organised on anything like a national scale. Old records indicate that before the time of Edward III there were 'stacks of wood set up in high places which were fired when the coming of the enemies was descried'; but during the era of Edward III, 'pitch boxes, as now be, were set up, and this properly a beacon' [40]. The new type of beacon was more practical for signalling and it became the custom to erect it around the coasts: the higher hills inland, where the signals were not called upon so often to function, retained the primitive bonfires. Froissart in 1386 wrote that the beacon sites ranged from the Humber to Cornwall. They were constructed of platforms set on casks fixed with sand from which the watcher could see 'seven leagues across the water' and could give orders for the beacon to be lit. The chief combustible materials used were tow or flax and pitch, wood and charcoal being added to give more substance [41].

Responsibility for the beacons was in the hands of the King who acted through Orders in Council sent to the appropriate Sheriffs. In addition there was a 'Commission for beacons' acting under the Great Seal, which was empowered to levy 'beaconage' for watching and maintenance, as well as for the 'hoblers' or horsemen who acted as messengers.

While burdensome for some the order to erect beacons was, for those families which received patronage for this task from the Crown, considered to be an honour. The Belknap family, for example, seems to have held their duty in high esteem and adopted a crest which comprised a fiery beacon proper Or, on a Griffin Vert.

The strategically important southern counties of England, including the Isle of Wight, were of course well provided with beacon sites. An Inquisition held in 1324 at Shide by Carisbrook recorded 31 beacon sites. The most important posts were on the high downs of Bembridge and Freshwater on the eastern and western sides of the island. Here the beacons were positioned in groups of three; elsewhere, in the coastal regions pairs of beacons were employed, and inland the beacons operated as single units.

The kingdom's beacon warning system was not the outcome of haphazard growth but had been systematically planned and executed. Carefully prepared procedures were available to enable the beacon stations to signal the strength and imminence of possible attacks. For the posts on the Isle of Wight, if more than ten enemy ships were sighted one beacon on each of the Bembridge and Freshwater downs was lit: if more than ten ships were seen, or if there was the threat of an actual enemy invasion two fires at each group of three were lit. These had to be answered by the lighting of single beacons throughout the maritime regions. If all three beacons of the groups were fired the danger of conflict was very great and the entire inland system was alerted.

In England the cost of the beacon stations varied a good deal in practice but the responsibility for the upkeep and repair of the beacons was shared by the sheriff and the JPs or the lieutenancy, and the high constables. The duty of keeping watch fell to each village in rotation and the JPs were expected to select four reliable, prosperous men to watch, two by day and two by night. Each received 8d (3.3p) for 24 hours duty and each had to be armed, on pain of three days' imprisonment. Once at his post the watcher had to spend an uncomfortable and tedious night scanning the horizon for ships that rarely appeared. For example, the constables had to build a new shelter for the watchman at one of the Kent beacons, after the old hut had fallen down: it was to be 'noe costlie house but only to save them from the stormy weather, without any seates or place of ease lest they should fall asleepe; only to stand upright in with the hoale towards the beacon' [42]. Not surprisingly beacon watching was an expensive and unpopular business: as a Hampshire Deputy Lieutenant observed at the time, 'the Countrey doth somewhat murmure' [43]. The season for watching normally extended from spring to autumn, beginning about March and ceasing in October or November. Kent was unburdened on 8th October 1570, and nine maritime counties on 23rd October 1592 because the long, cold nights made it 'very tedyous and troublesome to the county, and it maie be forbourne for the winter season without anie danger' [43].

The system of warning by beacons reached its zenith during the reign of Queen Elizabeth I when Philip of Spain was contemplating an invasion of England. The most famous instance of its use was at the time of the Armada, for then it was employed to call the whole nation to arms. Within a few hours the beacons had carried the news from the Channel coastline to London, Wales and the rest of England. In his stirring poem 'The Armada' Lord Macaulay told how the message of the sighting was sent out:

> From Eddystone to Berwick bound, from Lynn to Milford Bay,
> That time of slumber was as bright and busy as the day.
> For swift to east and swift to west the ghastly war flame spread.
> High on St Michael's Mount it shone, it shone on Beachy Head.
> Far on the deep the Spaniard saw, along each southern shore,
> Cape beyond cape, in endless range, those twinkling points of fire.
>
> .   .   .   .
>
> From Surrey's pleasant hill flew those bright couriers forth;
> High on black Hampstead's swarthy moor they started for the north;
> And on, and on, without a pause, untired they bounded still;
> All night from tower to tower they sprang; they sprang from hill to hill [44].

The beacon system at this time was highly organised and efficient. The ambassador to Charles V, the Holy Roman Emperor and King of Spain, wrote to his King telling him that between 25,000 and 30,000 men could be mustered within 24 hours. A Privy Council note records that the Harwich beacons could call up 17,000 men from the Eastern Counties, a further 16,000 could be assembled at Portsmouth, drawn from eight counties; 17,000 at Plymouth, and another 11,000 at Falmouth. The men of Kent and Surrey apparently gathered at Tilbury, and in all over 70,000 could be marshalled to defend the country in addition to the regular troops [40].

The beacon communication system was augmented by the horse and foot messenger posts which each parish had to provide, and also by the pinnaces which roved at large to give early warning of enemy ships. In 1534 the French ambassador noted that there were always five or more ships sailing on reconnaissance duty around the coasts ready to signal to the watchmen on land, 'so that no foreign vessel could show itself without the whole country being warned'[43]. Fortunately the services of all these men were never called upon, for as Elizabeth herself declared, 'the breath of the Lord has blown, and the enemies were scattered' [43].

A similar warning system was re-introduced in 1803 when the forces of Napoleon Bonapart threatened to invade the British Isles; but although one or two beacons were fired by mistake, the invasion never materialised and there was no spectacular firing as there had been when the Armada was sighted [45].

And so for nearly 3,000 years from the time of the fall of Troy the only effective method of transmitting a signal speedily over long distances was by the use of fire beacons, and then only a signal of extremely limited information content.

Good communications lead to good organisation. For the ruler it is the means for the maintenance of law and order; for the scholar it is the means for the ordering and increase of knowledge. The Romans considered the creation of some form of public post absolutely necessary to serve the military and administrative needs of the government.

The *cursus publicus* was the state postal and messenger service of the Romans, and originated sometime during the 3rd century BC. It was created wholly for the transmission of governmental and administrative correspondence and enabled the emperors of Rome to govern their far-flung empire from the capital by correspondence alone: it was not a general facility which could be employed by the public. Such a facility was not introduced until about the 15th century, although commercial companies and wealthy men had their own staff of tabellari [46].

R.J. Forbes [13], in his 'Studies on ancient technology', mentions:

The emperor Augustus completely organised and centralised such earlier services. The Ptolemaic system seems to have inspired him for he tried to finance it by making the cities and population along the roads pay for it. It was first confined to the neighbourhood of Rome but gradually grew to embrace the whole empire under Trajan. Gradually the old system of financing became impossible and Hadrian made it a state organisation. Private tabellarii [initially] were not allowed to use its facilities, which served officials only who had to get an official passport (diploma). However, soon private travellers were allowed its facilities as a favour.

The service was strictly regulated and well-organised. Along the highways there were places (mutationes), situated approximately one day's journey apart 23 miles (c. 37 km), where horses could be changed, and hostels (mansiones) where the couriers and messengers could be accommodated and fed. When rivers and open seas had to be crossed, ferries, and fast sailing vessels (dromones, or cursioriae) were provided. The diploma rigidly prescribed the type and extent of the transport that the holder was allowed, as well as the hospitality to be accorded to him in a mansione.

Forbes has estimated the average speed and daily distance travelled of the imperial post as 5 miles (c. 8 km/h) and 47 miles (c. 75 km) respectively. On special occasions when urgent dispatches had to be sent the distance covered was sometimes as much as 93 miles (150 km); and the maximum ever reported was 149 miles (240 km). These figures may be compared with the c. 38 km per day which Cicero achieved in 51 BC. Sir William Ramsay [47] has opined that the messengers could only cover, on average, about 50 miles (80 km) a day, so that Constantinople could be reached from Rome in 24 days, and Alexandria in 54 days.

According to Forbes [13]: 'The regularity, certainty and speed of travel of the *cursus publicus* were not surpassed until the days of Napoleon.' (Pertinently, as late as 1858 it took 40 days for the news of the Indian Mutiny to reach Trieste for onward transmission by telegraph to London.)

For millennia communications were dependent on messengers and riders. A well documented system began in the 13th century with the students' messengers of the University of Paris, and led by stages to the State controlled postal and communications authority of France of c. 1600 (see end of chapter Note 1).

In England the origins of the messenger/postal service date from the end of the 15th century. As in other countries the letters carried were those associated with State business. The first holder of the office of Master of the Posts was Sir

Brian Tuke who was appointed in 1512. He was responsible for the general supervision of the service, the procurement of the necessary men and horses, and their keep.

Of prime concern to Elizabeth I during the period when a Spanish invasion was seriously expected was the defence of the south coasts of England and Ireland. There was a need to communicate by means of posts [48] as rapidly and reliably as possible between the Court and the ships maintaining a defensive watch. That such means were provided is apparent from the records (in the British Museum) of Lord Burghley, the Lord Treasurer. His private list of the posts to the sea gives details of 'postes from London to Holyhead'; and of 'postes from London to Bristoll'. He noted that 'the postes were laid towardes Ireland . . . both for the conveying of packetts and expedition of messengers'; and he described plans for 'Stages where postes are appointed to be layd between the Court and Portsmouth 1595' [48]. Meticulous records of the distances between the posts and of the times taken were kept by Lord Burghley. The consequence was a postal system which was reliable, regular, and permanent, although the general public, which was excluded from using the post, exercised elaborate schemes of deception to circumvent the restriction. The number of letters carried 'on the King's business' became so suspiciously large during the reign of James I that a Proclamation was issued commanding that no person should carry letters without a clear authorisation from the Master of the Posts.

This situation persisted until 1635. In that year Charles I in an important Proclamation stated:

his Majesty hath bene graciously pleased to command his servant Thomas Witherings Esquire his Majesties Postmaster of England for forraigne parts to settle a running Post or twoe to run night and day betweene Edenburgh in Scotland and the citty of London . . . and to take with them all such letters as shal be directed to any post towne or any place near any post towne [48]. (See Fig. 1.8)

And so for the first time in England private letters could be conveyed by the State postal system – a system which had developed to fulfil a national need for speedy communications at a time when the integrity of the realm was gravely threatened.

At the beginning of the 17th century a scientific instrument was invented, which advanced scientific knowledge by enormous strides and led eventually to the realisation of the first practical, long range, communication system using electromagnetic waves. The invention was the telescope; its impact on signalling is considered in the next chapter.

(a)

(b)

Figure 1.8   (a) A pony messenger. (b) A town crier propagating the news, 1682 AD.
Source: AT&T.

Note 1

The messengers of the University of Paris [49]

In 1279, when France was at war with England and Flanders, Phillippe IV gave an assurance to the Flemish students who were studying at the University of Paris that they and their messengers would be safeguarded against assault. Later, in 1314, the university, which had for many years attracted students not only from Paris but from other parts of Europe, was granted permission by Louis X (1289–1316) – who became King in 1314 – to provide messengers for all its students. Slowly this type of service was adopted, but on a smaller scale, by a number of French universities. The messengers undertook to carry letters to and from students, and money and packages parents wished to send to their sons.

At the University of Paris, messengers were classed as 'grands' or 'petits'. The grands messengers, who numbered one for each diocese represented by the students, were persons of some distinction. They 'took their recognised place in formal processions and functions and were granted many personal exemptions'. As the system of communications improved so the term 'grands messenger' became honorific; it was a title eagerly sought by the principal citizens of Paris. The real functionaries were the petits messengers. At first they carried only students' letters and packages, but meeting with no opposition, undertook similar services for persons not connected with the university.

The system seems to have worked so well that when King Louis XI (1423–83) ascended the throne in 1461 he issued an edict, dated 19th June 1464, which led to the establishment of the Couriers de France. It was the first move towards the inauguration of a state controlled postal and communications authority.

A description of the operations of the Couriers de France has been given by W.B. Parsons [50] in his book *Engineers and engineering in the Renaissance*. There, it is mentioned that the head of the Couriers was the Grand Master, a member of the royal household, who had authority to appoint the local masters of couriers. During King Louis XI's reign the number of couriers was fixed at 230, but his successors Charles VIII (1483–1498) and Louis XII (1498–1515) determined that 120 was a sufficient number.

Stations where 'four or five horses of light build, well saddled to run at a gallop', were to be maintained were established on the main roads of the kingdom at intervals of four lieues (10.64 miles). The master couriers in charge were required to mount all persons bearing the passport of the Grand Master, and to carry over their respective sections all royal dispatches that bore the Grand Master's certification. To assure rapidity and regularity of transmittal, each master courier was to note on the dispatches the hour of receipt from the preceding master and of dispatch to the succeeding [master], thus providing a complete registry of expedition and fixing responsibility for delay, if any occurred.

The establishment was to be confined to the royal service, but might under certain conditions be used by the 'very holy Father the Pope and foreign princes with whom his Majesty maintains amity and alliance.'

On the speed of transmission of a dispatch, it has been recorded that during the reign of Louis XII a royal messenger took just three days to travel from Milan to the King's chateau at Amboise, near Tours; and that a regular six-day service, using relays, was maintained between Paris and Rome. These times improved under Charles IX. News, from Paris, of the massacre of St. Bartholomew (1572) was received in Madrid three days later; and when the King died in 1574 his death was reported to his successor Henry III, then King of Poland, in Warsaw, 12 days afterwards.

On his accession, Henry III continued the policy of his predecessors to curb the privileges, franchises, customs and rights of the University of Paris's messenger service, and to create a central state-controlled authority for the dissemination of important information. The dispute, which is beyond the scope of this book, dragged on for c. 150 years. Eventually, by the end of the sixteenth century the university's feudal system was abandoned, and the state had sole rights to the postal and communications services.

References

1 Quoted in 'Pioneers of electrical communication – Claude Chappe', by R. APPLEYARD, Electrical Communication (Macmillan, London, 1930)
2 HEINL, R.D. (Ed.).: 'Dictionary of Military and Naval Quotations' (United States Naval Institute, 1966)
3 Entry on Sun Tzu, 'Encyclopaedia Britannica' (Encyclopaedia Britannica Inc., Chicago, 1911)
4 Entry on Sun Tzu, 'Collier's Encyclopaedia' (Collier's, New York, 1994)
5 SUN TZU.: 'Ping Fa' (The art of warfare), English translation and introduction by R.T. Ames (Ballantine Books, New York, 1993)
6 CHERRY, C.: 'On communication before the days of radio', *Proc. IRE*, Section 17, Information Theory, 1962, pp. 1143–1145
7 WOODS, D.L.: 'A history of tactical communication techniques' (Orlando Division, Martin Company, Martin-Marietta Corporation, Orlando, 1965), chapter 1
8 Article on Greek literature, The New Universal Library (International Learning Systems Corp., London, 1967–1969), **6**, p. 373
9 GAMBLE, J.: 'Essay on the different modes of communicating by signal' (London, 1797)
10 ROSSI, L.: 'Trajan's column and the Dracian wars' (Cornell University Press, 1971), pp. 80–82
11 Ibid, pp. 73–74
12 SHORE, H.N.: 'Signalling methods amongst the ancients', *United Service Magazine*, 1945, pp. 166–174
13 FORBES, R.J.: 'Studies in ancient technology' (E.J. Brill, Leiden, 1955), vol. II, pp. 152–155
14 PARSONS, W.B.: 'Engineers and engineering in the Renaissance' (MIT Press, 1939), pp. 294–298
15 MERRIMAN, A.C.: 'Telegraphing amongst the ancients', Papers of the

Archaeology Institute of America, Classical Series, vol. 3, p. 2 (Cambridge University Press, 1890)

16   RAWLINSON, G.: 'The history of Herodotus' (London, 1880)

17   Ref. 9, p. 10, quoting Herodotus's Calliope, p. 585

18   Ref. 9, p. 2, quoting Herodotus's Urania, p. 562

19   Ref. 9, p. 13, quoting Cleomedes de Mundo, lib. ii in his description of the world

20   Ref. 9, p. 13, quoting Caesar's 7th book of commentaries

21   AESCHYLUS.: 'Agamemnon', vv. 282–286

22   VERRALL, A.W.: 'The Agamemnon of Aeschylus' (MacMillan, London, 1904), pp. xx–xxii

23   Ref. 15, pp. 17–32

24   DARMSTAEDTER, E.: 'Feuer – telegraphie im Alterturm' (Umschau, 1924), vol. 23, pp. 505–507

25   SOLYMAR, L.: 'Getting the message' (Oxford University press, 1999). [On p. 11, Solymar lists the beacons as being on Ida, Lemnos, Athos, Peparethus, Euboea, Messapium, Cithaeron, The, Megarid, and Arachneus (and has a map on p. 12)]

26   The Interpreter's Bible (Abingdon-Cokesbury Press, New York, 1952), The book of Jeremiah, ch. 6, v.1,

27   GUBER, R.: 'Signal fires of Lachish' (Massada, 1970). The Lachish letters, letter IV, lines 6–13

28   Ref. 26, The Book of Judges, ch. 20, v.38

29   Ref. 26, The Mishnah Rosh Hashanah, ch. 2, v.2–3

30   POLYBIUS.: 'The histories of Polybius', 'Methods of signalling', pp. 42–43. English translation by W.R. Paton (Harvard University Press, Cambridge, Mass., 1922–1927)

31   SARTON, G.: 'A history of science' (Harvard University Press, 1959), chapter xx, pp. 343–346

32   The Stratagems of Polyaenus, lib. 6, cap. 16; quoted in J Gamble (Ref 9), p. 28

33   Ref. 30, p. 43

34   Ref. 30, p. 44

35   Ref. 13, vol. 6, p. 172

36   AFRICANUS, S.J.: 'Kestoi', p. 77; quoted in R.J. Forbes (Ref 13), p. 176

37   PFLAUM, H.G.: 'Essai sur la Cursus Publicus sans le Haute Empire Romaine', (Imprimerie Nationale, Paris, 1950)

38   Ref.10, pp. 130–131

39   Apud Strivilling, 13 Die Octobris, AD MCCCCLV (Aeta Parliamentorum Jacobi II), section 1

40   YARHAM, E.R.: 'Britain's old-time fiery beacons', *Army Quarterly*, 1943, pp. 62–66

41   HARDY, T. D.: 'Rymer's Foedera', II, p. 502 (Public Record Office, London, 1972)

42   BOYNTON, L.: 'The Elizabethan Militia' (Routledge and Kegan Paul, London, 1967)

43   Ibid, pp. 132–138

44   MACAULAY, Lord.: 'Lays of Ancient Rome, with Ivry and the Armada' (Longmans, Green, London, 1874),

45 MOORHOUSE, S.: 'When the Sussex beacons blazed', *The Sussex County Magazine*, 1943, pp. 348–349

46 BAILEY, C.: 'The legacy of Rome' (Oxford University Press, 1924), pp. 151–152, 160–161

47 RAMSAY, A.M.: 'The speed of the Roman Imperial Post', *J. Roman Studies*, 1925, **15**, pp. 60–74

48 COLVILLE, R.: 'The Armada and the Post Office', *Chambers Journal*, 1939, pp. 503–505

49 BARNARD, H.C.: 'The Messageries of the University of Paris', *British Journal of Education Studies*, 1955, **4**, No. 1, pp. 49–56

50 PARSONS, W.B.: 'Engineers and engineering in the Renaissance' (MIT Press, 1939), pp. 294–298

Chapter 2

Semaphore signalling

Although the telescope – (the discovery of which in 1608 is usually attributed to Johannes Lippershey (d. 1619) of Middleburg) – caused vast waves of concern to be spread in philosophy and religion [1], it also advanced scientific knowledge by enormous strides and led eventually to the realisation of economically viable signalling systems. Galilei Galileo (1564–1642) heard of the invention and succeeded in 1609 in constructing his first telescope, an instrument that magnified by three diameters and which consisted of a convex lens and a concave lens fitted into the opposite ends of a small lead tube. Later he manufactured larger telescopes, (up to 4.45 cm in diameter), and with these discovered the mountains and craters on the moon's surface, the satellites of Jupiter, the starry nature of the Milky Way and the phases of Venus.

The invention of the telescope seems to have been considerably delayed for the 11th century Arab scientist Alhazen [2] had published the results of his experiments on parabolic mirrors and magnifying lenses, and these accounts had been translated into Latin by 1572. The first reference to eye glasses was that by Meissner (1260–1280) who stated that old people derived advantage from the use of spectacles. It is also known that Nicholas Bullet, a priest, used spectacles when signing an agreement in 1282 and that convex lenses of Murano glass were in use in Venice in the latter half of the 13th century.

Roger Bacon (c. 1220–c. 1292) appeared to recognise the usefulness of lenses in telescopes for he wrote [3] 'thus from an incredible distance we may read the smallest letters – the sun, moon, and stars may be made to descend hither in appearance – which persons unacquainted with such things refuse to believe'. If this statement is true, it is difficult to account for the delay in the general use of lenses for 'distant seeing' until the time of Galileo. Roger Bacon's statement is important for he correctly advanced an application of lenses which was to have a profound effect on the growth of signal communications.

In 1684, the English natural philosopher Dr Robert Hooke (1635–1703), who had succeeded Oldenburg as Secretary to the Royal Society, presented a discourse [4] to the Society, 'shewing a way how to communicate one's mind at great distances'. He said that he had considered this matter some years prior to 1677, 'but being [recently] laid by the great siege of Vienna, the last year, by the Turks, [it] did again revive in my memory'. Thus as on similar occasions, going back more than 2,000 years, at times of anxiety scientists and inventors gave some thought to the problems of signal communications.

Hooke's apparatus (Fig. 2.1) consisted of an elevated framework supporting a screen D behind which were suspended as many deal-board characters, or symbols, as there are letters in the alphabet. In daytime these characters were either exposed or drawn back behind D, as the occasion demanded, while during the night use was made of torches, lanterns or lights.

With this equipment Hooke thought

'tis possible to convey intelligence from any one high and eminent place, to any other that lies in sight of it, tho' 30 or 40 miles [48 or 64 km] distant, in as short a time almost as a man can write what he would have sent, and as suddenly to receive an answer, as he that receives it hath a mind to return it, or can it write down in paper.

By using a number of such devices at relay stations greater distances could be covered. For this purpose the natural philosopher had the benefit of 'a late invention, which we do not find any of the ancients knew', namely the telescope. Hooke considered the length of these for different station-to-station distances; 'for one mile [1.6 km], one foot [30.5 cm]; for two miles [3.2 km], two foot [61 cm] . . . for 10 miles [16 km], 13 ft [3.96 m] and so forward'. The characters he chose were selected so that communications could be made 'with great ease, distinctness and secrecy'. In addition to those representing the letters of the alphabet, Hooke devised others to signify certain meanings: ' "O" I am ready to communicate, ")(" I am ready to observe, "(" I shall be ready presently, ")" I see plainly what you show, "∪" Shew the last again, "∩" Not too fast'. Here, he was anticipating modern practice.

Hooke's paper clearly shows that at the time no scheme such as he described had been put into practice, but he was very optimistic about the outcome of his practical system:

with a little practice thereof, all things may be made so convenient, that the same character may be seen at Paris, within a minute after it hath been exposed at London, and the like in proportion for greater distances; and that the characters may be exposed so quick after one another, that a composer shall not exceed the exposer in swiftness.

His paper is important as it gives the earliest well-defined plan of telegraphic transmission. The scheme was certainly realistic as he described it although the closest that anyone came to using something of its nature appears to have been the occasion in 1689 when the Duke of Gordon attempted to relieve Edinburgh Castle [5]. A white board signified 'All well' and a blackboard the reverse. Messages were spelled out with black letters on large white boards. The transmission distance was probably considerable as a telescope was used.

Although Hooke's apparatus was the forerunner of all later mechanical methods of transmitting signals it was not the first to be advanced. In 1661 that

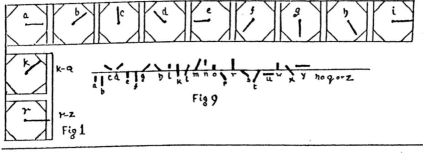

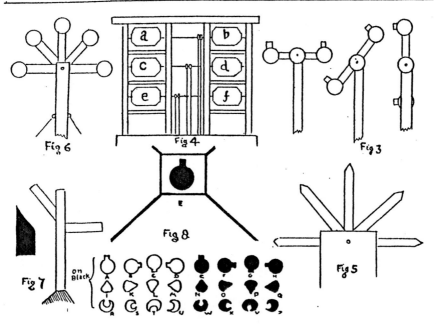

Figure 2.1   1. Ring dials – 'A century of Inventions', Marquis of Worcester, 1666. 2. Signal tablets – 'Philosophical experiments', R. Hooke, 1684. 3. Chappe's telegraph – 'Essay on modes of communication', Rev. J. Gamble, 1797. 4. Lord George Murray's signal apparatus – from an engraving, 1796. 5. Gamble's semaphore. 6. Goddard's semaphore – 'Observations on telegraphic correspondence', Admiralty. 7. Admiral Sir H. Popham's semaphore – from a lithograph, c. 1820. 8. Joseph Conolly's emergency signals, 1808, Admiralty. 9. The marquis of Worcester's cipher.

Source: *Journal of the Society for Army Historical Research*, **XXII**.

indefatigable inventor, the second Marquis of Worcester [6] (1601–1667), announced [7] that he had discovered

a method by which at a window as far as the eye can discover black from white, a man may hold discourse with his correspondent, without noise made or notice taken; being according to occasion given, or means afforded, ex re nata, and no need of provision beforehand: though much better if foreseen, and course taken by mutual consent of parties.

This method, he asserted, could be put into practice 'by night, as well as day, though as dark as pitch is black'.

Though Worcester printed his 'Inventions' he was evidently anxious that unauthorised copyists should not plagiarise it and therefore gave no details. His method, however, used a rotating pointer the position of which defined the letter of the alphabet to be transmitted, as shown in Fig. 2.2.

Figure 2.2   Chappe's synchronous telegraph being demonstrated on 2nd March 1791.
Source: A. Belloc, *La télégraph historique* (Paris, 1894).

Both Worcester's and Hooke's inventions gave inspiration to Claude Chappe whose semaphore system of 1792 was the first to be established on a national scale. Ignace Chappe paid tribute to Hooke and an examination of Claude Chappe's synchronised system of 1791 (Fig. 2.2) shows a remarkable amalgam of the ideas of the two British 17th century scientists [8].

Hooke was not alone in putting forward an idea which was subsequently developed into a practical form, even though at the time it was conceived it was incapable of a realistic solution. Kirchner [9] (1550) in his *Ars magna lucis umbrae* described a method for transmitting messages to a distance which was based on a form of magic lantern. The sender would write the message on a mirror and then, using the Sun as a source of light and a lens, the message would be conveyed by the light to the distant station. Kirchner's object 'was not merely to communicate the most secret thoughts of the heart to a distance but also to transport to the eyes of a friend at an enormous distance your profile or silhouette'. It seems that he was anticipating television.

The idea of recording signals and sending the recorded message to a distant station was first mentioned by John Wilkins, Bishop of Chester in the reign of Charles II, in the first edition of 'Mercury, or the secret and swift messenger, showing how a man, with privacy and speed, may communicate his thoughts to a friend at any distance' [10] (see Chapter 6).

In this work Wilkins described a method of telegraphing, using only three torches, or lights, to designate the 24 letters of the alphabet, which he ascribed to one Joachimus Fortius. These letters were according to Fortius's plan to be placed in three classes of eight each: one torch indicated class 1, two torches class 2, three torches class 3 and the number of the letter was to be shown by the number of times the torch was raised. This system was, of course, a modification of that described by Africanus in the third century AD and did not represent an improvement in communication.

Bishop Wilkin also mentioned a method of telegraphing by means of two lights attached to long poles, which he stated, 'for its quickness and speed is much to be preferred before any of the rest'. For long distance reception he suggested the use of the then newly invented telescope; which he called 'Galileus his perspective'.

The invention of the telescope by Lippershey in 1608 and Galileo in 1609 undoubtedly gave impetus to the development of signal communications in the 17th century, but in addition the publication of William Gilbert's great work *De Magnete* in 1600 possibly led to the interest that Famianus Strada (1617), George Hakewill (1627), Jean Leurechon (1628), Athanasius Kircher (1641), Sir Thomas Browne (1646) and others showed in the use of magnets in telegraph systems.

The dormant era of signalling development was now coming to an end but a national emergency would be required before sufficient government funds could be attracted to the large scale implementation of any one of the schemes being suggested. Actually, a national programme for the installation of a long distance communications system had to wait until 1792. Nevertheless there were in the 17th century many scientists and others who were aware that a solution to the problem of fast, reliable signalling could be achieved by an application of scientific principles and known experimental results to the task.

The earliest allusions to a magnetic telegraph were probably those put forward by Giambattista della Porta [11] (1540–1615), an Italian natural philosopher, who in his *Magiae naturalis* described a series of experiments with magnets for the purpose of communicating intelligence to the distance. 'And to a friend that is at a far distance from us and fast shut up in prison, we may relate our minds; which I doubt not may be done by two mariner's compasses, having the alphabet writ about them'. This idea was also suggested by Daniell Schwenter, Professor of Oriental languages at Altdorf, who in 1600 under the assumed name of Janus Hercules de Sunde described in his *Stegarologia et Steganographia* the means of transmitting intelligence to distance by utilising two compass needles circumscribed by an alphabet.

Famianus Strada [12], an Italian author and Jesuit priest, in 1617 also used the same principle in his *Prolusiones Academicae*.

If you wish your distant friend, to whom no letter can come, to learn something, take a disc or dial and write round the edge of it the letters of the alphabet in the order in which children learn them, and, in the centre, place horizontally a rod, which has touched a magnet, so that it may move and indicate whatever letter you wish. Then a similar dial being in the possession of your friend, if you desire privately to speak to the friend whom some share of the earth holds far from you, lay your hand on the globe, and turn the movable iron as you see disposed along the margin of all the letters which are required for the words. Hither and thither turn the style and touch the letters, now this one, and now that. Wonderful to relate, the far distant friend sees the voluble iron tremble without a touch of any person, and run now hither, now thither; conscious he bends over it and marks the teaching of the rod. When he sees the rod stand still, he, in turn, if he thinks there is anything to be answered, in like manner, by touching the various letters, writes it back to his friend.

Obviously neither della Porta, nor Schwenter nor Strada were able to demonstrate practically their proposals. As Jean Leurechon [13], (1591–1670), the confessor of Charles IV of Lorraine, said in his *La récréation mathématique* (1628): 'The invention is beautiful, but I do not think there can be found in the world a magnet that has such a virtue'. Sir Thomas Browne [14], an English physician was rather more scathing and pragmatic in his criticism of Strada's concept of the magnetic telegraph. In his *Pseudodoxia Epidemica, or inquiries into vulgar and common errors* (1646) he stated:

The conceit is excellent and, if the effect would follow, somewhat divine; whereby we might communicate like spirits, and confer on the earth width Menippus in the moon. And this is pretended from the sympathy of two needles, touched with the same lodestone, and placed in the centre of two abecedary circles or rings, with letters described round about them, one friend keeping one and another keeping the other, and agreeing upon the hour when they will communicate, at what distance of place soever, when one needle shall be removed unto another letter, the other, by wonderful sympathy, will move unto the same.

Browne to his credit tried the experiment which Strada and others clearly had not performed. As a result he found that 'though the needles were separated but half a span, when one was moved the other would stand like the pillars of Hercules, and if the earth stand still, have surely no motion at all'.

Still, della Porta, Schwenter and Strada put forward the idea of a 'dial' telegraph, and while the principle of operation was hopelessly inadequate for their purpose this concept was used later in some electrostatic telegraph systems, notably that of Francis Ronalds of 1816.

No advancement in long distance signalling appears to have taken place until 1767. In that year the ingenious Mr Richard Lovell Edgeworth (1744–1817), road builder, inventor, politician, educationist and writer, placed a bet with Lord Marsh, an inveterate gambler [15].

One day, Lord Marsh found he could not attend a particular race meeting at Newmarket but declared: '[I will] station fleet horses on the road, to bring me the earliest intelligence of the event of the race, and I shall manage my bets accordingly'. He thought the news could be given to him five or six hours after the race had ended, but Edgeworth told him he could have the result one hour later. This seemed to be an incredible boast, and one which should be tested. Marsh promptly laid a wager that it could not be undertaken. Though not a gambler Edgeworth was so sure his idea could be implemented he offered to support his claim by laying a bet of £500, which Sir Francis Delaval, FRS, one of Lord Marsh's friends, increased by a similar sum. That night Edgeworth explained to Sir Francis his idea for sending messages. A series of signalling stations, each comprising an ingenious apparatus, would be erected on high ground, at separation distances of c. 24 km: the signals would be observed by telescopes. The simplicity and feasibility of the scheme were such that Delaval doubled his bet, a move which was accepted by Marsh and his friends. However, when he was told mechanical contrivances would be utilised rather than swift horses Marsh and his friends withdrew their bets.

Unknown, presumably, to Lord Marsh, Edgeworth in 1767 had invented his tellograph. He proposed to 'make use of windmill sails instead of the hands or pointers [of Hooke's scheme]' and thereby 'to contrive not only a swift but an unsuspected mode of intelligence'. With a Mr Perrot, of Harehatch, Edgeworth had in 1767 observed that the windmill at Nettlebed could be seen with a telescope from Perrot's house even thought the separation between the two places was 16 miles (25.8 km). (See Fig. 2.3.)

Edgeworth's claim so inspired Delaval he insisted on an experiment being carried out.

Under my [Edgeworth's] direction Sir Francis erected an apparatus between his house in Downing Street and part of Piccadilly, an apparatus which was never suspected to be a telegraph. I also set up a night telegraph between a house which he occupied at Hampstead, and one in Great Russell Street. The nocturnal telegraph answered very well, but was too expensive for common use.

Edgeworth did not return to his invention until 1796, that is until after Chappe had demonstrated his method of semaphore telegraphy to the French Government.

To Claude Chappe (1763–1805) must be accorded the honour of having devised and implemented the first practical system of semaphore telegraphy. It was capable of sending messages over long distances and when it was finally superseded by electric telegraphy in 1850, France was covered by a network of 556 semaphore stations stretching over a total distance of 2,983 miles (4,800 km) [16]. Communications could be received in Paris from Lille in 2 minutes, from Calais in 4 minutes 55 seconds, from Toulon in 13 minutes 50 seconds, from Strasbourg in 5 minutes 52 seconds and from Brest in 6 minutes 50 seconds.

Claude Chappe was born on 25th December 1763 at Brulon in the Sarthe department [8]. He was one of ten children of whom seven survived. Of his early childhood little is known except that he was a pupil at the College de Joyease at Rouen and then later at a small school at La Fleche. He was trained for the church, becoming an Abbe Commendataire, but this did not require Chappe to carry out any religious duties. Two generous benefices

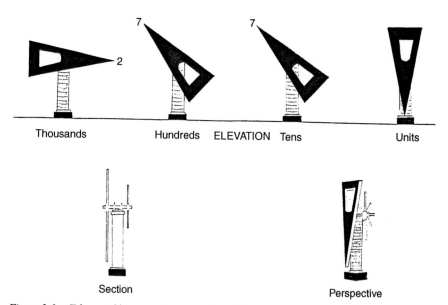

Figure 2.3   Edgeworth's semaphore type signalling apparatus.
Source: R.L. Edgeworth, 'An essay on the art of conveying secret and swift intelligence' (Royal Irish Academy, 1795).

enabled him to pursue his prime interests which were in physics and in mathematics, and by the age of 20 Chappe was submitting articles to learned society periodicals [16]. His researches came to an abrupt end on 2nd November 1789 when the benefices were terminated by the Legislative Assembly. He returned home and, with the assistance of his four brothers, Ignace Urbain Jean (b. 1762), Pierre Francois (b. 1765), Rene (b. 1769) and Abraham (b. 1773), who also were now unemployed, he applied himself to the problem of signalling.

From his investigations three different schemes were designed and operated, namely:

1. the synchronised system of 1791,
2. the shutter system of 1791, and
3. the semaphore system of 1792.

The synchronised system depended upon the use of two identical clocks (see Fig. 2.2) working synchronously, situated at the sending and receiving stations. Each clock was provided with a rotating seconds pointer which passed over divisions marked on the face of the clock, each division being indicated by a symbol. To transmit a message the operator at the sending end of the telegraph link struck a gong at the instant the pointer passed the symbol to be transmitted. Chappe used 16 characters on the dial face and all of these were included in Hooke's alphabet of 24 symbols which he devised for sending signals in code. Since there was no need for Chappe to use these particular symbols – he could just as well have used numbers and still maintained secrecy of message transmission – the possibility exists that he was aware of Hooke's work. Like Hooke, Chappe employed a code for interpreting the messages. The first experiment with this system was in 1790, the distance between two stations being 437 yd (400 m).

A disadvantage of the above scheme was that the transmission distance was limited by the initial intensity, and subsequent attenuation, of the sound from the gong. Chappe endeavoured to replace the sound signal by an electrical signal sent along a wire but the results were unsatisfactory because of the great difficulty, at that time, of insulating the conductor. Accordingly he turned his attention to a consideration of an optical system in which the precise moment of the transit of the pointer past the symbols was indicated by the appearance and disappearance of surfaces of different colours and forms. For his purpose Chappe used a pivoted wooden plate, painted black on one side and white on the other, placed at a height of 13 ft (4 m), on a supporting structure. On 2nd March 1791, before a group of a local officials, he sent messages from Paris to Brulon, a distance of more than 9 miles (15 km).

The method Chappe used to transmit messages over a long distance was that which had been proposed by Guillaume Amontons [17] and exhibited in 1704 before the royal family of France and members of the Academie des Sciences.

Let there be people placed in several stations, at such a distance from one another, that by the help of a telescope a man in one station may see a signal made in the next before him; he must immediately make the same signal, that it may be seen by persons in the station next after him, who are to communicate it to those in the following station and so on. These signals maybe as letters of the alphabet, or as a cipher, understood only by the two persons who are in the distant places, and not by those who make the signals. The person in the second station making the signal to the person in the third the very moment he sees it in the first, the news may be carried to the greatest distance in as little time as is necessary to make the signals in the first station. The distance of the several stations, which must be as few as possible, is measured by the reach of a telescope. (Fig.2.4.)

Another possibility is that Chappe knew of the ideas of Jean Jacques Bathelemy who, in 1788, had published in Paris the first edition of his *Voyage du Jenne Anarcharis*. In this work Bathelemy alludes to the possibility of

Figure 2.4   Amontons experimenting with his system of telegraphy before the Dauphin in the garden of Luxembourg, Paris, 1690.
Source: A. Belloc, op. cit.

telegraphing by means of clocks (pendules, not horloges) having hands similarly magnetised, in conjunction with artificial magnets. These were 'presumed to be so far improved that they could convey their directive power to a distance: thus, by the sympathetic movements of the hands or needles in connection with a dial alphabet, communications between distant friends could be carried on.' [17]

In a letter written in 1772 to Mme du Deffand he noted:

It is said that with two timepieces the hands of which are magnetic, it is enough to move one of these hands to make the other take the same direction, so that by causing one to strike twelve the other will strike the same hour. Let us suppose that artificial magnets were improved to the point that their virtue could communicate itself from here to Paris; you have one of these timepieces, we another of them; instead of hours we find the letters of the alphabet on the dial. Everyday at a certain hour we turn the hand, and M. Wiard [Mme du Deffand's secretary] puts together the letters and reads. . . . This idea pleases me immensely. It would soon be corrupted by applying it to spying in armies and in politics, but it would be very agreeable in commerce and in friendship [18].

There is certainly a similarity between Barthelemy's timepieces with the letters of the alphabet on the dial and Chappe's apparatus, although Chappe being pragmatic in outlook would have realised that magnets, whether permanent or artificial, could not have led to the anticipated result he desired.

Chappe's initial attempts to prove his system were not without incidents. After many disappointments, Chappe, with the influence of his brother Ignace who was a member of the Legislative Assembly [19], finally obtained permission to set up his synchronised system at the Etoile in Paris only to find a short time later, one morning towards the end of 1791, that it had been destroyed by a revolutionary mob who thought the equipment was being used to communicate with King Louis XVI, who at that time was imprisoned in the Temple.

Chappe's early experiments with the synchronised system led him to abandon it and try a shutter method. The equipment consisted of two rectangular frames, each frame being fitted with five shutters. These could be made to appear or disappear at will. The resulting experiments with the equipment – which was also smashed and burnt by the mob – demonstrated that elongated bodies were more clearly visible than shutters, and so Chappe in 1792 finally adopted semaphores for signalling.

Chappe presented an account of his semaphore system, which he called a tachygraphe, to the Legislative Assembly on the 22nd March 1792. (The word telegraph, from the Greek roots tele (far) and grafein (to write), was suggested by Miot de Melito, a classical scholar in the Ministry of War, in 1793.) He appealed for protection of his apparatus and for a fair and conclusive trial of it by competent investigators. However the Assembly itself was in an even less secure position than the apparatus they had been asked to defend and in September 1792 the Assembly was replaced by the National Convention [16].

Disaster followed upon disaster. When, a few months later, a telegraph station was set up in the Le Peletier Saint-Fargeau park at Menilmontant, the burghers seemed to believe that endeavours were being made to send messages to the Austrians and Prussians; rioters assembled in the park and fired the telegraph. Anxious to prevent further incidents Chappe, on 15th October 1792, wrote to the Convention and requested that official authorisation be given to the establishment of the Belleville station [16]. The Convention moved to delegate the issue to the Committee of Public Instruction.

Eventually, on 1st April 1793, following the outbreak of war in February/March 1793 between France and the coalition powers of England, Holland, Prussia, Austria, Spain and Sardinia, Gilbert Romme, the President of the Committee, brought Chappe's plans to the attention of its members. He reported favourably on the efforts of the Chappe brothers, and advised that, of the various systems of communication that had been offered to the government, only Chappe's telegraph appeared to merit any attention. Romme asked for the telegraph to be subjected to a trial experiment. Fr 6,000 were appropriated by the Convention from the general funds of the War Department for this purpose, and three members of the Committee of Public Instruction, the scientist Lakanal, the legislator Dannou, and the mathematician Arbogast, were appointed observers [20].

With the protection given by a decree published by the Convention, the Chappe brothers constructed three semaphore telegraph stations on a 22 yd (35 km) line between Le Peletier de Saint Fargeeau park at Menilmontant and Saint Martin-du-Tertre, via the heights of Ecouen. The apparatus, in its developed form (Fig. 2.5) consisted of a post 33 ft (10 m) high, at the top of which was attached a wooden beam, termed a regulator, approximately 5 yd (4.62 m) in length and 0.38 yd (0.35 m) in width, which turned in a vertical plane about its centre. At each end of the regulator there was an indicator – a moving arm 2.2 yd (2 m) long and 0.36 yd (0.33 m) wide. The indicators were balanced by thin iron counterweights and the mechanical system of regulator and indicators was connected by ropes and pulleys to the base of the post so that the three components could be moved to any of the many desired positions. These positions were exactly reproduced in the signalling room by a repeater. For good visibility the machine was painted black and the arms were of a balanced louvred construction to minimise wind resistance [21]. In devising his semaphore system Chappe was considerably aided by the noted clock and watch maker A.L. Breguet.

Each indicator could be set in one of seven positions, namely, 0°, 45°, 90°, 135°, 225°, 270°, and 315°, with respect to the regulator. The 180° position was not utilised because then the indicator would have appeared as an extension of the regulator. With the two indicators and the regulator (which could take one of four positions, namely, horizontal, vertical, and plus or minus 45° to the vertical), the total number of different combinations was $7 \times 7 \times 4 = 196$. Of these 98 were used for the handling of messages and 98 were employed for the regulation and policing of the line.

Delaunay, a cousin of the Chappe brothers, developed a code for use with the

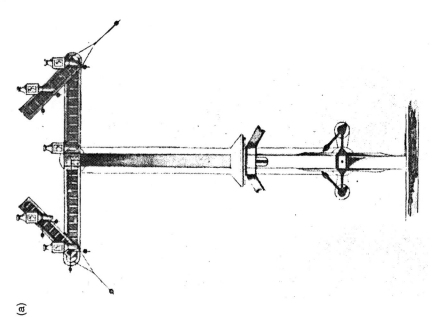

Figure 2.5 *(a) A schematic diagram of Chappe's semaphore. (b) A semaphore telegraph post erected before the battle of Condé, 30th November 1794.*

Source: A. Belloc, op. cit.

semaphores. As he had been on the French consular staff at Lisbon, he adopted as the basis of his code the one in service for diplomatic correspondence. The code, of 9,999 words, phrases and expressions each represented by a number, soon proved to be too slow and inconvenient and was replaced in 1795. It was revised again in 1830.

On 12th July 1793, the world's first telegraph line was demonstrated before the representatives of the Committee of Public Instruction, and several savants and artists. Lakanal and Arbogast observed from Saint Martin-du-Tertre and Dannou watched from Menilmontant. At 4.26 p.m. the Saint Martin-du-Tertre station signalled: 'Prepare', and in 11 minutes Dannou was able to send Lakanal the following message:

Dannou has arrived here; he announces that the National Convention has just authorised its Committee of General Security to affix the seals to the papers of the representatives the people.

Lakanal's reply [16] took just nine minutes to send:

The inhabitants of this beautiful region are worthy of liberty by their respect for the National Convention and its laws.

On 26 July 1793 Lakanal presented a eulogistic report to the Convention: 'What brilliant destiny do science and the arts not reserve for a republic which, by the genius of its inhabitants, is called to become the nation to instruct Europe.'

He stressed the accuracy of the semaphore system, and the secrecy of the dispatches which might be sent since these would be transmitted in code and only the terminal stations would have access to the code books.

The Convention approved the adoption of the semaphore apparatus for national communications; Chappe was given the title of ingeniure-telegraphe and was granted the pay of an engineering lieutenant [19]. The Committee of Public Safety was ordered to plan suitable telegraph routes and Chappe and his brothers Ignace and Pierre Francois were named as administrators of the telegraph line [16].

Realising the vital importance of establishing good communications the Comite de Salut, from August 1793, made decrees signed by such persons as Couthon, Barere, Herault, St Just, Thuriot and Robespierre for immediate extensions of Chappe's system. The inventor was authorised to place his machines in any belfries, towers or situations of his choosing; he had authority to cut down any trees that might interfere with the line of vision, and general provision was to be made by the Government for hastening the work. By mid-August 1794 a line was established between Montmartre, Paris and Lille in the face of great difficulties due to lack of cash, means of transport and work. This line included 15 stations. The first telegram was transmitted by Chappe from Lille to Monmartre on 15th August 1794 and told the Administration in Paris that Le Quesnoy had been recaptured by the French. Fifteen days later Chappe telegraphed the news that Conde had been recovered.

These messages were of great importance for the two towns had been captured by the Allies in 1793; Mainz had surrendered to them, and Valenciennes had capitulated. But in June 1794 Carnot had began a counter offensive, especially against the Austrians, and at the battle of Fleurus on 25th June 1794, the Allies were scattered towards Holland and the Rhine. Later the French under Jourdan regained the important area that included the fortresses of Landrecies, Quesnoy, Conde and Valenciennes. These gains heralded the eventual French victory and so the news transmitted by the Chappe telegraph was greeted with indescribable enthusiasm by the Convention. The importance of the telegraph in one form or another was now firmly established and was to be an essential feature for many years, in peace and war, of the communications systems of the developed countries.

Chappe had undertaken his work at a most propitious time and one of crucial significance to his country. It is perhaps doubtful whether his system would have been implemented in less troublesome times: certainly there was nothing of great originality in it. The system was not based on any particularly novel principles of mechanisms or of communications: indeed the combination of pulleys and levers used could have been devised many hundreds of years previously, and while the discovery of the telescope was essential for the rapid and efficient working of the communication channel there was nothing about the system as a whole that had prevented its introduction prior to 1792 and after 1609. The success of the Chappe scheme was primarily due to the need for good communications in France at a time when the country was in a desperate plight with her frontiers assailed by the Allied forces of England, Holland, Prussia, Spain, the Italian States, Hanover and Hesse, and with the country racked internally by the excesses of the Revolution. Thus, the times demanded a signalling method for the maintenance of good organisation and the transmission of messages between the forces in the field and the Administration in Paris. Without the telegraph the capture of the Dutch fleet by the French cavalry, with the support of a single battery of horse artillery, under Moreau, would not have been possible [8].

Following the news of the recapture of Conde and Quesnoy, further lines were planned and constructed in the face of immense difficulties due primarily to a lack of financial resources. 'Funds, funds, once more funds, otherwise we are not able to construct' [8], Chappe wrote when, in default of payment, construction work in 1796 became disorganised.

In an attempt to ease the situation caused by a lack of finance, Chappe suggested that his invention could be applied to commerce and the relaying of daily stock exchange news. Another of his ideas was to use the telegraph to provide a summary of the news of the day. None of the schemes were thought to be practicable, but Bonaparte did agree to the weekly transmission of the winning numbers of the national lottery [19]. This was quickly put into effect and as the ingenieur-telegrapher had predicted the lottery proposal paid for the greater part of the telegraph service.

Nevertheless the Chappe telegraph network became more and more extensive.

On the 31st May 1798, the 480 km trunk line, with 46 stations, between Paris and Strasbourg was opened; and the 870 km Paris–Brest trunk line with 55 posts was inaugurated on 7th August 1798 [19]. The extension of the Paris–Lille line to Brussels and Boulogne was completed in 1803. In 1805 Emperor Napoleon extended a line to Milan. Anvers and Flushing were linked to the network in 1809 and Amsterdam was connected in 1810. At the end of the Napoleonic era France had 224 stations distributed along 1,112 miles (1,790 km) of telegraph routes.

Chappe did not live to see the later development of his system. He killed himself by throwing himself down a garden well at the telegraph headquarters. One account states that he 'had long suffered from a painful ear complaint that may have deranged him'. Another account mentions:

On every side he met rivals claiming to have invented telegraphs and seeking a share of the credit for the invention. A watchmaker named Breguet and his associate Betancourt so discouraged Chappe by their quarrels with him over the priority of his ideas that he committed suicide on 23rd January in 1805 [16].

In his final message Chappe wrote: 'I kill myself because I am weary of a life that burthens me. I have nothing to reproach myself with' [22].

After his death the development and extension of the semaphore system was continued by his brothers Ignace, Pierre, Abraham and Rene. The first three were given the rank of Chevaliers of the Legion of Honour when Louis XVIII was restored to the throne of France in 1814, and when Ignace and Pierre retired in 1823 the King granted them pensions of 4,255 francs and 2,252 francs respectively. Thereafter the younger brothers Rene and Abraham remained in the service of the telegraph administration for several years [16].

Many lines were constructed during the period of the Bourbons. Calais was linked to Paris in 1816, and in 1821 the Lyons to Toulon line was in operation. The capital and Bordeaux were connected, in 1823, via Orleans, Poitiers and Angouleme, and in the same year telegraphic messages could be sent from Bayonne, near the French–Spanish border to Paris. The Avignon to Montpellier line, via Nimes, was operational in 1828 [16].

Among the later routes to be provided with the Chappe telegraph service there were:

1.   Avranches to Nantes (1832)
2.   Narbonne to Montpellier (1834), and to Perpignon (1840)
3.   Avranches to Cherbourg (1834)
4.   Bordeaux to Narbonne, via Toulouse (1834)
5.   Dijon to Bescancon (1840)
6.   Bayonne to Behobie (1847)

These telegraph lines were firmly controlled by the Government. After the July 1830 revolution another attempt was made to extend the use of the telegraph to non-military communications. A.M. Alexandre Ferrier was so persuaded of the benefits of the system to the railways, stock exchange transactions, trade and personal users that he constructed a non-government line between Paris and Rouen. It was opened in July 1833 against some opposition from Alphonse Foy, the Head of the Telegraphic Administration. The line closed before the end of the year.

Another entrepreneur from October 1836 established a line between Paris and Brussels but again Foy was of the opinion that the state should own and control all telegraphs. He persuaded the Minister of the Interior, de Gasperin, to put before the French Parliament a bill for a state monopoly in telegraphy. The bill was passed in March 1837 by the Lower House with 212 votes in favour and 37 against; and a month later by the Upper House with 86 votes in favour and two against. As a consequence the French parliament approved the measure that:

Anyone who transmits any signals without authorisation from one point to another one whether with the aid of mechanical telegraphs or by any other means will be subject to imprisonment for a duration of between one month and one year and will be liable to a fine of from Fr 1,000 to Fr 100,000.

In 1844 when the first electric telegraph line was established between Paris and Rouen the French signalling network linked, over a distance of more than 5,000 km, 29 of the largest cities and towns in France by means of 534 semaphore stations (see Fig. 2.6). The system had been in operation for 50 years and had provided France with an efficient and reliable means of transmitting essential intelligence which was much copied or adapted elsewhere. Chappe's drive and fortitude in the face of grave difficulties ensured the telegraph's early success. A few days after his death the French publication *Moniteur* opined: 'People rightly say that the signalling art existed long before him. What must be added to be just and impartial, is that he made of this art an application, so simple, so methodical, so sure and so universally adopted, that he can be regarded as an inventor'.

Reports of Chappe's telegraph reached England in the autumn of 1794. Soon afterwards the audience in one of London's music halls was regaled with the following ditty:

> If you'll just promise you'll none of you laugh
> I'll be after explaining the French telegraph!
> A machine that's endowed with such wonderful pow'r
> It writes, reads and sends news 50 miles in an hour.
> Then there's watchwords, a spy-glass, an index on hand
> And many things more none of us understand,
> But which, like the nose on your face, will be clear
> When we have as usual improved on them here [19].

The news stimulated R.L. Edgeworth [15] (see end of chapter Note 1); and Lord George Murray [23] (1761–1803) proposed a system of visual telegraphy to the British Admiralty. According to the 1797 edition of the Encyclopaedia Britannica the first description of Chappe's telegraph was sent from Paris to Frankfort-au-Maine by a former member of the Parliament of Bordeaux, who

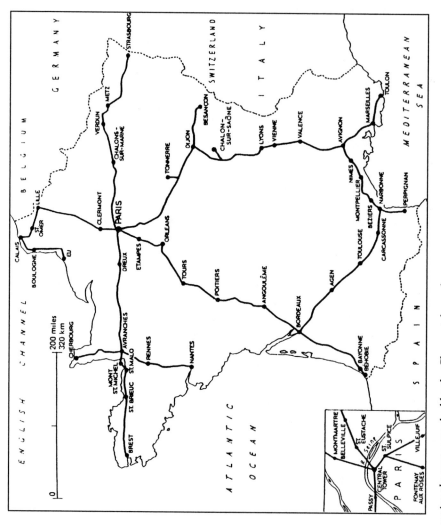

Figure 2.6   Map showing the routes worked by the Chappe telegraph.
Source: G. Wilson, *The old telegraphs* (Philimore, London, 1976).

had seen the telegraph post which was erected on the mountain of Belville. Subsequently two working models of the apparatus were executed, it was said, in Frankfort, and were sent by Mr W. Playfair to Frederick Augustus, the Duke of York (1763–1827), commander of the British Army in Flanders, who passed it on to his chaplin, John Gamble (d. 1811).

The Rev. J. Gamble, in his well-researched book of 1797 [24], *An essay on the different modes of communication by signals*, gives a somewhat different report:

The first intelligible account of this machine [Chappe's] was received from a prisoner – taken when the Duke of York's quarters were in Berlisum, the latter end of the same month, August 1794 – in whose pocket Captain Brinley, assistant adjutant-general, had found a rough drawing and description of the instrument, which he showed to me as a curiosity. Early the September following, Major Gordon, of the 11th, brought to His Royal Highness, while at Groosbeck, near Nimeguen, a complete working model, with the alphabet, which he had received from Liege. Of this, as I believe it was the 1st sent to England the following is the description ...

Arriving in England later in January 1795, and finding it had been a subject engaging much attention, I took the liberty of mentioning to Field Marshal the Duke of York, that I thought the French telegraph might be improved, or rather that a machine might be constructed to answer the purpose much better.

A model being completed early in February was left at York House, and there inspected by several of the first military officers. The latter end of the same month it was left with the Bishop of Lincoln, then on a visit in Downing Street; and its principles were approved by Mr Pitt and Mr Rose.

The British Army did not adopt Gamble's signalling method and passed it to the Admiralty in April 1795. However, the Admiralty preferred the system devised by Murray for, after receiving details of Gamble's signalling invention, Their Lordships replied that they were 'so well satisfied with the telegraph erected under the direction of Lord George Murray that they did not think it necessary to make any experiment with the radiated form'. Murray retired after five years and was appointed Bishop of St David's.

Murray, the fourth son of the third Duke of Atholl, matriculated from New College, Oxford in 1779 and graduated BA in 1782. In 1787 he was made Archdeacon of Man [23]. Then, 'applying his scientific skill and philosophical knowledge to that curious mechanical invention, the telegraph, he made many improvements in that machine'. Murray employed a large open wooden frame to which were fitted six boards pivoted about a horizontal axis. In view of the enmity existing between England and France at that time it may have been the case that Murray did not wish to adopt the Chappe system.

Figure 2.7 shows a view, and details of working, of the telegraph which was erected on the Admiralty Office at Charing Cross. Sixty-four changes could be made with the boards, by the most plain and simple mode of working. Of these, the 'all open' and 'all closed' shutter positions were reserved for the 'message ends' and 'ready to receive' signals; the letters of the alphabet and the ten numerals required 36 positions; the remainder were used for common phrases and commands.

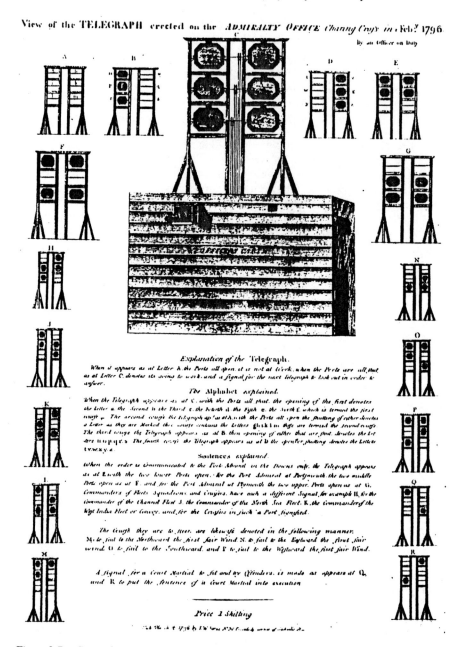

Figure 2.7   *Copy of a print, now in the Admiralty Library, of the telegraph of 1796, with an explanation of the telegraph.*
Source: The Admiralty Library.

Murray's system [19] was subjected to full scale trials on Wimbledon Common in September 1795. On 18th December 1795, Murray was granted an audience with King George III and as a result was created the first Director of Telegraphs in the Admiralty in March 1796. A chain of stations, 15 in all, was erected for the Admiralty between London and Deal (69 miles, 112 km) at a cost of nearly £4,000. Each station cost £230 to erect and was provided with an eight guinea clock and two £12.60 telescopes. A surveyor George Roebuck, was the contractor and in 1796 he was appointed the Superintendent of Telegraphs with a salary of £300 a year. On 27th January of that year a signal could be sent from London to Deal and acknowledged in two minutes.

Twenty-four stations, at a spacing of about 3 miles (4.8 km), were erected between London and Portsmouth (62.7 miles/101 km) and there was also a project in 1801 for a line between London and Yarmouth. However the Treaty of Amiens intervened and the scheme lay dormant. The outbreak of war in 1803 gave an added urgency to the question of communications and in October 1805, a few days before Admiral Nelson finally engaged the French Mediterranean Fleet, the Admiralty decided to link telegraphically the naval base at Plymouth with London. Since the line was not completed until 4th July 1806, it clearly could not be utilised to inform the Admiralty of the great victory at the battle of Trafalgar and the actual news of the event took 38 hours and 19 changes of horse to be sent from Falmouth to London. The Yarmouth scheme was revitalised in 1807 and by the middle of 1808 Roebuck had 65 stations under his supervision, all operated by the shutter system; it lasted for less than six years (Fig. 2.8).

The performance of Murray's signalling method was very much dependent on the weather and atmospheric conditions which prevailed along the lines of the stations. Loss of visibility could be caused by local fogs and the burning of coal. In the experience of the operators working on the London–Portsmouth line, poor visibility reduced the number of days when the telegraphs were functional by about 100 per year, and only intermittent communication could be maintained on a further 60 days: however, on the remaining 200 days the signals could be read throughout daylight hours.

Unlike Chappe's system, which used coded messages, the shutter scheme was worked with plain language messages. Routine signals could be passed at considerable speed by employing abbreviations, and by omitting the definite and indefinite articles, pronouns and prepositions, and self-evident vowels. This feature of the method stemmed from the highly redundant nature of the English language. With a code, where, say, a four character figure represents a word, phrase or sentence, an incorrect received character can completely change the meaning of the transmitted message; but with the use of a language containing a degree of redundancy the message may still be intelligible.

By the end of 1813 Napoleon Bonaparte's empire was constrained by the borders of France and three months later on 30th March 1814 the continental Allies were marching into Paris. Hastily the Admiralty gave orders on 4th May to discontinue the Yarmouth service and following the signing of the Treaty of Paris on 30th May 1814 the Admiralty, on 6th July 1814, issued instructions 'to

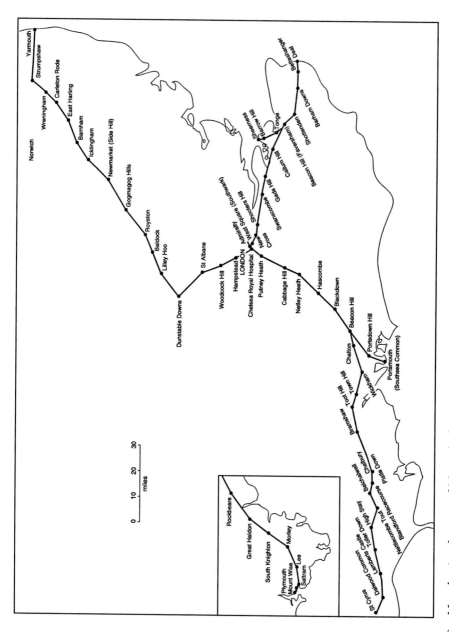

Figure 2.8   Map showing the route of Murray's shutter telegraph.
Source: G. Wilson, *The old telegraphs* (Philimore, London, 1976).

discontinue the line of telegraphs from London to Sheerness and Deal, and from London to Portsmouth the beginning of September next and to return the telescopes to their office.'

Some of the telegraph buildings were converted for use as homes by superannuated naval officers, and others were dismantled and the timber, and furniture, sold. Such was the speed at which the telegraph system was dismantled that in the spring of 1815, when Napoleon returned to France from the island of Elba, no telegraph was in an operational state in England. The Admiralty's wanton policy was now quickly overturned and 11 days after the Battle of Waterloo an Act of Parliament was passed for 'Establishing Signal and telegraph Stations'.

Perhaps somewhat curiously, Murray's scheme, which had given satisfactory service when operated in good atmospheric conditions, was now abandoned and a semaphore system [25], devised by Rear Admiral Sir Home Riggs Popham (1762–1820) was adopted (Fig. 2.1). The explanation for the change is that in November 1815 the Admiralty had instructed Popham to construct a semaphore line to Chatham and it was immediately obvious following its inauguration in July 1816 that the 8 ft (2.44 m) long arms on the 30 ft (9.14 m) mast had a greater visibility than the 5 ft (1.5 m) × 5 ft (1.5 m) rectangles or octagons of Murray's telegraph. Popham's telegraph seems to have been based on the three-arm Depillon semaphore telegraph used by the French coastal stations from c. 1803 to give warnings of British naval movements. These stations were known to the Admiralty for an 1808 report mentions that Captain Thomas Cochrane (1775–1860) had 'with his single ship the *Imperieuse* kept the whole coast of Languedoc in alarm destroying the numerous semaphoric telegraphs'. Also in May 1805 he had raided a station on the Ile de Re and had captured the station's code book. As a consequence many of the Admiralty's Coastal Signalling Stations, which communicated with ships off-shore, had, in the later stages of the war with France, replaced their signalling flags, pennants and balls with copies of Depillon's machine.

Popham found that just two arms were adequate for signalling. He described his 'telegraph and marine vocabulary' in a paper published in 1816 in the Transactions of the Society of Arts and received the Society's Gold Medal.

Many variants of the basic semaphore system were advanced following the success of the French system and during the first half of the 19th century mechanical type signalling had been adopted by many countries. Both day-time and night-time (lamp) methods were available (see Reference 19.)

The importance of good communications was certainly understood and appreciated by the end of the 18th century. By way of illustration in 1797 one reviewer wrote:

Were telegraphs brought to so great a degree of perfection, that they could convey information speedily and distinctly; were they so simplified that they could be constructed and maintained at little expense; the advantages which would result from their use are almost inconceivable. Not to speak of the speed with which information could be communicated, orders given in time of war, by means of which misfortunes might be prevented or instantly repaired, difficulties removed, and disputes precluded, and by means of which the whole kingdom could be prepared in an instant to oppose an invading enemy, it might be used by commercial men to convey a commission cheaper and speedier than an express can travel. The capitals of distant nations might be united by chains of posts, and the settling of those disputes which at present take up months or years might then be accomplished in as many hours. An establishment of telegraphs might then be made like that of the post; and instead of being an expense, it would produce a revenue. Until telegraphs are employed to convey information that occurs very frequently, the persons who are stationed to work them will never become expert, and consequently will neither be expeditious nor accurate, though, with practice, there is no doubt that they will attain both in a degree of perfection of which we can as yet have but little conception [26].

This was a far sighted, pragmatic view of the future of telegraphy. A source of revenue was clearly desirable, as Chappe had noted, and the suggestion made by the unknown writer above was eminently sensible. But before this proposal could readily be said to have been implemented two events were to take place: one was the detection of a capital crime, the other was the introduction of railways. Both of these events involved the electric telegraph rather than the mechanical semaphores of Chappe or the shutter boards of Murray: however, the speed of Chappe's system influenced the emergence of a practical method of electric telegraphy to an appreciable degree. This is considered in the next chapter.

Note 1
Edgeworth's telegraph of 1795

Edgeworth's 1795 method of telegraphy comprised, at all signal stations, four co-linear pillars, 16 ft to 20 ft (4.87 m × 6.09 m) in height, each carrying a triangular shaped pointer which could rotate about a horizontal axis. The individual positions of the four pointers represented the units, tens, hundreds and thousands of a four figure number. Any pointer could be set to one of eight positions, namely: upwards (0), horizontally to the right (2), downwards (4), horizontally to the left (6), and the four intermediate positions (1), (3), (5) and (7) (Fig. 2.3).

The figures denoted by the set of four pointers referred to words in a vocabulary; for example the number 2774 represented 'The Royal Irish Academy'. Edgeworth described his vocabulary as

composed of a large book with mahogany covers, framed, to prevent them from warping. Its size is 47 inches by 21 (119.4 cm × 53.34 cm). It consists of 49 double pages . . . The book is divided into seven parts, consisting of seven pages, by thin slips of mahogany, which serve to open it easily at each of these divisions. Every one of these seven divisions contains seven pages, and each page contains forty nine words [15].

Edgeworth found from experiments that six men – one at each of the pointers, one at the observing telescope, and another at the vocabulary – could send a word in 20 s. Otherwise, with just one man to work the station the rate of message transmission was one word per minute. With his son, Edgeworth, on 4th August 1795, successfully demonstrated before 'a vast concourse of people' [15] the sending of four messages across the channel between Donaghadee and Port Patrick.

In October 1796, following an attempted invasion by the French, Edgeworth petitioned the Lord Lieutenant of Ireland and offered to establish a telegraph system between Dublin and Cork, a distance of c. 180 miles (290 km), by means of 14 or 15 stations, at a cost of £100 per station. Initially, Edgeworth was much encouraged by T. Pelham, the Chief Secretary to the Admiralty, and various demonstrations were given in Ireland and in England to notable personages, including the Duke of York. Unfortunately for Edgeworth, his naivety in business matters was not conducive to either his well-being or the acceptance of his telegraph. As his biographer has written: 'The entire administration in Ireland seethed with nepotism and graft, and public expenditure on defence or any other branch of the service flowed into the pockets of a host of contractors and castle-hacks' [15]. Edgeworth became aware of this state of affairs for, in a letter to Lord Charlemont, the Commander-in-Chief, he opined: '. . . had I introduced my telegraph in the form of a lucrative job in which there might be good pickings for others, I might have increased the number of my friends, and have gratified those who are in power, by an opportunity of increasing patronage' [15].

Feeling much frustrated and bitterly disgusted by his treatment, Edgeworth sought to expose the chicanery of Pelham and the sophistry of others and, early in 1797, published a pamphlet entitled 'A letter to Rt Honourable the Earl of Charlemont on the Tellograph and on the Defence of Ireland'. It did nothing to advance Edgeworth's cause.

Shortly before the pamphlet's publication, on 15th December 1796 a French fleet, consisting of 17 warships and 13 frigates with 15,000 soldiers, sailed from Brest with the objective of invading Ireland. The intended invasion was at the behest of the United Irishmen – a patriotic organisation established in 1791 – who had entered into secret negotiations with the French Directory.

The impending assault on Ireland demanded a good system of communications so that rapid mobilisation of the country's defences could be effected. It was a situation for which Edgeworth's tellograph stations would have been invaluable, but none were available. There was a prospect that the small English garrison could be overwhelmed in the absence of reinforcements. Adventitiously, for the British Government, misadventure characterised the expedition. One ship sank in calm weather with the loss of 1,200 men, four ships collided, others went astray, and just half the fleet anchored in Bantry Bay on the south west coast of Ireland. The 7,000 men who reached the bay could not be landed because of strong gales, and the ships carrying the expedition's guns and money failed to arrive. After a few days, the fleet's commander decided to abandon his objective and return to Brest, and the small English garrison, with its poor communications to the main English field forces, escaped almost certain defeat. It would have been powerless to have opposed the full might of the invasion force. Incompetence, and not technology, saved the day; but elsewhere – during the invasion of Bavaria by Austrian forces, a few years later – the Chappe telegraph played a crucial role in the eviction of the invaders. The episode also inspired the development of the world's first telegraph based on electric currents.

References

1 KING, H.C.: 'The history of the telescope' (Griffin, London, 1955)
2 Ibid, p. 26
3 GUNTHER, R.J.: 'Early science at Oxford' (Dawsons, London, 1966)
4 HOOKE, R.: 'Discourse to the Royal Society, shewing a way how to communicate one's mind at great distances'; reproduced in Gunther, op. cit., pp. 658–664
5 FFOULKES, C.: 'Notes on the development of signals used for military purposes', *Journal of the Society for Army Historical Research*, **xxii**, pp. 20–27
6 A.F.P.: article on Edward Somerset, Dictionary of National Biography, **xviii**, pp. 640–645
7 WORCESTER, Marquess of.: 'Century of the names and scantlings of such inventions as at present I can call to mind to have tried and perfected' (London, 1666)
8 APPLEYARD, R.: 'Pioneers of electrical communication' (Macmillan, London, 1930)
9 MOTTELAY, P.F.: 'The bibliographical history of electricity and magnetism' (Griffin, London, 1922)
10 WILKINS, J.: 'Mercury, or the secret and swift messenger, showing how a man, with privacy and speed, may communicate his thoughts to a friend at any distance' (London, 1641)
11 PORTA, J.B.: 'Natural magic in xx books' (London, 1658)
12 STRADA, F.: 'Prolusiones Academicae' (Oxford, 1662)
13 LEURECHON, J.: 'La récréation mathématique', 1628, (English version published by T. Cotes, London, 1633)
14 BROWNE, T.: 'Pseudodoxia Epidemica, or inquiries into vulgar and common errors' (London, 1646)
15 CLARKE, D.: 'The ingenious Mr Edgeworth' (Oldbourne, London, 1965), chapter 10, pp. 137–151
16 KOENIG, D.: 'Telegraphists and telegrams in revolutionary France', *Science Monthly*, 1944, pp. 431–437
17 FAHIE, J.J.: 'A history of electric telegraphy to the year 1837' (F.N. Spon, London, 1884)
18 BATHELEMY, J.J.: 'Voyage du Jenne Anarcharis' (Paris, 1788)
19 WILSON, G.: 'The old telegraphs' (Phillimore, London, 1976)
20 Ref. 19, p. 122
21 Ref. 19, p. 124

22 Ref. 19, p. 133
23 A.F.P.: article on Lord George Murray, Dictionary of National Biography, (Oxford University Press, London, 1949), p. 1258
24 GAMBLE, J.: 'An essay on the different modes of communication by signals' (London, 1797), p. 70
25 POPHAM, H.: 'Telegraph signal' (London, 1800); and 'Telegraphic signals or marine vocabulary' (London, 1812)
26 ANON.: Article on the 'Telegraph', Encyclopaedia Britannica, (London, 1797), XVIII, Part 1, pp. 334–337

Chapter 3

The development of electric telegraphy from c. 1750–1850

Prior to 1753 no references occur in the literature to the use of electrified bodies for the purpose of signal communication, although the attractive power of amber when rubbed was known to Thales about 600 BC.

Essentially telegraph systems based on the science of electrostatics require a charge generator, a transmission path, and a charge detector for their successful working. Otto Guericke [1] (1602–1686) in 1675, was the first person to construct an electrical machine. He used a globe of sulphur as the substance to be excited and found that light bodies suspended within its sphere of action themselves became excited. Guericke was a contemporary of Boyle who in the same year had published his book on *Experiments and notes about the mechanical origin of electricity*. The publication of this work, following the *De Magnete* of W. Gilbert [2] (1544–1603) in 1600, marked a re-awakening of interest in the subject of electricity, and effectively heralded the birth of a new branch of physics.

Other machines followed Guericke's discovery and in 1729 Grey and Wheeler made an observation which completed the requirements for an electric telegraph. In that year they succeeded in producing, at a distance of 203 m, motion in light bodies, using pack thread and 'fictional electricity'. C.F. de C du Fay's experiments [3] (1733–1737) showed that wetting the pack thread increased its conducting power and G.M. Boze [4] (1710–1761) in 1741 introduced the 'prime conductor' suspended from silk threads. The classification of 'bodies' into 'electrics' (insulators) and 'non-electrics' (conductors) was made in an essay by J.T. Desaguliers (1683–1744) in 1742, and by 1747 Dr Watson [5], Bishop of Llandaff, had transmitted electricity through 933 yd (853 m) of wire and 2666 yd (2438 m) of water. At about the same time, in 1745, E.G. von Kleist and Muschenbrock [6] had simultaneously discovered the Leyden jar and during the next few years various applications of electricity had been put forward. Thus a Mr Maimbray [7] in October 1746 at Edinburgh electrified two myrtle trees for a month and found that they put 'forth leaves and blossoms sooner than those that had not been electrified'. In 1748 Benjamin Franklin [8] (1706–1790) had performed his celebrated experiments on the banks of the Schuylkill in North America and had concluded them with 'a picnic, when spirits were fired by an electric spark sent through the river, and a turkey was killed by the electric shock and roasted by the electric jack before a fire kindled by the electrifying bottle'. Finally, in 1748, J.L. Jallabert [9] (1712–1768), at Geneva, had 'entertained the idea of submitting some invalids to electrical treatment'.

With the component parts of an electric telegraph now available and with the considerable interest being shown in the discovery of new principles and concepts and the application of the science of electrostatics to everyday life, it was perhaps inevitable that someone would propose a scheme for an electric telegraph. This was mentioned in a most interesting and remarkable letter published in the Scots magazine [10] on 17th February 1753. The letter, sent from Renfrew, was signed by a writer known only by his initials, CM. Sir David Brewster [11], after some painstaking work, opined that it was very probable that the CM was identical with a Charles Morrison of Greenock. Another suggestion was that CM was Charles Marshall, 'a person of whom an aged lady says that he was a very clever man who had formerly resided in Renfrew and who could make lightning speak and write upon a wall' [12].

In his letter CM proposed that a set of wires, equal in number to the letters of the alphabet should be provided between two given places to enable a conversation with a distant friend to be carried out. From each wire at the receiving end a ball was to be suspended, and about a sixth or an eighth of an inch (4.23 mm to 3.17 mm) below the balls were to be placed the letters of the alphabet, marked on bits of paper that were light enough to rise to the electrified ball.

Having set the electrical machine a-going as in ordinary experiments, suppose I am to pronounce the word Sir; with a piece of glass, or any other electric per se, I strike the wire S, so as to bring it in contact with the barrel, then i, then r, all in the same way: and my correspondent, almost in the same instant, observes these several characters rise in order to the electrified balls at his end of the wires [10].

So far as is known CM's proposal was never tested, and it would have proved very expensive. CM was aware of the difficulty of the method concerning the suspension and insulation of the wires and suggested: 'At every 20 yards [18.3 m], let them be fixed in glass, or jewellers cement, to some firm body, both to prevent them from touching the earth or any other non-electric and from breaking by their own gravity' [10]. Again at the end of his letter he wrote:

Some may perhaps think that, although the electric fire has not been observed to diminish sensibly in its progress through any length of wire that has been tried hitherto; yet as that has never exceeded some 30 or 40 yards [27 or 36 m], it may be reasonably supposed that in a far greater length it would be remarkably diminished, and probably would be entirely drained off in a few miles by the surrounding air. To prevent the objection, and save longer argument, lay over the wires from one end to the other with a thin coat of jeweller's cement. This may be done for a trifle of additional expense; and as it is an electric per se, will effectually secure any part of the fire from mixing with the atmosphere [10].

Though CM's suggestion was entirely practical given suitable materials, the first real attempt to use fictional electricity for the transmission of signals between two rooms was made by G.L. Lesage [13] in 1774 in Geneva (Fig. 3.1). His apparatus consisted of 24 metallic wires insulated from each other and connected to separate electrometers formed of small balls of elder held by threads and each marked with different letters of the alphabet. In another system for communicating between buildings he proposed to employ a subterranean tube of glazed earthenware, divided at intervals of 6 ft (1.83 m) by partitions with 24 separate openings intended to hold apart that number of wires, the extremities of the wires being

arranged horizontally, like the keys of a harpsichord, each wire having suspended above it a letter of the alphabet, while immediately underneath upon a table were pieces of gold leaf, or other bodies that could be as easily attracted, and were at the same time easily visible [13].

This suggestion was submitted to Frederick of Prussia but was never implemented.

Lesage's ideas seem to have been improved by Lomond [14] who in 1787 is claimed to have reduced the number of wires to one: however, details of his scheme are somewhat sketchy.

Figure 3.1   Le Sage's telegraph.
Source: J. IEE, 1931, **69**(419).

You write two or three words on the paper; he takes it with him into an adjoining room and turns a machine in a cylinder case, on the top of which is an electrometer having a pretty little ball of pith of a quill suspended by a silk thread; the brass wire connects it to a similar cylinder and electrometer in the distant apartment, and his wife, on observing the movements of the corresponding ball, writes the words which it indicates. From this it appears that he has made an alphabet of motions. As the length of the brass wire makes no difference in the effect, you could correspond with it at a great distance, as for example with a besieged city or for objects of much more importance. Whatever be the use that shall be made of it, the discovery is an admirable one.

During the latter half of the 18th century many proposals for electric telegraphs were advanced, namely, those of:

1753	CM[15], 26 wires, and balls
1767	J. Bozolus[16] (unknown dates), similar to that of CM
1773	L. Odier[17] (1748–1617), unknown
1774	G.L. Lesage[18] (1724–1803), 24 wires, and electrometers
1782	S. Linguet[19] (1736–1794), 24 wires, and balls
1787	C.J.B. Lomond[20] (1749–1830), 1 wire and electrometer
1787	A. de Betancourt[21] (1760–1826), 1 wire
1794	M.R. Reizen[22] (unknown), 26 wires and illuminated letters
1795	M. Cavallo[23] (1749–1809), 1 wire and sparks
1796	D.F. Salva[24] (1747–1808), 42 wires and 22 men
1797	D.F. Salva[24], 1 wire
1816	F. Ronalds[25] (1788–1873), 1 wire and electrometer

Odier's telegraph is not known, but in a letter to a lady friend Odier, a distinguished physician of Geneva, captured something of the enthusiasm of the period:

I shall amuse you, perhaps, in telling you that I have in my head certain experiments, by which to enter into conversation with the Emperor of Mogul or of China, the English, the French, or any other people of Europe in a way that, without inconveniencing yourself, you may intercommunicate all that you wish, at a distance of four or five thousand leagues in less than half an hour! Will that suffice you for glory? There is nothing more real. Whatever be the course of those experiments, they must necessarily lead to some grand discovery; but I have not the courage to undertake them this winter. What gave me the idea was a word which I heard spoken casually the other day, at Sir John Pringle's table, where I had the pleasure of dining with Franklin, Priestly and other great geniuses [17].

Linguet had a similar vision: 'I am persuaded that in time it will become the most useful instrument of commerce for all correspondence of that kind' [19].

The most important advance in signalling using electrostatic methods was the reduction in the number of wires employed in the transmission path. Lomond was the first person to conceive such a system if the scanty evidence available can be relied upon. Cavallo's scheme depended on the number of sparks made by the apparatus to designate the various signals and in Salva's single wire method the details are not known with any accuracy.

The most ingenious and complete of the above methods, and one of the last of the telegraphs to depend on static electricity, was that proposed by Mr Francis Ronalds, in 1816, when he was 28 years of age. The details of the telegraph are fully described and illustrated in his *Description of an electrical telegraph and of some other electrical apparatus* which was issued in pamphlet form in 1823 and reprinted in 1871, and which was the first work on electric telegraphy to be published.

For his experimental line, Ronalds [26]

erected [in the garden of his London home in Upper Mall, Hammersmith] two strong frames of wood at a distance of 20 yards [18 m] from each other, and each containing 19 horizontal bars; to each bar he attached 37 hooks, and to the hooks were applied as many silken cords, which supported a small iron wire (by these means it was well insulated), which (making its inflections at the points of support) composed in one continuous length a distance of rather more than eight miles [12.9 km] (Fig. 3.2).

After making many experiments with this overhead line, he then laid one underground:

A trench was dug in the garden 525 ft [160 m] in length and four feet [1.22 m] deep. In this was laid a trough of wood two inches [50.8 mm] square, well lined on the inside and outside with pitch, and within this trough thick glass tubes were placed, through which the wires ran [27].

The transmitting and receiving apparatuses each consisted of an ordinary electric machine and a pith ball electrometer arranged in the following manner.

He placed two clocks at two stations; these two clocks had upon the second hand arbor a dial with 20 letters on it; a screen was placed in front of each of these dials and an orifice was cut in each screen, so that only one letter at a time could be seen on the revolving dial. The clocks were made to go isochronously; and as the dials moved round the same letter always appeared through the orifices of each of these screens. The pith ball electrometers were hung in front of the dials. The attention of the observer was called through the agency of an inflammable air gun fired by an electric spark. (Actually the clock dials moved synchronously and not just isochronously as stated by the above writer.)

When asked how he would deal with vandals who might damage his telegraph wires Ronalds responded: 'hang them if you can catch them, damn them if you cannot, and mend [the wires] immediately in both cases'.

Here was another example of a communications system that utilised the principle first put forward by Aeneas in 341 BC – the principle which Chappe had adopted in his first telegraph system. Ronalds' scheme was based soundly on experimental observations, although it would have been very slow in operation. Like other inventors such as Aeneas and Chappe he used a coding arrangement to transmit his messages.

Ronalds made strenuous efforts to bring his invention before the British Government but was met with little encouragement: 'Telegraphs of any kind are now wholly unnecessary and no other than the one now in use will be adopted', wrote Mr John Barrow, Secretary of the Admiralty, in 1816 [28]. (The one

(a)

(b)

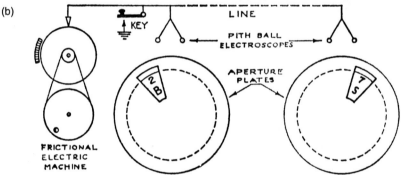

Figure 3.2   (a) The illustration shows Ronalds' apparatus in the foreground and the
structure which he used to support his line conductor. (b) Diagram illustrating
the principle of Ronalds' telegraph (see text).
Source: F. Ronalds, *Descriptions of an electric telegraph* (London, 1823).

alluded to was the semaphore telegraph between London and Portsmouth.)
'I felt', observed Ronalds, 'very little disappointment and not a shadow of
resentment because everyone knows that telegraphs have long been great bores
at the Admiralty.'

   Their response was similar to that given to R. Wedgwood (a member of the
famous family of pottery manufacturers), when in 1814 he submitted his telegraph proposals to Lord Castelereagh. Wedgwood was informed that the war
being at an end, the old system was sufficient for the country. Earlier in February
1813 J.R. Sharpe had 'exhibited' an experiment before The Right Honorable the

Lords of the Admiralty demonstrating 'the advantages to be obtained from the application of the certain and rapid motion of the electric principle through an extensive voltaic circuit to the purpose of the ordinary telegraph', but had been told that as the war was over and money scarce they could not carry it into effect [13].

Because of the Admiralty's uninterest Ronalds did not apply for a patent. He abandoned the subject of telegraphy and applied his intellect and energies to meteorology, becoming, from 1843 to 1852, Honorary Director of the Royal Observatory at Kew, and a Fellow of the Royal Society. In 1852 he received an annuity of £75 per annum from the Crown for his important discoveries in electricity and magnetism and subsequently spent much time completing and cataloguing his unique collection of more than 5,000 books – in various languages – relating to the literature in these fields. (He donated his library to the Institution of Electrical Engineers.) Ronalds had some reward for his enterprise: he was knighted for services to telegraphy, in 1870, the year before he died. For this he could 'thank a good constitution for not having been one among many benefactors of mankind whose services are only appreciated after death' [28].

The above examples illustrate the British Government's view on electric telegraphs in the first quarter of the 19th century but possibly Ronalds' system was doomed to be consigned to the scrap heap of unused inventions notwithstanding the Admiralty's opinion. His scheme necessitated the use of synchronous clocks at the transmitter and receiver; the pith ball electrometer detector was not capable of being easily replaced by a multiplier to extend the distance of working as was possible with later electromagnetic telegraphs; the transmitting apparatus was cumbrous compared to that which Volta's great discovery allowed; and it had a slow rate of message transmission. Nevertheless, it was a carefully worked out scheme and probably represented the best that could be achieved at that time using electrostatic principles. Its emergence came during an inopportune period for a new system: had Ronalds worked on his system 20 years earlier when Salva was carrying out his experiments and the Admiralty was contemplating telegraphic communications following the success of the Chappe method the position might well have been different.

The invention which effectively led to the emergence of electrical and electronic engineering was made known in a letter [29] dated 20th March 1800 written by A. Volta (1745–1827), a Fellow of the Royal Society, to Sir Joseph Banks, the President of the Royal Society. In it Volta announced his discovery of the voltaic pile, an arrangement consisting of discs of silver, zinc and moistened card placed in series. 'La Couronne de Tasses' was afterwards devised and used a circle of cups, each cup containing a solution of salt, silver and zinc, thus forming an electric battery. Progress in the new science of electrodynamics developed rapidly: batteries were improved by Wollaston, Trommsdorff, Dyekhoff, Behrens and others [30], so that by 1808 Napoleon was able to present a 'trough galvanic battery', of 600 pairs of plates, to the Polytechnic School in Paris, and by 1813 the Royal Institution's battery comprised of 2,000 pairs of zinc and copper plates, each having a surface area of 32 square inches (206 cm^2).

Following Volta's discovery, the electrolysis of liquids using a pile was discovered in 1800 by Nicholson and Carlisle, although the same phenomenon had been observed in 1797 by Pearson, an English physician, using the discharges from Leyden jars, and in 1789 by Deiman and van Troostwijk in Amsterdam by the use of a frictional machine. Much work on electrolysis was undertaken after 1800 by Henry, Hisinger and Berzelius, and Davy. This work led to the first use of electrolysis, as a practical detector of electric currents, in Soemmering's electric telegraph of 1809. As on previous occasions the development of the telegraph was hastened by an national emergency.

On the 9th April 1809 the Austrian army began to cross the river Inn and enter Bavaria. King Maximilian of Bavaria, on being informed of this, on the 11th, retired in haste with his family to the town of Dillingen, situated on the Western frontier of his kingdom. He took with him Baron von Monteglas who was head of the foreign and home departments – the two most important branches of administration – in Bavaria [31].

By means of Chappe's telegraph, which had been established from the French frontier to Paris, Emperor Napoleon learnt, on the 11th, of the Austrian invasion much sooner than it had been thought possible by the Bavarians. Without delay he departed from Paris for Bavaria to be with his army. He arrived so unexpectedly that he found King Maximilian in bed.

There is no doubt that the deliverance of Bavaria from the Austrians was due to the swift arrival of Napoleon in the midst of his army. Munich had been occupied from 16th April by the Austrian General Jellachich and his forces, but in less than a week, on the 22nd, he was obliged to withdraw and King Maximilian was again able to enter his capital.

This episode highlights the central role which good communications can play at a time of national emergency.

One of the witnesses to Napoleon's surprise arrival at Dillingen was Baron von Monteglas. Under his extensive administration had been placed the Munich Academy of Sciences, a member of which was Dr S.T. Soemmering (1775–1830), a well known anatomist and physiologist. Soemmering had often dined with Monteglas and on 5th July 1809, at one of their dinners, the Minister asked the anatomist to obtain from the Academy of Sciences proposals for a telegraph system. Monteglas had probably been so impressed with the Chappe system that he felt a suitable telegraph scheme would be an advantage to his country. Fortunately, Soemmering was aware of the discoveries that had been made in galvanism, as he had hoped that some of the problems of physiology would be solved by a study of the new science. The difficulty that faced Soemmering stemmed from an essential lack of knowledge of the subject as a whole: only the detection of a current by its decomposition of water into hydrogen and oxygen was known at the time.

Soemmering applied himself to Monteglas's request by endeavouring to determine whether this discovery could be used in telegraphy. On 6th August he noted in his diary: 'I tried the entirely finished apparatus, which completely answers my expectation [31].' He was able to work his telegraph through 724 ft

(220.7 m) of wire and 12 days later had increased the distance to 2,000 ft (609 m). On the 29th August, less than two months after the Bavarian Minister had made his request, Soemmering exhibited his telegraph before a meeting of the Academy of Sciences. Later, on 5th December 1809 the apparatus was presented to the National Institute in Paris and Biot, Carnot, Charles and Monge were appointed to report on it.

In his system [32] (Fig. 3.3) Soemmering used 35 wires (which represented the 25 letters of the alphabet and the 10 numerals) each of which terminated in a pin which projected through the base of a glass vessel filled with acidulated water. By connecting any pair of the wires to the extremities of a Volta pile, he was able to cause bubbles to rise from the appropriate pins. Later he reduced the number of wires to 27 and by 23rd August 1810 he was able, with the aid of a very ingenious mechanical contrivance, to use the device to ring an alarm bell to call the attention of the operator.

Actually, Soemmering's work had been anticipated by Salva who, in 1804, had appreciated that Volta's pile 'yields more fluid than the electric machine, and could be well applied to telegraphy, as the force can be obtained more simply and more steadily than in the static form.'[33] In a paper read to the Academy of Science in Barcelona in 1804 Salva [33] gave a full description of an

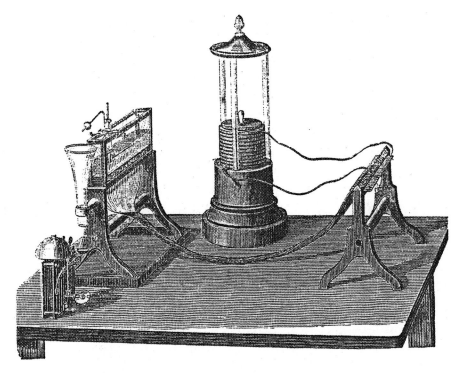

Figure 3.3 Soemmering's telegraph.
Source: A.R. von Urbanitzky, *Electricity in the service of man* (Cassell, London, 1886).

electro-chemical telegraph in which bubbles of hydrogen and oxygen were used for indicating purposes, but unfortunately for his prestige his paper was not published at that time and the method had to be re-invented by Soemmering.

Ten days before Soemmering's demonstration, Baron Pawel Lwowitsch Schilling von Canstatt (1786–1837), a nobleman of Wurtemburg origin, who was attached to the Russian Mission at Munich, had seen Soemmering's telegraph and had become very enthusiastic about the new art of signalling. He was so forcibly struck with 'the probability of a very great usefulness of the invention, that from that day galvanism and its applications became one of his favourite studies' [31]. He drew the attention of many persons to it including in particular, in 1811, Count Jeroslas Potocky, a Russian Colonel of the Engineer Corps. He took a telegraph made for him to Vienna and exhibited it on the 1st July to the Emperor Francis I, in the presence of the Archdukes Charles and John, with the result that the Emperor expressed a wish to have a telegraphic communication link established between Vienna and his country palace at Laxenburg, 9 miles (14.5 km) distant.

Baron Schilling was also the link by which the British envoy and Minister Plenipotentiary at Munich, the Honourable F.J. Lamb (the youngest brother of Lady Palmerston and subsequently the third Lord Melbourne) witnessed experiments with the apparatus on the 12th July 1816, being the first Englishman to do so [34].

Schilling maintained his contact with Soemmering for many years, by correspondence and visits, and in 1818 Soemmering wrote to him and said: 'Never has anybody except yourself at the first glance comprehended everything so clearly, and conceived how easily the telegraph might be applied on a large scale' [31]. Subsequently Schilling made a number of contributions to the art of electric telegraphy as mentioned later in this chapter.

Although by 15th March 1812 Soemmering had telegraphed messages through 10,000 ft (3040 m) of wire the telegraph was never practically used. Even by 1820 it was still only being demonstrated to visitors. The importance of Soemmering's achievement was to show that signalling could be developed using continuous electric currents, to serve as an inspiration for others, including Schilling and Gauss, and to enable the development of an economical and practical system to be achieved. It is likely that the anatomist's telegraph was not used because of the success of the Chappe apparatus, the difficulty in obtaining suitable insulated wires, and the costly use of 27 wires between the transmitter and receiver. The Chappe telegraph had given good service, it was reliable and in addition considerable capital sums had been expended on its development and installation: hence, until some revenue producing scheme had been suggested, it was not going to be replaced by an electric telegraph of uncertain reliability and limited range of working. In France the optico-mechanical apparatus of Chappe was in use until 1850, and in England the last of the semaphore stations was not closed until 31st December 1847.

In the spring of 1812, Baron Schilling carried out some experiments on devising a conducting cord which was so sufficiently insulated that it would transmit currents not only through wet earth but also through long stretches of water. The war then impending between France and Russia made Schilling anxious to find an insulated conductor which would serve for telegraphic communication purposes between fortified strongholds and the field and also for exploding powder mines across rivers. By the autumn of 1812 he had succeeded in exploding this type of mine from across the river Neva, near St Petersburg [35].

Later, in the 1830s, Schilling constructed an electromagnetic telegraph as did Wheatstone and Cooke in England, Gauss and Weber, and Steinheil in Germany, and Morse, Gale and Vail in America. A number of major discoveries, not available to Soemmering, led to these inventions and completely transformed the art of signalling in the first half of the 19th century.

For many years there had been some speculation as to the possible connection between electricity and magnetism, but it was not until the 21st July 1820 that H.C. Oersted (1770–1851) in Copenhagen published a circular giving details in Latin on: 'Experimenta circe effectum conflictus electrici in acum magneticam'. Dr Hamel, a member of the Imperial Academy of Sciences at St Petersburg, in the 19th century, has argued that Oersted must have known about G.D. Romagnosi's important discovery which he published on 3rd August 1802 in a newspaper at Trent and which began: 'Il signore Consigliere Giandomenico Romagnosi si affrettta a communicare ai Fisici dell' Europa uno sperimento relativo al fluido galvanico applicato al magnetismo'. Hamel did not explain why Oersted should have waited 18 years to make his announcement, though in a work published in 1807 Oersted proposed 'to try to whether electricity in its latent state will not affect the magnetic needle' [36].

Oersted's experiments were soon repeated by scientists in many countries and in the same year, on 2nd October 1820, A.M. Ampère [37,38] (1775–1836) proposed a telegraph based on Oersted's findings but did not actually construct the apparatus. From the point of view of coding Ampère's proposal was similar to CM's suggestion of 1753, and to Soemmering's experiments, since he considered using a separate wire and a separate needle, for observation purposes, for each letter of the alphabet.

Schweigger (1779–1857), a chemist of Halle, soon improved upon Oersted's discovery and in 1820 found that by taking a silk covered copper wire and coiling it around several times, the deflections of a magnetic needle suspended in it became 'extremely sensible'. Other galvanometers followed this observation but on 25th September 1820 Arago (1786–1853), Professor of mathematics at the Polytechnic School in Paris, stated to the Royal Academy of Sciences that 'he had ascertained the attraction of iron filings by the connecting wire of the battery, exactly as by a magnet'. Sturgeon (1783–1850), of Woolwich, used this knowledge to construct in 1825 the first electro-magnet.

Thus by the above date all the elements of an electromagnetic telegraph were available, namely: the battery, the magnetic needle, the coil or 'multiplier', and the electromagnet.

Schilling saw the possibility of employing Schweigger's multiplier as part of the indicating apparatus of an electric telegraph and in the decade from 1820 to 1830 he carried out many experiments with it. The horizontal needle of Schilling's detector was hung by a silken thread, within the multiplier, and was capable of rotating about a vertical axis. In order to ensure a motion without oscillations, a strip of platina plate was fixed to the lower extremity of the needle's vertical axis and immersed in a cup of mercury. In addition this suspension carried a circular disc of paper painted with a black horizontal line on one side and a black vertical line on the other. Schilling used five multipliers and six wires in his scheme, each letter or numeral being signalled by a particular combination of horizontal and vertical lines simultaneously visible to the receiving operator. As each indicator could take up one of two positions, Schilling could achieve $2^5 = 32$ different combinations with his five discs. He required just five wires, plus an earth return, instead of the 35 wires of Soemmering's system. Schilling was the first person to apply the important principle of the binary code to electric telegraphy. In his system the sides of the first disc represented 1 and 2, the sides of the second disc 3 and 4, and so on. Each combination of five numbers stood for a given word which was defined in a dictionary of numbers. The calling signal actuated an electromagnetic trip which released a clockwork alarm bell [39]. (Fig. 3.4 shows a six multiplier telegraph and its coding.)

In 1830 Schilling undertook a journey to China, but on his return in 1832 he again occupied himself with his invention [40]. The five needle telegraph was demonstrated to Czar Nicholas in that year in Berlin and he seems to have given Schilling much encouragement. The instrument was also exhibited in the section on natural philosophy and chemistry (over which Professor G.W. Muncke, of the University of Heidelberg, presided), at a meeting of the German Naturalists at Bonn on 23rd September 1835. Two years later Schilling ordered a submarine cable from a rope manufacturer in St Petersburg so that he could connect Kronstadt through the Gulf of Finland with the capital (St Petersburg) for telegraphic correspondence. The work was not completed as Schilling died in 1837 but his system of coding was used in a very similar form in the needle telegraph of Cooke and Wheatstone.

The development of the electric telegraph in England and its practical application was due almost entirely to the efforts of William Fothergill Cooke (1806–1879) and Charles Wheatstone (1802–1875). Since their combined efforts led to the establishment of electric telegraphy in the United Kingdom a few words on their backgrounds are apposite.

Cooke was born on the 4th May 1806 near Ealing, Middlesex, where his father was a surgeon. Later, the family moved to Durham following the appointment in 1833 of Dr William Cooke to the post of Professor of anatomy at the newly established University of Durham. Cooke was educated at Durham and at the University of Edinburgh, and at the age of twenty entered the East India Company's Army, in which subsequently he held a variety of posts. After about six years' service he returned to England and because of poor health resigned his appointment. He travelled to Paris in 1833 and in 1834 and attended lectures in anatomy and physiology, intending to follow his father's profession.

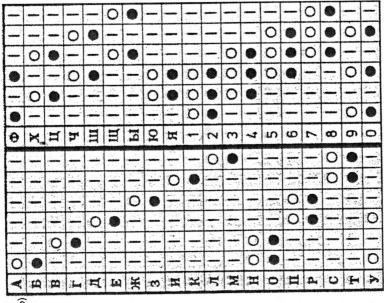

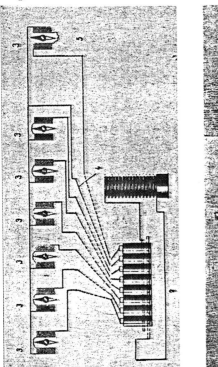

Figure 3.4 *(a) Schilling's telegraph, and (b) code.*
Source: R. Oberliesen, *Information, Daten und Signale* (Deutsches Museum, Rowohlt, 1982).

While in Paris, Cooke became adept at modelling anatomical dissections in coloured wax and in the spring of 1834 he returned to Durham and prepared a set of models that his father used to illustrate his lectures. The following year Cooke accompanied his parents on a tour of Switzerland and during their journey visited Heidelberg. Here Professor Tiedemann, the director of the Anatomical Institute, offered to assist Cooke in procuring the means to make his anatomical models. He availed himself of this offer and, following a visit to Berne, he returned to Heidelberg in November and took lodgings at no. 97 Stockstrasse. The house belonged to a brewer named Wilhelm Speyrer and bore an unusual inscription: 'Bierbraueri zum neuen Essighaus' ('Beer brewery at the vinegar house'). Cooke was not allowed to engage in wax modelling in this house and so rented a room at no. 58 Stockstrasse. Here, during the winter months of 1835–36, Cooke spent many industrious hours and was able to despatch four cases of models to his father in Durham.

On his period in Heidelberg Cooke has written [41]:

In the month of March 1836 I was engaged at Heidelberg in the study of anatomy, in connection with the interesting and by no means unprofitable profession of anatomical modelling, a self-taught pursuit to which I had been devoting myself with incessant and unabated ardour, working frequently 14 or 15 hours a day, for about 18 months previous. About the 6th March 1836, a circumstance occurred which gave an entirely new bent to my thoughts. Having witnessed an electro-telegraphic experiment exhibited about that day by Professor Muncke of Heidelberg who had, I believe, taken his ideas from Gauss, I was so much struck with the wonderful power of electricity and so strongly impressed with its applicability to the practical transmission of telegraphic intelligence, that from that very day I entirely abandoned my former pursuits and devoted myself thence-forth with equal ardour, as all who know me can testify, to the practical realisation of the electric telegraph, an object which has occupied my undivided energies ever since. Professor Muncke's experiment was at that time the only one upon the subject that I had seen or heard of. . . . His apparatus consisted of two instruments for giving signals by a single needle . . . the signals given were a cross and a straight line marked on the opposite sides of a disc of card fixed on a straw, at the end of which was a magnetic needle suspended horizontally in galvanometer coils by a silk thread. . . . Within three weeks after the day on which I saw the experiment, I had made, partly at Heidelberg and partly at Frankfurt, my first electric telegraph of the galvanometer form, which is now at Berne.

The apparatus that Cooke saw was Schilling's and not that of Gauss as he stated. Schilling had shown his telegraph at the German Naturalist's meeting in Bonn in 1835 [40]. Muncke, who from 1826 had been an honorary member of the Imperial Academy of Sciences at St Petersburg, was much pleased with Schilling's instrument, and he determined at once to get one for exhibition at his lectures. (Dr Hamel later discovered, in Heidelberg, the apparatus which had been made in imitation of the one exhibited by Baron Schilling at Bonn.) Cooke's telegraph had three needles and six wires in three separate circuits; with three keys and a rudimentary switch it was able to give 26 signals.

At about this time Cooke felt that he needed the expert assistance of an instrument/clockmaker and so returned to London [42]. He arrived in the capital on 22nd April 1836, immediately applied himself to the design and construction of an electric telegraph, and sought the aid of one of the experienced instrument makers of Clerkenwell – the centre of the clockmaking industry in London.

Cooke was quite decided on the uses that an electric telegraph system could have and in June drafted a prospectus of his invention, which he intended to issue as a pamphlet. (It was not printed until 1866.) The pamphlet was headed 'Plans for establishing a rapid telegraphic communication for political, commercial, and private purposes, in connection with the extended lines of railroads now in progress between the principal cities of the United Kingdom, through the means of electro-magnetism. By W.F.C.'.

The history of communications prior to the second quarter of the 19th century has shown that, notwithstanding the crucial importance of signalling in wartime and times of emergency, systems for transmitting intelligence tended to lapse into disuse in peacetime and new ideas remained under-developed. Fortunately for the advance of electric telegraphy the early part of the 19th century witnessed the emergence of the railway as a means of fast transport. Necessarily a complex pattern of railway lines required some form of signalling for its safe and efficient operation. Here surely was a use for the electric telegraph.

The first railway locomotive was made by Richard Trevithick in 1804, five years before Soemmering demonstrated his electro-chemical telegraph. Although one of Trevithick's locomotives was used to haul iron ore along the cast iron tramway from the Pendarren iron works to the Glamorgan canal – 'the first and only self-moving machine that ever was made to travel on a road with 25 tons (25,400 kg) at 4 m.p.h. (1.8 m/s) and completely manageable by only one man' – these early locomotives were very heavy and often broke the lines. George Stephenson made a number of improvements and in 1814 built his first locomotive, the *Blucher*. Later, as a result of the fame that he acquired as the builder of railway engines, he was asked in 1821 to undertake a survey for the first public railway in Britain. The survey for the Stockton and Darlington railway was completed in a fortnight and the first rail laid in May 1822. When the line was opened on 27th September 1825 Stephenson had commenced work on the second railway between Liverpool and Manchester. Later in 1830 George Stephenson's son, Robert, surveyed the route for the longest railway then projected. This was for the London and Birmingham Railway, which was to start at Euston, pass through Camden Town, Watford, Tring, Wolverton, Blisworth, Kilsby and then on through Rugby and Coventry to Birmingham. Work began in 1833 and in 1838 the line was opened. The task involved a tremendous effort: at different times between 12,000 and 20,000 men were involved and when the line was completed the cost was £5.5 million [43].

Cooke meantime had read Dr D. Lardner's book on the steam engine and had learnt that on a steep stretch of railway line it was necessary to have a spare engine to provide extra traction to passenger and goods trains. He saw at once that the expense and inconvenience of maintaining a locomotive in continuous readiness at the foot of an incline might be avoided by a communication system which gave a warning of the approach of a train. Furthermore, the value of the telegraph to Government would allow it 'in [cases] of disturbances [such as Peterloo, for example] to transmit [its] orders to the local authorities and, if necessary, send troops for their support; while all dangerous excitement of the public might be avoided' [42]. Additionally, Cooke foresaw the telegraph's application to commerce for sending information, such as the prices of stocks and shares, quickly from London to the provinces. Again, the telegraph could be of assistance to a family at a time 'when sickness appear[ed] [to be] hastening towards a fatal termination with such rapidity that a final meeting [was] without the range of ordinary means' [42].

On 18th November 1836, after many months of effort, frustration, and much expenditure – £361.73p by August 1836 – Cooke received from Moore, his Clerkenwell instrument maker, the finished apparatus. Alas, it did not satisfy his expectations. 'My instrument came home late last night but does not answer [my needs]' [42]. He felt in need of some scientific guidance and consulted Michael Faraday, the great 19th century experimentalist of electrical science. Faraday called to see the apparatus and averred that 'the principle was perfectly correct', and the instrument was probably capable 'when well finished of answering the intended purpose' [42]. Cooke was heartened by this encouragement.

In December, Cooke, who had been supported financially in his endeavours by his father and was anxious not to impose a further burden on his father's generosity, decided to seek funding for his work from independent capitalists. He wrote to his father and requested a letter of introduction to a Mr Walker, a friend, whose brother Joseph was associated with the Liverpool and Manchester Railway. Cooke travelled to Liverpool in January 1837, met the directors, and discovered that they had a problem with the Lime Street tunnel – a problem that possibly could be solved by the use of the telegraph.

During the early years of the railways, either because the locomotives could not haul trains up steep gradients from a standing start, or because the local residents objected to the noise and smoke of the engines near their homes, the terminal stretches of railways were provided with rope haulage. At Liverpool, the carriages ran downhill, under gravity, to the station via a 2,249 yd (2,057 m) long tunnel, and were hauled back up the gradient by a rope attached to a stationary winding engine at the summit. Clearly, for successful train movements, there was a need to communicate between the station and the engine house. A similar need arose with single line working to prevent collisions happening between trains travelling in opposite directions.

Unfortunately for Cooke, the Directors of the Liverpool and Manchester Railway had already decided to solve their problem by utilising a pneumatic telegraph. This was basically a tube provided with a whistle at each end. Consequently, the directors were disinclined to provide him with financial backing, though they had been sufficiently impressed to offer him some facilities for further experiment.

On this episode Cooke, in 1841, wrote:

The Directors of the Railway Company thought my instrument, which was calculated to give 60 signals, of too complex a nature for the purpose of conveying a few signals along a tunnel and therefore proposed that I should arrange one adapted for their purpose. I immediately designed and drew the second form of the mechanical telegraph, which was based on the same principles as the first, but being calculated to give fewer signals, was less complex. I returned to London immediately afterwards and directed four instruments of the simpler form to be begun; which were soon afterwards made and are extant. I had two of them working together at the close of April 1837 [42].

At this stage of his work Cooke faced a problem the solution to which was unknown. As he said: 'If I want the electromagnets to work at the end of a long line, should I use a great many turns of thin wire or a few turns of thick wire?' It would appear that Cooke posed this question at his meeting with Faraday but did not receive the definitive answer he required. He therefore decided to undertake an experiment which would resolve the matter. One of Cooke's friends was Mr Burton Lane, a solicitor, and he agreed to allow Cooke to set up his trial in a room of his chambers in Lincoln's Inn.

I tried last week [at the end of February 1837] an experiment upon a mile of wire, but the result was not sufficiently satisfactory to admit of my acting upon it. I had to lay out this enormous length of 1,760 yards [1.61 km] in Burton Lane's small office in such a manner as to prevent one part touching another; the patience required and the fatigue undergone in making this arrangement was far from trivial. From Monday evening till Thursday night I was incessantly employed, and by Friday morning at 10 o'clock was obliged to be removed . . . [42].

Dissatisfied at the results obtained, I this morning obtained Dr Roget's opinion, which was favourable but uncertain. [Dr P.M. Roget (1779–1869) was the Secretary of the Royal Society, the compiler of the first thesaurus, and the author of a treatise on electricity published in 1832. Roget suggested that he should visit Professor Charles Wheatstone[44] and this he did on 27th February 1837.] Next Dr Faraday's who, though speaking positively as to the general results formerly, hesitated to give an opinion as to the galvanic fluid acting on a voltaic magnet at a great distance when the question was put to him in that shape. I next tried Clark, a practical mechanician, who spoke positively in favour of my views, yet I felt less satisfied than ever, and called upon a Mr Wheatstone, Professor of chemistry at the London University, and repeated my queries. Imagine my satisfaction at hearing from him that he had four miles of wire in readiness, and imagine my dismay on hearing that he had been employed for months in the construction of the telegraph and had actually invented two or three with a view to bringing them into practical use. We had a long conference and I am to see his arrangement of wire tomorrow morning and we are to converse upon the project of uniting our plans and following them out together. From what passed my plan, if practicable, will, I think, have advantage over any of his, but this remains to be proved [42].

Wheatstone was born on 6th February 1802 at the Manor House in Barnwood, a village about 2 miles (3.2 km) from the centre of Gloucester, and was the second son of William and Beata Wheatstone. Though William and his father were cordwainers, members of the Wheatstone family had been publishers of music and makers of musical instruments from c. 1750. In 1806 William and his family moved to London: there William established himself as a manufacturer of musical instruments and as a teacher of the flute and flageolet – one of his pupils was Princess Charlotte, daughter of the future George IV. Charles attended various schools and seems to have been a precocious pupil. He excelled in mathematics and physics, was awarded a gold medal for proficiency in French, and was a voracious reader of books on which he spent most of his pocket money. His introduction to electrical science stemmed from his reading, in French, an account of Volta's experiments. These he repeated with the assistance of his brother in the scullery of their father's house.

Wheatstone had a great love of intricate mechanisms, automata and science. In particular he was fascinated by the science of sound and carried out many experiments on the transmission of sound and the vibrations of sound producing bodies. Possibly his father's musical instrument making business stimulated this interest, though Wheatstone, himself, once said: 'As an admirer of music ... I remarked that the theory of sound was more neglected than most of the other branches of natural philosophy, which gave rise in me ... the desire of supplying this defect' [44].

In 1823 Wheatstone's uncle Charles, who had a music seller's shop at 436 Strand, died. The young Wheatstone with his brother William Dolman Wheatstone took over the business which subsequently, following a redevelopment of the Strand and the demolition of no. 436, they operated from 20 Conduit Street. Wheatstone's profession was now that of a musical instrument maker. He continued his experiments not only on sound but also on light and optics and contributed various papers to several learned journals including the *Transactions of the Royal Society* and Thomson's *Annals of Philosophy*. These activities, pursued from a love of the subjects rather than from any pecuniary objective, led to Wheatstone being appointed in 1834, at the age of 32, to the chair of experimental physics at King's College, University of London. He was a brilliant experimenter and in February 1836 was given permission by the Council of the College to lay down 'a series of iron and copper wires in the vaults of the College for the purpose of trying some experiments in electricity on account and at the expense of the Royal Society' [44]. Wheatstone began experimenting on the rate of transmission of electricity along wires and determined that the velocity of propagation of an electric disturbance was c. 200,000 mps (322 km/s).

From this brief synopsis of Wheatstone's early life it is apparent that Cooke's introduction to Wheatstone in February 1837 was of seminal importance to the furtherance of electric telegraphy in the United Kingdom. With his undoubted skills as an instrument maker and as an investigative experimentalist, with his detailed theoretical knowledge of physics, and with his access to laboratory and workshop facilities Wheatstone was, prima facie, the ideal person to counsel and to assist Cooke.

In view of their mutual interests in the electric telegraph project, Cooke and Wheatstone decided to enter into a partnership – Cooke assuming the business responsibilities for their telegraph and Wheatstone the scientific and technical aspects of the apparatus. The terms of their agreement were that Cooke should receive 10 per cent of any profits for managing the enterprise and that the remainder should be divided equally between them. In addition it was agreed that Cooke should have the sole right to act as the contractor in the installation of any telegraph lines.

Cooke set about his task with vigour. On 27th June 1837 he met Robert Stephenson (1803–1859), one of the fathers of the British railway system, and was given much encouragement for his views. Apparatus was exhibited to Stephenson a week later; on the 12th July, Wheatstone and Cooke filed their first patent [45] 'for improvements in giving signals and sounding alarums, in distant places, by means of electric currents transmitted through metallic circuits'; and on 25th July the first full-scale experiment in England using their system – which was based on the use of an electro-magnetic escapement and rotary dial – was carried out between Euston and Camden Town stations on the London and North Western Railway company's line, a distance of less than 2 miles (3.2 km) [46].

Wheatstone's and Cooke's experiment using the Euston to Camden stretch of this line was a great technical success and was witnessed by Mr R. Stephenson, Mr I.K. Brunel, Mr C. Fox and Sir Benjamin Hawes. Cooke was at Camden Town and Wheatstone was in a small office, lit by a tallow candle, at Euston. After Wheatstone had sent a message and received a reply he was thrilled by the importance of what they had done: 'Never did I feel such a tumultuous sensation before as when, all alone in the still room, I heard the needles click, and as I spelled the words, I felt all the magnitude of the invention pronounced to be practicable beyond cavil or dispute' [47].

Unfortunately, in spite of this success, the reaction of the railway company directors was unenthusiastic and the line was dismantled. Cooke endeavoured to persuade the directors of the London and Birmingham Railway to approve a telegraph from London to Birmingham but without success. Undeterred, Cooke next, in 1838, conducted some extensive experiments at St Katherine's Docks, London in which he demonstrated the telegraphic transmission of signals through 110 miles (177 km) of no. 16 gauge insulated copper wire. At this stage Isambard Kingdom Brunel used his influence and in July 1839 the Great Western Railway Company agreed to install a line – consisting of pipes with five wires (later suspended by poles) – between Paddington and West Drayton, a distance of 13 miles (21 km): the cost was £3,270.30. Wheatstone's five needle telegraph was utilised (Fig. 3.5).

The apparatus contained five vertical needles pivoted on a horizontal axis and arranged across a diamond shaped dial marked with the letters of the alphabet. Each of the five needles carried on its axis a small magnet placed in a coil of wire behind the bar, and the keys in front of the instruments were so connected that each needle could be deflected at will to the right or to the left, the corresponding needle in the distant receiver undergoing the same deflection. The signalling of any given letter was achieved by the deflection of two of the needles in opposite directions, the interception of their directions indicating the required letter. The apparatus was direct reading, whereas with Schilling's system the position of the discs had to be converted, using a code, to words or the letters of

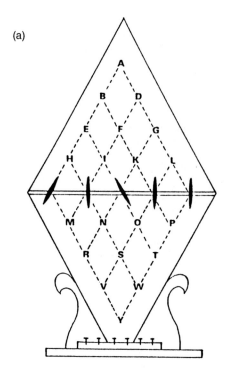

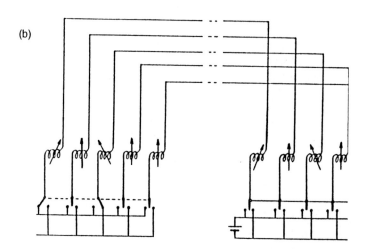

Figure 3.5   (a) The five needle telegraph, indicating the letter E. (b) Circuit of a pair of five needle telegraphs, sending the letter E.
Source: B. Bowers, *Sir Charles Wheatstone FRS 1802–1875* (IEE, London, 2001).

the alphabet. Wheatstone's and Cooke's apparatus was clearly an improvement on Schilling's telegraph and represented an ingenious combination of the binary code which used only two motions, (left or right), and the multi-place code used by C.M. Soemmering and Ampère which used separate indicators for each letter. (Fig. 3.6 summarises some of the early codes.)

Two years later the line was extended to Slough, although *The Times* [48] noted in 1839: 'It is the intention of the Great Western Railway Company to carry the tube along the line as fast as completion of the rails takes place, and ultimately throughout the whole distance to Bristol'. In the same report the newspaper observed: 'The space occupied by the case containing the machinery ... is little more than that required for a gentleman's hat box'; and: 'The machine and the mode of working it are so exceedingly simple that a child who could read would, after an hour or two's instruction, be enabled efficiently to transmit and receive information'.

1	2	3	4	5	6	7	8	9	10
1	970	E	\	•	•	•	—	•	•
2	730	N	/\\	••	—•	—•	••	••——	—•
3	610	R	///\	••	• ••	•—•	——	——•	•—•
4	550	I	//	•	••	••	•	—	••
5	430	S	//\/	•• ••	•••	•••	———••	—••	•••
6	430	T	/\ //	• •	—	—	—•	—•——	—
7	400	A	/	• •	•—	•—	—•—	•—	•—
8	310	G	\ //	•• •	———•	———•	••—•	•—••	——•
9	310	L	\\/	• ••	——	•—••	•——	••••	•—•
10	310	O	/\	•••	• •	•—•••	———	••	———
11	300	U	\/	✕	••—	••—•	•—•	—••	••—
12	250	B	\\	•• •	—•••	—•••	—••—	—•	—•••
13	250	D	\//	• •	—••	—••	—•	••—	—••
14	250	H	\\\	•••••	••••	••••	•—	•—	••••
15	250	M	\/\	•••	——	——	•••	•—•—	——
16	190	C	///	✕	•• •	—•—•	——•	•••	—•—•
17	190	K	///	•• •	—•—	—•—	——•	•———•	—•—
18	190	P	////	• •• •	•••••	•••••	•———•	—•—	•——•
19	190	V	✕	• •	•••—	•••—	—••	———•,	•••—
20	160	W	\///	• • •	•——	•——	•—•—	——••	•——
21	130	F	/\/	••	• —•	••—•	—••	•———•	••—•
22	130	Z	\\ //	•• • ••	••• •	•——••,	•• ——	————	——•••
23	80	Q	✕	✕	••—•	———•—	——•	•———•	——•—
24	80	Y	✕	✕	•• ••	——•••	✕	——	—•——
25	60	X	✕	✕	•—••	•••—••	✕	—••—	—••—

Figure 3.6   Columns 4 to 10 show the coding systems used by: Gauss and Weber; Steinheil; Morse, c. 1840; Gerke, c. 1848; Steinheil, 1849; Bain, 1843; Morse, 1851.
Source: R. Oberliesen, op. cit.

The telegraph was certainly reliable for the above newspaper stated in December 1839: 'The telegraph has now been in operation for nearly 12 months and not the least obstruction to its working by any of the wires, etc, becoming out of order has yet occurred'. Its value to non-railway interests was also known at this time: 'Merchants and others residing not only at the two extremities of the line but at any of the intermediate stations (at all of which dial plates will be fixed with competent persons stationed to work the telegraph) will then be enabled to avail themselves of the benefits and facilities of the invention' [49].

In addition to its daily use, an exhibition was staged at Paddington station to enable the public to see (for an admission fee of one shilling) this new marvel of science, which could transmit 50 signals a distance of 280,000 miles in one minute (7510 km/s) The *Morning Post* averred that the exhibition was 'well worthy of a visit from all who love to see the wonders of science'.

In 1840 Mr George Stephenson adopted the telegraph for the Blackwall Railway which was then worked by a stationary engine and rope; and in 1841 Sir Charles Fox ordered a telegraph for the Cowlairs and Glasgow railway. The telegraph was also employed in one of the Midland tunnels and on the Dublin and Dalkey Atmospheric Railway.

Cooke certainly thought that the telegraph had an important future for railway use and wrote and published a book *Telegraphic railways on the single way*, presumably to highlight the telegraph's significance and to promote discussion among the directors of the various railway companies. In addition, in 1842 a series of advertisements were published in the *Railway Times* and other papers drawing attention to the merits of Cooke's and Wheatstone's invention; and Cooke negotiated with the Admiralty for a line between London and Portsmouth. Some of his approaches were successful:

1. the Dalkey and Kingstown atmospheric railway telegraph was ordered in 1842; and
2. the telegraph line for the Norwich and Yarmouth Railway was opened in May 1844.

However, though the advantages of the telegraph were well-known and appreciated by scientists, commercial progress was not easily attained. 'At the beginning of 1843 we were at our lowest point of depression', Cooke wrote [42]. By February 1837 he had expended £385.42 on his experiments but by the end of 1843 this deficit had increased to £6,232.80.

Fortunately, there were two separate events that more than any others brought the telegraph to the notice of the general public.

On 6th August 1844 Queen Victoria's second son, Alfred Ernest was born at Windsor. By the aid of the telegraph the news was published in *The Times* within 40 minutes of the announcement and the editor acknowledged his 'indebted-[ness] to the extraordinary power of the electro-magnetic telegraph' [44] by which the information was so speedily transmitted. The telegraph was also used when, just before a banquet held at Windsor to celebrate the birth, the Duke of Wellington found that he had forgotten his dress suit. An urgent request was telegraphed to London and the suit was sent on the following train in time for the royal function.

The second event was the arrest, with the aid of the telegraph, of John Tawell, the murderer, at Paddington station on 3rd January 1845 [50]. A woman had been murdered in her cottage at Salt Hill, on the outskirts of Slough, and, following a report that a man had been seen leaving her home in rather suspicious circumstances some time before, enquiries were made at the Great Western station regarding this matter. The police learned that a man answering to their description of the suspect had boarded, a few minutes earlier, the slow train to London. The message was immediately sent by the electric telegraph to Paddington station:

A murder has just been committed at Salthill, and the suspected murderer was seen to take a first-class ticket for London by the train which left at 7.42 p.m. He is in the garb of a Kwaker, with a brown greatcoat on which reaches nearly to his feet. He is in the last compartment of the 2nd first-class carriage [42].

The word Kwaker, for Quaker (since the telegraph could not send the letter q) was not at first understood by the operator at Paddington but fortunately the delay in interpretation was insufficient to prevent the recognition of the man when the train arrived. He was followed and arrested. His trial attracted a great deal of interest and the event probably did as much, if not more, for the establishment of the electric telegraph of Wheatstone and Cooke than all their previous technical and business trials [42]. John Tawell was subsequently convicted and hanged; and it was widely reported that 'John Tawell had been hanged by the electric telegraph'.

Following their success with the five-needle telegraph, Wheatstone and Cooke on 21st January 1840 patented a step-by-step letter-showing telegraph [51]. Later, on 7th July 1841, Wheatstone described in a patent [52] an improvement to their apparatus so as to enable it to print the message directly on paper.

The 'A.B.C.' telegraph consisted of a disc, bearing around its periphery the letters of the alphabet, which was made to rotate by an electromagnet and armature mechanism so that when the armature was attracted to the magnet the disc was allowed to progress one step forward in its revolution and to expose to view, behind a small window, the given letter being signalled. A pallet and escapement wheel, similar to those employed in a clock, were used so that every pulse of current transmitted caused the disc to rotate one step or tooth [53].

This method used essentially a similar principle of operation to that adopted by Chappe in his synchronised system, and by Ronalds in his electrostatic telegraph. In his printing telegraph Wheatstone replaced the letter discs by the type wheel, that is a circular arrangement of flexible reeds each carrying a typeface, and the window was replaced by a hammer behind this wheel. When the correct position for a given letter was reached, a separate signal was used to release the hammer and cause it to strike the typeface against a sheet of paper placed on a drum – the latter moving with a linear and a rotary motion as in a typewriter. The apparatus was highly ingenious but at that time there was no demand for a printing telegraph and only two instruments were built. Wheatstone's printing telegraph was the subject of some controversy between himself and Alexander Bain [54] who claimed that Wheatstone had plagiarised his ideas.

The non-printing version of the A.B.C. telegraph was first used on the Paris to Versailles railway line. Various modifications were made to this form of instrument by Dujardin, Siemens, Brett, Poole, Highton, Nott and Allan.

Cooke and Wheatstone continued to improve their apparatus and in 1846 a two-needle telegraph was installed in the Octagon room of the House of Commons for the purpose of sending and receiving messages relating to parliamentary business.

Telegraph wires along railway lines became a familiar feature of the landscape. In Charles Dickens' *Hard times* Mrs Sparsil observed that the electric telegraph wires ruled the staves of 'a colossal strip of music-paper out of the evening sky'. By 1852 an estimated 4,039 miles (6,500 km) of telegraph lines had been erected in England, and by 1868 the total length of these lines had increased to c. 35,000 km. Railway locomotives and electric telegraphs formed a formidable partnership in Victorian Britain, speeding up transport and communications to a degree previously unattainable. They were referred to by *Punch Magazine* as 'the two giants of the time'.

Elsewhere, particularly in the United States of America, other types of telegraph systems were being established. The work of S.F.B. Morse is especially important.

Samuel Morse [55] (1791–1872) of the United States of America, and Edward Davey of Great Britain also produced, in 1837 and 1838 respectively, inventions for recording signals on paper. Morse, born in 1791, was a painter, of some distinction, of historical scenes. He spent the period from 1811 to 1815 in England, at London and Clifton, studying his profession and in 1829 went to Europe. In the autumn of 1832, after having lived for a year in Paris where he copied paintings in the Louvre, he returned to America on the packet ship, the *Sully*. One of the previous passengers on board was a Dr Charles T. Jackson [56] of Boston, who had attended some lectures on electromagnetism given by Pouillet at the Sorbonne, and during the voyage, which lasted from 8th October to 15th November, Jackson frequently conversed with Morse on the subject of electricity and the possibility of electromagnetic signalling. To Morse this was a new and ingenious idea and he became absorbed with the notion of using an electromagnet as the operative element in an electric telegraph.

Morse was not the first United States citizen to engage in the invention of an electric telegraph; that honour has been accorded to Harrison G. Dyar [57]. Sometime between 1826 and 1828, he attempted to devise an electrochemical system in which marks – dots and dashes – were recorded on chemically treated paper by the passage of electricity through the paper. It has been stated that Dyar's telegraph was tried out near a race track on Long Island by setting up poles with insulators to carry the wire that formed half the circuit and using the earth as a ground return for the other half. However, an unscrupulous partner, seeking a greater share in the expected gains, forced Dyar to drop his invention.

If this is correct, Dyar's actions were certainly most prescient. The use of electrochemical marking, dots and dashes, overhead lines, insulators and the earth return were all much in advance of their practical realisation by others.

A copy of Morse's notebook that records notes on his conversations with Jackson is held in the Museum of the History of Technology of the Smithsonian Institution. The notebook includes a sketch of 'what appears to be an electrochemical receiver similar to Dyar's and another sketch in which an electromagnet actuates an armature to move a style against a roll of paper'.

Morse's progress was slow for two reasons. First, his appointment in 1832 as Professor of painting and sculpture at the University of the City of New York (now New York University) left him with little time to pursue his ideas on telegraphy; and second, his ignorance of electrical science militated against a rapid practical implementation of those ideas. In November 1835 Morse became Professor of the literature of the arts and design in the university and as his duties left him with some free time he began to reduce his notions on electric telegraphy to practice.

Morse's first crude instrument, which did not work well, was constructed in 1835 and was based on the use of a port rule, or composing stick, onto which type was set (Fig. 3.7). This was cranked through an apparatus that opened or closed a circuit depending on the projections of the type. The receiver comprised a pendulum, carrying an ink pen, and a moving paper tape. In use the make and break of the electric current caused the pen, operated by an electromagnet energised by the current pulses, to record zig-zag lines on the paper strip as the pendulum moved to and fro across the tape, and as it was drawn continuously over a roller by a weight driven mechanism [58]. The early apparatus was not a success. In a demonstration given in January 1836 to Leonard Gale, Professor of chemistry at the university, the telegraph would not work through 40 ft (12.2 m) of wire. Morse's telegraph was capable of transmitting signals across the laboratory, but failed for longer distances due to the fact that he used an electromagnet which required a 'quantity' current [59]. Fortunately for Morse, Gale was familiar with the work of J. Henry, who was one of his friends, and suggested to Morse that he should wind his electromagnets with fine wire and use an 'intensity' battery. 'After substituting the battery of twenty cups for that of a single cup, we [Morse and Gale] sent a message through two hundred feet of conductors, then through 1,000 feet [305 m], and then through ten miles [16 km] of wire arranged on wheels in my own lecture room in the New York University in the presence of friends.'

The first published record that refers to Morse's invention is a letter written by his brother S.E. Morse and printed in the *New York Observer* of 15th April 1837. In this letter S.E. Morse mentions a telegraph which requires 26 wires. This response was probably stimulated by a resolution [60] passed by the House of Representatives of the United States on 3rd February 1837: 'Resolved, that the Secretary of the Treasury be requested to report to the House at its next session upon the propriety of establishing a system of telegraphs for the United States'. In compliance with this a circular was issued on 10th March 1837 by the

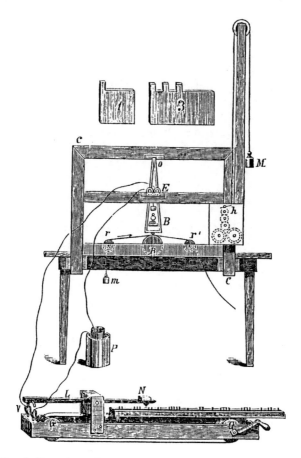

Figure 3.7   Morse's first telegraph.
Source: Urbanitszky, op. cit.

Secretary of the Treasury, Levi Woodbury, asking for this information: it was sent to 'certain collectors of customs and others desiring information with reference to telegraphic communication'. The circular alluded to communication 'by cannon or otherwise, or in the night by rockets, fires, etc.', and made no special reference to electric telegraphs.

S.E. Morse later, in April 1845 asserted that the plan [61] was his: however Professor L.J. Gale, who in 1837 lived in the same building as Samuel Morse, has written that in March/April 1837 his colleague consented to a public announcement being made of the existence of his invention.

By September 1837 several electric telegraph schemes – those of Schilling, Cooke and Wheatstone, Gauss and Weber, and Steinheil – had been described and demonstrated, and made known in the United States. Possibly because of the publicity given to these schemes Morse decided early in September to make his invention public. He wrote a letter, dated 4th September, to the New York *Journal of Commerce* and stated [62]:

You recently announced that I was preparing a short telegraphic circuit to show to my friends the operation of the telegraph; this circuit I have completed of the length of 1700 feet [518 m], and on Saturday, the 2nd, in the presence of Professor Gale and Dr Daubeny of the Oxford University, and several other gentlemen, I tried a preliminary experiment with the register. It recorded the intelligence sufficiently perfect to establish the practicability of the plan and the superior simplicity of my mode over any of those proposed by the professors in Europe. No account has reached us that any of the foreign proposed electric telegraphs have as yet succeeded in transmitting intelligible communications, but it is merely asserted of the most advanced experiment (the one in London), that by means of five wires, intelligence may be conveyed.

In a letter to the Secretary of the Treasury of the United States, dated 27th September 1837, Morse stated [63]:

About five years ago, on my voyage home from Europe, the electrical experiments of Franklin upon a wire some [6.4 km] in length was casually recalled to my mind in a conversation with one of the passengers, in which experiment it was ascertained that the electricity travelled through the whole circuit in a time not appreciable but apparently instantaneous. It immediately occurred to me that, if the presence of electricity could be made visible in any part of the circuit, it would not be difficult to construct a system of signs by which intelligence could be instantaneously transmitted.

The thought thus conceived took a strong hold of my mind in the leisure which the voyage afforded, and I planned a system of signs and an apparatus to carry it into effect. I cast a species of type which I devised for this purpose, the first week after my arrival home; and although the rest of the machinery was planned, yet from the pressure of unavoidable duties I was compelled to postpone my experiments and was not able to test the whole plan until within a few weeks. The result has realised my most sanguine expectations.

This lapse of time between the creation of his idea and its implementation did not deter Morse from making a rather startling claim: 'I assert to be the first proposer and inventor of electromagnetic telegraphy, namely on the 19th October 1832, on board the packet *Sully*, on my voyage from France to the United States . . . All telegraphs in Europe are, without one exception, invented later than mine' [64].

A chance encounter with Morse on 2nd September 1837 by Alfred Vail, then a young man, 16 years younger than Morse and of a mechanical turn of mind, enabled Morse to submit a proposal for a telegraph to the Government. Vail was fascinated by what he had seen of Morse's work, and, three weeks later, agreed, in exchange for a share of the rights of a potential patent, to construct, by 1st January 1838, at his own expense a telegraph based on Morse's design. Vail carried out his task at Morriston, New Jersey, where his father Judge Stephen Vail owned and managed the Speedwell Iron Plant. During this time Morse prepared a dictionary for use with the telegraph [65].

Morse's idea was not to produce on paper letters, or signs representing them, but to use 10 numerals for the 9 digits and 0 and, by means of a code dictionary, words. On 4th September 1837 the signal sent consisted of 62 zig-zags and 15 straight lines on the slip of paper. These characters represented the numbers 215, 36, 2, 58, 112, 04, 01837 and these in turn expressed the message: 'Successful experiment with telegraph September 4 1837'. Such a system of coding was undoubtedly cumbrous and restrictive and, in basic principle, was similar to that which Edgeworth had suggested in 1767. Following the demonstration Morse on 28th September 1837 filed a caveat in the patent office at Washington [66].

By the beginning of the New Year all was ready for a trial of the new equipment. On 6th January 1838 the three partners, Morse, Gale and Vail, found that their instrument could operate through 4.8 km of wire strung around the barn on the Vail estate. The zig-zag lines of the earlier apparatus were now replaced by dots and dashes and the motion of the armature was now vertical instead of horizontal [67].

Vail regarded Morse's earlier procedure as tedious and preferred to use a code in which a succession of symbols represent particular letters of the alphabet. The 'Morse code' that was devised, and subsequently extensively used, seems to have been the work of Vail for it has been stated [68]: 'Vail tried to compute the relative frequency of all the letters in order to arrange his alphabet; but a happy idea enabled him to save his time. He went to the office of the local newspaper in Morristown and found the result he wanted in the type-cases of the compositors. The code was then arranged so that the most commonly used letters were indicated by the shortest symbols – a single dot for an E, a single dash for T and so on'. The dots and dashes formed the elements of an alphabetic binary code and were the precursor of the later Morse code. There can be little doubt that Vail's practical experience, enthusiasm and financial resources aided the impecunious Morse in his endeavours. On 11th January 1838 the partners in Morristown publicly demonstrated their apparatus. Further demonstrations were held at New York University (24th January), and at the Franklin Institute (8th February), and, on 20th February the instrument was shown to the Committee of Commerce of the US House of Representatives. The chairman of the committee, F.O.J. Smith commented favourably upon the system. He soon became another partner in the Morse enterprise [69].

Though the Morse telegraph was based on the joint efforts of Morse, Gale and Vail, the portrait painter succeeded in persuading his partners that all inventions and developments in their work should be ascribed to him and that the system should be known as the Morse telegraph system [70]. (Because of this understanding it is difficult to determine the relative contributions of Morse, Gale and Vail and to say definitively who initiated the various improvements which were introduced into the system.) Morse seems to have had an overweening, egotistical craving to seek public acclaim which prevented him from giving credit to Gale [71] and Vail [72], and to the advice proffered by Henry [73]. Yet it was Henry who, by counselling Morse to abandon the employment of a 'quantity' battery and a 'quantity' coil and to use instead an 'intensity' battery and an 'intensity' coil, provided the solution which allowed Morse to be successful. Henry also invented the relay which was adopted by Morse to permit the telegraph signals to be transmitted over much larger distances than had been possible previously. And it was Henry who on 11th April 1837 had called on Professor Wheatstone at King's College where

he explained to us his plan of an electrical magnetic telegraph, and among other things, exhibited to us his method [the use of a relay] of bringing into action a second galvanic circuit; this consisted in closing the second circuit by the deflection of the needle. I informed him that I had devised another method of producing effects somewhat similar; this consisted in opening the circuit of a large quantity magnet by attracting a movable wire by a small intensity magnet, or magnet wound with a fine wire [74].

Morse's aspirations for the success of his telegraph were considerable. He, accompanied by Smith, arrived in London in June 1838 hoping to patent his invention but, perhaps unexpectedly for him, the patent application was opposed by Cooke and Wheatstone [75]. They argued that since Morse had already published his invention in the 10th February 1838 issue of the *Mechanics Magazine* this prior disclosure, according to English patent law, invalidated Morse's application. Morse's unhappiness with the decision, which he felt was unjust, was partly assuaged by his meetings with the English workers. His view of the telegraph situation in England is given in a letter [76] which he wrote in July 1838 on his way to Paris.

Professor Wheatstone and Mr Davy were my opponents. They have each very ingenious inventions of their own, particularly the former, who is a man of genius and one with whom I was personally much pleased. He has invented his, I believe, without knowing that I was engaged in an invention to produce a similar result; for, although he dates back into 1832, yet, as no publication of our thoughts was made by either, we are evidently independent of each other. My time has not been lost, however, for I have ascertained with certainty that the telegraph of a single circuit and a recording apparatus is mine . . . I found also that both Mr Wheatstone and Mr Davy were endeavouring to simplify theirs by adding a recording apparatus and reducing theirs to a single circuit.

Throughout his sojourn in Europe Morse endeavoured to obtain funding and patents but no backers were forthcoming, and various legal difficulties prevented him from taking out patents. He returned to the United States in 1839, and requested that his US patent be issued. It was finally granted on 20th June 1840 as US patent no. 1647.

Meantime Vail had devised a new form of transmitter which was manually operated instead of being worked automatically. This was the sending key a device later known to many thousands of electric telegraphy and wireless telegraphy Morse code operators around the world.

Vail was of great assistance to Morse [77]. In an editorial, published on 25th September 1858, headed 'Honor to whom honor is due' the editor and proprietor of the New York *Sun* wrote:

Alfred Vail entered into these experiments with his whole soul, and to him is Professor Morse indebted quite as much as to his own wit, for his ultimate triumph. He it was who invented the far-famed alphabet; and he too was the inventor of the instrument which bears Morse's name. But whatever he did or contrived went cheerfully to the great end.

Another writer, in an article in the April 1888 issue of the *Century Magazine* noted:

Prior to 1837 it [the telegraph] embodied the work of Morse and of Henry alone. From 1837 to 1844 it was a combination of the inventions of Morse, Henry and Vail; but . . . the elements contributed by Morse have gradually fallen into desuetude, so that the essential telegraph today, and the universal telegraph of the future comprises solely the work of Joseph Henry and Alfred Vail [77].

Unlike the situation in England, where the early railways played such an important part in the establishment of a telegraph network, 'the growth of the telegraph industry in America was dependent on political patronage and subsidy before any promise of commercial support could be obtained' [59]. During the period 1838–1843 Morse made repeated attempts to procure government assistance to set up a telegraph system, but without success. He wrote to Vail and opined that the lack of interest of the government was 'not the fault of the invention, nor [was] it [Morse's] neglect' [59].

In December 1842 Morse travelled to Washington hoping again to persuade Congress to provide the necessary funds. He demonstrated [78] his apparatus between two rooms in the Capitol building – sending messages back and forth – and eventually on 3rd March 1843 Congress passed the Telegraph Bill, by 89 votes to 83, which allocated $30,000 for a telegraph line along the Baltimore and Ohio railroad between Washington and Baltimore, a distance of 64 km. Seventy congressmen abstained from voting 'to avoid the responsibility of spending the public money for a machine they could not understand'. The first message [58] 'What hath God wrought' was transmitted on 27th May 1844 at a speed of about six words per minute and on 1st April the line was opened for public business [79].

On the Continent of Europe important work in the field of telegraphy was carried out by Carl Friedrich Gauss (1777–1855), Wilhelm Edward Weber (1804–1891), and Carl August Steinheil (1801–1870).

Gauss was born at Braunschwig and was educated at Göttingen. In 1807 he was appointed Professor and Director of the Observatory in that town. Weber, a native of Wittenberg had studied at Halle. He was introduced to Gauss by A. von Humbolt, and through his influence obtained the post of Professor of Physics at Göttingen.

In 1833 Gauss was engaged at the University of Göttingen upon a study of the earth's magnetic field. With his colleague Professor W. Weber he established a telegraph line between the observatory, where he carried out his researches, and the Cabinet de Physics [80]. The Göttingen *Gelehrten Anziegen* in reporting on Gauss's work in 1834 noted:

We cannot omit to mention an important and, in its way, unique feature, in close connection with the arrangements we have described (of the physical observatory) which we owe to Professor Weber. He last year stretched a double connecting wire from the Cabinet of Physics over the houses of the city to the observatory; [and] in this [way] a grand galvanic chain was established, in which the current was carried through about nine thousand feet

[2743 m] of wire. The wire of the chain was chiefly copper wire, known in the trade as no. 3. The certainty and exactness with which one can control, by means of the commutator, the direction of the current and the movement of the needle depending upon it, were demonstrated last year by successful application to telegraphic signalling of whole words and short phrases. There is no doubt that it will be possible to establish immediate telegraphic dictation between two stations at considerable distances from one another [58].

The coding [39] used depended on the use of right and left deflections of the needle of the galvanometer. In CM's, Soemmering's, and Ampère's systems each letter of the alphabet required a separate detector and communication channel but in Schilling's scheme, by the use of the binary code, the number of detectors and channels was reduced to five. Now, with the system (Fig. 3.8) of Gauss and Weber, in which one group of successively executed movements represented a letter or numeral, only a single indicator capable of two positions (right and left) and a single channel of communication were required. Their apparatus was not without a disadvantage. Apart from the use of a needle, having a mass of 25 lb (11.4 kg) (sic), whose deflections were so small they had to be observed through a telescope, and whose inertia seriously limited the speed of signalling to about seven letters per minute, the economy in the use of wire was obtained at the cost of signalling speed.

In 1836 Gauss and Weber replaced their apparatus's battery by a coil which could be moved up or down over a pair of large bar magnets. By utilising the principle of electromagnetic induction discovered five years previously by Faraday, motion of the coil in either direction generated a current which by means of a commutator deflected the needle either to the right or to the left.

In the same year the directors of the Leipzig–Dresden railway expressed an interest in the Gauss–Weber instrument since, as in England, railways were being developed in Europe and the use of the electric telegraph appeared likely to prove useful in transmitting information regarding the movements of the trains at remote places. Fortunately Gauss himself realised the deficiencies of the instrument and invited Steinheil to develop a simpler and more practical form of the telegraph [81].

Steinheil had studied science, initially, at Göttingen (under Gauss), and then at Königsberg. When he was 34 years of age Steinheil was appointed Professor of Physics and Mathematics at Munich. Later in 1849 he was invited, by the Austrian Government, to Vienna to superintend the establishment of telegraphic communications. There he prepared a complete plan for a telegraphic system for the various provinces of the empire. Upon returning to Munich, in response to the wishes of King Maximilian II, he founded in 1854 the optical and astronomical laboratories of the city and was elected a member of many scientific societies. He died in 1870.

Steinheil's first telegraph system was completed in 1837 and comprised three sections of line. The first, of 30,500 ft (9,296 m) in length, linked the Royal Academy of Munich with the observatory at Bogenhausen; the second, of 6,000 ft (1,830 m) connected the residence of Steinheil with the observatory in

(a)

(b)

Figure 3.8   Gauss's and Weber's sending (a) and receiving (b) apparatuses.
Source: Urbanitzky, op. cit.

the Lerchenstrasse; and the third, of c. 1,200 ft (366 m) joined the Academy with the workshop of the Cabinet de Physics. These stations could be connected in any required combination by means of a simple switching device. The instrument worked very well and could transmit at the rate of approximately six words per minute (Fig. 3.9).

In 1838 another telegraph was installed along the Nurnberg–Fuerther railway for a distance of about 5 miles (8 km). Steinheil's telegraph required only one wire as he discovered, in 1837, while experimenting on the line that the earth itself acted as a conductor [82]. This was a finding of major importance economically because it meant that just one telegraph wire was required in a communications link and so the cost of the overhead line was halved. Steinheil's discovery seems to have been received with considerable awe and surprise. In his book on the *Electric telegraph* (1855) D. Lardner wrote:

Of all the miracles of science, surely this is the most marvellous. A stream of the electric fluid has its source in the cellars of the Central Electric Telegraphic Office, Lothbury, London. It flows under the streets of the great metropolis; and, passing on wires suspended over a zigzag series of railways, reaches Edinburgh, where it dips into the earth, and diffuses itself upon the buried plate. From that it takes its flight through the crust of the earth, and finds its own way back to the cellars at Lothbury, London.

The instruments of Steinheil's Nurnberg–Fuerther telegraph included means for alerting the operators when a message was to be received.

When it was desired to telegraph [the sending end operator would send a signal which caused] the needle [of the distant instrument to] strike against either of two bells – the one needle striking one bell, and the other needle striking another differently toned.

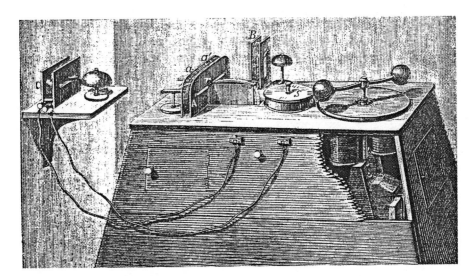

Figure 3.9 Steinheil's apparatus.
Source: Urbanitzky, op. cit.

When he required to permanently record the intelligence these needles were furnished with small tubes holding ink and, by their motions, dots were made on paper properly moved in front of them by a wound-up mechanism; one needle making dots in one line, and the other needle making dots in a line underneath the former [63].

Fig. 3.6 shows the code that Steinheil evolved [83]. In practice it was found that by substituting the bells for the ink pens the transmitted messages could still be read: his telegraph effectively became an acoustic telegraph. This facility was also noted, later, by the telegraph operators of the Morse system – much to Morse's displeasure.

Installation of Steinheil's telegraph was considered for the Munich–Augsburg railway but, notwithstanding the improvements that had been made, the railway operators decided the expense of the installation could not be justified.

Following the early work of the telegraph pioneers the electric telegraph developed rapidly, particularly in England. Here, by 1851 2,819 km of line, 11,750 km of wire and 198 stations had been installed. For the year 1878 the numbers of messages sent in England, France and Germany were 24.6 million, 14.4 million and 14.54 million respectively. This growth and some of the commercial and social aspects of it form the basis of the next chapter.

References

1   MOTTELAY, P.F.: 'The bibliographical history of electricity and magnetism' (Griffin, London, 1922), p. 126
2   Ref. 1, pp. 82–92
3   Ref. 1, pp. 161–162
4   Ref. 1, p. 166
5   Ref. 1, pp. 166–167
6   Ref. 1, pp. 173–175
7   Ref. 1, pp. 173–180
8   Ref. 1, pp. 193–199
9   Ref. 1, p. 189
10 C.M: a letter, *Scots' Magazine*, 17th February 1753
11 GORDON, Mrs.: 'The home life of Sir David Brewster' (Edinburgh, 1869),
p. 207
12 ANON.: *Cornhill Magazine* 1860, **ii**, pp. 65–66
13 FAHIE, J.J.: 'A history of electric telegraphy to 1837' (F.N. Spon, London,
1884)
14 Ref. 13, pp. 91–93
15 Ref. 13, pp. 68–77
16 Ref. 13, pp. 77–79
17 Ref. 13, pp. 79–80
18 Ref. 13, pp. 89–91
19 Ref. 1, p. 265
20 Ref. 13, pp. 91–93
21 Ref. 13, p. 318

22 Ref. 1, pp. 315–316
23 Ref. 1, pp. 243–245
24 Ref. 13, pp. 101–108
25 Ref. 13, pp. 127–145
26 DUNSHEATH, P.: 'A history of electrical engineering' (Faber and Faber, London, 1962) pp. 70–71
27 RONALDS, F.: 'Description of an electric telegraph and some other electrical apparatus' (London, 1823)
28 Correspondence and extracts on the electric telegraph, The IEE Archives. Also quoted in Ref. 26, p. 74
29 Ref. 13, pp. 191–192
30 Ref. 1, pp. 387–388
31 HAMEL, Dr.: 'Historical account of the introduction of the galvanic and electromagnetic telegraph', *Jour. of the Soc. of Arts*, 1859, **vii**, pp. 595–599 and pp. 605–610
32 Ref. 13, pp. 227–243
33 Ref. 13, pp. 220–227
34 Ref. 31, p. 599
35 Ref. 31, p. 598
36 See Refs. 13, pp. 257–262; 26, p. 53; and 31, pp. 605–606
37 Ref. 13. pp. 302–303
38 AMPERE, A.M.: 'Exposé des nouvelles découvertes sur l'electricité et le magnetisme de M.M. Oersted, Arago, Ampère, H. Davy, Biot, etc' (Paris, 1822)
39 OBERLIESEN, R.: 'Information, Daten und Signale' (Deutsches Museum, Rowohlt, 1982), p. 94
40 Ref. 31, pp. 607–608
41 COOKE, W.F.: 'The electric telegraph: was it invented by Professor Wheatstone', vol. 2 (W.H. Smith, London, 1866), p. 14
42 HUBBARD, G.: 'Cooke and Wheatstone and the invention of the electric telegraph' (Routledge and Kegan Paul, London, 1965)
43 SNELL, J.B.: 'Railways: mechanical engineering' (Longman Group, London, 1971)
44 BOWERS, B.: 'Sir Charles Wheatstone FRS, 1802–1875' (HMSO, London, 1975)
45 COOKE, W.F., and WHEATSTONE, C.: 'Improvements in giving signals and sounding alarms in distant places by means of electric currents transmitted through metallic circuits', British patent, no. 7390, dated 12th June 1837
46 COOKE, W.F.: letter 25th January 1847, in WFC papers, vol. 1, The IEE Archives
47 Quoted in Kingsford, P.W.: 'Electrical engineers and workers' (Edward Arnold, London, 1969), p. 49
48 ANON.: a report *The Times*, 2nd September 1839
49 ANON.: a report *The Times*, 5th December 1839
50 ROUTLEDGE, R.: 'Discoveries and inventions of the nineteenth century' (George Routledge, London, 1903), pp. 550–551
51 COOKE, W.F., and WHEATSTONE, C.: 'Electric telegraphs', British patent no. 8345, 21st January 1840

52 WHEATSTONE, C.: British patent no. 9022, 7th July 1841
53 Ref. 43, chapter 10
54 BURNS, R.W.: 'Alexander Bain, a most ingenious and meritorious inventor', *ESEJ*, 1993, **2** pp. 85–93
55 USHER, A.P.: 'Morse, Samuel Finley Breece', entry in 'Encyclopaedia Americana', 1992, p. 476
56 Ref. 31, p. 609
57 Ref. 13, pp. 155–161
58 URBANITZKY, A.R. von.: 'Electricity in the service of man' (Cassell, London, 1886), pp. 763–764
59 HARLOW, A,F.: 'Old wires and new waves', (Arno Press, New York, 1971 reprint edition), pp. 67–68
60 REID, J.D.: 'The telegraph in America, its founders, promoters and noted men' (Derby Bros., New York, 1879)
61 Ibid, chapter 8, 'The era of litigation', pp. 142–171
62 MORSE, S.F.B.: letter to the New York *Journal of Commerce*, 4th September 1837
63 HIGHTON, E.: 'The electric telegraph; its history and progress' (London, 1852), pp. 56–61
64 Ref. 31, p. 610
65 Paper 29, 'Development of electrical technology in the 19th century', Bulletin 228: Contributions from the [Smithsonian] Museum of History of Technology
66 MORSE, S.F.B.: US patent no. 1647, issued 20th June 1840 (filed 28th September 1837)
67 PRESCOTT, G.B.: 'History, theory and practice of the electric telegraph' (Trubner, London, 1860), chapter v, 'The Morse system', pp. 73–99
68 Quoted in Ref. 26, p. 81
69 Ref. 65, p. 298
70 Ibid
71 Ref. 60, p. 170
72 Ref. 60, pp. 75–76
73 Ref. 60, p. 169
74 Ref. 67, pp. 79–81
75 Ref. 44, p. 112
76 Quoted in Ref. 44, p. 113
77 Ref. 59, pp. 72, 75
78 Ref. 59, p. 83
79 Ref. 59, pp. 98–99
80 Ref. 58, pp. 756–757
81 SABINE, R.: 'The history and progress of the electric telegraph', (London, 1867), p. 37
82 Ibid, p. 42
83 Ref. 39, p. 101

Chapter 4

Electric telegraphy – commercial and social considerations

By 1845 Cooke's financial resources were declining rapidly; he had disbursed £71,719 in 1845 alone, compared with £31,279 for the entire period 1836–1844. According to his notes, receipts for the 1836–1845 period amounted to £96,974 excluding sums due on payment of works in hand [1]. The need for extra finance became crucial, a public company had to be formed. Cooke decided to enlist the support of G.P. Bidder whom he had met during the construction of the Blackwall Railway. Bidder procured the good offices of John Lewis Ricardo (1812–1862), the Member of Parliament for Stoke on Trent, the Chairman of the North Staffordshire Railway, and 'a man of extraordinary sagacity and great energy' [2].

It appears the objective of Bidder and Ricardo was to purchase part of the patent rights of Cooke's and Wheatstone's inventions with the intention of establishing a company to operate the electric telegraph. To further this plan Cooke's solicitor, R. Wilson, in August 1845, was instructed to list the various extant patents, licences and contracts [3]. Wheatstone agreed to exchange his royalty rights on all lines for £30,000. Subsequently, under a memorandum of agreement dated 23rd December 1845, Ricardo, Bidder and Cooke assented to the patents, together with Cooke's half-share in the telegraphs from London to Portsmouth, and to Slough, being valued at £160,000 and being apportioned as follows [3]:

J.L. Ricardo   12/32nds, at a price of £60,000
G.P. Bidder   11/32nds, at a price of £55,000
W.F. Cooke   9/32nds, at a price of £45,000

Interestingly, in January 1843, the English patents had been considered, by Cooke and Wheatstone, to be worth just £20,000.

In 1846 the House of Commons and the House of Lords considered 'An Act for forming and regulating "The Electric Telegraph Company", and to enable the said Company to work certain letters Patent'[4]. Alexander Bain opposed the formation of the company on the grounds that some of his patents would be infringed and gave evidence to Select Committees of both Houses. The Lords Select Committee was chaired by the Duke of Beaufort. Bain made such a favourable impression when he was examined that the chairman was inspired to indicate to the sponsors of the Bill that if they did not reach an agreement with Bain there was a possibility the Bill would not be supported. Bain was awarded £7,500 [5].

This was a substantial sum in 1846 as evidenced by the costs of goods and services at that time: for example, a labourer worked for approximately 75p per week, rent for a working class family was about 27p per week, a two bedroom house cost c £225, coal was 86p per ton in London, and an ounce of tobacco and a gallon of beer were 1p and 5p, respectively [6].

(An amusing anecdote may be related about this award. Bain, during the 1840s, sought patent protection for many of his inventions. A scrutiny of the patents shows that from his description of himself as a 'mechanist' in 1841 he advanced to the status of 'engineer' in 1844 and then to 'electrical engineer' in 1846. After the award Bain expressed himself simply as 'a Gentleman'. Bain's use of the title 'electrical engineer' in 1846 is noteworthy. The Institution of Electrical Engineers did not adopt its name until 1889 although The Society of Telegraph Engineers was founded in 1871.)

Towards the end of the 1840–1850 decade Bain and his family (Bain married in 1844) lived in Beevor Lodge, a large house in Hammersmith with an extensive garden. It is described in one of the volumes of 'A survey of London' and is also mentioned in several works on haunted houses. The 1851 Census Returns [7] show that Bain employed five servants and a resident teacher for his step-daughter. Clearly, the former poorly educated crofter's son, by force of much application and endeavour, had greatly improved his social, technical and financial position. Alas, the means – electric telegraphy – that had enabled him to reach this desirable state would lead to his downfall, and he would die almost penniless in 1877.

In 1846 the Electric Telegraph Company, which had been provisionally registered on 2nd September 1845 by Cooke and Wilson, was incorporated to purchase, by private subscription, the patents of Cooke and Wheatstone. It was the first company of its type to be formed in the United Kingdom. The first directors were J.L. Ricardo (Chairman), S. Ricardo, W.F. Cooke, G.P. Bidder and a Mr R. Till [8]. Cooke and Wheatstone were handsomely rewarded for their enterprise and endeavours. Apart from the £30,000 due to him, Wheatstone also received £3,000 for royalties then due, and was asked to be a scientific advisor at a salary of £700 per annum [9], though this was never confirmed: Cooke received £91,158 for his share of the patents, and his business as telegraphic engineer and contractor. This sum was dependent partly on shares and partly on profits.

After receiving their Act of Incorporation, The English Telegraph Company, under Ricardo's guidance, constructed their central station at the end of

Founder's Court, Lothbury, London, the capital for which had been provided by Sir Samuel Morton Peto. The handsome building was formally opened on 1st January 1848, at a time when 1,514 miles (2,436 km) of telegraph line had been erected or was in the process of erection [10].

In addition, ETC sought to convert Cooke's contracts, with the railway companies, from way-leaves into exclusive agreements for a long term, so as to enjoy a monopolistic control. This perspicacious policy was implemented, especially in the case of the leading railway companies and those operating the trunk lines which radiated from London to the north and west. By their Act the company had the authority to lay pipes and wires under the streets of towns and cities, and by 1st January 1848 ETC [11] had opened offices for receiving and transmitting public messages in London, Birmingham, Manchester, Liverpool and elsewhere, which could also communicate with the numerous smaller localities and railway stations previously linked.

The business was not an immediate financial success due to the high cost of transmitting a message [12]. Costs from London to various towns and cities are listed below:

Birmingham or Stafford	cost 3.9d (1.625p) per word
Derby, Norwich, Nottingham or Yarmouth	cost 4.5d (1.875p) per word
Liverpool, Leeds or Manchester	cost 5.1d (2.125p) per word
Edinburgh	cost 7.8d (3.25p) per word
Glasgow	cost 8.7d (3.625p) per word

Thus, a 20 word message to Leeds was priced at 102d (42.5p), which was more than a labourer earned in a week of hard toil. The consequences were inevitable. On 27th March 1848 ETC [13] discharged 80 per cent of its clerks, and by June the company was running at a loss of £3,220.40. Collapse seemed a possibility but fortunately Ricardo loaned the company money, and by December the actual loss had been reduced to £341.05. Income from messages was c. £100 per week. Slowly the position improved and in 1850 ETC's gross income from all sources was £43,524.19 out of which the profit was £10,075.61. Ten years later, in 1860 ETC's revenue had increased to £214,245.36 and its profits to £69,711.70.

The progress of the Electric Telegraph Company was initially gradual and determined. In 1850, it had 1,786 miles (2870 km) of line and 7,206 miles (11,600 km) of wire; but in 1860, the corresponding figures were 6,541 miles (10,500 km) and 32,787 miles (52,800 km), with 3,352 instruments; and in 1868 when the government decided to nationalise all telegraph operations in the United Kingdom some of the statistics relating to ETC were (approximately) [14]:

- 1,300 telegraph stations in Great Britain and Ireland
- 10,000 miles (16,093 km) of line
- 50,000 miles (80,465 km) of telegraph wire
- 8,000 sets of instruments
- 3,000 skilled personnel, and
- 3 continental cables under its control.

The instruments utilised during the above period can be classified as (a) the needle systems; (b) the dial telegraph; (c) the Morse system; and (d) the automatic apparatus: and the signals received were either visible or audible, the former being of a transient or permanent nature and differing in form, colour, or duration; while the latter, which were always transient, differed in tone and duration [15].

Initially, Cooke's and Wheatstone's five needle instrument was employed for sending messages but as this required five wires and was slow in operation (about 6 words per minute) it was replaced by the double needle instrument (Fig. 4.1a) which could be operated at 21 words per minute by the average operator, or 40 to 45 words per minute in expert hands. However, the temporary nature of its signals, and the very small difference between one letter and the next needed some expertise and good eyesight to avoid errors in transmission. For example: the letters X and Y of the double needle instrument were similar and could cause confusion. An anxious husband was probably much startled to learn that his wife had presented him with a 'box'. Also E, F and G were not too dissimilar; so that when a learned barrister had forgotten his wig his secretary was possibly mystified by a request to send him 'my wife in a bandbox'. And a station master, who had been asked to search for a 'black box' in a first class carriage, replied that he could not find the 'black boy' anywhere [16].

Cooke and Wheatstone continued to improve their apparatus and in 1846 the Electric Telegraph Company installed a two-needle telegraph in the Octagon room of the House of Commons for the purpose of sending and receiving messages relating to parliamentary business. Each letter was signalled by a sequence of two momentary, independent motions to the left or right which caused the two needles of the receiving apparatus to deflect to the left or right. By means of a code these deflections enabled messages to be transmitted: a trained operator could achieve a rate of sending of 22 words per minute.

The double needle instrument required two wires for its functioning and consequently was phased out and replaced by the single needle instrument (Fig. 4.1b) which needed just one wire for the same purpose. By 1872 only one double needle circuit remained in the Postal Telegraph Department (UK). Frequently with the single needle instrument good operators were able to reach speeds of 35 words per minute, but, of course, the rate was limited in practice by the quickness of the writing of the operators. Nevertheless, the same wires which in 1851 conveyed 17 to 20 words a minute could in the early 1870s transmit 50 to 80 words per minute [16].

The first type of recording instrument used in the UK was the Bain instrument. This displayed dots and dashes on chemically treated paper. Later, in 1862 the instrument was replaced by Morse's apparatus which recorded its characters, at a rate of about 20 words per minute, by means of indentations or embossments of paper (Fig. 4.2). However, in 1872 this type of instrument was succeeded by the Morse ink writer which had a writing speed of 35 to 40 words per minute. In 1872 there were 1,509 of these instruments in operation in the

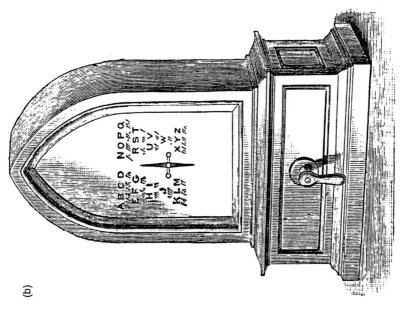

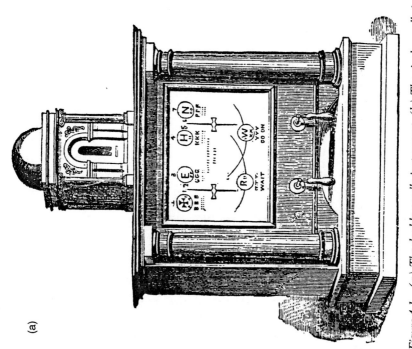

Figure 4.1   (a) The double needle instrument. (b) The single needle instrument.
Source: (a) R. Routledge, *Discoveries and inventions of the 19th century* (G. Routledge, London, 1903); (b) Urbanitzky op. cit. (Cassell, London, 1886).

Figure 4.2   The Morse embossing instrument.
Source: Urbanitzky, op. cit.

UK's Postal Telegraph Department dealing with messages having an average length of 34 words [17].

The instrument could record signals faster than a clerk could send or write and had few defects. Errors could arise from the tiredness of the operating staff, and more frequently from the transitory nature of the signals, and from the similarity of the signals. The letters H and S, P and G differed only by a dot and so, a gentleman who had telegraphed for his 'hack' (a horse) would probably have been surprised to learn that a 'sack' awaited his collection at the station. And the ordering of a 'gig' could lead to the recipient receiving a 'pig'. If a person telegraphed that he had forgotten his 'cocked hat' he would perhaps be irritated to receive a parcel containing 'cooked ham' [17].

Such errors sometimes arose because of the poor writing of the person sending the message. Furthermore, distortion could occur if little care was taken in punctuation and/or wording. Thus the physician who hurried to see a patient after having received the message 'Don't come too late' would have been spared his journey if the message 'Don't come; too late' had been sent; or if the message had been worded 'Too late – don't come', or 'Patient's dead: don't come' [18].

In practice in 1872 one error was evident from different causes in every 300 messages. Errors were inevitable, they could only be reduced to a minimum consistent with economical running of the telegraph service. If messages could not be transmitted with speed, cheap telegraphy would not have been possible: slow telegraphs were impracticable. Moreover, the telegraph signaller was not able to check the transmitted signal with the original message. He was in the same position as a person who wrote a message but was not able to see what had been received.

Both male and female staff were employed. In 1859 the directors of the District Telegraph Company advertised for female telegraph operators. They were surprised at the number of highly respectable and well-educated young women applicants. Some customers felt that female labour tended to lower the tone of the service although they agreed that the telegraph service was a most elegant employment for young women and they were very attractive to observe! Usually a female supervisor or matron was recruited to manage the operators.

The utilisation of automatic telegraphic apparatus enabled several advantages over manual working to be realised such as faster working, the reduction of errors caused by operator fatigue, and the elimination of delays caused by the need to repeat messages. Bain's apparatus was employed for a period in the 1860s, but in 1872 the machine most extensively used in the Postal Telegraph Department was that of Sir Charles Wheatstone (Fig. 4.3). It comprised the perforator, which – by means of a stiff paper tape having a series of holes punched in it – translated the message to a form suitable for the sending-end apparatus and the receiver which received the signal, as in the Morse system.

Three hundred messages per hour could be transmitted by the automatic equipment compared to the 30 to 35 per hour of the manual system. W.H. Preece, Divisional Engineer, Postal Telegraph Department, said in 1872 that 'the automatic apparatus [had] doubled the capacity of all our long wires and quad-rupled that of our shorter wires' [15]. This greater capacity was obtained at the expense of delay since the tapes had to be manually punched prior to being loaded into the sender.

And, of course, the need for a human operator did not lead to an error free system. Dots could be punched for spaces, spaces for dashes, and so on. During the proceedings of an important court case the following strange news report was noted: 'even Billy complained of the excessive heat, and the Court thereafter adjourned'. The Lord Chief Justice's indignation at the familiar, informal way he was mentioned was probably appeased when it was explained that the message should have been: 'eventually complained of the excessive heat' [19].

The most important advantage of the automatic system stemmed from its ability to work accurately and clearly for extended periods of time. In the transmission of news the same punched tape could be passed through several instruments in succession, and the same transmitter could actuate the receivers at several stations. By way of illustration: one transmitter at the Telegraph Street,

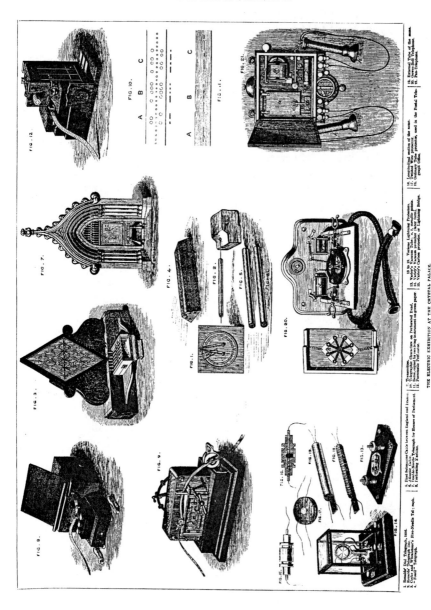

Figure 4.3 Electric telegraph apparatus exhibited at the Electric Exhibition, Crystal Palace, 1882.
Source: The Illustrated London News Picture Library.

London office worked the receivers at Bristol, Exeter, Plymouth, Gloucester, Newport, Cardiff and Swansea; another operated those at Nottingham, Sheffield, Leeds and Newcastle; and another actioned those at Birmingham, Manchester and Liverpool. The same tape could actuate all these machines, though, obviously, not simultaneously.

Some mention must be made of the dial telegraphs, such as Cooke's and Wheatstone's ABC instrument of 1840, (see Chapter 3). By 1872 these telegraphs had been much improved and there were 2,367 instruments of this kind in the Postal Telegraph Department (PTD). The advantage of them to the general public was the ease by which all persons who could read and write were enabled to send and receive messages. ABC telegraphs were to prove highly reliable and popular: indeed a few were still in use as late as 1920. Their rate of working was slow, rarely achieving 10 words per minute although experts could attain a speed of 20 words per minute [20]. Usually the rate was 4 to 5 words per minute. An advantage of the dial telegraphs was their freedom from batteries since the currents needed to work the senders were generated electromagnetically by the despatcher of the signal. This property was particularly advantageous for private telegraphy and, when combined with the dial telegraph's ease of use, enabled this form of apparatus to attain a deserved popularity. In 1862 there were 176 miles (284 km) of wire and 183 dial instruments for private telegraphy, but by 1872 these figures had increased to 7,190 and 1,767 respectively. The private wire department of the Post Office obtained an income of £48,790 in that year.

Printing telegraphs, in which the message was printed in bold Roman type upon a strip of paper that was then cut off the roll and handed to the recipient of the message, were employed from the 1850s. Several types were developed but only one form, the Hughes (Fig. 4.4) was practically adopted in England. The PTD had 23 in daily use in 1872. They were fast and could operate at over 60 words per minute, although 37 words per minute was about the norm [20].

Of the acoustic telegraphs, the Postal Telegraph Department had 394 bells and 211 sounders in use in 1872. With these a bell of one note signified a dot, and a bell of a different note a dash [21].

In the United States of America the growth of electric telegraphy (from the year, 1844, when Congress approved a liberal grant to establish the first telegraph line) was rapid. The aggregate length of the 57 telegraph lines, in 1852, exceeded 24000 km, of which:

- 48 lines totalling 1,305 miles (2,100 km) in length were worked on the system of Morse;
- 5 lines totalling 1,199 miles (1,930 km) in length used the Bain system; and
- 4 lines totalling 1,367 miles (2,200 km) in length utilised the apparatus of R. House (of New York).

In November 1852, 11 different companies had offices in New York, and there were in the United States between 20 and 30 joint stock electric telegraph companies.

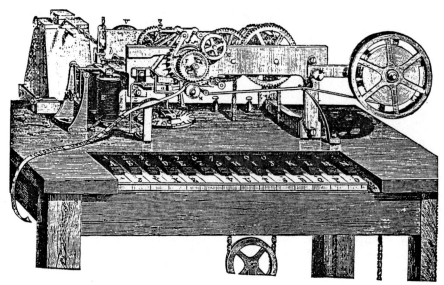

Figure 4.4   Hughes' type-printing telegraph.
Source: Urbanitzky, op. cit.

This rapid development was supported by the Federal Government, and promoted by individual States which passed general laws authorising the immediate construction of telegraph lines wherever they would be conducive to the good of the public; the laws afforded every facility to the founding of enterprises for that purpose. A c. 1852 special report [22], submitted to the British Parliament, on electric telegraphy in the United States noted:

If a company, or even a private individual, should propose to construct a telegraph line, and can show that it would be beneficial to the public (and as to this proof there is generally but very little difficulty), he may obtain an Act authorising him to proceed, as a matter of course; no private interests can oppose the passage of the line through any property; there are no committees, no counsel, no long array of witnesses and expensive hearings; compensation is made simply for damage done, the amount being assessed by a jury, and generally on a most moderate estimate. With a celerity that is surprising a company is incorporated, the line is built, and operations are commenced. Similar facilities are also afforded in many of the States, by general laws authorising the construction of rail roads.

In 1851, the Committee on Post Offices and Post Roads presented to the US Senate their report for an ambitious project which would link, telegraphically, San Francisco with Natchez on the Mississippi, and thence with the vast networks of lines of the Atlantic States [23]. The most distant points connected by telegraph lines in north America, in 1852, were Quebec and New Orleans, about 2,983 miles (4,800 km) apart, and the network of lines extended as far west as Missouri – approximately 500 towns and villages being provided with stations.

With the completion of the long distance lines from these States to California and Newfoundland it was considered that intelligence could be sent from the Pacific States to Europe and vice versa in six days.

The cost of erecting the telegraph lines obviously varied depending on the terrain being traversed, but an average cost was c. £22/km, this modest figure being attributed to the facilities afforded by the general telegraph laws for the formation of companies and the construction of lines.

Sometime towards the end of the 1840–1850 decade, Bain travelled to the United States and, on 18th April 1848, applied for a US patent on his electric telegraph but was not successful. The Commissioner of Patents in his report [24] (which runs to 20 closely printed pages) denied Bain a patent on the grounds that S.B.F. Morse had made a prior application. Essentially the Bain and the Morse claims were identical and so priority of invention depended upon the dates when the two inventors submitted their patents to the authorising bodies. The situation is summarised in Table 4.1 below.

Although Bain's UK patent was sealed on the 12th December 1846 the Commissioner argued that the pertinent dates were the 18th April 1848 (Bain) and 20th January 1848 (Morse). Hence, as the latter preceded the former, Morse had a prior claim.

After this disappointment Bain travelled to New York with the intention of returning to the United Kingdom. However in New York he met Henry O'Reilly. O'Reilly is an important figure in American telegraph history and so a few words of introduction are necessary [25].

He was born the son of a merchant in Carrickmacross, Ulster, on 2nd February 1806. Ten years later his father and family emigrated to the United States and established themselves in New York City. Eventually Henry O'Reilly became a journalist and after a period in this profession he ran, in 1836, for the Assembly as a Jacksonian Democrat. He was unsuccessful. Later in 1837 he was appointed Head of Rochester Post Office and in that year he became the editor of the *Albany Atlas*, a journal that advocated constitutional reform.

One day in 1845, while travelling on the night ferry to Albany, O'Reilly met a Mr John Butterfield. He was returning home from Washington where he had obtained an important telegraph contract. During his conversation with Butterfield, O'Reilly was persuaded that he too should engage in activity in

Table 4.1 Patent application dates

	Bain		Morse	
	UK dates	USA date		USA dates
Sealing	12.12.1846			
Enrolled	12.06.1847		Caveat	27.01.1847
Application		18.04.1848	Application (put in a secret archive)	20.01.1848

the new field of message transmission. Subsequently he participated in the construction of more than 12,800 km of telegraph lines in the USA.

At his meeting with O'Reilly, Bain was encouraged to press his case for a US patent, and in due course he was rewarded.

After this success O'Reilly, in 1848, announced [26] his intention to set-up telegraph lines in New England, using the Bain patent. The news was greeted with some enthusiasm. The *New York Express* reported: 'We are gratified to learn that the extortion which has so long been practiced on the telegraph line between [New York] and Boston is in a fair way to be abated.' A similar view was expounded by the *New York Journal of Commerce*: O'Reilly's arrangement with Bain would enable him

to work his lines with efficiency and success without infringing any other patent. [There-fore the public might expect] a healthy competition, a consequent reduction of prices, together with a more civil bearing of the telegraph owners and operators towards their customers.

Early in 1849 the New York and New England Telegraph Company was formed to operate a 'Bain' line between Boston and New York. This was ready for public use in 1850. The line was profitable [27]: receipts for the year ending March 1851 were $34,529.25 and for the year ending March 1852 were $41,521.30.

However this active opposition to the Morse interests was not conducive to commercial peace and as a consequence a bitter business war ensued. Telegraph rates were cut to two cents per word and then to one cent per word. Communication was greatly stimulated and this had the effect of enlarging the area of telegraphic utilisation. Whereas 20,000 messages had been transmitted in 1850, by 1851 the number had increased to 29,000, and in 1852 the total number of messages sent was 42,000.

The competition did not please Morse and his associates, who had adopted an aggressive policy to protect the use of the Morse patents, and as a result a suit against the use of the Bain patent was initiated in Philadelphia. It has been stated that the best leading talent in the country was engaged to sustain the Morse patent. And so, as it was likely that the suit would end disastrously for the Bain interests, a compromise between the two parties was arranged: 'There were strong arguments for peace'. An agreement to unite the two companies into a new organisation, the New York and New England Union Telegraph Company, was drafted and implemented.

During the period when the Bain patent was in operation in the USA, the Morse and House equipments were capable of transmitting 20 words per minute, but the apparatus of Bain (the electrochemical telegraph) was able to function at a much higher speed. According to Mr Whitworth's 1852 Special Report to Parliament [22]:

In Mr Bain's system, a weak current [was] found sufficient for very long distances; between New York and Boston, a distance of 270 miles [434 km], no branch or local circuit [i.e. relay circuit] [was] required. In some cases, where both Morse's and Bain's telegraphs [were] used by an amalgamated company in the same office, it [was] found convenient, in certain conditions of the atmosphere, to remove the wires from Morse's instruments, and connect them with Bain's, on which it [was] practicable to operate when communication by Morse's system [was] interrupted.

In November 1851 the North American Telegraph Company (which used Bain's patent) was defeated by the Magnetic Telegraph Company over a patent claim and welcomed the Morse company's offer of consolidation. As a sequel, its property, consisting of two wires between New York and Washington, was surrendered in January 1852 to its rivals, in exchange for an issue of $83,000 of Magnetic Telegraph Company stock. Six months later a similar merger was agreed between the Bain-operated New York and New England Company and Smith's New York and Boston Magnetic Company. When the New York and New England Union Telegraph Company was formed its joint capital was $300,000; two thirds was issued to the Morse company and one third to the Bain supporters. Later, on 7th June 1852, Faxton's New York, Albany and Buffalo Company purchased the Bain-operated New York State or Merchants State Telegraph Company for $65,379.20.

After 1852 no Bain lines were in operation. It seems that Bain personally lost a great deal of money in litigation against Morse.

The only other serious competitor to Morse, in the early days of electric telegraphy in the United States, was Royal House who was born and raised in Vermont. Little seems to be known about him. L. Turnbull in his 1853 book *The electro-magnetic telegraph* quotes [28] a letter, dated 23rd December 1850, that he received in reply to some inquiries regarding House.

Mr House is a self-educated man, and was engaged nearly six years in perfecting his instrument; he is decidedly scientific, but not learned. . . .

Such is the cast of his intellect, that he could form the entire object in his mind, and retain it there until he had completed its whole arrangement, without committing anything to paper; somewhat abstract in disposition, he is careless about money, little communicative concerning himself, capable of long protracted thought, and completely absorbed in his hobby, the telegraph. . . .

From some affection [sic] of the eyes, he was confined to his dwelling during most of the time spent in contriving his instrument; he resides at present in New York. An application was made for a patent in 1845 or '46, but it was refused on the ground that some of the specifications clashed with those of Mr Morse; one, however, was granted in October or November of 1848, to date from April 18, 1846.

House's invention was a letter printing telegraph in which the sending apparatus included a set of keys, rather like those of a piano; there was one key for each letter of the alphabet [29]. When a key was depressed a given number of electric impulses (corresponding to the particular letter) were sent to the receiving station. There, the received impulses advanced a type wheel until the letter that had been signalled by the sending end operator was reached. The letter was then printed by a pneumatic printing machine in which the valves were actuated by electromagnets excited by the received electric impulses. In use the machine could send and print at the rate of about 20 words per minute. As an indication of its complexity, the machine cost c. $250, the cost of the Morse register was c. $40.

The first telegraph line to utilise House's apparatus was completed in March 1849 [30]. It was owned by the New Jersey Magnetic Telegraph Company (later the New York and Washington Printing Telegraph Company) and linked Philadelphia to New York.

All classes of society used the electric telegraph. However, there was a system of precedence. Government despatches, and messages involving the life and death of any citizen, were entitled to priority; and then next in order of precedence were important press communications, but if these were not of especial concern they had to await their regular turn.

The tariff for press despatches was:

Under 200 miles (322 km)	1 cent per word
200 to 500 miles (322–805 km)	2 cents per word
500 to 700 miles (805–1126 km)	3 cents per word
700 to 1,000 miles (1,126–1,609 km)	4 cents per word
1,000 to 1,500 miles (1,609–2,414 km)	5 cents per word
over 1,500 miles (2,414 km)	6 cents per word

The press, of course, made great use of the various telegraph lines. For example: in 1852 the annual sum paid by the Associated Press of New York was c. $30,000. Hence, assuming an average cost of 3 cents per word, the total newsprint due to the telegraph amounted to about one million words per annum, or about 600 columns of a London newspaper of the largest size (that is nearly two columns per day). Usually the cost of the despatches was shared by several newspapers, so that for an association of six papers the annual charge per newspaper company, for two columns of intelligence per day, was just $5,000 (see Figure 4.5).

Business houses, too, used the electric telegraph to an appreciable extent. In 1852 the number of messages transmitted daily between New York and Boston varied from 500 to 600; and the sums paid, by some of the leading commercial houses, to the line's operator varied from $60 to $80 per month. On other lines the principal companies were estimated to pay from $500 to $1,000 per annum for telegraphic news.

A novel application of the electric telegraph for notifying fires was initiated in Boston [22]. The city was divided into seven districts, each provided with a powerful alarm bell. Associated with these districts were a total of 42 electric telegraph stations, each having an apparatus that was able to send a coded alarm message – which identified the station, and hence the approximate position of a fire – to a central telegraph office. On reception of a message the central office then notified an appropriate fire station.

Elsewhere, on the continent of Europe, the early growth of electric telegraphy was slower than in the United Kingdom and in the United States, as shown in Table 4.2, below [31].

THE ELECTRIC AND INTERNATIONAL
TELEGRAPH COMPANY.

INCORPORATED 1846.

The Charges for Messages not exceeding Twenty Words in Great Britain and Ireland :—

Within a Circuit of	50 Miles	....	**1s. 6d.**
Do.	do.	100	do.	....	**2s. 0d.**
Do.	do.	150	do.	....	**3s. 0d.**
Beyond a Circuit of 150	do.	....	**4s. 0d.**
To or from Dublin	....	**5s. 0d.**

No Charge is made for the Names and Addresses of either Sender or Receiver, or for Delivery within half a mile of the Company's Offices. The Company have

UPWARDS OF 360 STATIONS IN FULL OPERATION,

The whole of which are in

Direct Communication with the Continent,

Viâ the Company's

LINE TO THE HAGUE AND AMSTERDAM;

By which, under recent arrangements with the Continental Governments,

GREAT REDUCTIONS

Have been made in the charges, as shewn in the following list of

CHARGES TO

	£	s.	d.		£	s.	d.		£	s.	d.
Amsterdam	0	6	0	Copenhagen	0	12	0	Paris	0	11	0
Antwerp	0	7	6	Genoa	0	15	0	Riga	1	5	0
Berlin	0	11	0	Hamburg	0	10	0	Rotterdam	0	6	0
Bremen	0	8	6	Konigsberg	0	18	6	St. Petersburg	1	11	0
Brussels	0	7	6	Malta	1	11	6	Stockholm	0	18	0
Christiania	0	18	0	Memel	0	18	6	Trieste	0	12	0
Constantinople	1	18	6	Odessa	1	11	6	Vienna	0	12	0

For information as to number of words allowed, charges to other Stations, &c., &c., apply at any of the Company's Offices.

PRINCIPAL STATIONS IN GREAT BRITAIN.

Aberdeen	Falmouth	London	Sunderland
Birmingham	Glasgow	Lowestoft	Swansea
Bradford	Gloucester	Manchester	Truro
Bristol	Greenock	Newcastle-on-Tyne	Wakefield
Cambridge	Halifax	Norwich	Warrington
Cardiff	Haverfordwest	Oxford	Whitby
Carlisle	Holyhead	Perth	Wolverhampton
Darlington	Huddersfield	Plymouth	Windsor
Dublin	Hull	Preston	Wigan
Edinburgh	Leeds	Sheffield	Yarmouth
Exeter	Liverpool	Southampton	York.

Lothbury, London, June, 1853.

J. S. FOURDRINIER, Secretary

1-10-Lo.-11.

Figure 4.5   The charges (in 1853) for messages (1) not exceeding 20 words in Great Britain and Ireland, and (2) to certain overseas countries.

Source: Author's collection.

Table 4.2 Growth of the number of paid messages sent for the years 1853 and 1855

Country	Date of construction	1853	1855
Austria-Hungary	1850	120,001	209,208
Belgium	1851	25,420	35,635
Denmark	1854	–	26,380
France	1851	142,061	254,532
Germany*	1850	158,213	339,930
Italy (Modena)	1852	9,136	19,635
Netherlands	1852	26,087	78,508
Norway	1855	–	24,683
Portugal	1855	–	–
Spain	1855	–	2,085
Sweden	1853	851	60,607
Switzerland	1852	76,343	146,688
Electric Tel. Co. (United Kingdom)	1846	350,500	1,017,529

* Comprising Baden, Bayern, Braunschweig, Hamburg, Hannover, Mecklenburg, Bremen-Oldenburg, Prussia, Saxony, Schleswig-Holstein and Wurtemburg.

Unlike the situation that prevailed in the United Kingdom and in the United States, the electric telegraph systems of the major continental powers were operated almost from the start as state monopolies. The telegraph lines were considered primarily to be the means by which military and government intelligence could be transmitted quickly and no important lines were constructed by private enterprise. In France the first line was erected in 1845 by the government for its own use and it was not until 1850 that the public was permitted to use it, with priority for government messages. In July 1847 the Minister of the Interior, Lacave-Laplagne, declared in the Chamber of Deputies: 'The telegraph will be a political instrument, and not a commercial instrument' [32]. Essentially this view reflected the position held during the years of the Chappe semaphore telegraph. The telegraph was primarily for the use of the government and until 1878 the control and the responsibilty for its functioning rested with the Minister of the Interior, who was also the head of the nation's police force (Fig. 4.6).

Similarly, in Austria, following the experiments begun in 1846 along the Vienna to Brunn railway, Metternich declared the electric telegraph to be a monopoly of the state and it was not available for use by the public until June 1849 [33]. In Prussia, in 1846, a military commission was appointed to investigate the potentialities of the new mode of communication [32]. An experimental line was laid between Berlin and Potsdam, and subsequently several major lines were constructed during the ensuing years, but as in Austria, the general public was not allowed to utilise the state system until 1849.

Figure 4.6   Grande salle des télégraphistes au poste central de Paris.
Source: A. Belloc, *La télégraphie historique* (Paris, 1894).

Somewhat intriguingly, the state systems mentioned above were held, in the UK from c. 1850, to be superior to the overall system that had evolved in Great Britain. From this date arguments began to be advanced by private individuals and others for a nationalisation of the several British electric telegraph companies. The opening move – which subsequently led the Government, on 1st April 1868, to introduce the Telegraph Bill 'to enable the Postmaster General to acquire, maintain and work the Electric Telegraph in the United Kingdom' – was initiated by the Belgian Government. In 1851 it suggested that the transmission of international telegraphic communications should be regulated by a treaty signed by Belgium, France, Great Britain and Prussia [34]. This proposal was predicated on the assumption that the governments of these countries would control their telegraph systems. The issue was referred to the President of the Board of Trade, who, in a confidential memorandum, was advised: 'The time seems to have arrived for the Government to determine whether it will exercise any more systematic control over the telegraphic communication of the country than it has hitherto done' [34]. Telegraphs were undoubtedly important as a means of conveying intelligence, and, for the reasons put forward earlier for the Post Office being the government department responsible for the UK's mail, a case could be argued for the various electric telegraph companies being placed under state control or management.

In 1854 this question was raised in the *Quarterly Review*, a journal of intellectual non-conformity; and in the same year T. Allan, a promoter of the UKTC, argued that the Post Office should operate the telegraphs at a uniform rate of 1s(5p) per 20 words, irrespective of distance. Such a policy, he felt, would lead to an increased usage of the telegraph network. Further declarations of support for state control were made in 1856 by F.E. Baines, of the Post Office, and, perhaps surprisingly, in 1861 by J.L. Ricardo, who until 1858 had been Chairman of the Electric Telegraph Company [35].

Ricardo contended that 'it was imperative to transfer the telegraphs to a public department' and compared the situation in the UK with that on the continent. There, the telegraph was 'at once seen and understood as so powerful an engine of diplomacy, so important an aid to civil and military administration, so efficient a service to trade and commerce' that all the continental countries had instituted state telegraphic systems. Other voices were added to those of Ricardo, Baines and Allan but, by 1862, neither the Post Office nor the Government was persuaded that public opinion would accede to a move that would replace private enterprise by government control [36].

Meantime, the growth in the number of electric telegraph companies seeking, by means of Parliamentary Bills, and being granted by Parliament, approval to erect posts and wires along public highways was causing some unease to landowners and trustees of turnpikes, as well as to the general public [3]. The erection of telegraph poles in front of people's houses and the loss of public amenities gave concern that it was 'desirable to have a general measure, laying down some principles which would not deprive the public of the advantage arising from the competition of the various companies but at the same time would protect the public property' [3].

This issue was brought to the forefront by the United Kingdom Telegraph Company's Bill of 1862 which requested 'powers of a most extensive nature which if granted would enable it to use all public highways with imperfect protection to the rights and interest of property' [3].

The Government now acted and in 1863 introduced the Telegraph Act. Arbitration by the Board of Trade was introduced; the sale of a telegraph company could only proceed with the approval of the Board of Trade; sanctions could be imposed against any company which did not comply with the Act's provisions; and generally the extent of a company's aspirations were regulated by the powers devolved to the Board of Trade by the Act [37].

Further agitation for state control arose in 1865 when the telegraph companies in concert withdrew the uniform rate of 1s(5p) per 20 words (addresses free) between large cities which had been agreed in 1861. This action appeared to render somewhat doubtful the validity of the 1863 Act to satisfy the previous demands for state control. Clearly, further consideration had to be given to this question. In 1865 the Edinburgh Chamber of Commerce was stimulated to appoint a committee 'to consider the present conditions of telegraphic communication in the United Kingdom with a view to its improvement' [37]. The committee reported in October 1865. It found much that was wanting in the provision of a wide-spread, reliable, and efficient telegraphic system.

Among the complaints of the users of the electric telegraph were: the high charges of the companies, which militated against social correspondence; the frequent vexatious delays in the delivery of messages; the inaccurate reproduction of the messages, which sometimes caused confusion; the lack of telegraph facilities in many important towns; the remoteness of some of these facilities from the centres of business and of the population; and the short opening hours of some offices. The press, in its evidence to the committee, was vociferous in detailing the defects of the companies. News reports provided by them were often so inaccurate and unintelligible, due to the inexperience or carelessness of the staff, that 'in every newspaper office much valuable time was wasted in the irritating and wearisome occupation of deciphering and reducing to intelligible order' the news received. Taken together, these complaints were a severe indictment of the nation's telegraphs.

The committee's recommendations need not be enumerated and discussed here since in September 1865 the Postmaster General, Lord Stanley, directed F.I. Scudamore, the Assistant Secretary of the Post Office:

To enquire, and report, whether in his opinion, the Electric Telegraph service might be beneficially worked by the Post Office, and whether it would then possess any advantage over a system worked by the private companies, and whether it would entail a very large expenditure beyond the purchase of existing rights [38].

Scudamore reported to the Postmaster General in July 1866. He favoured the purchase of the telegraph companies by the state; the establishment of a uniform rate of message transmisssion of 1s(5p) per 20 words; and estimated that the whole of the property and rights of every description of the four principal companies could possibly be purchased for £2,400,000. His report was submitted to the Treasury and in January 1867 Scudamore drafted a bill 'To extend the facilities for the Transmission of messages throughout the UK' [39]. However, Parliament during the session 1866–1867 was heavily involved in debates on the Reform Bill and it was not until November 1867 when the Postmaster General announced his intention to seek powers to acquire the telegraph companies.

In February 1868 Disraeli became Prime Minister, following the resignation of Lord Derby, and on 1st April the new Chancellor of the Exchequer, Mr Ward Hunt, introduced the Telegraph Bill 'to enable the Postmaster General to acquire, maintain and work the Electric Telegraph in the United Kingdom'[39]. The Bill became law on 31st July 1868.

One effect of the agitation for state intervention of the telegraph industry, and the increasing prospect of nationalisation, was the upward movement of the share prices of the principal companies. The share price of the 'Electric' was £125 in 1865, £132 in January 1867, £145 in November 1867, £153 in January 1868, and £255 in July 1869 [40]. The share prices of the 'Magnetic' and of the UKTC doubled and quadrupled respectively during the same period. As a consequence Scudamore's 1866 estimate of £2,400,000 for the cost of nationalisation increased to £3,100,000 in February 1868, to £4,000,000 in April 1868 and to £6,000,000 in July 1868 when the Bill was enacted. Eventually the total cost reached £10,948,173 of which the telegraph and railway companies received £7,800,000; new extensions accounted for £2,100,000; and compensation for interests disturbed amounted to £190,000.

The vesting day when all the telegraph systems of the country were transferred to the General Post Office of HM Government was scheduled to be 30th June 1868 but for various reasons the actual date of the transfer was 28th January 1870. The receipts of ETC for the 13 month period ending on that date were £425,789.10 and its profits were £202,480.31. The shareholders received, in addition to certain interim dividends (which were limited by ETC's Act to 10% per annum) a sum of £2,938,826.45 divided among themselves, and a Trust Fund of £40,721.85 being equal to a dividend of £292.06 per cent on their capital [41].

Apart from the Electric Telegraph Company, several other operating companies had been formed prior to nationalisation. These (excluding the submarine telegraph companies) were [42]:

- the British Electric Telegraph Company (BETC) of 1849
- the English & Irish Magnetic Telegraph Company (EIM) of 1851
- the International Telegraph Company (ITC) of 1855
- the Electric & International Telegraph Company (the 'Electric') – an amalgamation of the ETC and the ITC – of 1855
- the British & Irish Magnetic Telegraph Company (the 'Magnetic') – an amalgamation of the BETC and the EIM – of 1857
- the London District Telegraph Company of 1859
- the United Kingdom Telegraph Company of 1850/1860

Together, these companies were predominantly responsible for a substantial increase in the electric telegraph network of the United Kingdom, as shown below [43]:

Year	Km of line	Km of wire	No. of stations	No. of messages
1851	2,819	11,750	198	48,490
1862	20,452	93,100	1,616	2,676,354
1872	40,225	141,000	5,179	15,500,000

In 1872 the uniform rate per word for the whole of the United Kingdom was 0.25p, a substantial reduction from the pre-nationalisation price of 2.5p per word. Moreover, whereas in the early days addresses were included in the message, by the 1870s they were sent free. This relaxation sometimes led to unusual addresses being given. As an illustration: a message was sent from Sherborne and delivered to: 'Jim Pierce, a little boy in charge of [a] horse and gig, in a field called "Eight Acres", just above [the] Railway station, Shepton Mallett, belonging to Mr Giles Farrington Gurney' [43].

The growth of the telegraph industry led to new occupations for both males and females. In 1849 the number of instruments at the Central Station (Figures 4.7 and 4.8), Lothbury – the main telegraph station in the City of London – was eight, with four or five boys working them – boys by day and men by night. In 1872 there were 409 instruments, worked by 649 females and 709 males, needing c. 17,000 battery cells (compared with 408 in 1849) (Fig. 4.7). Other statistics that highlight the enormous development of the telegraph industry are given in Table 4.3.

The development of electric telegraphy in the UK may be further appreciated by noting certain statistics relating to electric telegraphy in Europe, which were published by the International Telegraph Office in Berne. For the year 1875 some of these figures are given in Table 4.4.

Table 4.3 The growth of the telegraph industry in the UK, 1870–1872 [44]

	1870	*1872*	*Increase*
No. of telegraph offices	2,932	5,179	2,247
Length of wire in network (miles)	66,000	87,719	21,719
No. of instruments	2,175	8,284	6,109
No. of messages sent per office	43	49	6
Average charge per message (inclusive of porterage)	10.75p	5.69p	*

(* i.e. a reduction of 6d (2.5p) per message on the tariff, and 6d per message on the porterage charges – which had nearly disappeared.)

Table 4.4 Telegraphy in Europe in 1875 [45]

Country	*No. of paid inland messages (millions)*	*No. of messages per 100 persons*	*No. of telegraph offices*	*Average expenditure per message*	*% Profit / (loss)*
England	18.5	58	1 per 5,640 persons	4.79p	25
Austria-Hungary	4.5	12	1 per 11,556	7.5p	<(38)
Belgium	2.0	36	1 per ?	2.4p	<(15)
Denmark	<1.0	22	1 per ?	3.65p	<(15)
France	7.0	19	1 per 8,463	5.0p	18
Germany	8.25	19	1 per 7,980	5.3p	<(38)
Holland	1.5	38	1 per 115,449	4.27p	<(38)
Italy	4.25	16	1 per 15,522	4.38p	25
Norway	<1.0	27	1 per ?	5.18p	<(15)
Russia	<1.0	3.7	1 per 50,188	15.8p	19
Spain	<1.0	?	1 per ?	11.98p	<(38)
Sweden	<1.0	?	1 per ?	?	6
Switzerland	2.0	77	1 per 2,664	2.4p	12

THE ILLUSTRATED LONDON NEWS, Dec. 19, 1874. — 576

THE   CENTRAL   POST   OFFICE   TELEGRAPH   ESTABLISHMENT.

TRACING TELEGRAMS.

THE BATTERY-ROOM.

MESSAGE PRESSES.

THE SYMPATHETIC CLOCK.

THE CHRONOPHER, OR TIME-SENDER.

THE LINES TEST-BOX.

Figure 4.7   The Central Post Office Telegraph Establishment in 1874.
Source: The Illustrated London News Picture Library.

In a lecture on 'Telegraphy: its rise and progress in England', delivered at the Albert Hall on 18th July 1872, Mr W.H. Preece, Divisional Engineer, Postal Telegraph Department, said:

The public itself is becoming familiarised with telegraphy. The feelings of fear and alarm which once accompanied every telegram, are now more frequently converted into those of surprise and joy, for the wires have become the vehicle of communications as much for our little household joys as for the great concerns of State . . . [15]

The telegraphs of this country belong to the Nation. Every person has an interest in them. The prosperity of this property can be seen by any one every week in the papers, wherein the total number of weekly messages, is inserted. 126,000 messages at the transfer [nationalisation] have now risen to 308,000, and since every 1,000 messages produces £57.35, the money they bring in can readily be calculated. Indeed, it may be said that in the year that is coming, the telegraphs will unquestionably put £1,000,000 into John Bull's purse.

This enthusiasm of the public for the electric telegraph may be compared with the usage of the telegraph in France and Germany, as illustrated by the statistics, for the year 1878, given below [45].

Country	Population	No. of messages sent
England	33,799,000	24,600,000
France	36,900,000	14,400,000
Germany	42,700,000	14,540,000

The only cases in which the proportionate number of messages was larger than in Great Britain were those of Belgium and Switzerland. However, in both these countries the total number of messages transmitted included the number of foreign telegrams passing through them to destinations other than in Belgium and Switzerland.

A further point bearing on the public use of the system concerns the differences in the lengths of the messages in England, France and Germany. In the UK the average length, including the address was 28.18 words, in France it was 15.16 words, and in Germany 11.6.

These messages or telegrams could be classified into two categories: those relating to business or personal matters, and those whose content was of a general nature and of interest to the public at large. The latter category contained 'news': either universal news of events at home or overseas describing or commenting upon, for example, military campaigns, political changes and scientific discoveries; or specialist commercial news, such as commodity and stock market prices. The dissemination of such news by telegraph led to the establishment of news agencies. In Paris the Havas agency was formed in 1835 – before the first practical electric telegraph – and was given privileged access to the French semaphore system. Of the agencies set up during the era of electric telegraphy the most important were the Associated Press of New York (1848), the Wolff Bureau of Berlin (1849), Reuters of London (1851), the Press

THE ILLUSTRATED LONDON NEWS, Nov. 28, 1874. — 504

SKETCHES AT THE CENTRAL TELEGRAPH ESTABLISHMENT, GENERAL POST OFFICE.

Figure 4.8   Sketches taken at the Central Telegraph Establishment, General Post Office. The single needle, the automatic, the double needle, the Hughes' type printing, and the ABC telegraphs are shown being used.

Source: The Illustrated London News Picture Library.

Association of London (1868), the Central News of London (1871) and the Exchange Telegraph of London (1872).

Julius de Reuter (1816–1899) [46], a German Jew by birth but an Englishman by adoption, had had some experience of news collecting before he opened his London office. At one time he had provided a carrier pigeon service between the terminus of the Prussian telegraph line in Aachen and the French telegraph line in Brussels [47]. He chose an opportune time to start his London agency. The Dover to Calais submarine cable (see Chapter 5) had just been laid and so the telegraphic system that was rapidly expanding in the United Kingdom was connected to the lines being erected in the German States, France and elsewhere. Reuter soon developed a network of agents who despatched news messages by wire or cable. Where these did not exist the messages were sent by ship, railway or mail to a principal telegraph centre such as Trieste, Marseilles, Liverpool, Plymouth and Southampton.

Of prime importance to the general public was the association of the news agencies and newspapers. In October 1858 Reuter, after seven years of endeavour, succeeded in persuading the London newspapers to give his service a trial. It was immediately successful, and was followed by the syndication of overseas news whereby this news was channelled through Reuters to the London and provincial papers. The editors soon appreciated that by using the service provided they had access to more numerous and extensive sources than their own newspapers could provide. Essentially 'Reuters', as the agency was called, enabled the editors to publish a much greater and more widespread coverage of news than their special resident reporters could collect for themselves. For an annual fee of £200 the *Belfast Newsletter*, the *Glasgow Herald*, and the *Manchester Guardian* purchased news reports averaging 6,000 words per day when Parliament was in session and 4,000 words daily at other times of the year. The telegraph companies, too, realised the importance of Reuters. Both the 'Electric' and the 'Magnetic' contracted with the agency, for a fee of £800 per annum, for the exclusive rights to supply foreign telegrams to all towns in the UK [48].

Reuters' success was founded not only on the rapidity of its news gathering operations and the interest of its reports, but also on the trustworthiness and impartiality of the facts it disseminated. The agency was often the first with some news but its accuracy and reliability were never compromised. These qualities enhanced Reuter's reputation so that by 1861 one journal wrote that 'all our earliest information from Australia, India, China, and the Cape is derived from this gentleman's telegrams' [49]. His name became a hall mark of excellence in news reporting and he was accepted as an important figure in Victorian society – all from his exploitation of the electric telegraph.

Charles Dickens (1812–1870), who had been a journalist by training before pursuing his work as a novelist, portrayed Mr Reuter in a humorous address, given at a dinner of the Newsvendors' Benevolent Association in 1865. Dickens was referring to the situation that would ensue if the newsvendors went on strike [49].

Imagine all sorts and conditions of men dying to know the shipping news, the commercial news, the legal news, the criminal news, the foreign news and domestic news . . . Why, even Mr Reuter, the great Reuter whom I am always glad to imagine slumbering at night by the side of Mrs Reuter, with a galvanic battery under the bolster, telegraph wires to the head of his bed, and an electric bell at each ear, even he would click and flash those wondrous despatches of his to little purpose, if it were not for the humble, and by comparison, slow activity, which gathers up the stitches of the electric needle, and scatters them over the land.

The growth of the agencies' contacts and reporters revolutionised communications, initially throughout Europe and the United States of America, but later the whole world. No longer was the rate of news transmission dependent on the speed of a horse, a mail coach or a ship. Now, information that formerly had taken days or weeks could be sent in minutes or hours. *The Times* newspaper of 9th January 1845, for example, included a report of an event that had occurred approximately eight weeks previously. News from South America, from New York, and from Berlin was usually about six weeks, four weeks and one week old respectively when published. But with the development and laying of submarine telegraph cables and the expansion of the land lines news could be available from a far off land in a few hours. By way of illustration: the news of the death of President Lincoln on 15th April 1865 was not published in the London newspaper until 12 days later; but when President Garfield was shot in 1881 the event was printed in the British papers within 24 hours [49].

The news that was most in demand by the press and the public was bad news rather than good news. In 1883 Reuter's agents received a circular which clearly specified what was required:

Fires, explosions, floods, inundations, railway accidents, destructive storms, earthquakes, shipwrecks attended with a loss of life, accidents to British and American war vessels and to mail steamers, street riots of a grave character, disturbances arising from strikes, duels between, and suicides of persons of note, social or political, and murders of a sensational or atrocious character. It is requested that the bare fact be first telegraphed with the utmost promptitude, and as soon as possible afterwards a descriptive account, proportionate to the gravity of the incident. Care should, of course, be taken to follow the matter up [50].

In 1870 Reuters and the other news agencies mentioned above combined together in a 'ring' – to prevent expensive and possibly cut-throat competition – according to which news gathering and dissemination throughout the world was shared between them. Havas was allocated most of western and southern Europe and the French empire; Reuters covered the British empire, China and Japan; and Wolff had eastern and northern Europe. The USA was considered to be a neutral region where any of the agencies could collect news, but in practice Associated Press supplied most of the news from America.

Nevertheless, the cost of the 'electric' news was high, particularly when submarine cables had to be utilised. At first, the rate per word for telegrams sent via the first trans-Atlantic cable was £1, and the rate for transmission to Australia, in 1872 via the newly laid cable, was 10s. per word [49]. Necessarily overseas newspapers could not afford to take lengthy cable messages from the agencies. However, the practice of using cable-ese soon developed. Abbreviations, code-words, and the omission of conjunctions became commonplace. Football and cricket results could be sent to and from London by means of resourceful coding.

Occasionally though, contraction of a message and the inadvertent intro-duction of errors could lead to gross distortion. A delightful example of this was given by Mark Twain (Samuel Clemens, 1835–1910) in his book *A tramp abroad* (1880). A telegram had been received by Reuters in London which read: 'Governor Queensland twins first son'. Enquiry revealed that the Governor was Sir Arthur Kennedy and so the telegram on reception was expanded to: 'Lady Kennedy has given birth to twins, the eldest being a son'. Unfortunately the putatively welcome news was not greeted warmly by the Governor's family and friends in England because they knew him to be an elderly widower who had not remarried. Further enquiry showed that the message referred to the construction of a new railway and should have read: 'Governor Queensland turns first sod' [50].

Bloomers such as this could be amusing, but when commodity or share prices were subjected to error the consequences could be serious. Fortunately the press agencies made surprisingly few faux pas and, of course, crucially important despatches could always be repeated.

Notwithstanding the high costs of transmission, the principal newspaper editors greatly valued the inclusion of reports from their overseas reporters and correspondents, and from the news agencies. Exclusivity and the rapidity by which news from distant places appeared in the dailies were factors that affected the sales and the standing of the papers; and so the leading newspapers were willing to expend appreciable sums on gathering information the editors felt the general public required.

Wars, particularly those near at hand or those involving British troops, were primary news sources for reporters in the field. When the Franco-Prussian war of 1870–1871 was being fought, the *Daily News* and the *Daily Telegraph* were spending approximately £70 to £80 per day on telegrams. This was at a time when a labourer earned 85p per week, rent for a working class family was 30p per week, income tax averaged about 3p in the £, and a 2-bedroom house cost c. £250 [6]. Apparently, the expenditure was profitable for the circulation of the *Daily News* increased from c. 50,000 to c. 150,000 copies per day. And the sales of the *Daily Telegraph* (which had been established just a few years previously in 1855) were c. 190,000 per day in 1870 – or three times those of the well-established newspaper *The Times*. This adverse situation stimulated the manager to instruct his war reporter to follow the example of the *Daily News*' war correspondents [49].

I beg you to use the telegraph freely. After any very important event go yourself with all speed to the nearest telegraph station that has communication with London. Send by the wires, not a scrap of a few lines, but a whole letter. This is what the correspondents of the *Daily News* have been doing frequently.

Reuter's success was not fortuitous or wholly determined by the growth of electric telegraphy. His achievement came about during the mid-Victorian period when Great Britain stood out as the only truly world power. It possessed the largest empire the world had ever seen – an empire, it was said, 'on which the sun never [set]'. And so, with many of the country's citizens scattered throughout the colonies and dominions as administrators, settlers, law enforcers, traders and the like, there was much interest at home in their well-being and the well-being of the countries they were serving. 'It would be fatal to say "Discuss home matters, but not foreign ones" ... every issue of an English journal speaks to the whole world', opined *The Times*, 'that is its strength; it lives by its universality; that idea imparts conscious power' [46]. A similar sentiment was advanced by J. Grant, a London editor, in his book *The Newspaper Press* (1871): '[The mission of the press was] to enlighten, to civilise and to morally transform the world. These are the grand purposes which providence has in view in relation to our race.' But none of this could be accomplished – at least easily and speedily – without the network of submarine cables which were laid down around the world from c. 1850. This topic forms the basis of the next chapter.

Note 1

Standard time and the distribution of time signals

Mention has been made in Chapter 3 to the application of the electric telegraph to railway signalling. Railway companies were required to produce timetables for the benefit of their passengers but the times given were 'railway times' and not necessarily the 'local solar times' of the passengers' destinations. Since local solar time is a function of longitude – a change of 15° of longitude leads to a change of solar time of one hour – confusion and difficulties of railway scheduling could arise. With the rapid development of railway transportation a need for a 'standard time' arose. This was particularly so in large countries, such as the United States of America, where some routes passed through places that had local solar times differing by several hours.

Moreover, not only was a standard time desirable, there had to be a means of distributing time signals. In the United Kingdom, one of the first services to be provided for this purpose was introduced by John Pound, the Astronomer Royal (1811–1835) [51]. He employed his adopted son, John Henry, to carry a chronometer showing Greenwich time to the principal clock and watchmakers in London. Previously, on a given day of the week, these artisans had had to send their own men to Greenwich, or make transit observations of a star, to have their timepieces calibrated. Pound's successor, Sir George Airy, who was the Astronomer Royal from 1835 to 1881, in his Annual Report of 1849 noted: 'The general utility of the Observatory will be increased by the dissemination throughout the kingdom of accurate time signals by an original clock at Greenwich' [51].

In 1841 Alexander Bain invented an apparatus that would permit the distribution of time signals. His ideas on electric clocks were incorporated in five patents dating from 1841 to 1852. According to C.K. Aked of the Antiquarian Horological Society, 'most of the applications of electricity to horology [were] anticipated by his 1841 patent'; it 'shows quite clearly that Bain was a genius', and justifies Bain being given the title of 'Father of electrical horology' [5].

Among the points covered by the patent are: the application of electromagnets and pendulums, having operating contacts, to make or break an electric circuit, to drive a master and slave clocks; and the application of a master clock to impulse / to wind up / to regulate / to set the hands of any number of slave clocks. In May 1846 Bain demonstrated his system for distributing time signals when he transmitted, telegraphically, time signals from one of his clocks, placed in Edinburgh railway station, so as to synchronise a timepiece situated at the Glasgow terminus of the Edinburgh and Glasgow railway. A contemporary account describes the invention 'as being one of the greatest importance, as by its introduction the great evil of variation of time, in distant situations, will be entirely avoided' [51].

Bain's enterprise was not immediately progressed but in 1864 the Magnetic Telegraph Company initiated the control of public clocks by electric time signals transmitted from the Glasgow Observatory. This was followed one year later by a similar service provided by the Liverpool Observatory; and in 1869 Greenwich Observatory was reported by Airy to be sending synchronising signals via a telegraph line to the Lombard Street (London) Post Office clock. The Greenwich Observatory service to the Post Office expanded rapidly, as did those of private telegraph companies. For example, the firm of Barraud and Lund in 1877 established in the capital a time signal service using electric currents regulated by a timepiece that was corrected by signals from Greenwich. This service, eventually, was maintained by the Standard Time Company [51].

In North America the great lengths of the rail routes made the problem of timetable scheduling much more difficult than in the UK. One of the persons who pondered on this problem was Sir Sandford Fleming, a Canadian railway planner and engineer. He advocated, in the late 1870s, a plan for a worldwide time standard. A few years later (in 1884) delegates from 27 nations met in Washington, DC, and agreed on a plan of 24 time zones, which is basically the same as that now in use. During the conference, international agreement was reached that the Greenwich meridian should be the prime meridian of longitude, i.e. 0°, and in Great Britain Greenwich Mean Time became the standard time.

The method of distributing time signals by electric telegraphy began to be superseded by wireless telegraphy from 1905. On the 9th August of that year the first ever wireless time signals were transmitted from the Navy Yard at Boston, Massachusetts. Five years later, on 25th June, the first French radio time signal was radiated by equipment installed in the Eiffel Tower, Paris. British radio time signals, the 'pips', date from shortly after the British Broadcasting Company was formed in 1922 [52].

References

1 BOWERS, B.: 'Sir Charles Wheatstone FRS, 1802–1875' (HMSO, London, 1975), p. 136
2 BRIGHT, C.: 'Address of the President', *J. STE*, 1887, **16**, p. 18
3 KIEVE, J.L.: 'The electric telegraph. A social and economic history' (David and Charles, Newton Abbot, 1973)
4 Lords Journal, **12**, session 1846
5 BURNS, R.W.: 'Alexander Bain, a most ingenious and meritorious inventor', *ESEJ*, 1993, pp. 85–93
6 PRIESTLY, H.: 'The what it cost the day before yesterday book, from 1850 to the present day' (Kenneth Mason, Hampshire, 1979)
7 Census returns, H O 107/1469, folio 484, Public Record Office
8 CLARK, L.: Inaugural Address, *J. STE*, 1875, **4**(4), pp. 1–22
9 Ref. 3, p. 4
10 Ref. 8, p. 12
11 Ref. 2, pp. 7–41
12 Ref. 2, p. 21
13 Ref. 8, p. 12
14 Ref. 2, p. 22
15 PREECE, W.H.: 'Telegraphy: its rise and progress in England', *J. STE*, 1872, **1**, pp. 228–236
16 Ref. 15, p. 231
17 Ref. 15, p. 232
18 Ref. 15, p. 233
19 Ref. 15, p. 234
20 Ref. 15, p. 235
21 Ref. 15, p. 236
22 WHITWORTH, Mr.: 'Special report', 'Electric telegraphy', chap. xi, Lords Journal, **78**(9) Vict., 10 Vict., pp. 28–33
23 TURNBULL, L.: 'The electromagnetic telegraph' (A. Hart, Philadelphia, 1853), p. 145
24 ANON.: 'Decision of the Commissioner of Patents in the matter of inter-ference of S.F.B. Morse and Alexander Bain, applicants for patents for certain improvements in electric telegraphs', US Patent Office Report, 1848, pp. 1124–1143
25 W.M.B.: 'O'Reilly, Henry', in 'Dictionary of American Biography', **14** (Scribner's and Sons, New York, 1934), pp. 52–53
26 THOMPSON, R.L.: 'Wiring a continent: history of the telegraph industry in the United States, 1832–1866' (Princeton University Press, Princeton, 1947)
27 REID, J.D.: 'The telegraph in America, its founders, promoters, and noted men' (Derby Bros. New York, 1879), chap. 25, p. 301
28 Ref. 23, p. 116
29 PRESCOTT, G.B.: 'History, theory and practice of the electric telegraph' (Trubner and Company, London, 1860), pp. 111–126. Also Turnbull, op. cit., pp. 115–125
30 Ref. 23, p. 125

31 OBERLIESEN, R.: 'Information, daten und signale' (Deutsches Museum, Rowohlt Hamburg, 1982), p. 115
32 Ref. 3, p. 46
33 Ref. 3, p. 47
34 Ref. 3, p. 122
35 Ref. 3, p. 119–120
36 Ref. 3, p. 121
37 Ref. 3, p. 125
38 Ref. 3, p. 128
39 Ref. 3, pp. 136–138
40 Ref. 3, p. 167
41 Ref. 8, p. 13
42 Ref. 3, chap. 3, pp. 46–72
43 Ref. 15, p. 229
44 Ref. 15, p. 230
45 Graves, R.: 'A decade in the history of electric telegraphy', *J. STE.* 1880, **9**, pp. 249–272
46 READ, D.: 'The impact of "Electric News" 1846–1914. The role of Reuters', a paper in 'Semaphores to short waves', (Royal Society of Arts, London, 1998), ed. by F.A.L. James, pp. 121–135
47 Standage, T.: 'The Victorian internet' (Weidenfeld and Nicolson, London, 1998), p. 61
48 Ref. 3, p. 7
49 Ref. 46, pp. 125–127
50 Ref. 46, pp. 130–131
51 AKED, C.K.: 'Electrifying time' (The Antiquarian Horological Society, Wadhurst, 1976),
52 DALE, R.: 'Timekeeping' (The British Library, London, 1992), p. 57

Chapter 5

Submarine telegraphy

In 1840 the Minutes of the House of Commons Select Committee on Railway Communication were printed and published [1]. Wheatstone was one of the experts consulted and during his examination he was asked, on 6th February 1840, by Sir John Guest, MP, whether he had 'tried to pass the [electric telegraph] line through water'. Wheatstone responded by saying that 'there would be no difficulty in doing so, but the experiment [had] not yet been made'.

At this time (1840) Cooke was in dispute with Wheatstone over his (Cooke's) involvement in their 1837 patent on electric telegraphs and this led to a formal arbitration in 1840–1841. Cooke averred that [2]: 'The invention [of 1837] at once became a subject of public interest; and I found that Mr Wheatstone was talking about it everywhere in the first person singular . . . At length, in 1840, I required that our positions, relative to the invention and to each other, should be ascertained by arbitration.' No documents were published concerning the quarrel until December 1854, when, after a renewal of the controversy, Cooke published a pamphlet entitled *The Electric Telegraph: Was it invented by Professor Wheatstone?*

In his January 1856 'Answer' to Mr Cooke's pamphlet, Wheatstone [3] referred to his 1840 evidence to the Select Committee and wrote (in the third person):

Shortly after this, having been furnished with the necessary hydrographic information by his friend Sir Francis Beaufort, and received much useful counsel from the late Captain Drew of the Trinity Board, Captain Washington, and other scientific naval friends, he prepared his detailed plans [for a submarine cable from Dover to Calais], which were exhibited and explained to a great number of visitors at King's College, among whom were the most eminent scientific men and public authorities. He also made the subject known in Brussels . . .

Mr Wheatstone's plans were also shown in 1841 to some of the most distinguished scientific men in Paris, who came to see his experiments at the College de France.

However, as Wheatstone's biographer [2] has stated, it is not clear what

Wheatstone's plans for a submarine telegraph were in 1840: they were not published in his lifetime though they were shown to some of his scientific acquaintances. A few months after his death in 1875 the Society of Telegraph Engineers published a letter [4] from his son-in-law, Robert Sabine, together with a copy of Wheatstone's drawings for the submarine cable (Fig.5.1). They show that Wheatstone had given considerable thought to the difficulties that would be experienced in manufacturing and laying a submarine cable. This would consist of seven copper wires each lapped with hemp twine saturated with boiled tar; these were to be laid up together and lapped overall with similarly treated twine.

It is known that Wheatstone carried out very limited experiments in the field of submarine telegraphy in 1843 and 1844. In June 1843 Prince Albert visited King's College to inaugurate the George III Collection of Scientific instruments which was displayed in a 'spacious and well-lit apartment' [5]. After the formal opening and a tour of various demonstrations, the Prince was accompanied by Wheatstone to the terrace in front of Somerset House, where, according to *The Times* [6]: 'It had been intended that some experiments should be tried by the electric fluid from the top of the shot manufactory across the river to the terrace, but the experiment was not exhibited, as it was understood that the string or wire had been cut or broken by which the communication was to be effected.'

Wheatstone described this experiment in a lecture, delivered on 23rd May 1843 to the Society of Civil Engineers, which was reported in the Literary Gazette of the 3rd June 1843.

In [1843] he [Wheatstone] formed a communication between King's College and the shot-tower upon the opposite side of the river; the communicating wire was laid along the parapets of Somerset House, and Waterloo Bridge, and thence to the top of the tower, where one of the telegraphs was placed; the wire then descended, and a plate of zinc, attached to its extremity, was plunged into the mud of the river; a similar plate was attached to the extremity at the north side, and was immersed in the water. The circuit was thus completed by the entire breadth of the Thames, and the telegraphs acted as well as if the circuit were entirely metallic.

The following year, on 28th and 29th August 1844, Wheatstone carried out some experiments, in Swansea Bay, 'on the passage of the electric current through sea water' [7], but nothing was published about this work until Sabine's 1876 letter to the Society of Telegraph Engineers.

Although, in some accounts of the history of submarine telegraphy, Wheatstone is given some credit and priority for his efforts in this field, Alexander Bain [8] had conducted a systematic investigation of the conduction of electricity in water in 1842. He had found, in 1841, from some experiments with an electromagnetic sounding apparatus,

that if the conducting wires were not perfectly insulated from the water in which they were immersed, the attractive power of the electromagnet did not entirely cease when the circuit was broken. For the purpose of investigating the nature of this phenomenon, a series of experiments took place with great lengths of wire, in the reservoir of water at the Polytechnic Institution, when similar results were obtained.

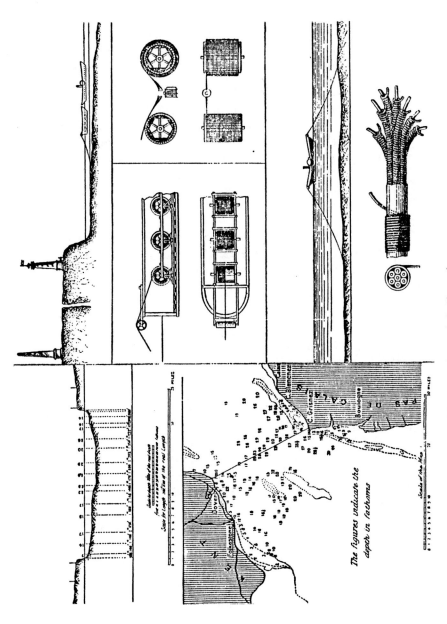

Figure 5.1 Wheatstone's plans for a submarine cable across the English Channel.
Source: *J. IEE,* 1876, **5**.

Bain felt that this effect might militate against the practical application of electric telegraphy and sought to pursue the matter. He was given permission by His Royal Highness the Duke of Sussex 'to commence a series of extensive and varied experiments on the Serpentine River, in Hyde Park'. The tests showed clearly that a volume of water could be placed in series with a conducting wire and items of apparatus without any disadvantage, and that dissimilar metal plates, for example of copper and zinc, when placed in moist earth or water acted as a voltaic cell. Bain's results were described and published in 1843, and his earth/water battery was patented in 1843 [9].

From 1840 Bain, whom *The Times* [10] later described as 'a most ingenious and meritorious inventor', was engaged in a prolonged dispute with Wheatstone whom he accused of plagiarising his inventions of the electromagnetic clock and the electromagnetic printing telegraph. Their quarrel was widely publicised and formed the subject of a 127 page tract by John Finlaison [11]. Finlaison states that on 21st August 1843 Wheatstone wrote to the Directors of Waterloo Bridge for leave to repeat Bain's experiments. This was granted on 3rd September 1843.

Elsewhere, in 1842 in the United States of America, Samuel F.B. Morse [12] conceived his subaqueous plan which, in December 1844, he submitted to the US House of Representatives. His first experiment [13] to establish the practicability of the plan was undertaken in the autumn of 1843 when, at New York, Morse laid some insulated wires in the water between Governor's Island and Castle Gardens, a distance of approximately 1.6 km. His initial experiment was soon brought to an end when one of his wires was caught and cut by a ship's anchor. 'The Professor, however, persevered, and next so arranged his wires along the banks of the canal, as to cause the water itself to conduct the electricity'

Morse wrote: 'Having ascertained the general fact, I was desirous of discovering the best practical distance at which to place my copper plates, and not having the leisure myself, I requested my friend Professor Gale to make the experimental facts for me.' Gale, late in 1844, determined that the length of each line should be three times the width of the water channel. Subsequently, the method was applied 'under the direction of my [Morse's] able assistants, Messrs Vail and Rogers, across the Susquehanna river, at Havre-de-Grace, with complete success, a distance of nearly a mile'.

Actually, the discovery that water can conduct electricity cannot be ascribed to either Wheatstone, or Morse or Bain. On 14th and 18th July 1747 Dr William Watson, an eminent English scientist, directed a series of experiments in which electricity was conducted from the Thames bank at Lambeth to the opposite bank at Westminster and then by wire across Westminster bridge. Later on the 24th of the same month at New River, Stoke Newington, Watson and his associates – who included the President of the Royal Society and Lord Cavendish – sent an electric charge through 800 ft (244 m) of water and 2,000 ft (610 m) of land, and then through 2,800 ft (853 m) of land and 8,000 ft (2438 m) water [14]. Watson's experiments were repeated by others, notably Franklin, in 1748, across the Schuykill at Philadelphia; by Deluc, in 1749, across Lake Geneva; and by Winckler, in 1750, at Leipzig.

Further work was carried out by Giovanni Aldini [15], on 27th February 1803, when (it is said) he transmitted a current, from a battery of 80 silver and zinc plates, from the West Mole of Calais harbour to Fort Rouge by means of a wire supported on the masts of boats, and made the current return through 200 ft (61 m) of water.

And during the same year the transmission of electricity through water was also demonstrated by F.H. Basse, of Hamel, and by Erman, of Berlin using the rivers Weser and Havel respectively [16]. The use of an earth return was shown to be possible by Steinheil [17] in 1838 although his finding was not published.

Contemporaneously with the endeavours of Wheatstone, Morse and Bain, James Bowman Lindsay [18] in 1843 conceived the notion of a submarine telegraph to America by means of an *uninsulated* wire and earth batteries, 'after having proved the possibility by a series of experiments'. He was the first person to suggest this combination. Subsequently, Lindsay carried out many experiments on telegraphing through water from 1843 to 1860 although this activity was but one of his many interests. He was such an extraordinary person – 'a genius in humble life' – that a few words about his life are apposite.

Lindsay [19] was born at Carmyllie, near Arbroath, Scotland, on 8th September 1799, and, but for his delicate state of health, would probably have become a farmer rather than as he did a linen weaver. He had such a passion for reading that in 1821 his parents arranged for him to be admitted to the University of St Andrews. Lindsay appears to have excelled in the mathematical and physical sciences, and, after completing the four year course, subsequently enrolled as a student of theology at Divinity Hall, St Andrews but did not present himself for a licence, although he completed his studies. In 1829 he obtained a position as a science and mathematics lecturer at the Watt Institution, Dundee. A former pupil, A. Maxwell, has said of Lindsay:

When I was [at the institution], I attended classes that were taught by Mr Lindsay, a man of profound learning and untiring scientific research, who, had he been more practical, less diffident, and possessed of greater worldly wisdom, would have gained for himself a good place amongst distinguished men. As it was, he remained little more than a mere abstraction, a cyclopaedia out of order, and went through life a poor and modest schoolmaster.

His acquaintance with languages was extraordinary. In 1828 he began the compilation of a dictionary in 50 languages, the object of which was to discover, if possible, by language the place where, and the time when, man originated. This stupendous undertaking, which occupied the main part of his life's work, he left behind in a vast mass of undigested manuscript consisting of dissertations on language and cogitations on social science – a monument of unpractical and inconclusive industry. In 1845 he published 'A Pantecontaglossal Paternoster', intended to serve as a specimen of his fifty tongued lexicon.

In 1858 he published 'The Chron-Astrolabe', for determining with certainty ancient chronology and in 1861 'A Treatise on Baptism' which is a curious record of his philosophical knowledge'.

His other pursuits included work on electric lighting and electric telegraphy.

Lindsay's main employment was as a teacher in Dundee Prison – on a salary of £50 per annum – a position he held from March 1841 to October 1858. A few months earlier, in July 1858, on the recommendation of Lord Derby, the Prime Minister, Queen Victoria granted Lindsay an annual pension of £100 per annum, 'in recognition of his great learning and extraordinary attainments'. This enabled him to pursue his literary and scientific interests until his death on 29th June 1862.

Lindsay outlined his ideas for a submarine telegraph in a letter published in the *Northern Warder*, of Dundee, in June 1845. In this he gave details of the processes to be used for joining the lengths of wire, and for protecting them from corrosion. Some costing was also given.

Between 1845 and 1853 Lindsay remained silent on the Atlantic project, but in March of 1853 he gave a lecture on 'Telegraphic communication', during which he stated that submerged wires, 'such as those now used for tele-graphic intelligence between this country and Ireland and France, were no longer necessary' [18]. His claim was illustrated on a small scale by means of a water trough. The following year, on 5th June, he patented his method, shown diagrammatically in Fig. 5.2. Lindsay wrote:

Now, I have found that if each of the two distances CD and EF be greater than CE and DF, the resistances through CE and DF will be so much less than that through the water between C and D, that more of the current will pass across the water, through the opposite wires [EBF], and recross at F, than take the direct course CD; or, more correctly speaking the current will divide itself between the two courses in inverse ratio to their resistances [18].

Lindsay's first public trials were across the Earl Grey Docks at Dundee, and then across the River Tay at Glencarse, where the river is nearly 0.75 miles (1.2 km) wide. Further experiments were conducted in 1854, first, with the assistance of W.H. Preece, in the gutta-percha testing tank at Percy Wharf on the Thames, and second, across the mill dam, (c. 450 m wide), Portsmouth. These experiments were repeated at intervals and at various places whenever and wherever he had the opportunity: his greatest achievement was to signal across the Tay from Dundee to Woodhaven, a distance of almost two kilometres.

In September 1859 Lindsay read a paper, 'On telegraphing without wires', at the British Association Meeting at Aberdeen, which earned a special com-mendation from Lord Rosse, the President of the section. Additionally, Faraday and Airy (the Astronomer Royal) were said to have approved the opinions expounded.

Further experiments in 1859 across the Tay showed that the current that traversed the river could be increased in four ways: '(1) by an increased battery power; (2) by increasing the surface of the immersed sheets; (3) by increasing the coil that moves the receiving needle; and (4) by increasing the lateral distance of the sheets' [18]. In most of this work the lateral distance was approximately

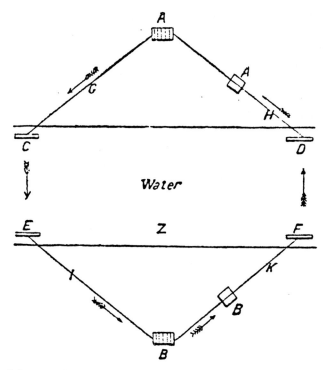

Figure 5.2   Schematic diagram of Lindsay's method of communicating across a river. A shows the positions of the battery and telegraph instrument for sending a message; B indicates the positions of the battery and needle instrument for receiving the message.

Source: J.J. Fahie, *A history of wireless telegraphy* (Blackwood, Edinburgh, 1899).

twice the transverse distance (i.e. the width of the river), but in one experiment on the Tay CD was c. 0.5 mile (0.85 km) while CE was c. 0.75 mile (1.2 km).

Lindsay was well aware of the limitation of his method of submarine telegraphy, and in his address to the BA Meeting he was reported as saying: 'In telegraphing by this method to Ireland or France abundance of lateral distance could be got, but for America the lateral distance in Britain was much less than the distance across' [18].

Further successful experiments were carried out across the river Dee, (in the presence of Lord Rosse, Professor Jacobi of St Petersburg, and others), and elsewhere, and to the end of his life Lindsay remained convinced of the correctness of his ideas and of their final success.

He was not alone in his views: J.W. Wilkins (1845), Dr W. O'Shaunessy Brook (1849), E. & H. Highton (1852–1872) and G.E. Dering (1853) were proponents of signalling through water using either the water itself or uninsulated wires for their purpose. Their work is considered in the Notes at the end of this chapter.

Of those persons whose ideas in the above field can be construed as unworkable or strange, mention may be made of J. Haworth (1862), J.H. Mower (1868), M. Bourbouze (1870), and M. Loomis (1872) [20].

1. Haworth's patent was described by Cromwell Varley, formerly electrician of the Electric and International Telegraph and Atlantic Telegraph Companies, as 'unintelligible: it is a jumble of induction plates, induction coils, and coils of wire connected together in a way that can have no meaning'.

2. Mower claimed to have transmitted messages for two hours across Lake Ontario, from a position near Toronto to a site on the coast of Oswego County, New York – a distance of 138 miles (222 km) – using his machine and a new discovery. The report of this unsubstantiated achievement stated: 'The upshot of the discovery – on what principle Mr Mower is not yet prepared to disclose – is, that electric currents can be transmitted through water, salt or fresh, without deviation vertically, or from the parallel of latitude. The statement of the discovery is enough to take away one's breath'.

3. Bourbouze's plan for signalling across a stretch of water was conceived during the investment and siege of Paris by the German forces in the winter of 1870–1871. His proposal was 'to send strong currents of electricity into the river Seine from a battery at the nearest approachable point outside the German lines, and to receive in Paris through a delicate galvanometer such part of these currents as might be picked up by a metal plate sunk in the river'. However, by the time that Bourbouze had obtained the necessary equipment the Seine had frozen over: two weeks later the armistice was proclaimed and the project was abandoned. Later occasional notices of some experimental work were 'all provokingly silent on the point of actual results over considerable distances'.

4. Loomis, an American dentist, proposed to signal *over* water by utilising the electricity of the higher atmosphere. He seems to have held the opinion that the charge distribution of the upper atmosphere was stratified and that the electric charges 'are in some way independent of each other, and that the electricity of any one stratum can be drawn off without the balance being immediately restored by a general redistribution of electricity from the adjacent strata'. A New York newspaper (5th February 1873) outlined his aerial telegraph plan to its readers, after cautioning them 'not to laugh too boisterously at it'. 'We are told that, from two of the spurs of the Blue Ridge Mountains, [20 miles, 32 km] apart, he sent up kites, using small copper wire instead of pack-thread, and telegraphed from one point to the other.' And in the *Electrical Review* (1st March 1879) it was reported that 'with telephones in this aerial circuit he (Loomis) can converse a distance of twenty miles [32 m]' A want of public demonstration militated against Loomis's claims being taken seriously by the technical press [20].

Fahie, from whose excellent book *A history of wireless telegraphy, 1838–1899* the above quotations have been taken, relates a delightful anecdote about the irrational fears that some people have with regard to new, or supposedly new, technologies. One inhabitant of New York felt the aerial telegraph was being used for illegitimate purposes and felt it to be his duty to expose the perpetrators.

By means of a $60,000 battery, he said, they transmitted the subtle fluid through the aerial spaces, read people's secret thoughts, knocked them senseless in the street; ay, they could even burn a man to a crisp, miles and miles away, and he no more know what had hurt him than if he had been struck by a flash of lightning, as indeed he had!

Still, he had the solution to such actions:

it had occurred to him that by appearing on the streets in a robe of pea-green corded silk, gutta percha boots, and a magenta satin hat with a blue-glass skylight in the top of it, he would be effectually protected from injury during his daily perambulations [20].

The prime problem to be solved before submarine telegraphy using *insulated cables* could become a practical reality concerned the insulation of the submerged conducting wire. With overhead telegraph lines it was necessary only to insulate the conductor from its surroundings at the points from which it was supported, whereas with undersea cables the conducting wires had to be protected from marine biological activity and had to be proof against the ingress of high pressure water into the insulation. Rubber was an unsuitable material for this purpose but providentially in c. 1843 Dr Montgomerie, a Scottish surveyor with the East India Company, brought to England a sample of a new gum, gutta percha [21].

The gum is the solidified juice of a tree, Isonandra percha, which abounds in Borneo and Malacca. The trunk of the tree grows to a diameter of c. 6 ft (1.8 m) but the timber is useless. When an incision is made through the bark and into the wood a milky juice flows out which quickly solidifies. Gutta percha is a very tough substance and differs from India rubber in being inelastic at room temperature, but when softened by gentle heating it can be easily moulded. In 1843 its properties were brought to the attention of the Royal Society of Arts and it was examined by Michael Faraday. The excellent insulation properties of gutta percha were recognised and its use as an insulator for submarine cables was advanced when in 1847 Charles Hancock at his rubber works at Stratford-le-Bow, London devised and patented a machine for processing the raw material. Tests carried out at the Birmingham Waterworks showed that 'the power of gutta percha tubing to resist pressure [was] quite extraordinary, and far beyond what would be supposed' [22]. Tubes, having internal diameters of 0.75 in (19.1 mm) and 0.125 in (3.2 mm), were subjected to pressures of up to 337 lbf/in^2 (2320 kPa) without any damage. The earliest uses of gutta percha as a coating for cables seem to have been made by Dr W. Siemens when he laid cables across the River Rhine and Kiel Harbour in 1847 and 1848 respectively. Subsequently gutta percha was employed extensively in the submarine cable industry. Some indication of its importance may be inferred from the imports of the material into the United Kingdom. In 1850 11,000 cwt (559,000 kg) were imported but in 1872 the figure was 46,000 cwt [22] (2,337,000 kg): indeed the demand for gutta percha led to reckless tree felling in Malaya [23].

In England, in 1849, C.V. Walker, Superintendent of Telegraphs to the South Eastern Railway Company, who had employed gutta percha coated wire in wet railway tunnels, conducted the first full-scale experiment to determine the desirability of using the new cable in a submarine environment [24]. On 9th January 1849 the small ship *Princess Clementine* sailed from Folkstone Harbour and paid out 2 miles (3.2 km) of no. 16 gauge copper wire covered with gutta percha. The shore-end of the cable was connected to the local telegraph office of the company and communications were established between London and the *Princess Clementine*, a distance of more than 79 miles (128 km).

In the meantime Jacob Brett, of Hanover Square, London, had patented his 'Subterranean and oceanic printing telegraph' by which he and his elder brother John, in 1845, hoped to link the Capital telegraphically with the British Colonies and the Channel Islands. John Watkins Brett had been an antique dealer and was quite wealthy; he could provide much of the finance needed to further his brother's aspirations. He had a more dominant personality than Jacob and had the ability to negotiate with government officials. To further their objective Jacob Brett registered the General Oceanic Telegraph Company on 16th June 1845 [25]. At this date Brett was not aware of Wheatstone's English Channel project. He remarked later:

Had these facts then been known to me, I cannot say how far they might have damped my determination to devote my whole time and means to establish and promote the submarine telegraph, and, if possible, to bring this country into instantaneous communication with India and America, then the sole object of my thoughts.

In July the Bretts wrote to the Prime Minister, Sir Pobert Peel, about their proposal and were referred to the Admiralty. Their Lords Commissioners were now persuaded of the advantages that technology could bring to communications and the Bretts received a more receptive and helpful response than had been accorded to Ronalds when he had sought the Admiralty's approval for his telegraph. A new company, the General Oceanic & Subterranean Electric Printing Telegraph Company, was registered in 1846 'to establish a telegraphic communication from the British Isles across the Atlantic to Nova Scotia and Canada' [25]. The Bretts intended to adapt R. House's printing telegraph, which was used in the United States of America, for their purpose [26], and for which in November 1845 an application for a patent had been made.

The brothers also went to Paris to obtain the consent of the French Government for a submarine telegraph between England and France [24]. After much negotiation their request was granted by King Louis Philippe in 1847 and they secured an exclusive concession. Their agreement was later rescinded but was renewed by Louis Napoleon on 10th August 1849 for a period of ten years subject to the condition that the link had to be established by 1st September 1850. Another company, the English Channel Submarine Telegraph Company with a capital of £2,000, was formed. The venture capitalists who each provided £500 were John W. Brett, Charles Fox, railway engineer, Charles J. Wollaston, engineer, and Francis Edwards.

The 25 nautical mile (46.3 km) cable had a single conductor of No. 14 gauge unannealed copper wire covered by a 0.5 in (12.7 mm) layer of gutta percha and was manufactured in 100 yd (91.4 m) lengths at the City Road, London works of the Gutta Percha Company. These were jointed together by a bell-hanger's twist and by soldering the conductors, each joint then being encapsulated with hot gutta percha pressed on by means of a wooden mould. Following manufacture, the cable was wound onto a large wood and iron drum (having a horizontal axis) and placed on the deck of the SS *Goliath*, a small Thames steam tug (Fig. 5.3).

At Dover the paying out of the cable commenced at 10.00a.m. on 23rd August 1850 in fine weather. The *Goliath* was escorted by HMS *Widgeon*, a government surveying vessel, which had previously marked the route with buoys. To ensure the cable lay on the bed of the English Channel, lead masses, of 14 to 24 lb (6.35 to 10.9 kg) were bolted on at a spacing of 100 m [27]. Prior to this operation the two shore ends had been laid: these consisted of a copper wire 0.065 in (1.65 mm) diameter, covered only with cotton and a solution of india-rubber, enclosed in a very thick tube of lead. On the English side this extended from a horse–box in the yard of the South Eastern Railway terminus to a structure of piles which had been constructed for the new Admiralty pier in Dover harbour. The French shore end, encased in a very thick lead tube, stretched from Cape Gris Nez, a headland near Calais, to a rocky ledge situated a considerable distance away [24].

Success was soon achieved. By late evening, with the aid of a single needle instrument, signals (including one to Louis Napoleon) were exchanged between huts on the beaches at Dover and Cape Gris Nez [28]. Regrettably the success was short-lived. After a few hours communication completely failed due, it is thought, to a break in the cable rather than to an insulation failure. According to Charles Bright, a Boulogne fisherman had – accidentally or otherwise – raised the cable to the surface and, on finding 'gold' inside the unusual 'seaweed', had cut out an appreciable length.

Notwithstanding this setback the efforts of Brett, C. Wollaston, the engineer in charge of the laying, and W. Reid, the contractor who had equipped the *Goliath* with the necessary gear, were not in vain. The practicability of communicating across more than 20 miles (32 km) of sea had been definitely proved; undeniably the operation had been an epoch-making event.

Four months later, on 19th December 1850, the Bretts were granted a new ten year concession (to be effective from 1st October 1851) from the French Government – the concession to be valid only if the new submarine link was in working order by that date. Disaster nearly struck this further attempt to connect the two countries but, fortunately, T.R. Crampton, a railway engineer and inventor, helped to raise the required capital of £15,000 and invested £7,500 of his personal fortune [29]. The new cable, fabricated by Messrs Newall & Company, was of a very substantial construction. It comprised four no. 16 gauge copper wires, each coated with a double layer of gutta percha, laid up with tarred hemp filling the interstices, followed by a lapping of tarred spun yarn. An armouring of ten No. 1 gauge galvanised iron wires was subsequently

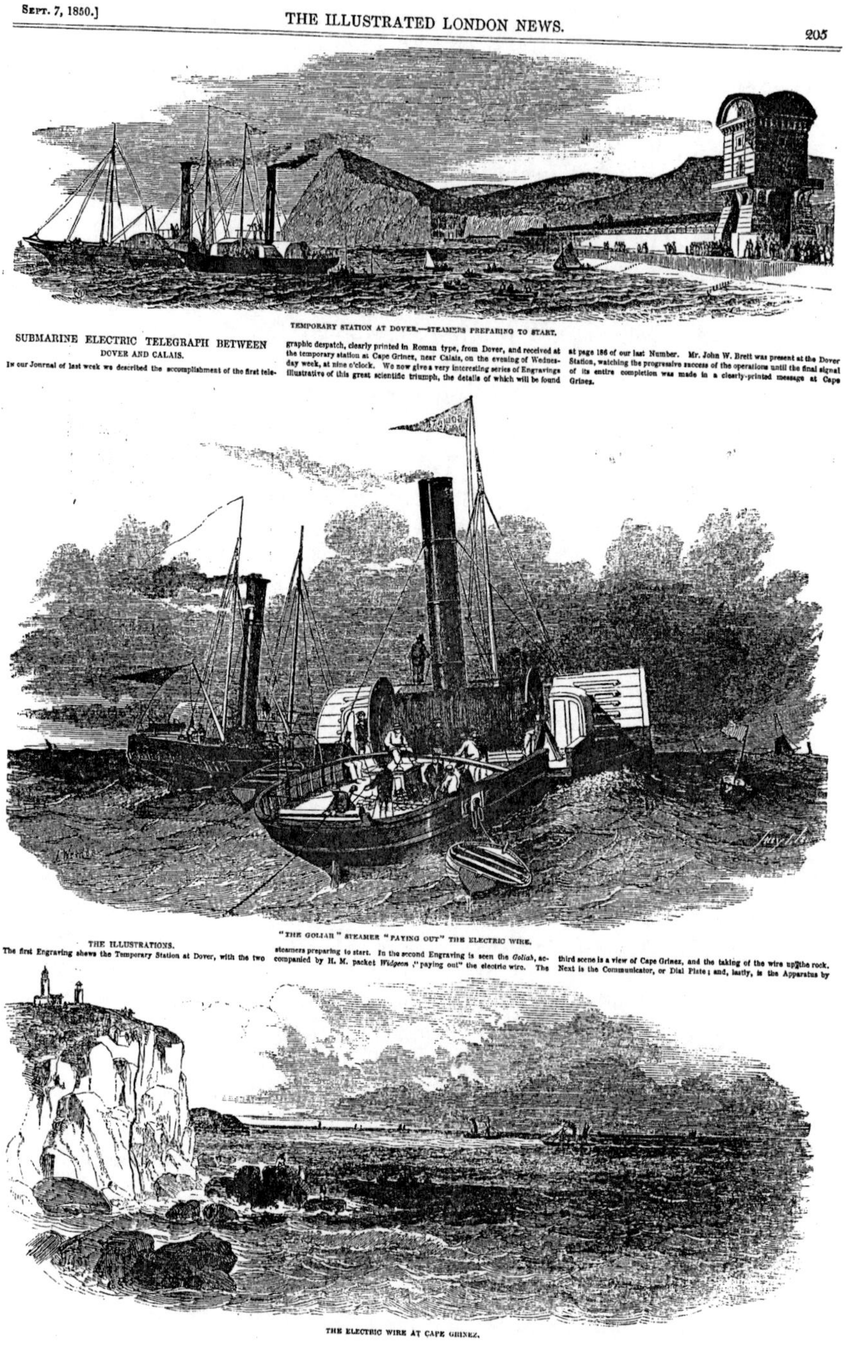

Sept. 7, 1850.] THE ILLUSTRATED LONDON NEWS. 205

TEMPORARY STATION AT DOVER.—STEAMERS PREPARING TO START.

SUBMARINE ELECTRIC TELEGRAPH BETWEEN DOVER AND CALAIS.

In our Journal of last week we described the accomplishment of the first telegraphic despatch, clearly printed in Roman type, from Dover, and received at the temporary station at Cape Grinez, near Calais, on the evening of Wednesday week, at nine o'clock. We now give a very interesting series of Engravings illustrative of this great scientific triumph, the details of which will be found at page 186 of our last Number. Mr. John W. Brett was present at the Dover Station, watching the progressive success of the operations until the final signal of its entire completion was made in a clearly-printed message at Cape Grinez.

"THE GOLIAH" STEAMER "PAYING OUT" THE ELECTRIC WIRE.

THE ILLUSTRATIONS.

The first Engraving shews the Temporary Station at Dover, with the two steamers preparing to start. In the second Engraving is seen the *Goliah*, accompanied by H. M. packet *Widgeon* ,"paying out" the electric wire. The third scene is a view of Cape Grinez, and the taking of the wire up the rock. Next is the Communicator, or Dial Plate; and, lastly, is the Apparatus by

THE ELECTRIC WIRE AT CAPE GRINEZ.

Figure 5.3 Laying the submarine cable between Dover and Calais, 1850.
Source: The Illustrated London News Picture Library.

applied with a long lay to provide the mechanical strength which had been so deficient in the first cable. The completed cable had a mass of c. 4400 kg/km [30].

On 25th September 1851 the cable was laid from the South Foreland lighthouse across to France. An anxious period had occurred during the lay when, due to stormy weather and strong currents, the cable ship, (a hulk, the *Blazer*), lent by the Government, was found to be out of position, with the result that when the ship was about 1.6 km from the French coast no more cable was left on board. Three gutta percha covered wires were temporarily installed to fill the gap and on 13th November 1851 the telegraph service was opened to the public. The ECSTC obtained its ten year concession on 23rd October 1851 and direct communication was established between London and Paris in November 1852 via the lines of ECSTC and the European & American Telegraph Company [31]. The new cable gave excellent service. Notwithstanding the enormous traffic of shipping in the Dover Strait the cable was seldom damaged, and on those occasions when repairs were necessary they were easily carried out. A report mentions that in 1867 the insulation was still in a perfect state.

The success of the Dover–Calais cable laying enterprise led to the execution of further operations. In 1852 three unsuccessful attempts were made by the Magnetic Telegraph Company to establish a cable link between, first, Holyhead and Howth (near Dublin), second and third, Portpatrick and Donaghadee. However, a fourth endeavour in 1853 between the latter places was completed successfully by the use of a heavier cable having a mass of seven tons per 1.6 km. Also in 1852 a cable was laid from Dover to Ostend, and further Anglo-Dutch and Anglo-German cables were laid so that by 1857 the Electric Telegraph Company was in direct communication with Holland, Germany, Austria and St Petersburg [32].

By 1867, 74 cables had been constructed in England and telegraphic links had been accomplished between Denmark and Sweden; Italy and Corsica and Sardinia; Sardinia and Africa; Malta and Alexandria (a cable distance of 1,565 miles (2,518 km)), as well as between England and Europe [33]. In this great expansion of submarine telegraphic communications the London firm of Glass, Elliott & Company was pre-eminent. It laid 24 cables with a total length of 3,677 miles (5,916 km) from 1854 to 1861.

Meantime the possibility of a telegraphic link between Europe and North America was being explored. The Western Union Company (which had laid a vast network of overhead lines in the United States of America) now planned to construct a land-line from the USA to Europe via British Columbia, Alaska, the Aleutian Islands, the Bering Sea and Siberia. Much construction work was undertaken, despite many difficulties, and a considerable section of the total line length was completed when proposals for a trans-Atlantic submarine cable brought the enterprise to a premature end – at a loss of c. $300,000.

In 1852 an English engineer, F.N. Gisborne, and an American syndicate were granted an exclusive concession to unite St John's, Newfoundland with Cape Roy in the Gulf of St Lawrence, by an overhead telegraph line. A company, the Electric Telegraph Company of Newfoundland, was formed and the objective was to obtain news from ships plying the North Atlantic route, from London to St John's, and then forward messages to the press, the shipping companies and others. A few miles of cable, manufactured in England, were laid from Prince Edward Island to New Brunswick; and, after Gisborne had surveyed a route for a Newfoundland overhead line, 40 miles (64 km) were erected before a shortage of funds brought the work to an end [34].

In 1854 Gisborne had talks with C.W. Field on the feasibility of connecting the two continents of Europe and North America by a cable. Field was an important person in its subsequent laying and so a few words of introduction to his background are appropriate.

Field was born in 1819 in Massachusetts and was the eighth child of a congregational minister and his wife [35]. At the age of sixteen years he left home with eight dollars given to him by his father and set out for New York. Here he intended to make his fortune. Initially, he was employed as an errand boy, at a salary of $50 per annum, in the warehouse of a large dry goods merchant, but later served a three year apprenticeship as a clerk. 'My ambition', he wrote, 'was to make myself a thoroughly good merchant. I tried to learn in every department all I could, knowing that I had to depend entirely on myself.' Field's conscientiousness and ability must have made a favourable impression on his peers and superiors for, when he resigned at the age of nineteen to advance his position, he was presented with a diamond tie pin by the company and was given a farewell dinner by his fellow clerks.

For a while Field worked as an assistant to his brother who had his own paper factory in Massachusetts: then he obtained employment with a firm of wholesale paper merchants in New York. A year later the company went bankrupt and Field had to deal with the firm's creditors. He promised to repay them even though he was not legally obliged to do so. Soon a new business, Cyrus W. Field & Company, wholesale paper dealers, with offices near the East River docks, was registered with his brother-in-law. Field worked tirelessly to build up a flourishing trade. He has stated:

In 1841 I was not worth a dollar. What money I had made had gone to pay the debts of the old firm. My business was conducted on long credit; we did a general business all over the country. I built up a first rate credit everywhere. All business entrusted to me was done quickly and promptly. I attended to every detail of the business and made a point of answering every letter on the day it was received [35].

Field's industry paid off; by 1853 he was worth more than $250,000.

There was no luck about my success which was remarkable. It was not due to the control or use of large capital, to the help of friends, to speculation, or to fortunate turns of events, it was by constant labour and with the ambition to be a successful merchant; and I was rewarded by seeing a steady, even growth of business.

In this year he retired from his firm – although he left $100,000 in the company – and embarked with a friend upon a six months exploration of South America. He was just 34 years of age [35].

Gisborne's discussion with Field was seminal to the trans-Atlantic submarine cable project. As a paper merchant Field had some knowledge and experience of the newspaper industry and was aware of the great demand of the New York papers for news from Europe. This demand had been exemplified by the success of the New York to Halifax (Nova Scotia) telegraph. When ships from Europe berthed in the port, agents of the New York press eagerly sought information about European affairs to telegraph to their New York offices. From his conversation Field felt that a cable link would generate so much telegraphic traffic that its viability would be assured. If it could be realised news transmission would be almost instantaneous rather than be delayed by the passage time of several days by ship.

Before the end of 1855 the practicability of joining the extreme west cost of Ireland and Newfoundland, almost 2,000 miles (3,218 km) apart, by cable became more evident. Experiments conducted in October 1856 by Charles Tilston Bright [36] and Edward O.W. Whitehouse – whom Bright described as 'a gentleman of very high intellectual power, and a most ingenious and painstaking experimenter' – with underground wires (between London and Manchester) and submarine cables connected together so as to form a continuous circuit exceeding 2,000 miles (3,218 km) in length showed that signals could be transmitted at rates of 210, 241 and 270 per minute [37]. This experiment was demonstrated on 9th October to Morse, the electrician of the New York and Newfoundland Telegraph Company, which had Field as its deputy chairman. Furthermore, soundings of the sea bed (which had been carried out by Lieutenant O.H. Berryman USN in the USS *Arctic* and by Commander J. Dayman RN in HMS *Cyclops*) had indicated that a gently undulating plateau of great breadth extended for almost the whole of the intervening distance between Newfoundland and Ireland. The depth of the bed varied from 1,700 to 2,300 fathoms (16,400 to 22,200 km), and upon its surface microscopic shells had been found which were exceedingly fragile [38]. This was an important finding for it meant that no currents existed at that depth which could move and possibly cause some abrasion of a submarine cable.

As part of his plan for a trans-Atlantic cable, Field, with the support of a powerful group of bankers, lawyers and merchants, formed the New York, Newfoundland and London Telegraph Company. Samuel Morse was appointed Vice-President and J.W. Brett joined the syndicate. Gisborne sold his company's concession to Field and he persuaded the Newfoundland Government to grant him, for 50 years, exclusive rights – on very favourable terms – to land cables in the State. This monopoly was then extended to include New Brunswick, Cape Breton Island, and the shores of the State of Maine.

In the summer of 1855 the company endeavoured to lay a cable from the mainland to Newfoundland. During the attempt the 85 mile (137 km) long cable, fabricated in England by Messrs Glass, Elliott & Company, was towed by a paddle steamer from Nova Scotia across the Cabot Strait. Unfortunately there was some confusion about the route to be followed and, when a storm blew up and the cable had to be cut to save the ship, the attempt had to be abandoned.

The following year a cable was successfully laid and St John's and New York were in telegraphic connection. With the completion of this stage the main task of linking Newfoundland with Ireland could commence.

In July 1856 Field came to England to raise capital for the venture and to hold talks with the country's technical experts. With Morse, he had discussions with the British Foreign Secretary and was accorded a favourable reception. The British Government offered a subsidy of £14,000 per annum in return for future use of the cable. Field also met Michael Faraday, and Robert Stephenson, and when Isambard Kingdom Brunel took Field to Millwall to view the enormous ship – which became the SS *Great Eastern* – being built Brunel indicated that this was the ship to lay the cable. Again, Field found so much support from leading merchants and bankers in Liverpool and London that, with Brett, E.O.W Whitehouse, and C.T. Bright, he formed the Atlantic Telegraph Company [39]. It was registered on 20th October 1856 and the 350 shares of £1,000 each became available for purchase from 12th November 1856. Meeetings were held in Liverpool, the headquarters of the British & Irish Magnetic Telegraph Company (the 'Magnetic'), Manchester and Glasgow. Many of the shareholders of the 'Magnetic' were shipowners and merchants who could discern the immense benefit to trade which would come about as a consequence of the two continents being united telegraphically. The principal individual shareholders were John Watkins Brett and Cyrus Field who initially each acquired 25 shares but who before the final allotment reduced their holding. Of the board members, nine were directors or shareholders of the 'Magnetic' (including Brett and Sir John Pender): others included Professor William Thomson, of the University of Glasgow, and George Peabody. Charles Bright and Wildman Whitehouse were appointed chief engineer and electrical engineer respectively. Of these persons Pender (1815–1896) subsequently became chairman of nine telegraph companies and acquired the sobriquet 'the cable king'.

Manufacture of the first trans-Atlantic cable began in February 1857 and was shared between Glass, Elliott & Company of Greenwich and Messrs R.S. Newall & Company of Birkenhead. The Gutta Percha Company supplied the core. The conductor comprised seven No. 22 gauge copper wires, having a mass of 107 lb (48.6 kg) per nautical mile (1.852 km), insulated with three coats of gutta percha, to a diameter of 0.375 in (9.525 mm), of mass 261 lb (118.4 kg) per nautical mile. This was covered with hemp saturated with a mixture of Stockholm tar, pitch, linseed oil and wax to form a bedding for the armouring, which comprised 18 strands of no. 22 gauge bright iron wire. The cable was then drawn through a compound of tar, pitch and linseed oil. Its overall mass was one ton per nautical mile and it had a breaking strength of 3.25 tons force. Both shore ends were protected with additional armouring which consisted of 12 no. 0 gauge iron wires [40]. The contract price for the complete cable was £225,000, the core costing £40 per mile and the armour £50 per mile.

Remarkably, for such an ambitious project, only four months were allowed for the fabrication of the 2,500 nautical mile (4630 km) cable, even though this involved the drawing of 17,500 miles (28,600 km) of copper wire (from 116 tons

(118,000 kg) of copper), over 315,000 miles (507,000 km) of iron wire (from 1,687 tons – 1,714,000 kg – of charcoal iron), and the preparation and application of 250 tons (254,000 kg) of gutta percha. Moreover, the paying-out machinery had to be designed and manufactured, and the two former battleships that had been commissioned for the difficult task of laying the cable had to be considerably adapted to make them fit for their purpose. It seems that an inappropriate time scale was insisted upon by Field – against the objections of Bright and Whitehouse who urged, in vain, for more manufacturing time to ensure greater care in the fabrication of the cable – because of business interests in the USA. Actually, Bright argued that the design of the cable was unsatisfactory: he advocated a cable having a conductor cross sectional area three times as large with a much greater thickness of insulation. Field's unreasonable haste meant that no storage tanks or facilities could be built in the available time to accommodate the finished cable, which had to lie exposed to the sun after manufacture. The consequence was that sections of the cable suffered damage and had to be cut out. This haste would prove to be calamitous [41].

The ships made ready for the enterprise were the 91-gun battleship HMS *Agamemnon* (3,200 tons – 3,251,000 kg) and the steam frigate USN *Niagara* (5,000 tons – 5,080,000 kg); each would carry half the complete cable. On 6th August 1857 the two ships sailed from Valentia, Ireland (Fig. 5.4). The *Niagara* having spliced its cable on to the shore end began to lay at a speed of 1.03 m/s. Before the first hour had elapsed disaster struck. The cable slipped off the sheaves, jammed and broke. The cable was grappled, repairs were made and on the following day the pay-out recommenced. Good progress was accomplished for the next three days and 335 miles (540 km) of cable had been laid when at 03.00 a.m. – when Bright, who was in charge of the laying, was taking some rest – the cable broke in a heavy sea. Arrangements had been implemented to allow for the rise and fall of the stern of the ship under such conditions, and so prevent excessive stress being imposed on the cable, but the braking system required constant alertness. Unfortunately, the operator left in charge was surprised by the speed at which the cable raced away and applied the brake at the wrong moment, as the stern was rising.

The expedition returned to Plymouth, where the cable was taken ashore, well tarred and left to dry, and stored. New paying out gear was designed and constructed and in due course trials were conducted in the Bay of Biscay, at depths of 1,800 fathoms (3290 m), both in paying-out and grappling.

The following year a second attempt was made to lay the Atlantic cable. Whitehouse was unable to join the *Agamemnon* and Professor W. Thomson, at the request of the directors, undertook the post of electrician – without any recompense. A decision had been taken to change the pattern of the laying. Instead of the two ships setting out from Valentia it was decided that they should proceed to mid-ocean and, after the two cables had been spliced [42], should then begin to pay-out the cable in opposite directions. Throughout the voyage Thomson used his mirror galvanometer, together with 75 Daniell's cells, for continuity testing and for signalling to shore. According to Thomson's

Figure 5.4   Re-shipment of the Atlantic telegraph cable on board the Agamemnon *and*
Niagara *in Keyham Basin (1858).*
Source: The Illustrated London News Picture Library.

biographer: 'The work which he undertook for it was enormous; the sacrifices he made for it were great. The pecuniary reward was ridiculously small' [43].

On 16th June 1858, after having endured on route a tremendous storm that rolled the ships by 45° and recorded waves of 40 ft (12.2 m) amplitude, the splice was completed and the two ships parted company, *Agamemnon* laying towards Valentia and *Niagara* towards Newfoundland. Three times – on the 26th, 27th, and 29th – the cable broke and each time the operation had to be recommenced (with the loss of 540 miles – 869 km – of cable), but on 5th August 1858 the cable was landed at Trinity Bay, Newfoundland and on the same day the *Agamemnon*, having endured heavy seas, entered Valentia Bay. A test was soon made and Trinity Bay reported 'very strong currents of electricity throughout the whole of the cable from the other side of the Atlantic': Thomson handed over his responsibility as 'electrician' to Whitehouse [44].

The first official message transmitted was: 'Glory to God in the highest, on earth peace, goodwill towards men' [45]. And the first press communique, sent from New York, read:

Emperor of France returned to Paris Saturday. King of Prussia too ill to visit Queen Victoria. Her Majesty returns to England August 31. Settlement of Chinese question. Chinese empire open to trade; Christian religion allowed; foreign diplomatic agents admitted; indemnity to England and France. Gwalior insurgent army broken up. All India becomes tranquil.

Great enthusiasm, on both sides of the Atlantic, greeted the news of the successful cable operation. *The Times* commented that 'since the discovery of Columbus, nothing [had] been done in any degree comparable to the vast enlargement which [had] thus been given to the sphere of human activity'. In the United States there was much jubilation – so much so that one torchlight procession in New York set the Town Hall alight. Queen Victoria sent a message of congratulations to the President of the United States, James Buchanan, and Charles Bright, who was just 26 years of age, was honoured with a knighthood. It was the first honour to be bestowed on anyone associated with the telegraph industry [46].

Alas, the triumph was short-lived. An insulation fault occurred about 483 km from Valentia, which caused the signals to become confused, and then on 1st September to become unintelligible. From 20th October the line, after a total of 732 messages had been propagated from 5th August, was closed to the transmission of public traffic from Newfoundland [47]. Fortunately the British Government had time to send a signal countermanding an order for two regiments to leave Canada for England, thereby saving c. £50,000. This signal, probably more than any other sent via the cable, confirmed the enormous advantages to be gained from swift communications – a conclusion amply demonstrated by the Chappe system of semaphore towers more than half a century earlier.

The cause of the failure was never fully explained, although some suspicion fell on Whitehouse's testing method. The cable had lain in the sun for lengthy periods after fabrication and may have suffered insulation damage; and it had been coiled and uncoiled many times. In addition, Whitehouse in testing the cable had insisted on applying very high voltages – from an induction coil five feet (1.52 m) in length – across the insulation 'in defiance of Thomson's tested conclusions' [48]. As his biographer has said: 'Thomson's self-abnegation and forbearance throughout this unfortunate affair are almost beyond belief. He would not suffer any personal slight to interfere with his devotion to a scientific enterprise' [43]. Whatever the origin of the breakdown £500,000 had been lost and it would be eight years before another cable could be laid. On the credit side the feasibility of telegraphic transmission across the Atlantic Ocean had clearly been demonstrated and this fact gave encouragement for the future.

At the time of the failure the United Kingdom had a vast empire of dominions and colonies. Good governance demanded good communications. But, since the empire stretched around the world, time delays between policy decisions agreed by the government in London and their implementation in a far off colony or dominion could be many days or weeks in the pre-submarine cable era. These delays could lead to unfortunate actions or misadventures when commands had to be sent to British military forces operating against insurgents and rebellious communities.

By this year the British Government was fully aware of the importance of electric telegraphy, so that when Gisborne in 1857 promoted the 'Red Sea and India Telegraph Company', to establish a link between the UK and the East Indian possessions, the government was ready to assist. (Gisborne prior to its formation had obtained powers from the Turkish Government to carry a telegraph line across Egypt and to lay a cable in the Red Sea.) The particular event which stimulated the Government to act was the Indian Mutiny which began on 10th May 1857 in Lucknow. News of the mutiny and a request for assistance was transmitted from Lucknow to Calcutta, then: by overland transport to Bombay, by ship to Suez, by overland transport to Alexandria and finally by ship to Trieste, the nearest telegraph station. From the besieged Indian town the message took 40 days to reach the telegraph station. The Government's assistance took the form of an annual subsidy of £36,000 [49].

The line, 4896 km in length, between Suez and Karachi, was divided into two sections. The first, from Suez to Aden (1,358 nautical miles – 2515 km), with intermediate landings at Kosseir and Suakin, was laid in 1859 but soon broke down. Bright, in his definitive account, mentions that the sections 'were all laid very taut', and 'in some places, owing to the tightness and high speed of laying, the elongation of the iron wires had "nipped" the gutta percha; and, in others, the sheathing was much stretched and broken' [50]. The second section, from Aden to Karachi (1,685 miles – 3121 km), with intermediate landings at Hallania Island and at Muscat, was laid during 1860. Again, faults developed in all three sections; 'and the company, having neither specially qualified men, nor the necessary materials, for carrying out repairs, was obliged to abandon the line, before any commercial use had been made of it'. Bright states: 'This was a most unfortunate line in every way. Report has it that a complete message was never got through the entire length, but only through each section separately. It ultimately failed altogether' [50].

The failures of the Atlantic Telegraph Company and the Red Sea and India Telegraph Company – the joint losses of which amounted to more than £1,000,000, and to which the Treasury had given a continuous guarantee – stimulated the British Government, in 1859, to investigate thoroughly the entire question of submarine telegraphy. A Joint Committee, appointed by the Lords of the Committee of Privy Council for Trade and the Atlantic Telegraph Company, of eight members was constituted with four members (Douglas Galton, Chairman, Charles Wheatstone, William Fairbairn and George Bidder) from the Board of Trade and four members (Edwin Clark, Cromwell F. Varley, Latimer Clark and George Saward) from the Atlantic Telegraph Company. The Committee's terms of reference included inquiring into the question of construction of the cable, and of the best methods of 'laying and maintaining submarine telegraph cables' [51].

Twenty two sittings were held from 1st December 1859 to 4th September 1860 to interrogate engineers, electricians, professors, physicists, seamen and manufacturers who had taken part in the various branches of submarine cable work, and whose knowledge and experience would be helpful to the committee. Wide ranging investigations were initiated 'concerning the structure of all cables previously made or in course of manufacture, and the quality of the different materials used as to special points arising during manufacture and laying, on the routes taken, electrical testing, and on sending and receiving instruments, [and on] speed of signalling' [50]. Experiments were also carried out to determine '(1) the effects of temperature and pressure on the insulating substances employed; (2) the elongation and breaking strain of copper wires; of iron, steel, and tarred hemp' [51].

The Committee's Report (of over 500 pages) was published in April 1861. In its analysis of the evidence submitted on the Atlantic cable the Committee stated:

We attribute the failure of this enterprise to the original design of the cable having been faulty owing to the absence of experimental data, to the manufacture having been conducted without proper supervision, and to the cable not having been handled, after manufacture, with sufficient care. We have had before us samples of the bad joints which existed in the cable before it was laid; and we cannot but observe that practical men ought to have known that the cable was defective, and to have been aware of the locality of these defects, before it was laid [51].

The Committee affirmed its 'conviction that submarine telegraphy might be as sure and remunerative in the future as it had been speculative in the past, provided that the specification, manufacture, laying, and maintenance of the cable were proceeded with on the lines laid down in the report', which has been called 'the most valuable collection of facts, warnings and evidence ever completed concerning submarine cables' [52].

Meantime the Atlantic Telegraph Company survived and Cyrus Field and his associates never lost hope of bridging the Atlantic ocean with a reliable commercial cable. Several schemes and new routes were proposed, including The Grand North Atlantic Telegraph (1860), which was based on four cable sections (to reduce the length of a continuous lay) by laying the cable via Iceland and Greenland and thence to Labrador. A detailed survey led to the abandonment of the plan. Among the practical steps taken before a new cable could be laid, the Atlantic Telegraph Company prevailed upon the British Government to despatch two vessels to examine further the ocean floor 300 miles (483 km) out from the coasts of Ireland and Newfoundland, and in 1862 HMS *Porcupine* explored the western region of the Atlantic ocean in about the same latitude as Ireland.

The raising of new capital for the third attempt was not easy. The American civil war between the north and the south raged from 1861–1865 and was not conducive to venture capitalism. Field worked vigorously on both sides of the Atlantic to seek funds for the new enterprise. It is said, according to Bright [53], that he crossed the Atlantic 64 times – suffering sea sickness on each occasion. In 1863, for example, he was in New York and Boston but was assured of only £70,000. Financial merchants told him that because of the shortage of money he would have to postpone the laying of the third Atlantic cable until after the war was over.

Apart from this difficulty, the directors of the 'Atlantic' required some assurance that the new cable would have the most desirable electrical and mechanical properties for the purpose. They sought the assistance of prominent and independent scientists and established a scientific consultative committee. This committee comprised Captain D. Galton, RE, FRS; W. Fairbairn, FRS; Professor W. Thomson, FRS; Professor C. Wheatstone, FRS; and J. Whitworth, FRS. Thomson [43] (1824–1907), later Lord Kelvin, played a most important part in the work of the committee and in the subsequent laying of the third cable and so a few words on his background are apposite.

He was born on 26th January 1824, in College Square East, Belfast, into a comfortable middle class family and was the second of six children. Unfortunately his mother died when he was just six years of age and the children were brought up by their father, their grandparents and a nurse. When William Thomson was eight years of age his father, Dr James Thomson, was appointed professor of mathematics at the University of Glasgow.

Much of young William's education was provided by his father; he taught his children English, geography, mathematics and classics. William was a quick, highly intelligent boy and at the age of ten he matriculated at the university and commenced a course in arts subjects. It was the custom at Glasgow university for the students to select by voting the persons who should be awarded class prizes. As a consequence William and his brother James won the prizes available for Latin, Greek, mathematics, astronomy and natural philosophy (including electricity) every year during their time as students. At home the two brothers were given a room where they could make apparatus and experiment. Here, William (from the age of 12 years) and James made an electrical friction machine, voltaic piles and galvanic batteries; they stored electrical energy in Leyden jars – and gave members of the family electric shocks – and experimented with metals and fluids.

During his last year at Glasgow (1840–1841) he communicated to the Cambridge Mathematical Journal, under the signature 'P.Q.R', an original paper 'On Fourier's expansions of functions in trigonometrical series', which was a defence of Fourier's work against some criticisms of Kelland. Thomson was only 16 years of age at that time.

In 1841 William was entered at Peterhouse, University of Cambridge where he soon distinguished himself in mathematics and geology, and became an accomplished oarsman and musician. His prowess in mathematics was extraordinary. When 17 he submitted a paper to the above mentioned journal on 'The uniform motion of heat in homogeneous solid bodies, and its connection with the mathematical theory of electricity'. His Tripos results caused no surprise; he was placed second Wrangler and first Smith's prizeman, and was elected to a fellowship worth £200 per annum. One year later, at the age of just 22 years, he was elected Professor of Natural Philosophy at the University of Glasgow.

Thomson was a popular, inspiring and somewhat unorthodox lecturer. During his lectures on ballistics he would fire a rifle at a lead pendulum to illustrate the law of the conservation of momentum; and while lecturing on acoustics he would demonstrate various principles by playing on his French horn – much to the joy of his class who would cheer his activities. His first lecture to the natural philosophy class was based on notes that, it is said, he used for 50 years and never reached the end of, even though the interest of his students was such that the one-hour lectures often extended to two or three hours. Thomson remained at Glasgow for 53 years and refused a professorship at Cambridge on three occasions. His output was prodigious; between 1841 and 1908 he wrote 661 papers.

In 1857 the Atlantic Telegraph Company issued a pamphlet in which it stated: 'the scientific world is particularly indebted to Professor W. Thomson, of Glasgow, for the attention he has given to the theoretical investigation of the conditions under which electrical currents move in long longitudinal wires; and Mr Whitehouse [a retired medical practioner who was the electrician on the *Agamemnon* [35]] has had the advantage of this gentleman's presence at his experiments, and counsel, upon several occasions'. This appreciation stemmed from Thomson's study of the conditions that affect the charging and discharging processes of a long telegraph cable. He showed that the rates of these processes was inversely proportional to the square of the length of the cable. Whitehouse questioned this 'law of squares' at the 1856 meeting of the British Association held in Cheltenham, when he delivered a paper entitled 'The law of the squares – is it applicable or not to the transmission of signals in submarine circuits', and professed to have refuted it by experiments. He declared that if the law were true Atlantic telegraphy was hopeless. Whitehouse concluded [54]: 'I believe nature knows no such application of that law and I can only regard it as a fiction of the schools, a forced and violent adaptation of a principle in physics, good and true under other circumstances, but misapplied here.' Thomson gave his reply in two letters published in *The Athenaeum* and mentioned that success was dependent primarily on the conductor of the cable having an adequate cross sectional area. In one of his letters he suggested appropriate dimensions for a trans-Atlantic cable and calculated that it would enable three words per minute to be propagated along a 2,400 mile (3860 km) long cable. These dimensions were subsequently adopted and as a consequence of his interest in their project the Atlantic Telegraph Company had sought his assistance.

Following the failure of the first Atlantic cable and the establishment of the Joint Committee, Thomson had given evidence before the committee of inquiry on 17th December 1859. In his testimony, which comprised 18 pages of text with many diagrams, Thomson reaffirmed his inverse square law and put forward the formula by which the capacitance of the cable can be determined from the relative permittivity of the gutta percha and the ratio of the conductor and dielectric diameters. He also described his marine galvanometer, which consisted of a very light steel magnet cemented to the back of a minute mirror – the mass of the whole assembly being less than 1.5 grains – and which was attached, initially, to a platinum wire suspension but later to a silk suspension. The instrument was designed to be unaffected by the motion of a ship and incorporated a zero control.

The findings of the scientific consultative committee led to the adoption of the specification of the new cable put forward by the Telegraph Construction and Maintenance Company, which had been formed in April 1864 by an amalgamation of the Gutta Percha Company with Messrs Glass, Elliot and Company; the chairman was John Pender. It would comprise seven copper wires, each of No. 18 gauge, having a mass of 300 lb per nautical mile (73.5 kg/km), and coated with four layers of gutta percha alternating with four of Chatterton's compound, the dielectric having a mass 400 lb per nautical mile (98 kg/km). The centre wire would also be covered with Chatterton's compound 'to prevent small air-bubbles remaining in the interstices of the wires when the gutta percha was put on, as well as to prevent the percolation of water along the cable from a loose, buoyed, or lost, end' [24] (See Fig. 5.5.).

According to Bright, 'the primary cause of the failure of the first Atlantic cable was the fact of the core, especially the conductor, being insufficiently large, coupled with its low specific conductivity' [24]. The purity of copper at that time was 'so low that an electrical conductivity of 40% [of pure copper] was as much as was ordinarily obtained for telegraphic purposes'. Moreover, as Thomson pointed out, the conductivity varied very considerably depending upon the origin and refining of the copper ore. For the new cable the conductivity of the copper was specified to be at least 85 per cent of that of chemically pure copper. This improvement and an increase in the cross sectional area of the conductor by a factor of three would, it was estimated, give a working speed of transmission of seven words per minute.

The core would be covered with a cushion of tarred jute and protected by ten

Figure 5.5 Making the steel wires for the Atlantic telegraph cable of 1865.
Source: The Illustrated London News Picture Library.

iron wires of no. 13 gauge. Each wire would be lapped with yarns of Manilla hemp previously soaked in a mixture of tar, india-rubber, and pitch to prevent corrosion. The complete mass of the cable would be 1 ton 15.75 cwt per nautical mile (981 kg/km) in air and in water 14 cwt per nautical mile (384 kg/km); the overall diameter would be 1.75 in (44.5 mm); and the breaking force would be 7 ton 15 cwt force (7875 kgf) or 11 miles (17.7 km) of its weight in water [55].

TC&M gained the contract for the manufacture of the cable, and payment to them was made partly by £250,000 in cash, partly by £250,000 in 8 per cent preference shares, and partly by £100,000 in mortgage debentures. A condition of the contract was the allocation of a further sum of £137,140 in unguaranteed stock of the Atlantic Telegraph Company – to be paid by instalments – as long as the cable worked satisfactorily. The capital of the newly formed ATC was £600,000 of which TC&M had provided more than half: Field was able to raise just £70,000 in the USA [56].

The cable core was manufactured at the Wharf Road Gutta Percha works in 4,500 yd (4115 m) lengths, each of which was kept for 24 hours under water at 75° F, and then carefully inspected. The sheathing was carried out at TC&M's works at Morden Wharf, Greenwich. After fabrication the cable was coiled into light iron tanks, filled with water, and tested daily [57].

Since experience of the laying of the first Atlantic cable had shown the inconvenience of employing two ships simultaneously, it was decided to attempt the laying of the new cable using one ship. The only ship then available that could accommodate the entire length of the new cable was I.K. Brunel's mighty SS *Great Eastern*, a vessel of 22,500 ton (22,860,000 kg) (Fig. 5.6). Originally named the *Leviathan*, the *Great Eastern* had been lying idle for almost ten years unable to obtain a suitable cargo. Brunel suggested the ship should be used for the cable laying and so in 1864 The Great Eastern Steamship Company, under the chairmanship of D. Gooch (later Sir Daniel Gooch, Bart., M.P., previously locomotive superintendent of the Great Western Railway) was chartered for the project. Gooch became an ardent supporter of the enterprise and eventually became the chairman of TC&M.

Because it was not practicable to moor the *Great Eastern* off the works at Greenwich, the cable had to be cut into lengths and coiled on two pontoons before being loaded into three vast water-tight iron tanks constructed in the holds of the ship.

On 23rd July 1865, off Valentia Island, the cable was joined up with the shore end cable (Figure 5.7), of 30 miles (48 km) in length, (manufactured by W.T. Henley at North Woolwich), of which 26 nautical miles (48 km) had been stowed on the SS *Caroline*. Afterwards the *Great Eastern*, escorted by two British men-of-war, the *Terrible* and the *Sphinx*, began paying out the cable at a speed of six knots [58].

Soon a fault developed. On the second day out the electric insulation was found to be inadequate. Inspection of the cable showed that a small piece of iron wire had been forced into the outer covering and gutta percha surrounding the inner conductor. When this section was cut out and a new splice was made

228 THE ILLUSTRATED LONDON NEWS Sept. 8, 1866

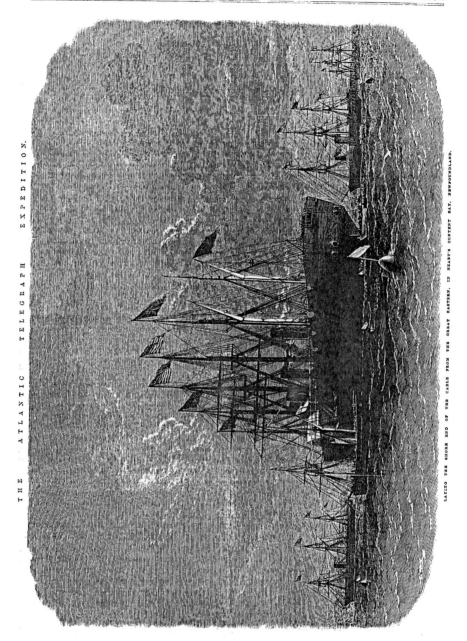

Figure 5.6 Laying the shore end of the cable from the Great Eastern, *in Heart's Content Bay, Newfoundland.*
Source: The Illustrated London News Picture Library.

the integrity of the cable was restored [59]. Again a total loss of electric insulation was detected on the 29th July and once more a section had to be cut out and a repair effected. Then, on 2nd August, when the ship was 636 miles (1023 km) from Valentia, and 1,028 miles (1654 km) from Newfoundland disaster struck – the cable broke. A graphic, first hand, account of this saga was written for *The Times* by a Mr Russell in his 'Diary of the cable'. It tells of some of the difficulties experienced during this epic undertaking – difficulties from which valuable lessons would be learnt for future cable laying. An extended extract from Russell's diary now follows [60].

The picking-up [see Fig. 5.8] was, as usual, exceedingly tedious, and one hour and 46 minutes elapsed before one mile was got on board; then one of the engine's eccentric gear got out of order; next, the supply of steam failed, and when the steam was got up it was found that there was not water enough in the boilers, and so the picking-up ceased altogether for some time, during which the ship forged ahead and chafed against the cable.

After two miles of cable had been picked up, the *Great Eastern* was forced to forego the use of her engines because the steam failed, while her vast broadside was exposed to the wind, which was drifting her to larboard or the left-hand side, till by degrees an oblique strain was brought to bear on the cable, which came up from the sea to the bows on the right side. Against one of the hawse pipes at the bows the cable now caught, while the ship kept moving to the left, and thus chafed and strained the cable greatly against the bow, for now it was held by this projection, and did not drag from the V-wheel in front. The *Great Eastern* could not go astern lest the cable should be snapped, and without motion some way there is no power of steerage. At this critical moment, too, the wind shifted, so as to render it more difficult to keep the head of the ship up to the cable. As the cable then chafed so much that in two places damage was done to it, a shackle chain and a wire rope belonging to one of the buoys were passed down the bow over the cable and secured in a bight below the hawse pipes. These were hauled so as to bring the cable, which had been caught on the left-hand side by the hawse pipes, round to the right-hand side of the bow, the ship still drifting to the left; while the cable, now drawn directly up from the sea to the V-wheel, was straining obliquely from the right with the shackle and rope attached to it. . . .

The cable and the wire rope together were now coming in over the bow in the groove in the larger wheel, the cable being wound upon a drum behind by the machinery, which was once more in motion, and the wire rope being taken in round the capstan. Still, the rope and cable were not coming up in a right line, but were being hauled in, with a great strain on them at an angle from the right hand side, so that they did not work directly in the V in the wheel. Still, up they came. The strain was shown on the dynamometer to be very high, but not near breaking-point. At last, up came the cable and wire rope shackling together on the V-wheel in the bow. They were wound round on it, slowly, and were passing over the wheel together, the first damaged part being inboard, when a jar was given to the dynamometer, which flew up from 60 cwt (3050 kg) – the highest point marked – with a sudden jerk, 3.5 in (89 mm). In fact, the chain shackle and wire rope clambered, as it were, up out of the groove on the right hand side of the V of the wheel, got on the top of the rim of the V-wheel, and rushed down with a crash on the smaller wheel, giving, no doubt, a severe shock to the cable to which it was attached. The machinery was still in motion, the cable and the rope travelled aft together, one towards the capstan, the

THE   ATLANTIC   TELEGRAPH   EXPEDITION.

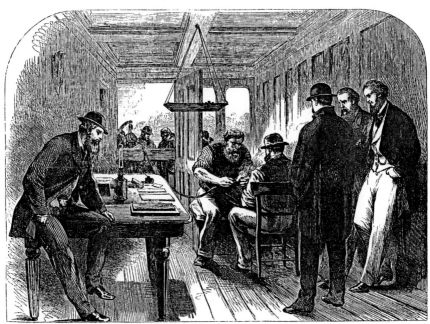

a)   MAKING A JOINT TO SPLICE THE CABLE WITH THE IRISH SHORE END.

b)   THE TELEGRAPH STATION IN HEART'S CONTENT BAY, NEWFOUNDLAND.—SEE PAGE 238.

Figure 5.7   a) Making a joint to splice the cable with the Irish shore end.
*   b) The telegraph station in Heart's Content Bay, Newfoundland.*
Source: The Illustrated London News Picture Library.

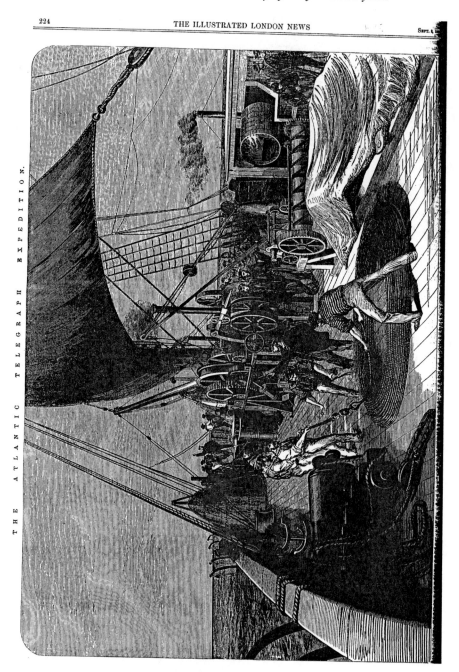

Figure 5.8 Preparing to grapple the broken cable.
Source: The Illustrated London News Picture Library.

other towards the drum, when, just as the cable reached the dynamometer, it parted, 30 feet (9.1 m) from the bow, and with one bound leaped, as it were, into the sea.

This is the scene depicted in Fig. 5.9 by the artist, Mr R. Dudley, who accompanied the expedition to sketch its many incidents for the *Illustrated London News* [61]. Several attempts were made to recover the cable but the effort had to be abandoned when the supply of grappling rope ran out, and on 11th August the fleet of ships parted company with much despondency. Approximately 5,500 miles (8850 km) of cable had been fabricated for the Atlantic cable project from 1857 to 1865; now c. 4,000 miles (6440 km) lay on the ocean floor and £1,250,000 had been expended. But much had been learnt. Although the picking-up gear had failed, the paying-out machinery had worked well; the feasibility of recovering cables by grappling from c. 12,000 ft (3660 km) had been demonstrated; the SS *Great Eastern* had proved to be an excellent ship

Figure 5.9   The cable breaks!
Source: J. Timbs, *Wonderful inventions* (Routledge, London, 1870).

for trans-oceanic cable laying; and, of course, the viability of sending signals across the Atlantic ocean had been proven. All of this gave encouragement to the promoters, and so, with the knowledge and experience gained from the first and second unsuccessful attempts, another venture was felt to be worthwhile.

To secure the necessary capital the Atlantic Telegraph Company was amalgamated with a new company, the Anglo-American Telegraph Company, to raise £600,000 of additional capital [62]. As on previous occasions Cyrus Field was actively involved in finding sponsors. Daniel Gooch promised to subscribe £200,000, and Thomas Brassey agreed to bear one tenth of the total cost of the undertaking. The Telegraph Construction & Maintenance Company subscribed £100,000 and each of the ten directors put forward £10,000. Further sums were obtained from firms that were likely to participate in the new enterprise as sub-contractors. Of the sum raised the TC&MC would receive £500,000 for the new cable and an additional sum of £100,000 if the venture succeeded. If both cables were successfully utilised the TC&MC would collect £737,140.

The objective of the enterprise was not only to lay a new cable between Ireland and Newfoundland but also to splice a length of cable to that which lay on the bed of the Atlantic ocean. With the unused length of cable from the second attempt and a further 1,600 miles (2,575 km) of new cable from TC&MC a total length of 2,770 miles (4,457 km) would be available for these purposes – 1,960 miles (3,154 km) for the third trans-Atlantic crossing and 697 miles (1,220 km) to complete the cable run of the second attempt: the reserve would be 113 miles (182 km). The only change in the cable itself would be the use of galvanised wire for the armouring.

On 30th June 1866 the *Great Eastern* arrived once more at Valentia. On board there were Willoughby Smith the chief electrician; Professor Thomson the consulting electrical adviser to the Atlantic Telegraph Company; Daniel Gooch, Cyrus Field, Samuel Canning who was in charge of the project; the same representatives acting on behalf of the contractors as in 1865; and an artist and a historian who would provide first hand illustrations and reports.

The laying of the new cable was successfully completed in just 14 days – from 13th July to 27th July – and telegraphic communications became possible between the old and the new worlds. It was a triumph of private enterprise, science and engineering. The jubilation of the achievement was compounded a few weeks later when, following the grappling of the 1865 cable and the splicing of it with the new cable on board the *Great Eastern* – an operation that had occupied the period between 31st August and 2nd September, the second cable link between Europe and north America was established [63].

In commemoration of this epic achievement Queen Victoria conferred the honours of two baronetcies and four knighthoods to six of the principal participants, namely: Sir Daniel Gooch, MP, and formerly the locomotive super-intendent of the Great Western Railway; and Sir Curtis Miranda Lampson, deputy chairman of the original Atlantic Telegraph Company; and Sir Richard Atwood Glass of the firm who designed and fabricated the 1865 and 1866 cables; Professor Sir William Thomson; Sir Samuel Canning, engineer-in-chief of the TC&MC; and Captain Sir James Anderson, commander of the *Great Eastern*, respectively [64].

Soon after the laying, Mr Latimer Clark arranged for the two Newfoundland ends of the cables to be joined together: then, connecting one of the Valentia ends to a silver thimble containing sulphuric acid and a tiny zinc plate and the other to a mirror galvanometer, he showed that signals could traverse the Atlantic ocean twice and still give a strong deflection of the spot of light of 12 in (30.5 cm) or more. It was a dramatic demonstration of long distance electric telegraphy.

The success of the 1866 Atlantic cable effectively brought to an end the early developmental period of submarine cable technology. Now, the proven engineering designs and techniques of cable fabrication, cable instrumentation and testing, cable laying and jointing, cable recovery from the depths of an ocean and ship-borne cable storage and handling, could be applied to the task of laying submarine cables around the world. The task would be the fulfilment of Puck's prophesy – in Shakespeare's *A midsummer night's dream*:

> I'll put a girdle round about the earth
> in forty minutes.

The pioneer work of British cable companies and engineers placed them in a dominant position for future submarine cable laying enterprises. According to W.H. Preece, when presenting, in 1894, his Presidential Address to the Institution of Engineers:

The form of cable has remained unaltered since the original Calais cable was laid in 1851. Various sizes of core and armour, and various modes of protection from decay, have been used to suit different routes, but the cable of today may be said to be typically the same as that used in the English Channel in 1851, and in the Atlantic in 1865.

The expertise and experience of the UK's engineers in the field of electric telegraphy led to the foundation, in 1871, in London of The Society of Telegraph Engineers [65], the first President of which was Dr William Siemens. In his Inaugural Presidential Address Siemens referred to this point and observed:

London . . . is the principal centre of the telegraphic enterprise in the world, and musters consequently the greatest number of telegraph engineers. It is a remarkable fact that the manufacture of insulated wire, and of submarine cables, is almost entirely confined to the banks of the Thames.

This state of pre-eminence continued for many years. The growth in the number, and length, of submarine cables world-wide was extraordinary. In 1852 the total length was 87 nautical miles (161 km); in 1893 it was 139,594 nautical miles (258,500 km) of which 125,115 nautical miles (231,700 km) had been laid by private companies, and 14,479 nautical miles (26,820 km) by Government Administrations (see Tables 5.1, and 5.2, in Notes 2 and 3 at the end of this chapter).

By 1898 the total length of submarine telegraph cable laid was almost 170,000 nautical miles, representing an investment of c. £ 50,000,000. There were c. 1,500 separate cables varying in length from a quarter of a nautical mile to over 2,700 nautical miles (5,000 km). Fig. 5.10 shows, for 1898, a map of the cable routes of the Eastern and Associated Telegraph Companies. This was by far the largest cable conglomerate in the world for many years (see Appendix 1). By stages this organisation, with several additional companies became, in 1934, Cable & Wireless Ltd. Appendix 2 lists the principal primary cables laid (from 1868 to 1928) for the companies that finally merged into the British company Cable and Wireless Ltd.

Of the c. 170,000 nautical miles (c. 315,000 km) of submarine cable, nearly 90 per cent had been provided by private enterprise, and c. 120,000 nautical miles (c. 222,000 km) had been manufactured and laid by just one firm of contractors, namely, the Telegraph Construction and Maintenance Company (TCM) of London.

Meanwhile, during the last quarter of the 19th century, a new invention – the telephone – was being exploited, which would eventually replace all forms of telegraphy. This topic is discussed in the next chapter.

Note 1

Some early attempts at telegraphing through water [66]

1. J. W. Wilkins was led to consider telegraphing through water at a time when, in 1845, he was 'engaged on the only long line of telegraph then existing in England – London to Gosport' and when, as he said, 'gutta percha was not imagined, or the ghost of a proposition for a submarine wire existed'. He put forward his thoughts for a submarine electric telegraph link between England and France in a paper published in the *Mining Journal* of 31st March 1849. His proposal was essentially the same as Lindsay's but it did contain one new feature. In describing how 'the electricity diffuses itself in radial lines' from the submerged plates, Wilkins opined: 'These rays of electricity may be collected within a certain distances – focused as it were – by the interposition of a metallic medium that shall offer less resistance than the water or earth'. The principle of using *uninsulated* submarine wires to augment the conductivity of the path between the sending and receiving apparatuses featured in the schemes of O'Shaughnessy, Highton, and Dering.

2. O'Shaughnessy seems to have been the first person to attempt to establish a practical, rather than an experimental, electric telegraph link across an expanse of water. In 1849 he laid a bare iron rod under the waters of the broad (4,200 ft – 1,280 m) river Huldee, India, with the associated batteries and delicate needle instruments in connection on each bank. Signals were transmitted, but 'it was found that the instruments required the attention of

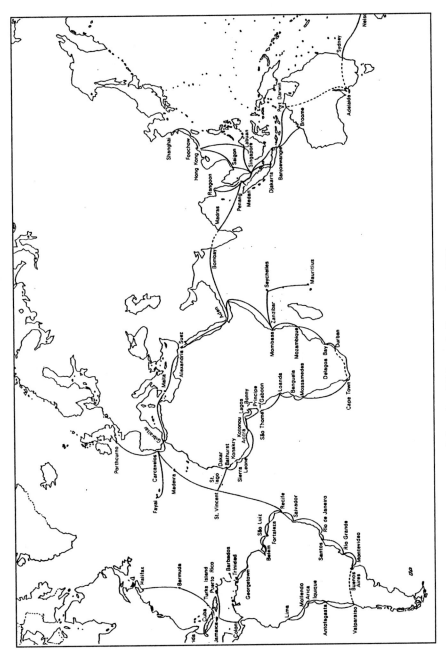

Figure 5.10   The cable routes of the Eastern and Associated Telegraph Companies, 1898.

skilful operators, and that in practice such derangements occurred as caused very frequent interruptions'.

He repeated the experiment without the iron rods but, though he succeeded in observing intelligible signals, he concluded the 'battery power' would be so enormous, and so prohibitively expensive, as to preclude the method being used in practice.

3. The Highton brothers, Edward and Henry, tried many experiments on 'transaqueous communication' in 1852. 'Naked wires [were] sunk in canals, for the purpose of ascertaining the mathematical law which governs the loss of power when no insulation was used. Communications were made with ease over a distance of about 402 m. The result, however, has been to prove that telegraphic communications could not be sent to any considerable distance without the employment of an insulated medium.'

4. Dering patented his scheme for a 'transmarine telegraph' on 15th August 1853. In his patent he stated:

To carry out my invention, I cause two uninsulated or partially insulated wires to be placed in the water or in the earth, at a distance apart proportionate to the total length of the circuit, the wires being insulated when they approach one another to communicate with the instruments. . . . In practice I find that from one-twentieth to one-tenth the length of the line-wires is a sufficient distance. . . . 'Thus an oblong parallelogram of continuous conductors is formed, having for its longer sides the uninsulated conductors, and for its shorter sides the insulated wires along the coasts.

The patent includes speculations on the feasibility of linking Holyhead and Dublin, and England and America by means of his method. Possibly the many failures of the cables laid between 1850 and 1860, mainly because of poor insulation, stimulated Dering and others to persist with the notion of telegraphing with bare wires.

Dering's experiments were carried out, in 1853, across the river Mimram at Lockleys, Herts, with parallel galvanised wires of no. 8 gauge laid at a distance apart of c. 30 ft (9.1 m), i.e. about one-tenth of the width of the river. Using 'a small battery power of only two or three Smee cells the signals were easily readable'. Indeed, so impressed were the Chairman and Directors of The Electric Telegraph Company of Ireland at one demonstration that they agreed 'there and then' to adopt the system for the Portpatrick to Donaghadee line.

On 23rd September 1853 the necessary wire was shipped to Belfast but, during the laying operation, the wire proved to be so unreliable, due to frequent breaks caused by its brittleness and the poor factory welds, that the enterprise was abandoned.

Previously, two unsuccessful attempts, using cable similar to that of the 1851 Dover–Calais cable, had been made to link Great Britain and Ireland. However, when the third attempt, in June 1854, between Portpatrick and Donaghadee was successful no further work was carried out by Dering.

Note 2

Table 5.1 Submarine cables laid worldwide by private companies – 1893 [24]

Company	No. of cables	Lengths of cables (nautical miles: 1 nautical mile = 1.852 km)
Direct Spanish	1	708
Spanish National	5	1,163
India Rubber, Gutta Percha & Tel. Co.	3	145
West African	12	3,015
Black Sea	1	346
Indo-European	2	15
Great Northern	24	6,948
Eastern	76	25,376
Eastern and South Africa	9	6,645
Eastern Ext. Australasia, & China	25	15,130
Anglo American	14	10,400
Direct United States	2	3,100
Compagnie Français du Télégraphe de Paris a New York	4	3,496
Western Union	8	7,743
The Commercial Cable	6	6,908
Halifax and Bermudas	1	850
Brazilian Submarine	6	7,369
African Direct	7	2,746
Cuba Submarine	4	1,049
West India and Panama	22	4,557
Soc. 21. Français des Tel. Sous-marins	14	3,754
Western and Brazilian	15	45,408
River Plate Telegraph	1	32
Mexican Telegraph	3	1,559
Central and South American	12	4,898
West Coast of America	7	1,699
Co. T-T del Plata, Co. T. del Rio de la Plata	2	56
Totals	289	125,115

Note 3

Table 5.2 Government cables worldwide – 1893 [24]

Administration	Number of cables	Lengths (nautical miles)
Great Britain	115	1,599
Austria	31	105
Belgium	2	54
Denmark	58	209
France	54	3,450
Germany	47	1,762
Greece	48	452
Holland	20	60
Italy	39	1,068
Norway	255	248
Russia (in Europe)	8	213
Spain	10	441
Sweden	13	94
Turkey	15	339
Senegal	1	3
Russia (in Asia)	1	70
Japan	31	215
Cochin China and Tonquin	2	795
India (British)	83	1,927
India (Dutch)	4	483
Australia (South)	5	50
Queensland	13	162
New Caledonia	1	1
New Zealand	3	196
America (British)	1	200
Bahama islands	1	213
Brazil	22	35
Argentine Republic	3	35
Totals	886	14,479

References

1 House of Commons Sessional papers, 1840 (xiii)
2 BOWERS, B.: 'Sir Charles Wheatstone' (HMSO, London, 1975), pp. 130–131
3 Quoted in Ref. 2, p. 144
4 SABINE, R.: a letter, *J. STE*, 1876, **5**, pp. 86–89
5 Ref. 2, p. 147
6 ANON.: a report, *The Times*, June 1843
7 Ref. 2, p. 148

8 BURNS, R.W.: 'Alexander Bain, a most ingenious and meritorious inventor' *ESEJ*, 1993, pp. 85–93
9 BAIN, A.: 'Electric time-pieces and telegraphs' British patent no. 9745, 1843
10 ANON.: *The Times*, April 1844
11 FINLAISON, J.: 'An account of some remarkable applications of the electric fluid to the useful arts by Mr Alexander Bain with a vindication of his claim to be the first inventor of the electromagnetic printing telegraph and also of the electromagnetic clock' (Chapman and Hall, London, 1843)
12 USHER, A.P.: article on S.F.B. Morse, Encyclopaedia Americana', 1992, p. 476. Also: ANON.: article on S.F.B. Morse in the 'Encyclopaedia Britannica', 1910, p. 874
13 TIMBS, J.: 'Wonderful inventions' (Routledge, London, 1870), p. 361
14 MOTTELAY.: 'Bibliographical history of electricity and magnetism' (Griffin, London, 1922), pp. 175–177
15 Ref. 14, pp. 304–306
16 Ref. 14, p. 384
17 URBANITZKY, A.R. VON.: 'Electricity in the service of man' (Cassell, London, 1886), translated and edited by R. Wormell, p. 759
18 FAHIE, J.J.: 'A history of wireless telegraphy' (Blackwood, Edinburgh, 1899), pp. 13–32
19 MILLAR, A.H.: 'James Bowman Lindsay and other pioneers of invention', (MacLeod, Dundee, 1925
20. Ref. 18, pp. 55–77
21 WYLDE, J. (ed.).: 'The industries of the world' (London Printing and Publishing Company, London, c.1886), II, pp. 503–504
22 ROUTLEDGE, R.: 'Discoveries and inventions of the nineteenth century' (Routledge, London, 1903), pp. 728–730
23 CLAPHAM, J.H.: 'An economic history of modern Britain' (Cambridge University Press, 1963), vol. 2, p. 43
24 BRIGHT, C.: 'Submarine telegraphs' (London, 1898). Also: Ref. 13, p. 363
25 KIEVE, J.L.: 'The electric telegraph. A social and economic history' (David & Charles, Newton Abbot, 1973), pp. 102–103
26 SMITH, W.: 'Resume of the earlier days of electric telegraphy', *J. STE*, 1881, **10**, pp. 312–333
27 ANON.: a report, *Illustrated London News*, 31st August 1850, p. 186
28 ANON.: a report, *Illustrated London News*, 7th September 1850, p. 206
29 Ref. 25, p. 104
30 DUNSHEATH, P.: A history of electrical engineering' (Faber and Faber, London, 1962), p. 211
31 ANON.: a report, *Illustrated London News*, 6th November 1852
32 Ref. 26, pp. 320–321
33 LARDNER, D.: 'Electric telegraph', revised and rewritten by E. B. Bright (London, 1867)
34 Ref. 25, pp. 103–104
35 KINGSFORD, P.W.: 'Electrical engineers and workers' (Arnold, London, 1969), pp. 62–64

36   BRIGHT, E. and C.: 'The life story of the late Sir Charles Tilson Bright, Civil Engineer, with which is incorporated the story of the Atlantic Cable and the first telegraph to India and the Colonies' (London, 1898)

37   BRIGHT, C.: Inaugural Address, *J. STE*, 1887, **16**, p. 29

38   Ref. 33, p. 110

39   Ref. 25, pp. 106–107

40   Ref. 30, p. 215

41   Ref. 24, p. 36

42   ANON.: a report, *Illustrated London News*, 3rd July 1858, pp. 5–6

43   THOMPSON, S.P.: 'The life of Lord Kelvin' (MacMillan, London, 1910), 2 vols.

44   ANON.: reports, *Illustrated London News*, 10th July 1858, p. 31; 17th July 1858, 53; 31st July 1858, pp. 111, 114, 116; 7th August 1858, p. 125; 14th August 1858, p. 149

45   Ref. 35, p. 67

46   ANON.: reports, *Illustrated London News*, 28th August 1858, p. 208; 4th September 1858, p. 227

47   ANON.: reports, *Illustrated London News*, 9th October 1858, p. 329; 23rd October 1858, p. 380

48   Ref. 35, p. 68

49   Ref. 24, p. 57

50   Ref. 24, p. 58

51   Report of the Joint Committee appointed by the Lords of the Committee of Privy Council for Trade and the Atlantic Telegraph Company to enquire into the Construction of Submarine Telegraph Cables, 1861, Parliamentary Papers

52   Ref. 37, p. 61

53   Ref. 24, 80

54   SOLYMAR, L.: 'Getting the message' (OUP, 1999), p. 79

55   Ref. 30, p. 220

56   Ref. 24, p. 82

57   ANON.: a report, *Illustrated London News*, 15th September 1865, pp. 256, 263

58   ANON.: a report, *Illustrated London News*, 5th August 1865, pp. 122–124

59   ANON.: a report, *Illustrated London News*, 19th August 1865, p. 151

60   Quoted in Ref. 13, pp. 375–376

61   ANON.: a report, *Illustrated London News*, 2nd September 1865, p. 223

62   Ref. 25, p. 114

63   ANON.: reports, *Illustrated London News*, 14th July 1866, p. 35; 21st July 1866, p. 51; 8th September 1866, p. 238; 15th September 1866, p. 263

64   Ref. 13, p. 388

65   APPLEYARD, R.: 'The history of the institution of Electrical Engineers' (Institution of Electrical Engineers, London, 1939)

66   Ref. 18, pp. 32–55

Chapter 6

The telephone

Although the transmission to a distance of voice signals by means of electricity was achieved by A.G. Bell in 1875, allusions to the possibilities that sounds could be propagated from one person to another, and even recorded, date from the 17th century.

In 1667 Dr Robert Hooke (1635–1703), the Secretary to the Royal Society, suggested that the communication of sounds could be achieved with the aid of a wire. He wrote:

And as glasses have highly promoted our seeing, so it is not improbable that there may be found many mechanical inventions to improve our other senses – of hearing, smelling, tasting, touching. . . . Tis not impossible to hear a whisper a furlong's distance, it having been already done; and perhaps the nature of the thing would not make it more impossible though that furlong should be ten times multiplied. And though some famous authors have affirmed it impossible to hear through the thinnest plates of Muscovy glass, I know a way by which it is easy to hear one speak through a wall a yard [0.91 m] thick. It has not been examined how far acoustics may be improved, nor what other ways there may be of quickening our hearing, or conveying sound through other bodies than the air, for that is not the only medium. I can assure the reader that I have, by the help of a distended wire, propagated the sound to a very considerable distance in an instant, or with as seemingly quick a motion as that of light, at least, incomparably swifter than that which at the same time was propagated through the air; and this not only in a straight line, or direct, but in one bended in many angles [1].

Approximately a quarter of a century earlier the Bishop of Chester, John Wilkins, in the first edition of 'Mercury, or the secret and swift messenger, showing how a man, with privacy and speed, may communicate his thoughts to a friend at any distance' hinted at the possibility of making a contrivance for recording sounds. In this work Wilkins reported:

There is another experiment . . . mentioned by Walchius, who thinks it possible so to contrive a trunk or hollow pipe that it shall preserve the voice entirely for several hours or days, so that a man may send his words to a friend instead of writing. There being always a certain space of intermission, for the passage of the voice, betwixt its going into these cavities and its coming out; he conceives that if both ends were seasonably stopped, while the sound was in the midst, it would continue there till it had some vent. . . . When the friend to whom it is sent shall receive and open it, the words shall come out distinctly, and in the same order wherein they were spoken. From such a contrivance as this (saith the same author) did Albertus Magnus make his Image, and friar Bacon his Brazen Head, to utter certain words [2].

Another allusion to sound recording was narrated in the April 1632 issue of the *Courier Veritable*, a monthly periodical in which unusual imaginations were frequently aired.

Captain Vosterloch has returned from his voyage to the southern lands, which he started on two years and a half ago, by order of the States-General. He tells us, among other things, that in passing through a strait below Magellan's, he landed in a country where Nature has furnished men with a kind of sponge which holds sounds and articulations as our sponges hold liquids. So, when they wish to dispatch a message to a distance, they speak to one of the sponges, and then send it to their friends. They, receiving the sponges, take them up gently and press out the words that have been spoken into them, and learn by this admirable means all that their correspondents desire them to know [3].

A few years later Savinien Cyrano de Bergerac, a witty French writer, in his curious work 'Histoire comique des états et empires de la lune', described a strange box which would now be called a gramophone.

On opening the box, I found a number of metallic springs and a quantity of machinery resembling the interior of our clocks. It was, in truth, to me a book, indeed, a miraculous book, for it had neither leaves nor characters, and to read it, one had no need of eyes, the ears alone answering the purpose. It was only necessary to start the little machine, whence would soon come all the distinct and different sounds common to the human voice [4].

By the beginning of the 19th century various devices comprising megaphones, speaking tubes and the like for conveying speech to a distance had been demonstrated; and the young Wheatstone had undertaken some experiments on the transmission of musical sounds through solid rods [5]. Some of these experiments were developed into public demonstrations. In one of these – the enchanted lyre – visitors would see in his father's shop in Pall Mall an instrument in the form of a large ancient lyre – suspended by a brass wire that passed through a hole in the ceiling – surrounded by, but out of contact with, a velvet covered hoop supported by three vertical rods resting on the floor. At its upper end the wire was connected with the sound boards of musical instruments in the room where unseen players performed on the harp, piano and dulcimer (Fig. 6.1). The reproduction of these sounds by the playerless lyre elicited much surprise, and evoked some speculation on the possible uses of the phenomenon [6]. An anonymous report in the 1821 issue of *Ackerman's Repository* envisaged music being transmitted across London from one concert hall to another, or to a person's house, and parliamentary debates being heard contemporaneously instead of being read the next day only.

*Figure 6.1   The enchanted lyre demonstration which brought Wheatstone some
recognition.*
Source: B. Bowers, *Sir Charles Wheatstone FRS, 1802–1875* (The IEE, London, 2002).

In March 1830 Wheatstone's work led to a Royal Institution Discourse, by Faraday, and an 1831 published paper 'On the transmission of musical sounds through solid linear conductors, and on their subsequent reciprocation'. In this Wheatstone observed [7]: 'Could any conducting substance be rendered perfectly equal in density and elasticity so as to allow the undulations to proceed with a uniform velocity without any reflections and interferences, it would be as easy to transmit sounds through such conductors from Aberdeen to London as it is now to establish a communication from one chamber to another'.

A few years later in 1837 Professor Page [8,9] of Washington discovered that an iron bar, when magnetised and demagnetised at brief intervals, emits sounds, which he called 'galvanic music'. The discovery of galvanic music led experimenters, almost simultaneously in several countries, to investigate the acoustical effects produced by magnetisation. Among these experimenters were Marriam, Beatson, Gassiot, Matteuci, Guillemin, Wertheim, Wartmann, Janniar, Joule, Laborde, Legat, Reis, Poggendorf, du Moncel, Delzenne and others as well as de la Rive [8]. De la Rive, of Geneva, in 1843, increased the audible effect by using a long, stretched wire which passed through a solenoid. This finding became the basis of the telephone receiver of Reis.

The first scientist who avowed a belief in the possibility of sending speech signals from one place to another was the French scientist Charles Bourseul [10]. He pointed out in 1854 that sounds are caused by vibrations and suggested that if a person spoke near a flexible disc the vibrations of the disc might be used to make and break the current in an electric circuit. Similarly he thought the current variations could be employed to produce vibrations in a receiving disc and so reproduce the speech of the sender. There is no evidence that Bourseul ever carried out any experimental work on his notions, but a few years later Reis effected a rudimentary form of telephone based on a vibrating diaphragm.

Philipp Reis [11] was born on 7th January 1834 at Friedrichsdorf, near Hamburg. He received a good elementary education and at the age of 16 years entered a business house, but his love of the sciences led him to devote his leisure time to the study of mathematics, chemistry and physics at the local commercial institute. Subsequently he left the business and was admitted to Dr Poppe's establishment at Frankfurt to train and qualify as a teacher. It seems that some time towards the end of the 1850–60 decade he began to experiment on telephones by following the researches of Wertheim, Marian and Henry on the production of sounds by electricity.

The first apparatus that Reis constructed consisted of a beer barrel in the bung hole of which there was placed a cone covered at its smaller end with an animal membrane upon which a tiny platinum strip or wire was fastened by means of sealing wax. The receiver consisted of a violin, upon which a knitting needle, having a coil wound around it, was fixed. Later the receiver was made in the form of the human ear. In Fig. 6.2a the platinum wire f was attached to the membrane M, and to the spring R, by sealing wax, and a platinum contact L was placed opposite f. A screw V adjusted the spring. The wires P P' connected the apparatus to the battery. In operation, sound waves impinging on the

(a)

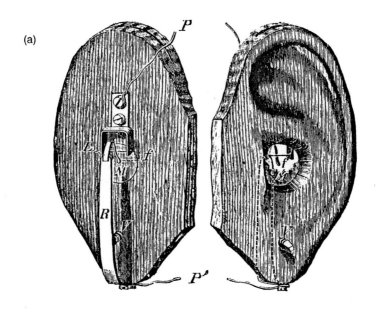

(b)

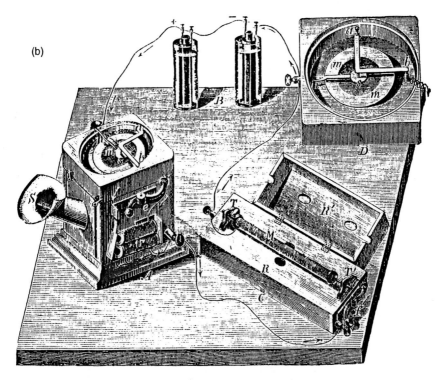

Figure 6.2   (a) Reis's 'ear'; (b) Reis's system.
Source: A.R. von Urbanitzky, *Electricity in the service of man* (Cassell, London, 1886).

membrane caused it to vibrate, thereby establishing a make-and-break type of contact between f and L and an intermittent current in the circuit comprising the battery and 'ear'.

Reis's system (Fig. 6.2b) was formed from the sender A, the battery B, and the receiver C. The upper portion of A, shown separately at D, included the stretched membrane mm having attached to it the platinum strip ns, which could make interrupted contact with the L-shaped piece of metal ab, the ends of which dipped into a mercury cup, when the membrane vibrated. The receiver C consisted of a needle, 8.5 in (21.5 cm) long and 0.03 in (0.9 mm) diameter, around which the coil M was wound. Sound waves incident on mm produced an interrupted current that led to the rapid magnetisation and demagnetisation of the needle and hence, following Page's discovery, the production of sounds. These had the same frequencies as the applied sound waves although the apparatus because of the make-and-break nature of the current could not be responsive to the amplitudes, and hence the quality, of the sounds.

Reis showed his telephone for the first time to the Physical Society of Frankfurt in 1861. Though much has been written on whether his instrument could transmit and reproduce words as well as sounds, Reis in a letter of 1863 to a Mr F.J. Pisko noted [12]: 'The apparatus gives whole melodies in any part of the scale between C and c''' well, and I assure you, that if you come and see me here, I will show you that words also can be made out.' At that time Reis's invention was treated as a toy but Reis was well aware of its potential. He remarked to a friend that 'he had shown to the world a road to a great discovery, but left it to others to follow it up' [12]. Reis died in 1874 just two years before Bell patented his telephone.

The 'others' who tried to modify the apparatus for practical use included S. Yeates (1865), Wright (1865), C. Varley (1877), C. and L. Wray (1876), E. Gray (1874), Van der Weyde, and Pollard and Garnier [13]. Of these experimenters the efforts of Gray of Chicago are notable.

Gray's early patents show that he envisaged applying his invention to telegraphy by replacing the Morse code alphabet of dot and dash signals by tones, of different frequencies for the various letters of the alphabet [14]. These tones could be generated more rapidly in succession, and their duration could be shorter, than the printed marks of the Morse code. Of course, if this code were still to be required the dots could be represented by one tone and the dashes by another. In subsequent British patents of 1875 and 1876 Gray described the application of his instrument to multiplex telegraphy by which several distinct messages could be transmitted simultaneously, in the Morse code, along a single wire [15]. He envisaged that each telegraph receiving instrument would be 'tuned so as to be affected only by its corresponding transmitter at the sending station, and thus the receiving instruments along a line of wire [would] have the power of selecting those messages intended for themselves and letting all others pass'.

A similar method was employed by P. Lacour. He used tuning forks, of differing resonant frequencies, at the sending end and a series of corresponding tuning forks in the receiving instruments.

Gray's invention was successfully operated in 1877 on the lines of the Western Union Telegraph Company from Boston to New York and other places but did not achieve commercial success. It has been stated that with it up to four messages could be sent simultaneously over 2,400 miles (3,860 km) of the company's lines.

Gray, in 1876, also patented a telephone system (Fig. 6.3a). In this the underside of membrane of the sender B is joined to a metal rod t which dips into a vessel G containing a poorly conducting fluid and a terminal t'. The receiver comprises the vessel B' which is closed at one end by the membrane m', to the centre of which is attached a piece of soft iron. In operation the membrane of B vibrates and causes the distance between the end of t and t' to vary thereby altering the resistance, and hence the current, of the circuit formed from B, B', the line wire l, and the battery plates L and L'. This varying current produces a varying force between the electromagnet e and the soft iron armature of the receiving instrument and in consequence the membrane m' vibrates and reproduces speech [16].

Unfortunately for Gray his caveat or 'notice of invention' was filed at the Patent Office on 14th February 1876, the same day that A.G. Bell submitted the complete specification of his invention to the Patent Office [17]. Bell had had his invention notarised in Boston on 20th January 1876 and so the honour of inventing the telephone has been accorded to Bell. His priority of invention, based on his fundamental discovery of 2nd June 1875, was later challenged, without success, in hundreds of court actions.

Bell's telephone was not the outcome of some fortuitous finding; rather, it stemmed from nearly 20 years of investigations of speech and hearing, subjects that had greatly interested his father and grandfather. Since the importance of the telephone cannot be overstated a few words on Bell's background are apposite.

Alexander Graham Bell was born on 3rd March 1847 at 16 South Charlotte Street, Edinburgh, the second of three sons of Alexander Melville Bell (1819–1905), a speech therapist, and his wife Eliza Grace, the daughter of Samuel Symonds, a surgeon in the Royal Navy, and his wife Mary [18]. Bell was educated initially at home and then at James Maclaren's Hamilton Place Academy, Edinburgh (1857–1858), and at the Royal High School, Carlton Hill, Edinburgh from 1858. In 1862, at the age of 15, he left Edinburgh to stay in London with his grandfather, Alexander Bell (1790–1865), who was a well-known elocutionist and the author of *The practical elocutionist* and *Stammering, and other impediments of speech.*

From 1863 to 1864 Bell was a pupil teacher at Weston House, Elgin, Moray, where he taught both music and elocution and received instruction in Latin and Greek. These disciplines were further studied at the University of Edinburgh from 1864 to 1865. He returned to Weston House as an assistant master in 1865, and in 1866 became a master at Somerset College, Bath. The following year he joined, as an assistant, his father who was continuing his own father's work in London and who was also a notable elocutionist; he had written *A new*

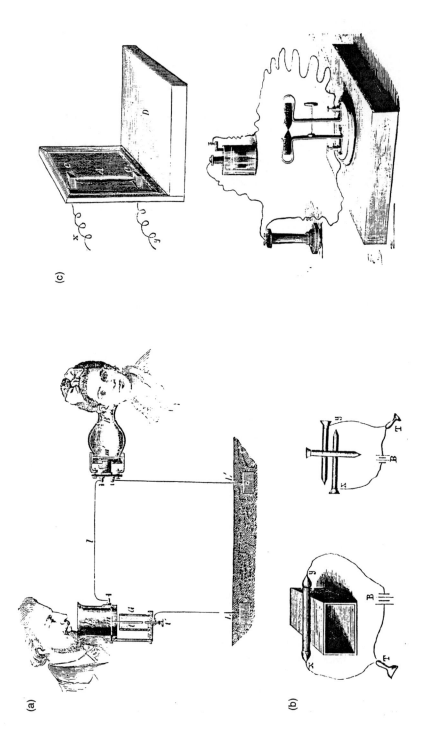

Figure 6.3   (a) Gray's telephone; (b) Hughes' microphone without carbon; (c) Hughes' carbon microphone.
Source: A.R. von Urbanitzky, op. cit.

elucidation of the principles of speech and elocution (1849), and *Principles of speech* (1863). As a consequence of these associations, and his keen interest in the work of his father and grandfather, Bell became adept at teaching the deaf. As a boy he had developed a flair for practical experiments and at 18 years of age had studied the formation of vowel sounds in his own mouth and the effects of cavity shape on resonance.

Bell's early experimental work was influenced by A.J. Ellis, a phonetician of London, who had been converted to A.M. Bell's system of 'visible speech'. In 1866 Ellis introduced Bell to the work of Helmholtz [19] and his electric tuning fork apparatus which could produce composite vowel-like sounds. Ellis suggested that Bell should repeat Helmholtz's experiments [20]. Bell found that he could read Helmholtz in French but could not understand the work because of a lack of knowledge of electrical science and technology. He decided to make good this deficiency and was soon absorbed in experiments involving electromagnets, tuning forks, electric telegraphy, and related devices and subjects with the objective of generating vowel sounds artificially.

Bell matriculated at the University of London in 1868 and attended physiology and anatomy classes from October of that year. Then in 1870 he and his parents emigrated to Brantford, Ontario, Canada. A.G. Bell continued his work for the deaf in the USA, in Boston and Northampton, Massachusetts, and in Hartford, Connecticut, and in 1873 he was appointed Professor of vocal physiology and elocution at Boston University. He became a naturalised American citizen in 1874.

At Brantford Bell extended his experimental activities. He soon conceived the notion of a multiple, or harmonic, telegraph, using tuned reeds, which would permit a number of telegraph messages to be sent simultaneously over a single wire by means of interrupted tones of different frequencies. This work continued for several years [21] (Fig. 6.4a).

In 1874 Bell saw demonstrations of Koenig's manometric flame apparatus and Scott's phonautograph. Both apparatuses utilised membrane diaphragms. In Koenig's manometric capsule a gas flame vibrates by the action of sound waves produced by the human voice. The flickering of the gas flame can be viewed as a continuous wavering band of light in a revolving mirror. Scott's French patent of 1857 was the first to contain a description of an instrument, the phonautograph, which could visually record sound waves. It consisted of a horn terminated by a diaphragm to which was attached a pivoted lever carrying a single bristle. This bristle lightly contacted the surface of a strip of paper, covered in lamp black, which was wrapped around a brass drum. When this was rotated by a handle a lead screw caused the drum to move axially under the bristle. Sound waves incident on the horn vibrated the diaphragm and bristle assembly, thereby causing a white tracing, on the lamp-blacked paper, of the waveform of the sound. Using either of these devices Bell felt that he could teach deaf-dumb persons to speak by comparing their attempts at speech with previously recorded speech patterns. The deaf would observe the effects of changes of the organs of speech – the lips, the tongue, and the soft palate – on

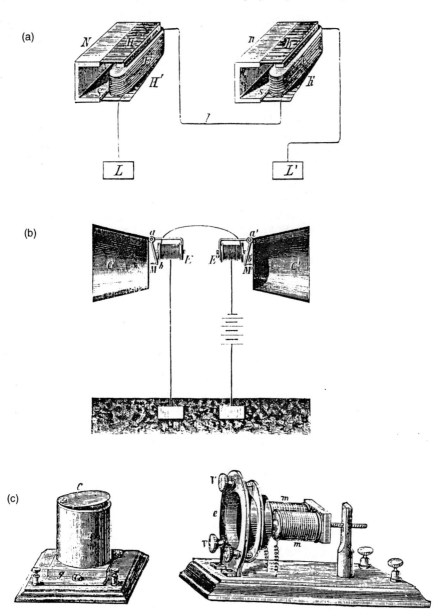

Figure 6.4   (a) Bell's electric harmonica; (b) Bell's second telephone; (c) Bell's third telephone.
Source: A.R. von Urbanitzky, op. cit.

the generation of sounds and so be taught to pronounce words distinctly. 'Visible speech' was demonstrated by Scott at the 1859 British Association meeting held in Aberdeen. As previously noted Reis's 'telephone' of 1861, which was well known by 1874, also employed a membrane diaphragm at the transmitter.

In the course of his work Bell, using apparatus put at his disposal in the Institute of Technology, Boston compared the responses of the phonautograph and the manometric capsule for the same vowel sound and found that the two outputs did not match. This caused him to commence searching in other directions for a possible teaching aid for the deaf. However, while pondering this matter Bell was struck by the similarity of the phonautograph to the human ear. He reasoned that a phonautograph modelled on the human ear probably would lead to more accurate tracings of speech vibrations than the instruments he was using. Soon afterwards, at Brantford in the summer of 1874, following a suggestion [22] by Dr C.J. Blake, (who became a renowned American otologist), Bell fabricated his ear phonautograph. This used a human ear, with its eardrum but with the stapes removed. He moistened the membrane-tympani and the ossiculae with glycerine and water to make them supple and used a small piece of hay 0.98 in (c. 25 mm) in length for the bristle. When he spoke into the ear, the piece of hay vibrated in accordance with the voice sounds. Tracings were made of the sound waves using smoked glass.

While experimenting with this human ear Bell was continuing his work on the problem of transmitting musical sounds, by a telegraphic instrument, based on the interruption of an electric current.

I had dreams that we might transmit the quality of a sound if we could find in the electric current any undulations of form like these undulations we observe in air. I had gradually come to the conclusion that it would be possible to transmit sounds of any sort if we could only occasion a variation in the intensity of the current exactly like that occurring in [the] density of the air while a given sound is made. . . .

I had obtained the idea that theoretically you might, by magneto electricity, create such a current. If you could only take a piece of steel, a good chunk of magnetised steel, and vibrate it in front of the pole of an electromagnet, you might get the current we wanted . . . It struck me that the bones of the human ear were very massive, indeed, as compared with the delicate thin membrane that operated them, and the thought occurred that if a membrane so delicate could move bones relatively so massive, why should not a thicker and stouter piece of membrane move my piece of steel. And the telephone was conceived [23].

During the autumn and winter of 1874, and subsequently, Bell, with his young assistant, Thomas A. Watson, spent more and more time on his experimental work. Their efforts were rewarded when, on 2nd June 1875, Bell's first membrane diaphragm telephone transmitter produced crude speech sounds. Watson has described the occasion [24].

On that hot June day we were in the attic, hard at work experimenting with renewed enthusiasm over some improved piece of the apparatus. About the middle of the afternoon, we were retuning the receiver reeds, Bell in one room pressing the reed against his ear one by one as I sent him the intermittent current of the transmitters, from the other room. One of my transmitter reeds stopped vibrating. I plucked it with my fingers to start it going. The contact point was evidently screwed too hard against the reed and I began to readjust the screw while continuing to pluck the reed when I was startled by a loud shout from Bell and out he rushed in great excitement to see what I was doing. What had happened was obvious. The too-closely adjusted contact screw had prevented the battery current from being interrupted as the reed vibrated and, for that reason, the noisy whine of the intermittent current was not sent over the wire into the next room, but that little strip of magnetised steel I was plucking was generating by its vibration over the electromagnet, that splendid conception of Bell's, a sound-shaped electric current.

We spent the rest of the afternoon and evening repeating the discovery with all the steel reeds and tuning forks we could find and before we parted, late that night, Bell sketched for me the first electric speaking telephone, beseeching me to do my utmost to have it ready to try the next evening. And, as I studied the sketch on my way to Salem on the midnight train, I felt sure I could do so.

The model of the telephone that Watson constructed became known as the 'gallows' telephone. It consisted of a wooden frame onto which was mounted one of Bell's harmonic receivers, a tightly stretched parchment membrane to the centre of which the free end of a reed was fixed, and a mouthpiece that directed the voice sound waves against the membrane. The expectation was that the movement of the reed-membrane would induce a voltage in the associated iron-cored solenoid which would follow the wave-shape of the speech signals.

Bell tested this device on 3rd June 1875. He used one of his harmonic telegraph receivers for listening and though no intelligible words were transmitted, nevertheless speech sounds were heard and Bell felt that he was on the right track to success [25]. Months of tedious experimenting now followed the significant events of the 2nd and 3rd of June in an attempt to improve the apparatus to the stage where intelligible speech could be transmitted using a second stretched membrane device as the receiver. Also during this period Bell spent much time preparing a patent application for his electric speaking telephone. Some delay was experienced because Bell was sick for several weeks, and also because he wanted his patent application to be filed in the United Kingdom and in the United States of America simultaneously. Eventually, on 14th February 1876 Bell's future father-in-law, Gardiner Greene Hubbard, filed the US patent. It was granted on 3rd March, and issued on 7th March 1876, as US patent no. 174,465 (Fig. 6.4b).

In January 1876 Bell set up a workshop at 5 Exeter Place, Boston. He rented two rooms in the attic of a boarding house at that address for $4 per week. The back room was established as a laboratory/workshop, the front room was used partly as a bedroom and partly as a laboratory, and a wire was installed between the two rooms. Bell's move was necessitated because he had heard that strangers were visiting the workshop of Charles Williams – a manufacturer of telegraph apparatus where Bell's devices were being constructed – and were examining his apparatus with unusual curiosity [26].

By the early spring of 1876 Bell had designed and Watson had fabricated a new type of transmitter in which a wire attached to a membrane was inserted into acidulated water contained in a metal cup. Vibrations of the membrane by speech signals altered the resistance between the wire and the cup and consequently the resistance of the circuit that embraced the transmitter, the receiver and the connecting line and battery. This new variable resistance telephone transmitter had followed some work on telegraphy in which Bell had used a variable water resistance to bridge the contacts of his telegraph key.

On the evening of 10th March 1876 the new device was tested. Watson has described the occasion.

When all was ready I went into Bell's bedroom and stood by the bureau with my ear to the receiving telephone. Almost at once I was astonished to hear Bell's voice coming from it distinctly, saying, 'Mr Watson, come here, I want you.' We had no receiving telephone at his end of the wire so I could not answer him, but as the tone of his voice indicated he needed help, I rushed down the hall into his room and found he had upset the acid of the battery over his clothes [27].

On entering Bell's room Watson exclaimed: 'Mr Bell, I heard every word you said distinctly' [27]. In their elation the spillage of the acid was quickly forgotten: the first ever intelligible sentence had been sent over a telephone circuit – it had been an historic event.

Further development work led to Bell's third system (Fig. 6.4c). The tuned-reed receiver was replaced by one that consisted of a thin iron diaphragm which moved under the influence of an electromagnet excited by the line current; and the transmitter comprised a non-magnetic membrane, to which a piece of iron was attached, that caused a voltage to be induced in an associated coil when voice signals were incident on the membrane. This system was described and demonstrated before the American Academy of Arts and Sciences in Boston on 10th May 1876 [28].

In a lecture given to the Society of Telegraph Engineers, London, Bell stated [29]: 'It was my original intention, and it was always claimed by me, that the final form of telephone would be operated by permanent magnets in place of batteries, and numerous experiments had been carried on by Mr Watson and myself privately for the purpose of producing this effect.' Bell was at pains to mention this point since he

doubt[ed] not that numbers of experimenters have independently discovered that permanent magnets might be employed instead of voltaic batteries. Indeed one gentleman, Professor Dolbear, of Tufts College, not only claims to have discovered the magneto-electric telephone, but I understand charges me with having obtained the idea from him through the medium of a mutual friend.

During the summer of 1876 the Philadelphia Centennial Exposition was held to commemorate the 100th anniversary of the founding of the United States of America. Bell's two membrane transmitters, his liquid transmitter, and his iron-box receiver were exhibited and on 25th June 1876 Bell demonstrated his system to the Exposition judges. Another noteworthy test was carried out in August when articulate speech was, for the first time, transmitted and received between places that were separated by miles of space, namely, the eight miles (12.9 km) between Brantford and the small town of Paris.

And, in his opening address to Section A of the British Association meeting in Glasgow, the illustrious Sir William Thomson (later Lord Kelvin) described his delight in using the telephone receiver for the first time [30].

In the Canadian Department I heard 'To be or not to be . . . there's the rub', through an electric telegraph wire; but, scorning many monosyllables, the electric articulation rose to higher flights, and gave me passages taken at random from the New York newspapers . . . All this my own ears heard, spoken to me with unmistakable distinctness by the circular disc armature of just such another little electromagnet as this which I hold in my hand. The words were shouted with a clear and loud voice by . . . Professor Watson. . . . This, the greatest by far of all the marvels of the electric telegraph, is due to a young countryman of our own, Mr Graham Bell.

In the meantime Bell sought protection for his methods and apparatus and applied for additional patents. His patents nos. 174,465 and 186,787 of 7th March 1876 and 30th January 1877 respectively became the fundamental telephone patents. Over a period of 18 years they were tested in around 600 separate court actions.

Interestingly, in 1876 Bell's telephone patent rights were offered to Western Union Telegraph Company for the very modest sum of $100,000. Western Union declined the proposal, their view being that the telephone was a toy and would never be of practical use. They quickly realised their error and within a year began to acknowledge its commercial importance. Soon both Western Union and the Western Electric Manufacturing Company were fabricating apparatus based on patents acquired from Gray, Edison, Dolbear and others. This caused much unease to Bell and his supporters and during the summer of 1878 a suit was brought against the Western Union Telegraph Company for infringing the Bell patents. In November 1879 Western Union settled out of court and, from 10th November 1879 for a period of 17 years, agreed to disengage from the telephone business, grant rights to the National Bell Telephone Company (successor to the Bell Telephone Company which instituted the suit) on all patents they had obtained, and to sell to the Bell company all of the telephone systems that Western Union and their subsidiaries had installed. In return the Bell company agreed not to compete with Western Union in the public-message telegraph field, and to purchase the telephones (totalling more than 50,000) that Western Union and its subsidiaries had already made.

Meantime, a policy of further tests, demonstrations, developments and publicity of Bell's invention was pursued. On 26th November 1876 telephonic signals were sent over 15.9 miles (25.7 km) of telegraph line. And on 12th February 1877 Bell gave a lecture and demonstration in the Lyceum Hall, Salem during which Watson, in Bell's laboratory in Boston, was heard speaking by the audience of 600 people. For this demonstration Bell used his latest permanent magnet telephones with metallic diaphragms: there was no battery in the line circuit. Bell had the vision to see the great potential of telephone communications in commerce and presumably engaged in lecturing and demonstrating to further his views. He was not to be disappointed. *The Boston Globe* reported Bell's Lyceum Hall lecture and stated:

This special [report] by telephone to the Globe has been transmitted in the presence of about twenty who have been witnesses to a feat never before attempted – that is, the sending of a newspaper despatch over the space of eighteen miles by the human voice – and all this wonder being accomplished in a time not much longer than would be consumed in an ordinary conversation between two people in the same room [31].

This news item seems to have stirred newspaper editors into an awareness of the benefits of their journalists transmitting current affairs reports by telephone for, by March 1880, there were in the United States 138 telephone exchanges in operation with c. 30,000 subscribers.

The first line to be employed for regular telephone use was installed between Charles Williams' Boston workshop and his house in Somerville, about three miles (4.8 km) away: it was inaugurated on 4th April 1877. The first commercial application of the principles of telephony, as they are known today, has been accorded to G.W. Coy who, from 28th January 1878, operated the District Telephone Company of New Haven with a central switchboard which could interconnect 21 subscribers [32].

Previous to these events, while Bell was working on his telegraph system, T. Sanders, a leather merchant and friend, had made Bell a verbal offer of help to finance his endeavours, in return for a share in whatever patent rights might accrue. Soon afterwards, Gardiner G. Hubbard, a Boston attorney and Bell's future father-in-law, made a similar offer. Both Hubbard and Sanders had deaf daughters – who had been taught by Bell – and had been impressed by Bell's ability as a teacher of the deaf. (On 11th July 1877 Bell married Mabel Hubbard, who had been deaf from early childhood. They had two daughters, Elsie and Marian.) The offers were accepted, and on 27th February 1875 the Bell Patent Association was formed. This organisation became the Bell Telephone Company (from 9th July 1877), and then the National Bell Telephone Company (from 13th March 1879).

The elimination of competition from Western Union was of course a major commercial victory for the Bell company. It led to a vast increase in business, but the great deal of capital necessary to implement the means to handle this business, and to buy the Western Union telephone equipment, posed another problem for the Bell interests. A further reorganisation was necessary. On 20th March 1880 the American Bell Telephone Company was formed: (its Certificate of Incorporation was filed on 17th April 1880). By a special Act of the Massachusetts Legislature, signed on 19th March 1880, the new company was enabled to permit the 'manufacturing, owning, selling, using and licensing others to use electric speaking telephones and other apparatus and appliances pertaining to the transmission of intelligence by electricity, and for that purpose constructing and maintaining by itself and its licensees public and private lines and district exchange' [22]. The capitalisation of the new company was limited to $10,000,000.

Shortly afterwards, on 26th November 1881, the American Bell Telephone Company purchased a controlling interest in the Western Electric Manufacturing Company. The company was renamed the Western Electric Company and on 6th February 1882 it officially became the manufacturing unit of the Bell System, thereby providing a dependable source of instruments and apparatuses which would be coherent in system use and also be of consistently reliable quality.

Three years later the American Bell Company became the American Telephone and Telegraph Company (incorporated on 3rd March 1885). It developed into a prestigious and mighty organisation which 100 years after Bell's invention had a budget of $704,000,000 and a staff of c. 16,000 [33].

Before its incorporation into the Bell System, Western Electric had for a number of years been interested in the sale and manufacture of communication equipment abroad. Its overseas business continued after the merger and branches of Western Electric existed throughout the world by 1918. At that time, its foreign interests were so extensive that it was considered desirable to concentrate them into a new subsidiary, the International Western Electric Company. This organisation grew rapidly. At the same time a huge expansion of the telephone network in the USA was taking place and the Bell management felt that it would be commercially desirable to avoid any possibility that these foreign activities would interfere with the domestic (US) business. Hence, in 1925, the International Western Electric Company was sold to the International Telephone and Telegraph Company (IT&T) and its name changed to International Standard Electric Corporation (ISEC). Neither of these companies had any corporate connection with the Bell System [34].

By 1924 the technical inquiries of the Bell System had increased so much in scope and size, and in the number of personnel employed, as to suggest the formation of a single new organisation to deal with most or all of these activities. Such an organisation was established on 27th December 1924, and commenced operations on 1st January 1925 as the Bell Telephone Laboratories, Incorporated; but sometimes referred to simply as the Bell Labs. The Labs. had a dual responsibility – to the American Telephone and Telegraph Company for fundamental researches and to the Western Electric Company for the application of the results of these researches to designs suitable for manufacture.

At the date of its incorporation, the personnel in the Labs. numbered approximately 3,600, of whom about 2,000 were members of the technical staff, and comprised engineers, physicists, chemists, metallurgists and mathematicians. The Bell Labs. occupied space in the building at 463 West Street, New York City: this was to remain one of the major locations of the Labs. for more than 40 years.

Returning now to the 1880s, a novel application of the telephone was demonstrated in 1881. In that year a French engineer named C. Ader filed a patent, titled 'Improvements of a telephone equipment in theatres'. Essentially, it described a method of stereophonic reproduction (Fig. 6.5).

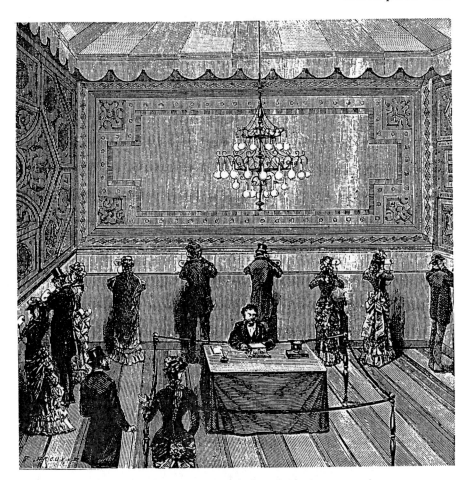

Figure 6.5   Ader's 'stereophonic' listening telephone being demonstrated.
Source: A. Belloc, *La telegraphie historique* (Paris, 1894).

The transmitters (i.e. the telephone mouthpieces) are distributed in two groups on the stage – a left and a right one. The subscriber has likewise two receivers [headphones], one of them connected to the left group the other to the right one . . . This double listening to sound, received and transmitted by two different sets of apparatus, produces the same effect on the ear that the stereoscope produces to the eye [35].

Ader's notions were tried out during the Paris Exposition of 1881, when sound signals were transmitted from the stage of the Paris Opera to the homes of subscribers. However, the poor quality of the received sound led to the suspension of the service and soon afterwards the idea of stereophony became dormant. Some mention of Ader's musical telephone was made in the Paris press, not because of its stereophonic characteristic, but because the telephone permitted home listeners to 'attend' the opera dressed in their bathrobes and pantoufles. The press editorialised at length about the impropriety of such attire.

In the United Kingdom the commercialisation of telephony proceeded with some difficulty [36]. The earliest telephone companies were The Telephone Company, of 1878, formed to acquire Bell's patent, and The Edison Telephone Company of London, of August 1879, established to work T.A. Edison's telephone patents (Fig. 6.6). In that month The Telephone Company opened the first telephone exchange in London at a site in Coleman Street. Both companies commenced negotiations with the General Post Office for the sale of their patents to the British Government; both were unsuccessful.

A portent of the difficulties that telephone companies generally were likely to face in their dealings with the GPO occurred when Parliament debated the Telegraph Bill of 1878. During the passage of the bill the Postmaster General (PMG) tried to insert a clause that would subsume with the word telegraph 'any apparatus [including the telephone] for transmitting messages or other communications with the aid of electricity, magnetism, or any other like agency'.

Although unsuccessful, the PMG persisted with his monopolistic policy for communications in the United Kingdom by instituting in 1879 proceedings against The United Telephone Company (the company formed in June 1880 by the amalgamation of The Telephone Company and The Edison Company to end their mutually damaging rivalry) which wished to commence telephone operations in London. The PMG argued that this action infringed his monopoly rights under the Telegraph Act of 1869. His view was supported by a Mr Baron Pollock and Mr Justice Stephen who ruled that the telephone was a telegraph within the meaning of the Act, and that telephone exchange business could not legally be carried on except by the PMG or with his consent [37]. The ruling also covered any future invention relating to 'every organised system of communication by means of wires according to any preconcerted system of signals'. During the hearing Mr Baron Pollock enquired if the voice of a speaker using the telephone could be identified. The Attorney-General replied that he might distinguish 'a rough voice from a tenor voice, but it comes out more like a voice you would hear in a Punch and Judy' [37].

The aspirations of the telephone companies at this time were causing some concern to the Postmaster General. In December he wrote to the Lords of the Treasury and expressed his belief that the object of the companies was not so much to meet a public need as to create a telephone system which, at a future date, the Government might have to purchase; (cf. the position of the Government and the telegraph companies). 'I propose, then, that the Post Office should at once establish a telephone exchange system of its own, and leave the companies no time to set up vested rights and a practical monopoly' [37].

Their Lordships acceded to the PMG's proposal, 'on the understanding that its object [was], by the establishment of a telephonic system to a limited extent, by the Post Office, to enable your department to negotiate with the telephone companies in a satisfactory manner for licences' [38].

When the United Telephone Company appealed against the decision, the PMG in 1881 partially relaxed his intransigent stance and agreed to grant

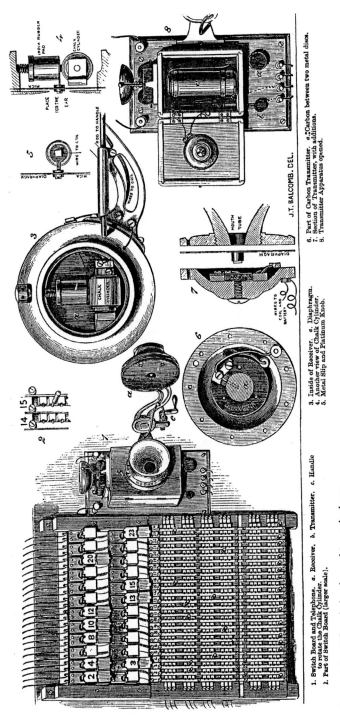

Figure 6.6 Edison's loud-speaking telephone.
Source: The Illustrated London News Picture Library.

licences for restricted area telephone networks in London and in the provinces. The conditions though were far from generous. In the capital the restricted area was limited to about 5 miles (8 km) in radius, and in the provinces to about 2 miles (3.2 km) in radius. Moreover, potential licensees were to be precluded from opening public call offices and from laying trunk lines from one town to another. The licences were for 31 years, expiring in 1922 without any recompense, and were subject to a minimum royalty payment to the GPO of 10 per cent of the gross revenue. These inhibiting limitations led The United Telephone Company to constrain its activities to London only: subsidiary companies operated in the provinces.

In 1882, the PMG announced he would no longer issue licences unless the licensees agreed to sell telephones to the GPO [37]. The consequence was that only eight companies – out of more than 70 which had applied – obtained or accepted licences. One year later, the proposed policy of the GPO became clear: it wished to compete with the telephone companies. The policy did not find favour with the Treasury, which argued that the government should supplement and not supersede private enterprise.

These moves by the General Post Office did not have a popular appeal. The GPO's attitude was not conducive to the rapid growth of a national telephone business, and so public opinion attempted to prevail upon the PMG to issue new licences which would be applicable throughout the realm. The Postmaster General acceded to the views presented and agreed that:

1. the restrictions on the areas within which the telephone companies could operate would be abolished;
2. the licensees would be allowed to open public call offices – but not to receive or deliver written messages;
3. the licenced companies would be permitted to erect trunk wires;
4. the new licences would terminate in 1911 without any provision for purchase or compensation in that year, though with the option to the government to purchase the plant of the licensees in 1890, or 1897, or 1904 at a price to be determined by arbitration; but
5. the royalty of 10 per cent would continue, and
6. the GPO reserved the right to compete either directly or by granting other licences, and it would not be under any obligation to grant way-leaves.

To further its operations the United Telephone Company in 1884, 1885 and 1888 asked Parliament for rights of way in streets (to lay cables) but was refused on each occasion [37]. Necessarily, the company could only install overhead wires when way-leaves had been granted.

With the restriction on the erection of trunk wires removed the various telephone companies decided that the development of a national telephone service would be eased if standardised systems and procedures were adopted, and that economies would be effected if the management of the overall system was centralised. The various companies therefore combined in 1889 as the

National Telephone Company. It fared no better than its predecessors in attempting to seek permission to lay wires underground. In 1890, 1892, 1893 and 1894 the company's applications were rejected by the London County Council, Parliament, and some Local Authorities.

Undoubtedly, the progress of telephony in the USA proceeded more rapidly than in the UK. The figures given below show the numbers of instruments and exchanges in service in America for the years ending 1877, 1880, 1883 and 1885, together with the situation in 1885 in Canada and the United Kingdom [39].

	1877	*1880*	*1883*	*1885*
No. of instruments (USA)	780	60,800	249,700	325,574
No. of exchanges (USA)		100		782
No. of instruments (Canada)				c. 18,000
(population > 4,000,000)				
No. of instruments (UK)				c. 13,000
(population > 31,000,000)				

Elsewhere, the numbers of subscribers in various European cities were: Berlin, 4,248; London, 4,193; Paris, 4,054; Stockholm, 3,825; Rome, 2,054. As W.H. Preece, the Assistant Engineer-in-Chief, GPO, said when he presented this information in a paper to the Society of Telegraph Engineers [38]: 'These figures give us clear indications that from some cause or other the development of the telephone in England has not been so brilliant or so successful as some of us could wish'. He intimated that the reasons for this state of affairs were not to do with the charges or with any restrictions in the UK – both of which, he said, were greater in America – or of the apparatus utilised. Rather, it was the difficulty of comprehending what was heard at the end of a line which had inhibited progress; 'it [was] a difficulty with the conductor, and with its environment'. Preece mentioned, 'in America [people] have frequently talked to a distance of 1,000 miles [1610 km]', but in London he had not yet succeeded 'in finding an inventor, or in finding an instrument' that would enable him to speak over the Post Office's 4 mile (6.4 km) test circuit in London because of the disturbances that manifested themselves. These were due to induction, resistance, mutual induction, earth currents, atmospheric agencies, the contiguity of telegraphic circuits and of electric light circuits, the currents used for electric tramways, and for the transmission of electric power. 'All these disturbances cause a blurring or a muffling of the sounds reproduced' [40].

Preece had a number of amusing anecdotes to relate about the vagaries of the early telephones. At a meeting of a learned society he invited a distinguished member of the audience to speak into the telephone and then to listen to the reply given by one of Preece's assistants. Seemingly at a loss, on the spur of the moment, to say anything of great profundity the distinguished member said: 'Hey diddle-diddle'. After a pause he delightedly told the audience the reply was: 'The cat and the fiddle'. Preece felt this response was out of character and later questioned his assistant about it, but was told that he (the assistant) had simply asked the member to repeat his remark [41].

Fortunately, there was a simple, though costly, solution to the problem of bad reception. This was to use a pair of twisted wires, instead of a single wire and earth return, as the transmission line. The various induction effects and those arising from contiguous wires were 'entirely eliminated'. Preece found that the practical distance to which speech could be conveyed was 100 miles (161 km) (in 1886). The invention of the twisted metallic pair has been ascribed to D. Brooks, of Philadelphia, but in a US patent interference action it was determined that A.G. Bell, as early as 1876, had suggested the method. In a patent of 1881 Bell, with customary clarity and insight wrote: 'With twisted wires the relative distance of the wires is of little or no consequence so far as obviating inductive disturbance is concerned, since by the twist the wires of each pair are brought alternately to the same position relative to the other wires' [42].

Much fundamental research work on speech and hearing was undertaken by staff at the Western Electric Laboratories prior to the incorporation of the Bell Telephone Laboratories. There had been numerous investigations, over a 50 year period, of hearing sensitivity, beginning with those of Toepler and Boltzmann in 1870 and Rayleigh in 1877 but, as Fletcher and Wegel of Western Electric showed in 1922, the results were too widely disparate to be of practical use. There had been no appreciation of the great variation in the sensitivity of hearing of people with supposedly 'normal' ears. This variation could be as much as 1,000 fold in terms of acoustical energy. One of the difficulties that the early workers faced was the lack of suitable instrumentation for their measurements. Fortunately, following the necessary development of radio communications during the First World War, valves and radio components became easily attainable from c. 1920. Electronic oscillators and amplifiers enabled reliable measurements to be obtained, and when an oscillator and amplifier were combined with a capacitor microphone, a thermal receiver and a calibrated attenuator, a new instrument, the audiometer, became available for hearing tests.

Among the inventions that rendered long distance telephony practicable, that of hard-drawn copper wire of high conductivity, by T.B. Doolittle in 1877, and its introduction in 1883, was of the greatest importance. The wire became universally used for such service. Prior to its introduction, telephone wires were, following telegraph practice, of iron or steel, sometimes galvanised and sometimes untreated. However, these lines were troubled with problems stemming from corrosion, from rust causing loose and noisy connections, as well as from high attenuation due to the high resistivity of iron and steel. It was known that copper had a resistivity about 10 per cent that of steel but the commercially available 'soft' copper wire was unsuitable for overhead lines because of the 'excessive and permanent sag' that occurred under load: in its 'hard' state the wire was brittle. Fortunately, the drawing and annealing process developed by Doolittle greatly increased the tensile strength of the copper wire [43].

In 1890 the Post Office (UK) had the option (under the 1884 licencing conditions) to purchase the plant of the telephone licensees but the Government announced in June 1890 it would not do so. The Postmaster General accordingly prepared a scheme for active competition between the telephone undertakings and the Government. Again the Government procrastinated and it was not until the end of 1891 that the new PMG obtained the Chancellor of the Exchequer's approval for what Preece had urged seven years previously, namely, the acquisition of the companies' trunk lines by the State [37].

The Telegraph Act of 1892 authorised both the purchase (subsequently £500,000) and the extension of the trunk wires from the National Telephone Company, which had a practical monopoly of such wires, and the actual transfer commenced in April 1896.

In 1895 some clarification of the position of the companies and of the Post Office with regard to certain aspects of the telephone service seemed desirable. As a consequence, in 1895, a Select Committee of the House of Commons (with the PMG as Chairman) was appointed. Its terms of reference were [37]: 'To consider and report whether the provision now made for the telephone service in local areas is adequate, and whether it is expedient to supplement or improve this provision either by the granting of licences to local authorities or otherwise'. Unanimity of counsel being unattainable the committee did not submit a report but merely presented to the House the evidence it had received.

Three years later, in 1898, another Select Committee was appointed 'to consider whether the telephone service [was] calculated to become of such general benefit as to justify its being undertaken by municipal and other local authorities, and if so under what conditions' [37]. The committee reported, on the 9th August, that the telephone service was unlikely to become of general benefit 'so long as the present practical monopoly in the hands of a private company shall continue', and observed [44]:

Our trunk system, which is in the hands of the Government and is worked on the toll system, is the most extensive in Europe. The exchange service, which is almost wholly in the hands of the company, and is chiefly confined to subscribers is much behind that of some continental countries. . . .

Within the London area, containing a population of over six million persons, there are only 237 call offices open to non-subscribers for the transmission of messages. In Stockholm there are over 700 for a population of only a quarter of a million. . . .

Competition appears to be both expedient and necessary, in order, firstly to extend and popularise the service, and next to avoid a danger which is by no means remote, if no alternative scheme is in operation, that a purchase of the company's undertakings at an inflated price may be forced upon the government of the day. . . .

In areas where the company have already an exchange, municipal competition, if permitted, should be conducted, so far as possible, on equal terms. . . .

[Finally a] general, immediate, and effective competition by either the Post Office or the local authority is necessary . . . a really efficient Post Office service affords the best means for securing such competition (Fig. 6.7.).

Two years after Bell's invention of the telephone, Bell invented an apparatus that he considered was the 'greatest invention he had ever made, greater than the telephone'. It seemed at the time to offer the prospect of telephony without wires. This invention – the photophone – and optical communications generally are considered in the next chapter.

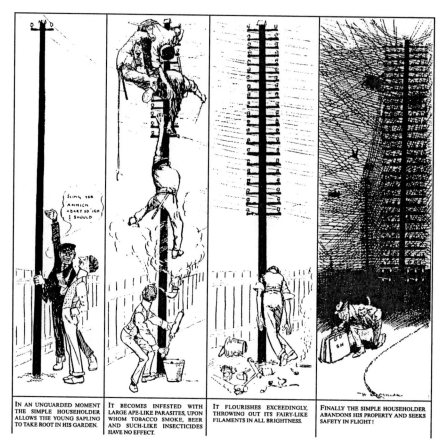

Figure 6.7   A cartoon: 'A suburban tragedy'.
Source: The Bodleian Library.

References

1   Quoted in Mottelay, P.F.: 'Bibliographical history of electricity and magnetism' (Griffin, London, 1922), pp. 142–143
2   Ref. 1, pp. 119–120
3   Ref. 1, p. 171
4   Ref. 1, p. 103
5   BOWERS, B.: 'Sir Charles Wheatstone' (HMSO, London, 1975), pp. 14–31
6   Ref. 5, pp. 7–8
7   WHEATSTONE, C.: 'On the transmission of musical sounds through solid linear conductors, and on their subsequent reciprocation', *J. Roy. Inst.*, 1831, **2**
8   PAGE, C.G.: 'The production of galvanic music', *Silliman's Journal*, 1837, **32**, p. 396. See also p. 354; and **33**, p. 118

9 URBANITSKY, A.R. von.: 'Electricity in the service of man' (Cassell, London, 1886), pp. 178–179

10 ROUTLEDGE, R.: 'Discoveries and inventions' (Routledge, London, 1903), p. 582

11 Ref. 9, p. 659–660

12 Ref. 9, p. 661

13 Ref. 9, p. 662

14 PREECE, W.H.: 'The telephone', *J. STE*, 1876, **5**, pp. 525–530

15 ANON.: 'Bell's articulating telephone', *J. STE*, 1876, **5**, pp. 519–525

16 Ref. 9, pp. 665–666

17 Ref. 10, p. 590

18 BURNS, R.W.: article on A.G. Bell, 'New Dictionary of National Biography' (Oxford University Press, Oxford, 2003)

19 HELMHOLTZ, H.: 'Die Lehre von dem Tonempfindungen' (English translation by A.J. Ellis: 'Sensations of tone') (Longmans, London, 1895)

20 BELL, A.G.: 'Researches in electric telephony', *J. STE*, 1877, **6**(20), pp. 390–391

21 Ref. 20, pp. 387–403

22 FAGEN, M.D. (ed.).: 'A history of engineering and science in the Bell System. The early years (1875–1925)' (Bell Telephone Laboratories, 1975)

23 Quoted in Ref. 22, p. 5

24 Quoted in Ref. 22, pp. 7–8

25 Ref. 22, p. 8

26 Ref. 22, pp. 9–11

27 Quoted in Ref. 22, p. 12

28 Ref. 22, p. 13

29 Ref. 20, p. 410

30 Ref. 15, p. 524

31 Quoted in Ref. 22, p. 17

32 Ref. 22, pp. 18–20

33 ANON.: 'Facts about Bell Laboratories' (Bell Laboratories, 1976)

34 BURNS, R.W.: 'The life and times of A D Blumlein' (IEE, London, 2000)

35 ANON.: 'The telephone at the Paris Opera', *Scientific American*, 31st December 1881, p. 422

36 GARCKE, E.: article on the 'Telephone', 'Encyclopaedia Britannica', 1910 edition, p. 554. Also, see BAKER, E.C.: 'Sir William Preece FRS. Victorian engineer extraordinary' (Hutchinson, London, 1976), p. 180

37 GARCKE, E.: op. cit., p. 555; and BAKER, E.C.: op. cit., p. 183

38 BAKER. E.C.: op. cit., p. 184

39 PREECE, W.: 'Long distance telephony', *J. STE*, 1886, **XV**, p. 277

40 Ref. 38, pp. 278–279

41 BAKER, E.C.: op. cit., p. 185

42 Ref. 22, p. 212

43 Ref. 22, p. 203

44 BAKER, E.C.: op. cit., pp. 192–194

Chapter 7

Optical communications

Towards the end of the 19th century some proponents of submarine cable telegraphy argued that a trans-Pacific cable from the western sea-board of Canada to the British dominions of Australia and New Zealand would have a great strategic value for the protection of the integrity of the British empire. Though the financial viability of such a cable was questioned by opponents of the scheme, the Pacific Cable Board was established to plan the cable. The PCB comprised representatives from Great Britain, Canada, Australia and New Zealand, and received financial sponsorship from these Commonwealth countries.

An essential requirement of the scheme was that all the terminal and inter-mediate telegraph stations should be located on British controlled territory. This condition could be met south of the equator where the Fiji islands and Norfolk island could be used as staging posts, but north of the Fiji islands to British Columbia all the suitable islands – the Sandwich group – were owned by the United States of America. However, there was an uninhabited atoll called Fanning island, and as no other landing place for a cable existed the British Government took formal possession of this remote place. Its location would entail a cable length of 3,458 nautical miles (6404 km) from Bamfield, British Columbia – a length much longer than that of any existing submarine cable.

Nevertheless following a commission of enquiry the Telegraph Construction & Maintenance Company of Greenwich contracted to supply and lay the cable [1]. A new cable ship that could accommodate the complete length of the Bamfield–Fanning cable was ordered, and in 1902 the *Colonia* (8,000 ton (8,130,000 kg)) was launched.

The completion of the 'All Red' route submarine cable was an outstanding achievement: it meant that Puck's desire to 'put a girdle around the earth' had been fulfilled. In just over 50 years communications between countries of the world and within individual states had been transformed: the days and weeks that had been necessary for inter-state and trans-continental message transmission in the first half of the 19th century had now been reduced to minutes or a few hours. The overall system of land-lines and submarine cables constituted the world's first electric information highway. Governments could now exercise greater strategic control and command over their possessions; businesses could trade more quickly and efficiently with their overseas suppliers and customers; and the public could become more knowledgeable about world affairs and national events.

These attributes were desirable in peacetime; during wartime they were essential – a point exemplified by Napoleon's statement, quoted in Chapter 1, 'Do not awake me when you have good news to communicate; with that there is no hurry. But when you bring bad news, rouse me instantly, for then there is not a moment to be lost'.

The electric telegraph was, of course, of very great strategic and tactical benefit during a war, and where the lines of conflict between the two opposing sides were reasonably clearly defined – as on the Western Front in the First World War – the systems of electric telegraphy were usually reliable and efficient. However this was not always the position with highly mobile engagements, or where guerilla activity was strong, or where the terrain was difficult to secure. A.P. Finley [2] in his 1888 book *Recent improvements in the art of signalling for military and commercial purposes* identified 12 problems inherent in military electric telegraphy. These were associated with (1) bad installations; (2) breaks in the telegraph lines; (3) enemy interception; (4) imperfect conditions; (5) cutting or grounding of the lines; (6) delays in the construction of lines; (7) hidden breaks difficult to find; (8) replacement by the enemy of line conductors by lengths of non-conducting line; (9) destruction of lines by ignorant members of one's own forces; (10) exposure of telegraph lines to the enemy; (11) possibility of the sounder in the receiver being masked by the noises of battle; and (12) liability of losses of material during a retreat. Furthermore the laying of the land-lines was labour intensive and costly.

Fortunately, there was one method of military signalling that could be used in the above mentioned situations which was not susceptible to these problems and conditions, namely optical telegraphy based on the use of mirrors and sunlight. The instrument that made this possible – the heliograph – was due to Mr (later Sir) Henry C. Mance (1840–1926) of the Persian Gulf Telegraph Department of the Government of India, which he had joined in 1863 [3]. He had assisted in the laying of submarine cables in the Persian Gulf and had devised the method of detecting and localising defects in such cables. In 1869 he was stationed at the Jask telegraph station in Baluchistan, and adapted the principles of the heliostat and the heliotrope to the invention of the heliograph: it was adopted by the Indian Government in 1875. Among the instrument's advantages were portability, low cost, great range, secrecy of signalling except to observers directly in the signalling path, ease of setting-up and convenience in operation in difficult terrain or enemy infested territory.

The earliest account of the utilisation of reflecting surfaces appears to be that given by Herodotus. He tells of a signal flashed in 480 BC from Athens to Marathon, by means of a burnished shield, when the Greeks were about to engage Darius's army. Xenophon mentions that a similar signal was given by Lysander before the battle of Aegospotamos, and King Demetrius of Macedonia has been said to have signalled the start of the battle at Salamis in Cyprus by displaying a polished shield [4].

Among the fantasies that have been recorded, Henry Cornelius Agrippa, (a magician of the 15th century) in his *de Van Scient* (*Vanity of Sciences*) relates of Pythagoras 'that he had a method of writing, or drawing in blood upon a speculum whatever he chose, when, throwing it upon the surface of the full moon, it became visible upon her disc, to all the persons standing behind the mirror' [5]. F. Risner in his book *Optics* (Cassel, 1606), in an article on reflected vision states:

Many persons were persuaded that, by means of reflecting specula, it was easy to discover what was doing, in the interior of houses, in the streets, or in any part of the earth illuminated by the sun. And that also by a certain artifice, painted images or written letters, might be placed in such a manner opposite the full moon, on a clear evening, that the rays being thrown upon the surface of the moon would thence be reflected back, so that you might see or read at Paris the views sent from 'Constantinople'. But he concludes, that these miraculous relations have not the demonstration of experiment [6].

Sun flashing seems to have been used by: the Moors in Algeria in the 11th century; by some Native North Americans – particularly the Dakota tribes who employed small pieces of silica and mica for their purpose; by the Russians at the siege of Sebastopol [7]; and also by Alexander the Great on his return from India, when his fleet is said to have been guided along the Persian Gulf by flashing mirrors. In these instances the signalling does not seem to have been reduced to a system of codes for general message transmission; rather the signalling appears to have been somewhat arbitrary in nature. Again no instrument was devised to facilitate the process.

Finlay, who made a comprehensive study of the origin of the heliograph, cites the use of sun-flashing during the trigonometrical survey of the British Isles [8]; and towards the close of the 18th century General Roy, when engaged in relating the meridians of London and Paris, employed this method. At night during the survey Bengal lights or Argand lamps were burned to permit the determination of the bearings of distant points but these sources gave a very limited range and were not wholly satisfactory in practice [3]. These difficulties led to Captain Drummond of the Royal Engineers inventing a lamp, which bore his name: ranges of 30 to 40 miles (48 to 64 km) were obtained. Later, in 1822, Colonel Colby, RE, designed an apparatus for signalling by flashing the sun's rays which enabled greater ranges to be achieved. Drummond improved Colby's design by the invention of the heliostat – a somewhat complex instrument (later simplified) comprising an adjustable mirror worked in conjunction with a combination of telescopes [3]. Professor Gauss, when undertaking the survey of Hanover, also employed a similar apparatus. The heliostat, as a surveying instrument, could be worked over very extended ranges. By its aid triangles having sides of more than

100 miles (160.9 km) in length were incorporated in trigonometrical surveys; for example, in England, that formed by Sca Fell in Cumberland, Slieve Donard in Ireland, and Snowdon in Wales, the sides of which are respectively 111, 108, and 102 miles (178.6, 173.7 and 164.1 km) in length [3].

The Mance heliograph is illustrated in Fig. 7.1. When in use the instrument is fixed to a tripod A, and a tangent screw E enables the mirror B, supported by two arms C, to rotate to any bearing with respect to the arm L. The inclination of the mirror from the vertical is determined by the vertical rod J and the clamping screw K. A small unsilvered spot in the centre of the mirror (that normally has a diameter of c. 5 in (c. 12.7 mm) allows the instrument to be sighted onto the distant station with the aid of the sighting vane P. In use the heliograph is aligned by turning the horizontal and vertical adjustment screws until the 'shadow spot' cast by the unsilvered centre of the mirror appears on the vane. In this position the sun's rays are reflected to the receiving station and the instrument is ready for use. Signals, based on the Morse code, are transmitted by altering the vertical angle of the mirror by the depression and release of the collar I which, with the pivoted arm U-V, acts as a telegraph key. A 'duplex

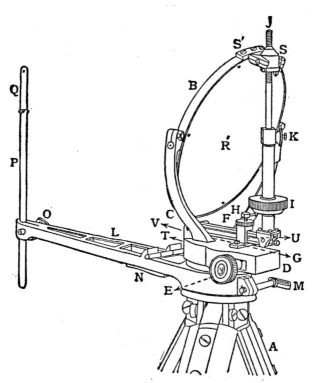

Figure 7.1   The Mance heliograph.
Source: HMSO.

mirror' is employed when the sun makes an angle of more than 120° with the mirror [9].

The first recorded use of the Mance heliograph in a tactical situation was that of the Bengal Sappers and Miners during the Jowaki Expedition in India in 1877 and 1878; and the first actual trial of the instrument was carried out during the Second Afghan War [10] (1878–1880). Subsequently, the heliograph was extensively utilised in India, and during the Waziri expedition of 1881 communication was maintained over a record distance of 70 miles (113 km) with a 5 in (12.7 cm) heliograph [9]. It was of major tactical importance in the 1879 Zulu War [10] in Africa, and was much used by both sides in the Boer war (1899–1902). Messages could be sent at c. 16 words per minute [11]. Apart from being able to communicate vital information, the heliograph also had – for some – rather strange powers. During the Afridi campaign the superstitious rebels viewed the heliograph as a mystic instrument of their sun god and became convinced the British were in alliance with their gods against them. Their belief seemed to be supported when snow, which normally fell in November, did not fall until February [10] (Fig. 7.2).

Secrecy, speed of deployment and light weight 4.46 lb (c. 2.3 kg) were three of the major advantages of the optical telegraph. Its major disadvantages stemmed from the need for good sunlight and the general restriction to day-time working, although some success was reported with a derivative instrument – the celinograph – which used the moon's rays. The celinograph had a range in India of

Figure 7.2   Signallers using a heliograph.
Source: Imperial War Museum, Q.9191.

approximately 12 miles (19.3 km) when previously set-up and tested in daylight, since the sighting and location of a distant station would have been impossible otherwise.

In his 1888 work Finley noted that the following nations had established signal corps: Algeria, Austria, Belgium, China, Denmark, France, Great Britain, Germany, Greece, Holland, Italy, Japan, Norway, Portugal, Roumania, Russia, Serbia, Spain and Switzerland [2]. Of these states heliographs were employed in the military services of Algeria, Afghanistan, Austria, Egypt, France, Great Britain, Holland, Spain, and Switzerland [12].

In the United States of America the heliograph was much used in campaigns by the US Army against the Native North Americans [13] – particularly the Apache tribe. By April 1886 abortive attempts to subdue the Apaches led to the appointment of Brigadier General N. Miles as commander of the army forces pitted against them. He was the nation's most successful Indian fighter and had gained much experience in unbeaten actions involving the Kiowas, Comanches, Sioux and Nez Perce. Miles immediately strengthened his forces by the appointment of an extra 500 Indian scouts, and organised his command into 25 small detachments which he sent out into the Arizona desert on reconnaissance missions. Additionally, he established a 'flying column' of troops, and introduced the then latest signalling means – the heliograph.

Arizona's atmosphere was clear and bright, and experiments showed that messages could be flashed through it for 50 miles (80.5 km) or more. The use of the heliograph entailed the establishment of 27 stations on mountain peaks from 25 to 30 miles (40 to 48 km) apart. So expert did the heliographers become that once they transmitted a message 800 miles (1290 km) over inaccessible mountain peaks in less than 4 hours. They handled 2,264 messages during the months from 1st May to 30th September 1886 . . .

It was the heliograph system which really was the decisive factor. Flashing all day long from mountain to mountain top, the mirrors kept [Captain H.W. Lawton of the 4th Cavalry] and the other commanders continually informed of the progress of the Apaches. Geronimo had not a moment's rest. Ceaselessly he had to keep on the move. As he shook off one pursuing detachment of soldiers or scouts, another would cut his trail. Twist and dodge as he would, he was never free [14].

The major difficulty in employing the heliograph was deciding whether a flash represented a dot or a dash. Experiments were conducted in which dots and dashes were indicated either by obscurations of the reflected sunlight, or by appearances of the reflected sunlight. The former method was less fatiguing to the eye, but learners could read signals more easily by the latter method. Experience showed that a dash should be three times the length of a dot for good signalling.

Until 1880 the use of reflected sunlight was limited to optical telegraphy, that is the transmission of messages which had been coded into a binary, or 2-state (on-off), form; but on 19th February 1880 A.G. Bell and his co-worker S. Tainter demonstrated optical telephony by an instrument which they called a photophone. Their invention was made possible by the 1873 discovery, by Willoughby Smith, Chief Engineer of the Telegraph, Construction & Maintenance Company, and his assistant May, of the photo-conductive effect of selenium. This very important discovery also initiated a great deal of interest, and subsequent activity, in 'seeing by electricity' or 'distant vision' and must be considered in some detail.

Selenium, which belongs to the sulphur and tellurium family, is a non-metallic element and was first discovered by Berzelius in 1817 in a red deposit found at the bottom of sulphuric acid chambers when pyrites containing selenium was used. Like sulphur it exists in several modifications, being obtained as a dark red amorphous powder, as a brownish black glass mass, as red monoclinic crystals or as a bluish grey, metal-like crystalline mass. In its natural state selenium is almost a non-conductor of electricity, its specific conductivity being forty thousand million times smaller than that of copper, but Knox, in 1837, found that on being annealed it became a conductor having a large resistivity compared to that of copper [15].

It was this property that led to its use in certain experiments by Willoughby Smith and to the discovery which he reported in a letter to Mr Latimer Clark, then Vice-president of the Society of Telegraph Engineers [16].

Wharf Road
4 February 1873

My Dear Latimer Clark
Being desirous of obtaining a more suitable high resistance for use at the Shore Station in connection with my system of testing and signalling during the submersion of long submarine cables, I was induced to experiment with bars of selenium – a known metal of very high resistance. I obtained several bars, varying in length from 5 cm to 10 cm, and of a diameter from 1.0 mm to 1.5 mm. Each bar was hermetically sealed in a glass tube, and a platinum wire projected from each end for the purpose of connection. The early experiments did not place selenium in a very favourable light for the purpose required, for although the resistance was all that could be desired – some of the bars giving 1400 MΩ absolute – yet there was a great discrepancy in the tests, and seldom did different operators obtain the same result. While investigating the cause of such a great differences in the resistance of the bars, it was found that the resistance altered materially according to the intensity of light to which they were subjected. When the bars were fixed in a box with a sliding cover, so as to exclude all light, their resistance was at its highest, and remained very constant, fulfilling all conditions necessary to my requirement; but immediately the cover of the box was removed the conductivity increased from 15 to 100 per cent, according to the intensity of the light falling on a bar. Merely intercepting the light by passing the hand before an ordinary gas-burner, placed several feet from the bar, increased the resistance from 15 to 20 per cent. If the light be intercepted by a glass of various colours, the resistance varies according to the amount of light passing through the glass.

To ensure that the temperature was in no way affecting the experiments, one of the bars was placed in a trough of water so that there was about an inch of water for the light to pass through, but the results were the same; and when a strong light from the ignition of a narrow band of magnesium was held about 9 in [22.9 cm] above the water the resistance immediately fell more than two-thirds, returning to its normal condition immediately the light was extinguished.

I am sorry that I shall not be able to attend the meeting of the Society of Telegraph Engineers tomorrow evening. If, however, you think this communication of sufficient interest, perhaps you will bring it before the meeting. I hope before the close of the session that I shall have an opportunity of bringing the subject more carefully before the Society in the shape of the paper, when I shall be better able to give them full particulars of the results of the experiments which we have made during the last nine months.

Yours faithfully
Willoughby Smith

Smith's letter was published in the *Journal of the Society of Telegraphic Engineers* and at the meeting at which it was read the chairman remarked that he thought it indicated a very interesting scientific discovery, and one about which it was probable they would hear a good deal in future. The chairman had himself witnessed the experiments and could confirm all that Smith had stated. Selenium's sensibility to light, the chairman remarked, 'was extraordinary, that of a mere lucifer match being sufficient to affect its conductivity' [16]. The chairman's statement on it was quite prophetic: Willoughby Smith's discovery gave rise to much attention and speculation in the scientific world at that time, particularly in the fields of picture telegraphy and distant vision (television). (See chapter 9.)

Smith forwarded a further letter [17] to the Society of Telegraph Engineers on 3rd March 1876. An extract from this letter is given below because it relates to the reason for Smith's use of selenium.

While in charge of the electrical department of the laying of the cable [from] Valentia to Hearts' Content in 1866 I introduced a new system by which ship and shore could communicate freely with each other during the laying of the cable without interfering with the necessary electrical tests. To work this system it was necessary that a resistance of about one hundred megohms should be attached to the short end of the cable. The resistance which I first employed was composed of alternate sheets of tin foil and gelatine, and, although they answered the purpose, still the resistance was not constant enough to be satisfactory. While searching for a more suitable material the high resistance of selenium was brought to my notice, but at the same time I was informed that it was doubtful whether it would answer my purpose as it was not constant in its resistance. I obtained several specimens of selenium and instructed Mr May, my Chief Assistant at our works at Greenwich, to fit up the system we adopt on shore during the laying of cables, using selenium as the high resistance, and employ the spare members of the staff as though they were on shore duty and report to me on the subject. . . . It was while these experiments were going on that it was noticed that the deflections varied according to the intensity of light falling on the selenium. . . . During the laying of the 1873 and 1874 Atlantic cables, the Lisbon and Madeira, Madeira and St Vincent, St Vincent and Pernambuco, and the Australian and New Zealand cables, I have with success adopted selenium bars protected from the action of light.

In a further communication [18] to the Society of Telegraph Engineers, dated 23rd May 1878, Willoughby Smith reported that he had heard the action of a ray of light falling upon a bar of selenium by listening to a telephone in circuit with it. 'With the assistance of the microphone one can hear the footsteps of a fly as loudly as if it were the trampling of a horse on a wooden bridge, but it strikes me as much more wonderful that by means of a telephone I can hear a ray of light falling on a metal plate.'

The notion of including a selenium bar in series with a telephone occurred to several persons in 1878, including a Mr W.D. Sargent of Philadelphia, a Mr J.F.W. of Kew, and a Mr A.C. Brown [19] of London. In a letter to *Nature*, dated 13th June 1878 J.F.W. wrote [20] that he had looked in vain for any account

of experiments with the telephone or phonoscope, inserted in the circuit of a selenium (galvanic) element. One is inclined to think that by exposing the selenium to light, the intensity of which is subject to rapid changes, sound may be produced in the phonoscope. Probably by making use of selenium, instead of the tube-transmitter with charcoal, of Professor Hughes, and by exposing it to light as above, the same result may be obtained.

This letter may have stimulated Alexander Graham Bell, for in a lecture before the Royal Institution, in 1878, he announced the possibility of hearing a shadow fall upon a piece of selenium included in a telephone circuit [21].

In September/October 1878, A.C. Brown, of the Eastern Telegraph Company submitted to Bell a confidential note giving 'the details of a most ingenious invention of his'. Bell later opined that 'the honour of having distinctly and independently formulated the conception [mentioned above], and having devised apparatus, though of a crude nature, for carrying it into execution' was due to Brown [22].

Bell and his co-worker Sumner Tainter from 1878 undertook a careful experimental investigation of photophones – a device which allowed sounds, both musical and vocal, to be transmitted to a distance by the agency of a beam of light of varying intensity and a selenium cell.

In the articulating photophone [23] (Fig. 7.3) a mirror M, reflected a beam of light through a lens, L, and, if desired for the purpose of experimentally cutting off the heat rays, through a cell, A, containing alum-water, and casting it upon the transmitter, B. This comprised a small disc of thin glass, silvered on the front, about the size of the diaphragm of a telephone and mounted in a frame with a flexible india rubber tube approximately 16 in (40.6 cm) long leading to a mouthpiece. A second lens, R, interposed in the beam of light after reflection from the transmitter rendered the rays parallel. The receiver consisted of a parabolic mirror, C, which served to concentrate the beam and to reflect it down upon a selenium cell, S, which was placed in a circuit of a battery, P, and a pair of telephones, T.

In operation the lenses were so adjusted that when the mirror, B, was flat, (i.e. not vibrating), the projected beam was focused onto the receiving instrument. However, when a speaker spoke into the transmitting apparatus the mirror disc, B, was set into vibration thereby making it alternately slightly convex and concave and caused the focus of the transmitted beam to vary. With careful adjustment of the apparatus it was hoped that the received electric current wave shape would be a replica of the sound variations at the transmitter.

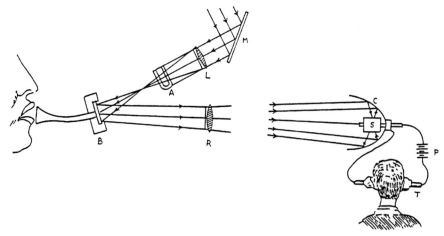

*Figure 7.3   Bell's photophone enabled speech to be transmitted by the agency of a light
beam. It was demonstrated on 19th February 1880. Bell considered the
photophone to be his greatest invention.*
Source: *Scientific American,* 1881, **44**(1).

Bell and Tainter achieved their desired objective on 19th February 1880 in
Bell's laboratory at No. 1325 L Street, Washington [24]. The inventor of the
telephone seems to have been greatly elated by his success for he wrote to his
father [25]:

Can imagination picture what the future of this invention is to be . . . We may talk by
light to any visible distance without any connecting wire. . . . In warfare the electric
communications of an army could neither be cut nor tapped. On the ocean communication may be carried on . . . between vessels . . . and lighthouses may be identified by
the sound of their lights. In general science, discoveries will be made by the photophone
that are undreamed of just now. . . . The twinkling stars may yet be recognised by the
characteristics sounds, and storms and sun spots detected in the sun.

By 26th March 1880 Bell and Tainter had transmitted audible effects over a
distance of 269 ft (82 m) and on 1st April Tainter successfully sent a message 698.7
ft (213 m) from the top of the Franklin School to Bell in the L Street laboratory:
'Mr Bell, if you hear what I say, come to the window and wave your hat' [26].
In the same year Bell and Tainter devised not only a variety of selenium
cells and a parabolic reflector but also more than 50 methods of varying a light
beam [27]. These included magnetic fields that affected polarised light, lenses
of variable focus, variable apertures and adaptations of Koenig's manometric
capsule. Some of these ideas were to be used in future distant vision equipments.
Bell and Tainter also devised a simple process for transforming selenium into the
required crystalline state in a few minutes instead of the elaborately accepted
procedure which took 40 to 60 hours [28].

As early as May 1880 Bell had considered the photophone to be sufficiently well developed for him to offer it to the National Bell Telephone Company for the sum of $2,000. William H. Forbes, the President, accepted it for the company but noted rather reservedly: 'Whether this discovery ever approaches the telephone itself in practical importance or not, it is no less remarkable and a thing which we should be glad to possess.' Forbes's caution was justified by subsequent events. The photophone could only be employed in clear air conditions and its range was small.

Nevertheless Bell continued his work and in 1893, at an exhibition held in Chicago, showed a thermophonic (i.e. an infrared) version of his basic system [29]. The transmitter, as in the photophone, comprised a thin silvered mirror diaphragm, but the receiver now consisted of a small glass bulb in which charred pieces of cork had been placed, and an ear tube. The apparatus was capable of transmitting speech over a distance of c. 328 ft (100 m).

Four years later in June 1897 Bell and H.V. Hayes (Head of the American Bell Telephone Company's Boston laboratory) discovered the speaking arc [15]. They showed that, if a battery and microphone were suitably coupled to a direct current arc, the arc would audibly reproduce voice signals and that at the same time the light intensity of the arc would vary with the speech (Fig. 7.4). This finding was applied to photophony and Hayes demonstrated the equipment over ranges up to several miles in length [30]. At an exhibition of electrical novelties held in the Madison Square Gardens, New York, in 1899 the apparatus was

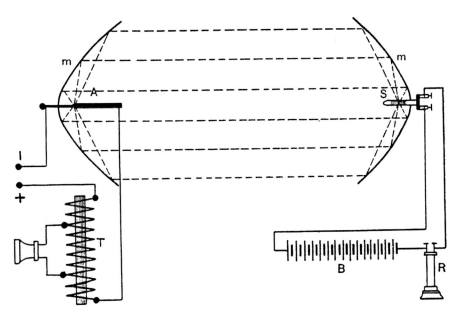

Figure 7.4 *Ruhmer's photophonic transmitter and receiver: m, mirrors; A, arc; T, transformer; B, battery; S, selenium cell; R, telephone receiver.*
Source: E. Ruhmer, *Wireless telephony* (Crosby Lockwood, London, 1908).

shown working over a 393.7 ft (120 m) optical link [31]. Nevertheless, the unavailability of a sensitive detector and the means for amplifying the very small received signals severely constrained the maximum distance over which the system could be operated.

Bell in 1898 held the photophone to be his greatest invention, and in 1921, less than a year before his death and in an age of inter-continental radio communications, he told an interviewer: 'In the importance of the principles involved I regard the photophone as the greatest invention I have ever made; greater than the telephone.'

Shortly after his death, General Carty received a letter from Mrs Bell (who was deaf) in which she wrote: 'I really believe that the reason Dr Bell did not follow up his invention of the photophone – or radiophone, as it became afterward . . . and the reason he took up aviation instead was that I could not hear what went on over the radiophone but that I could see the flying machine' [24].

In Europe work on photophony was undertaken by H.T. Simon, and by E. Ruhmer. At Nuremberg, in September 1901, Simon transmitted speech over a distance of 0.72 miles (1.2 km) using a Schuckert's searchlight with a parabolic mirror (of 35.4 in – 90 cm diameter), and a receiver comprising another parabolic mirror (of 11.8 in – 30 cm diameter) and a selenium cell. Later a range of 1.5 miles (2.5 km) was achieved in a demonstration at Gottingen [32].

The most comprehensive researches in the field of optical telephony of the European workers were those undertaken by Ruhmer. He described these in a copiously illustrated book [29] titled (in the English translation) *Wireless telephony*. Ruhmer began his work on the speaking arc in 1900 and was able in 1902 to transmit speech over a sender–receiver link of c. 4.9 miles (8 km). During the summer of that year a motor boat exhibition was held on the Wannsee, near Berlin, at which the Hagener Accumulator Company were showing their electric motor boat the *Germania* which was fitted with a Schukert torpedo boat searchlight of 13.7 in (35 cm) aperture. The boat and searchlight were placed at Ruhmer's disposal for use as a mobile transmitter, and his 19.6 in (50 cm) aperture receiver reflector was mounted on a jetty near the Wannsee Electricity Works. Many experiments were carried out in variable weather conditions but the maximum range was never greater than about 4.9 miles (8 km). Ruhmer reported: 'The transmission was in all cases good, and in [the] last experiment surprisingly loud and clear' [33].

Further investigations [34] were undertaken in Berlin in 1902 between the works of the Schukert Company, in Kopeicker Road and the parish school in the Baumschulweg approximately 1.5 miles (2.5 km) distant. Again a Schukert's searchlight 'with a very perfect parabolic mirror of 60 cm diameter was used as the transmitter, while the receiver was the same as that utilised on the Wannsee except that a larger mirror of 60 cm diameter was fitted to it'. Additional searchlights were employed to enable two-way transmissions of speech to take place. According to Ruhmer: 'The transmission was excellent here also, particularly as regards the loudness of the enunciation, which, both by day and in wet weather, was as good as could be desired'.

PROFESSOR RUHMER, of Berlin, working on the lines of Bell's photophone, has brought light telephony to considerable perfection. Bell used a plane mirror, reflecting a beam of light upon a selenium cell in circuit with a common telephone-receiver at the receiving end. The mirror, vibrating to the voice, altered the intensity of the beam playing on the selenium (which alters its electrical resistance under the action of varying light), and thus reproduced the vibrations on the diaphragm of the receiver. This, with important modifications and improvements, is Ruhmer's method. The great merit of his invention is that it is now possible to telephone without wires in the daytime. Other advantages are that messages can be sent more rapidly than at present, that replies to the messages can be sent instantaneously, and that perfect secrecy is ensured. The German Government has placed £2000 at the disposal of the inventor for further experiments.

RUHMER sought to make the Bell system of commercial value, and was aided by two important original discoveries. He found that selenium is sensitive to other than red and yellow rays, and also that by increasing the size of his mirror he could increase the distance at which the message would be audible. During his experiments on the Wannsee, Berlin, the inventor first succeeded in sending his messages a mile; afterwards he spoke to a distance of nine miles.

Figure 7.5   Talking on a ray of light: experiments in wireless telephony.
Source: The Illustrated London News Picture Library.

From these, and other, experiments Ruhmer concluded: 'The experimental results ... prove clearly that light-telephony is of practical importance, more particularly in the Navy, for which in the spring (25th to 28th May) of 1903 the author carried out some experiments on board the warships *Neptun* and *Nymphe*, using their ordinary searchlights for the purpose of light-telephony' [34] (Fig. 7.5).

Later investigations in 1904, using the electric barge *Teltow* on the Griebnitz-see (near Neu-Babelsberg) and on the Havel, finally determined the factor [35] that limited the range and hence the usefulness of optical telephony at that time.

The large Schukert searchlight, with its almost mathematically perfect parabolic glass mirror, silvered on the back, served the purpose excellently; and yet with this, even though a very small arc was used, the divergence at great distances was very considerable ...

There are no further factors which require consideration in the transmitter; *only the discovery of a new source of light with a greater specific brilliancy could lead to a further advance.* (Author's italics.)

Such a new source would not be realised until 1960 when, following the classic 1958 paper [36] by Townes and Schalow on the possibility of laser action, Maiman demonstrated the principle in a ruby crystal [37].

Ruhmer failed to increase the range of working of his system because, as he stated in his book, the diameter of the searchlight beam was as much as several hundred metres at the ranges he was using. Given a divergence of the beam of 3° the spread of the beam at 1.8 miles (3 km) is c. 164 yds (150 m). Thus, if the diameter of the light gathering device is, say, 11.8 in (30 cm), only about four millionths of the light flux from the searchlight will be incident on the device. The need for a light source – a laser – that produces a collimated beam of radiation was obviously essential for light telephony.

At this stage in the history of communications a return must be made to the topic of wire communications. Not only were telegraph lines used to transmit 'news' information, they were also, for brief periods, utilised to send 'picture' information, that is two-dimensional representations of portraits, scenes and documents. Indeed, the birth of facsimile transmission began in 1843, almost contemporaneously with the commercial exploitation of electric telegraphy.

References

1 LAWFORD, G.L., and NICHOLSON, L.R.: 'The Telecon story' (TCM Co., London, 1950), p. 84
2 FINLEY, A.P.: 'Recent improvements in the art of signalling for military and commercial purposes' (American Helio-Telegraph and Signal Light Co., Washington, DC, 1888)
3 ANON.: a report, *The Times*, 29th September 1875
4 HARLOW, A.: 'Old wires and new waves' (Arno Press, New York, 1936), pp. 9–11

5 GAMBLE, J.: 'An essay on the different modes of communication by signals' (London, 1797), p. 56

6 Ibid, p. 57

7 Ref. 4, p. 10

8 WOODS, D.L.: 'A history of tactical communication techniques' (Orlando Division, Martin Co., Martin-Marietta Corporation, Orlando, 1965), pp. 150–151

9 ANON.: article on 'Signal', 'Encyclopaedia Britannica', 1910, pp. 70–73

10 Ref. 8, pp. 152–153

11 Ibid, p. 160

12 Ibid, p. 158

13 WELLMAN, P.I.: 'The Indian Wars of the West' (MacMillan, New York, 1935), 2 vols

14 Quoted in Ref. 10, p. 154

15 BARNARD, G.P.: 'The selenium cell: its properties and applications' (Constable, London, 1930)

16 SMITH, W.: a letter to Latimer Clark, *J. STE*, 1873, **2**, pp. 31–33

17 SMITH, W.: a letter, *J. STE*, 1876, **5**

18 SMITH, W.: a letter, *J. STE*, 1878, **7**

19 BELL, A.G.: 'Upon the production and reproduction of sound by light', *J. STE*, 1880, **9**, pp. 404–426

20 J.F.W.: a letter, *Nature*, 1878, **18**, pp. 169–170

21 Ref. 19, p. 413

22 Ref. 19, p. 415

23 ANON.: 'Bell's photophone', *Sci. Am.*, 1881, **44**(1), pp. 1–2

24 BRUCE, R.V.: 'Bell' (Gollancz, London, 1973), p. 336

25 Ibid, p. 337

26 Ibid, p. 338

27 Ibid, p. 419

28 Ibid, pp. 415–418

29 RUHMER, E.: 'Wireless telephony', trans. E. Murray (Lockwood, London, 1908), pp. 10–11

30 Ref. 24, p. 342

31 Ref. 29, p. 30

32 Ref. 29, pp. 34–35

33 Ref. 29, p. 49

34 Ref. 29, pp. 49–59

35 Ref. 29. P. 64

36 TOWNES, C.H., and SCHALOW, A.L.: 'Infra-red and optical lasers', *Phys. Rev.*, 1958, **112**, 1940

37 MAIMAN, T.H.: 'Stimulated optical radiation in ruby', *Nature*, 1960, **187**, p. 493

Chapter 8

Images by wire, picture telegraphy (1843–c. 1900)

By the early 1840s electric telegraphs had been invented that could either print marks on paper tape (as in Morse's apparatus) or print the letters of a received message on a sheet of paper (as in Bain's printing telegraph). The recording of transmitted signals led Bain to ponder on the possibility of sending and receiving the signals that delineated an image drawn or printed on paper. For this purpose the two-dimensional image would have to be analysed at the sending end of the telegraph link to generate a sequence of electric pulses, the pulses would have to be transmitted, and, finally, a two-dimensional image would have to be synthesised at the receiving end from the received pulses. In addition the two instruments that carried out these processes would have to have their scanning actions synchronised.

The first proposal for transmitting facsimiles electrically from one place to another was contained in a British patent [1] dated 27th November 1843. In this Alexander Bain, a Scottish clock and instrument maker, described 'Certain improvements in producing and regulating electric currents and improvements in electric time pieces and in electric printing and signal telegraphs'. His patent was comprehensive and included seven different ideas for developments in electric telegraphy: the sixth of these related to his 'improvement for taking copies of surfaces, for instance the surface of printer's types at distant places'.

Bain was born in October 1810 at Houstry, in the parish of Watten in the county of Caithness [2]. He was one of 11 children of John Bain, a crofter, and his wife Isobella Waiter.

Alexander Bain received only a very basic education. He was employed as a herdsboy during much of a year and attended school in the winter. Some time after he left school – the date is not known – he became an indentured apprentice to John Sellar, a watch maker of Wick, but did not complete his apprenticeship.

The turning point in Bain's early life stemmed from a lecture on 'Light, heat and electricity' which he attended in January 1830 in Thurso. He must have been keen to hear the speaker for after the meeting he had to walk, in bitterly cold weather, 13 miles (21 km) back to his father's cottage and then, next morning, walk 8 miles (13 km) to Seller's shop in Wick.

In the 1830s in Caithness opportunities for advancement were minimal. Bain decided to travel to London to seek work and to avail himself of the educational facilities accessible in the capital. Arriving there in 1837 he found employment as a journeyman clockmaker in Clerkenwell and soon began to attend lectures, exhibitions and demonstrations at the Adelaide Gallery of Popular Science and at the Royal Polytechnic Institution. By 1838 Bain had begun to contemplate how a clock could be operated from an electric battery. His ideas progressed during the next two years and in 1840 he showed a model of his electric clock to one of his colleagues. It was the first electromagnetic clock ever invented. He also devised at about this time an electromagnetic printing telegraph [3]. Later, an amalgam of his general notions on electric clocks and electric printing telegraphs led to his invention of apparatus for sending black and white images from one place to another.

Bain's many diverse inventions seem to have been soundly based and eminently practicable for the period in which they were advanced. In a report [4], published in April 1844, on one of his telegraphs *The Times* noted: 'The results have proved highly satisfactory, and established the rapidity and accuracy of communication and the simplicity of the means by which it is accomplished. Mr Bain has proved himself a most ingenious and meritorious inventor of a very novel and efficacious instrument.'

In April 1850 he demonstrated his electrochemical telegraph (in which the image producing feature was comparable to that of the facsimile apparatus) in the Elysee Palace before the President of the Republic and some notable figures of the French Government. During the exhibition 'as an instance of the extraordinary powers of the telegraph' a despatch containing 1327 letters 'was conveyed between Lille and Paris in the space of a 55 seconds, being at the rate of nearly 1500 letters per minute' [2].

Highton [5] an early writer on electric telegraphy, mentioned that, around 1850, Bain's telegraph was one of the three of the most commonly used in America, coming after Morse's in general use, although in rapidity of signalling it was the fastest. And Schaffner [6] who wrote *The telegraph manual* (1859) referred to 'the many ingenious contrivances invented by Mr Bain' and said: 'He was not a commercial man but his inventive powers were most wonderful. He has given the world some invaluable inventions'.

After Bain's 1843 patent on picture telegraph apparatus was enrolled many inventors put forward various devices and systems to further the progress of this application of electrical science but, despite some attempts at commercialisation in the 1860s and the first decade of the 20th century, permanent picture telegraphy services did not commence until the 1920s. Then several schemes were introduced and facsimile transmission became a feature of modern communications.

R.H. Ranger was one of the engineers who participated in the design of the RCA system. His 1925 paper on 'Transmission and reception of photo-radiograms' includes the following acknowledgement [7] to the work of Bain:

The transmission of pictures electrically had its inception almost at the same time as straight telegraphy for in 1842 [sic] Alexander Bain, an English physicist [sic] first proposed a device to send pictures from one place to another by electric wires. His plan is so basically correct that it is only right, at the start, to show the simplicity of his plan and how, generally, we are all following in his footsteps.

Fig. 8.1, taken from Bain's 1843 patent, shows the arrangement that he submitted for sending a copy of the surface of printer's type. Essentially, the oscillatory motion of the pendulum combined with the vertical controlled motion of the metal frame caused the stylus to scan, indirectly, the surface of the type. The transmitting and receiving instruments, which were similar in construction, were synchronised by arranging that the two pendulums actuated an electric circuit so that if one preceded the other by a small amount in its swing it was held until the other had reached the same position, when both then started a new stroke. The two pendulums were thus the basic synchronisers of the system. On each swing the frame descended by a given constant amount so that the whole surface was scanned uniformly.

At the transmitter the metal frame was filled with short insulated wires, parallel to each other and at right angles to the plane of the frame, so that they made contact with the raised surface of the metal type on one side and the moving stylus, attached to the pendulum, on the other side. Consequently, as the stylus moved across the frame, an electric circuit containing the stylus, the frame and type was continually made and broken according to the arrangement of the type.

The receiving frame held two thicknesses of damp paper which had been previously saturated with a solution composed of equal parts of prussiate of potash and nitrate of soda. At the back of the paper there was a smooth metal plate that pressed the paper into contact with the ends of the parallel wires which filled the frame, as in the transmitting frame. By chemical action it was intended that the making and breaking of the current in the circuit should discolour the paper at the receiver to give a copy of the original surface.

This, then, was Bain's invention: it did not contain any radically new discovery nor even a new electrical principle, other than that of scanning, but it was based on a sensible application of the technology available at the time to the solution of a new problem. His proposals represented a natural development of the science of electric telegraphy and were made apparently realistic by the advances that had occurred previously in this field. Thus, his use of electro-chemical marking followed the practice put forward in a patent [8] in 1838 by Edward Davy for a chemical marking telegraph. In his scheme 'three wires were to be used, and the points of the metal wires were to be caused to press, by means of the motions of magnetic needles, upon chemically prepared fabric at the distant or receiving station'. The fabric to be used was calico or paper, and it was moistened with a solution of hydriodate of potass and muriate of lime.

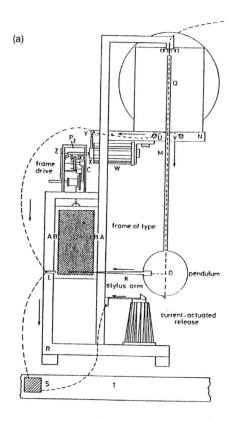

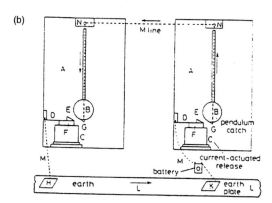

Figure 8.1   *(a) Bain's facsimile apparatus as shown in British patent no. 9745, dated 27th November 1843. The frame of type was scanned in two dimensions by the linear motion of the frame drive and the sinusoidal motion of the pendulum. (b) The transmitter and receiver of Bain's apparatus were synchronised at the ends of every swing. The use of line synchronising pulses in modern television can be traced back to Bain's notions.*

Source: British patent no. 9745, 27th November 1843.

Davy described the operation in the following way:

The motion of a needle to the right [should cause] a mark to be made on one part of the fabric, and the motion of the same needle to the left [should cause] a mark to be made on another part of the fabric; and the same for each needle attached to the respective wires. Thus the single or combined marks [can be made] to express letters, or other desired symbols.

Although Davy's patent was bought [9] by the old Electric Telegraph Company it was never utilised. It seems likely that Bain knew of Davy's idea, for much work on electric telegraphs was being carried out at that time. Also, because Bain made several applications for patents prior to 1843, he was probably aware of the patent literature on the subject.

The novel concept incorporated into Bain's invention was undoubtedly the principle of automatically scanning a two-dimensional array and transmitting, automatically, signals dependent upon some variable characteristic of the surface. Scanning is a vital requirement in all facsimile and, *mutatis mutandis*, television systems, but had not been proposed prior to Bain's 1843 patent. In their book [10] *Engineers and electrons: a century of electrical progress* J.D. Ryder and D.G. Fink state: '[Bain's] concept embodied all the geometric and timing methods of the modern television system'.

It is rather surprising that Bain did not extend his invention to include the transmission of drawings, maps and the like. This was left to F.C. Bakewell [11] to accomplish in 1848, and as a consequence some controversy took place in 1850 as to who was actually the first to suggest the facsimile transmission of handwritten letters. In a letter [12] to *The Times* dated 17th November 1850 Bain wrote:

My copying telegraph is capable of transmitting not only manuscripts written with all the characters of the autograph, but also of delineating at a distance any figure whatever which can be traced by drawing, stamping, etc. Thus a paper profile of a fugitive could, by its means, be transmitted in a few moments to all parts of the kingdom to which telegraphic wires extend.

Bain's 1850 letter to *The Times* is interesting as it shows his invention had not been put into practice because it required greater accuracy in the mechanism and more perfect insulation of the wire than had yet been realised [13]. A further factor was possibly the high cost of sending telegraphic signals. Highton gives the rates charged, in 1850 in England, for a telegraphed message of 20 words as follows [14]:

London to Birmingham	112 miles (180 km)	32.5p
London to Hull	200 miles (322 km)	47.5p
London to Glasgow	420 miles (676 km)	50.0p

Assuming the same speed of signalling for facsimile transmission as for a telegraph message and a *pro rata* increase in charges, the cost of sending a page of a letter or a diagram of similar size would have been many pounds.

The first instrument to be practically demonstrated was that constructed by F.C. Bakewell and patented [11] by him on 2nd December 1848. The patent was well presented and differed from that of Bain in a number of important points. First, the message to be transmitted was written with a non-conducting liquid, such as varnish, on tin foil, and the tin foil then wrapped around the cylinder of the transmitting equipment. On the receiver's cylinder the paper, thoroughly moistened with a solution that was readily decomposed by an electric current, was placed. Marks were produced on the paper whenever the electric circuit, which comprised the transmitting and receiving cylinders and associated apparatus, was completed. The solution preferred by Bakewell consisted of a mixture of one third part of muriatic acid, one third part water, and one third part of a saturated solution of prussiate of potash.

Both cylinders were rotated at equal, uniform speeds by means of weights and a clockwork-type mechanism, and each cylinder was traversed by a metal style which was carried in a traversing nut mounted on a lead screw. Hence, whenever the transmitter style pressed on the exposed tin foil, the circuit was closed through the moistened paper and a mark was recorded.

The preparation of the master surface was much simpler in Bakewell's apparatus than in Bain's and, furthermore, his use of rotating cylinders and associated linearly moving styles was a forerunner of many 20th century facsimile machines.

Again, whereas Bain had suggested a mechanism to check the motion of his pendulums at each swing, Bakewell made use of freely swinging pendulums. He was aware that such a simple control might not have the desired effect and utilised a reference or guide line on the cylinder of the transmitting instrument. 'By means of this "guide line" ', he wrote, 'the person in charge of the receiving instrument is enabled to regulate it exactly in accordance with the transmitting instrument by regulating the pendulum and adjusting the weight' [11].

The problem of synchronising two non-mechanically linked mechanisms was to exercise the minds of inventors in the fields of still picture transmission and television for very many years. Bakewell soon encountered this difficulty, for a newspaper report on his system noted [15]:

The chief difficulty with which he has had to contend lay in making the revolutions of the two cylinders correspond exactly, and this he has endeavoured to overcome by means of an electromagnetic regulator, which acts upon the receiving instrument and checks its motion so as to keep pace with the other. The machine is still in an experimental state, and evidently short of perfection: but sufficient success has been secured to render the practical result aimed at almost certain and it is impossible to witness the delicacy and ingenuity of the process without feelings of surprise and delight.

Neither Bain's nor Bakewell's designs were subsequently used for a regular service, but in a letter [16] to *The Times* in July 1894 Mr Armytage Bakewell observed that 40 years previously his father's copying electric telegraph had successfully transmitted autographic messages between Brighton and London.

Invisible dispatches that could be rendered legible by the recipient had also been sent by the system.

Great interest was taken in Bakewell's invention by the Prince Consort and the inventor had the honour of exhibiting his instruments and of explaining their mechanical and electrical principles to His Royal Highness at Buckingham Palace. The copying electric telegraph was later exhibited at the Great Exhibition of 1851 and received the highest award, the Council Medal.

The Abbe G. Caselli, of Florence, in the provisional specification for his 1855 patent, gave a possible explanation for the non-use of the invention [17]: 'The principle on which facsimile copies of messages may be produced through the medium of the electric current is well known; but it is the rapidity of transmission that is required to render this principle of practical value.' Later, in the complete specification of the patent, he stated another view: 'The principle barrier to success in a machine of this nature is to obtain a perfect synchronism of motion in the machine which transfers the dispatches, and that at the opposite end of the line which receives and fixes them on the paper.'

A third reason was given by T.A. Dillon in his patent [18] of 1879. Here he put forward the suggestion that the original document to be copied and transmitted should be enlarged before it was scanned and that correspondingly at the receiver the copy should be reduced. 'Because of the enlarged nature of the letters, the practical difficulties which operated against the Bain, Bakewell, Caselli and Bonetti automatic copying systems are obviated', he claimed.

Actually, the non-implementation of either the Bain or the Bakewell schemes was probably due to a combination of all three factors mentioned above, plus an overriding limitation based on economic grounds, as events were to show in the latter half of the 1860–70 decade.

Caselli's pantographic telegraph, which bore a considerable resemblance to that of Bain, was the first to be utilised, on a regular basis, anywhere in the world (Fig. 8.2).

The first notification in England of the application of Caselli's invention was contained in the foreign intelligence column [19] of *The Times* for 22nd February 1862:

A new system of telegraph has been submitted to the [French] Emperor, to which its inventor, M. Caselli has given the name of 'pantograph' [pantelegraph]. This telegraph has been already worked at Florence and Leghorn. It transmits autograph messages and drawings with all the perfections and defects of the originals. [See Fig. 8.3.] An inhabitant of Leghorn wrote four lines from Dante and they appeared in the same handwriting at Florence. A portrait of the same poet was painted at Leghorn, and was reproduced at Florence line for line and shade for shade. A bill of exchange was drawn in the same manner, and its authenticity admitted. The Emperor was much pleased at the trial made in his presence, and he proposes to establish it in France.

The first dispatch was sent from Lyons to Paris on 10th February 1862 [20]. Later, Le Corps Legislatif ordered the installation of the pantelegraph on the railway between these two cities and from 16th February 1863 the public was

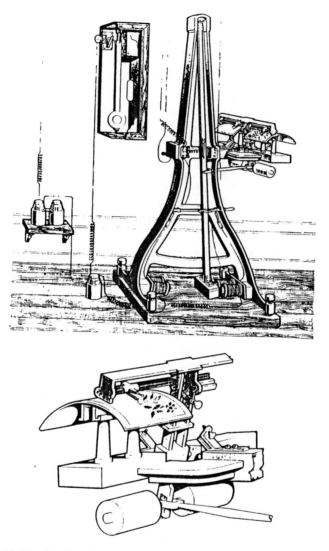

Figure 8.2   L'Abbe Caselli in 1862 demonstrated an apparatus, for transmitting facsimiles, which was based on the principles advanced in 1843 by Bain.
Source: J.M.J. Boyer, *La transmission télégraphiques des images et des photographies* (Paris, 1864).

able to forward messages. In 1867 the Director of Telegraphs, a Monsieur de Vougy, sanctioned the setting up of a second line on the Marseille to Lyons route and his department provided the necessary metallised paper. The cost was 0.20 Fr for each square centimetre of image transmitted. Unfortunately, the public did not appreciate the importance of the service and after a few years the State abandoned its enterprise.

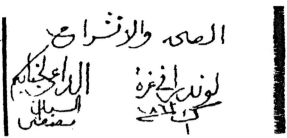

Message in Persian by Caselli's pantelegraph

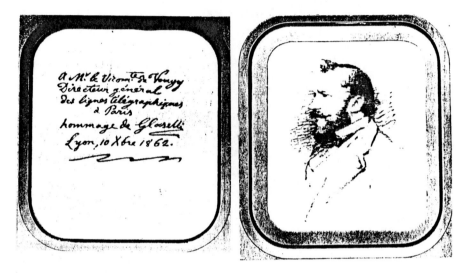

Figure 8.3   Caselli's pantelegraph was able to transmit line drawings and messages in Roman and non-Roman script.
Source: Boyer, op. cit.

A similar system [21] was employed by a French telegraph engineer named Meyer, except that he used synchronously running metal cylinders, much the same as those employed by Bakewell. Meyer's apparatus (Fig. 8.4) was put into operation between Paris and Lyons in 1869, but after a short period it, too, was taken out of service.

Another French engineer, d'Arlincourt, was not discouraged by the lack of success of the Caselli and Meyer projects for he carried out some experiments between Paris and Marseille in 1872 with a comparable scheme (Fig. 8.5). Although it was favourably commented upon at the Vienna Exhibition of 1873, it was quickly cast aside like its predecessors [22].

Nevertheless, the pioneer thoughts and work of Bain (1843), Bakewell (1848), Caselli (1855), Meyer (1869), d'Arlincourt (1872), and, later, Cowper (1879), Senlecq (1879), Edison (1881), Gray (1893) and Amstutz (1893) showed the way to a more satisfactory solution of the problem and possibly inspired others to

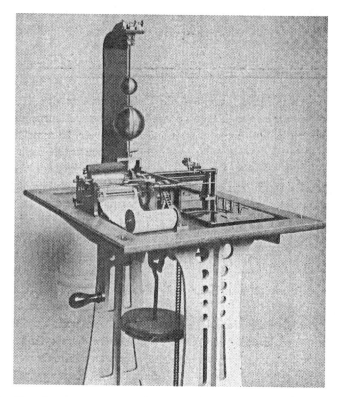

Figure 8.4   Meyer's apparatus was used commercially between Paris and Lyons in 1869.
Source: Boyer, op. cit.

make attempts at the realisation of an efficient, reliable and economic facsimile service. Their endeavours highlighted the areas where new ideas could be applied:

1   conversion of the tonal gradations of the drawing, picture, message or photograph to an electrical signal;
2   synchronisation of the transmitting and receiving instruments;
3   conversion of the received electrical signals to produce visible impressions, corresponding faithfully with those on the original document, on a sheet of material at the receiver;
4   scanning of the transmitted and received surfaces.

Apart from improvements in synchronisation techniques, further developments in picture/facsimile transmission were dependent upon the utilisation of recent discoveries which would allow intermediate tones to be transmitted. Bain's, Bakewell's and Caselli's apparatuses worked by sending pulses of current along the propagation path; the signals were essentially telegraphic rather than telephonic in character: either a signal was present or it was absent. This meant that although line drawings, diagrams and letters could be 'faxed' from one

Figure 8.5   D'Arlincourt in 1872 carried out some experiments in picture transmission between Paris and Marseilles. His apparatus was not a commercial success.
Source: Boyer, op. cit.

place to another, it was not possible to reproduce, electrically, portraits which comprised graded tones. Fortunately, the development of photography was proceeding contemporaneously with the advancement of electric telegraphy and picture transmission (Fig. 8.6).

Four years before Bain put forward his notions for taking copies of surfaces, L.J.M. Daguerre in France and W.H. Fox Talbot in England had publicised the first practical techniques for creating permanent images by the agency of light. The really important factor in their work was that they had each discovered and published a way of developing a latent image so that it became visible on paper or on a plate.

The problem that had faced artists and scientists using the camera obscura during the early years of the 19th century had been how to fix the image they had obtained by the action of light, without having to trace it on to translucent paper. Clearly, a light sensitive chemical was required that was capable of being developed and fixed. Berzelius in his *Textbook of chemistry*, published in 1808,

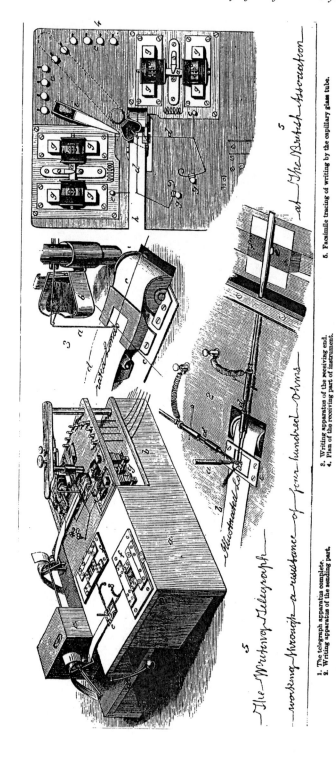

Figure 8.6 Cowper's writing telegraph.
Source: The Illustrated London News Picture Library.

had listed more than one hundred substances that had their chemical or physical structure altered by the light, and indeed the influence of light on silver nitrate had been reported by an Italian physician named Angelo Sala in 1614. Not surprisingly, several workers in Britain, Europe and America experimented with these substances in the late 18th and early 19th centuries in the hope of obtaining permanent pictures.

This activity was probably spurred on by the demand for inexpensive naturalistic pictures, particularly portraits, which existed towards the end of the 18th century. A simple way of reproducing pictures was by means of silhouettes, made by tracing the outline of the projected image of the face and filling it in with black paint. G.L. Chretien in 1786 invented a machine, the physiontrace, in which the projected image of a head was traced by a stylus, and by a pantograph arrangement an engraving tool could cut a copper plate that could be inked and printed. Aloys Senefelder invented lithography in 1798, but although it was introduced in Paris in 1802 it was not until 1813 that it became a success and a fashionable hobby.

Thomas Wedgwood, son of the famous Josiah Wedgwood, and Sir Humphrey Davy achieved some fame with their use of silver nitrate and silver chloride in 1802, and Nicéphore Niepce in 1822 triumphed in making a heliographic copy of an engraving on a glass plate coated with bitumen. The first permanent camera picture was taken by Niepce in 1826 using his asphalt process on a pewter plate. The exposure was inordinately long, about eight hours on a bright summer's day, and hence the shadow and intermediate tone effects recorded on the plate represented a distortion of the scene at a given instant of time.

Another person who was experimenting with silver salts at this time was L.J.M. Daguerre, a painter. In December 1839 Niepce and he formed a partnership. Subsequently, a full description of their invention and methods was presented at a joint meeting of the Académie des Sciences and of the Académie des Beaux Arts by Francois Arago, on 19th August 1839.

Meanwhile, Fox Talbot had been working with light sensitive substances from 1835 and had used paper coated with silver chloride. He publicised his process in 1839, after the first announcement of the daguerreotype but before the official description of it to the two Academies. Two years later he patented the calotype process which used silver iodide as a sensitive material together with silver nitrate and potassium iodide. The sensitivity of Talbot's paper was further increased by treatment with gallic acid, the sensitising properties of which had been discovered in 1837 by J.B. Reade.

Another important advance was made in 1851 when F.S. Archer introduced the wet collodion process, in which silver salts were coated on glass in a film of collodion. The plates were exposed while still wet and gave very clear glass negatives. Talbot believed that his patents covered the collodion process, but following a lawsuit in 1854 (Talbot versus Laroche) a favourable verdict was given to Laroche and the use of Archer's process was thenceforth free for general utilisation, as its inventor had intended. The British patent of the daguerreotype expired in 1853 and, as Talbot did not renew the calotype patents, all forms of photography could be used in Britain, without restrictions, by amateurs and professionals.

Thus when Caselli patented his pantographic telegraph in 1855, the art of photography was well established. And yet the transmission of photographic images by electric telegraphy was not advanced until Amstutz put forward his ideas in 1893–21. The actual development of a suitable method was made in 1907 when A. Korn of Munich and the Belgian inventor H. Carbonelle published their independent processes.

The major difficulty that faced the early inventors in this field was the lack of a photo-electric cell. Prior to the discovery [23] of the photo-conductive property of selenium in 1873, the only known relationship between the production of an electrical effect and its optical cause was the disclosure made by Edmund Becquerel in 1839 [24]. He had noted that an electric current was established in a cell containing two dissimilar liquids when exposed to light. The finding, however, was not really appropriate for the electrical transmission of images because a sensitive galvanometer was needed for the observation of the current. This meant that the obvious method of transmitting a photographic image by electrical means – the projection of a scanning beam of light through a photographic transparency on to a photoconductive cell – was not available until the arrival of the selenium cell.

A consequence of this limitation was that only photographic methods that created or were capable of creating images in relief could be considered for use in facsimile systems.

When Caselli carried out his experiments the technique for producing a photographic image in relief was well known. In 1852 Fox Talbot had perfected and patented a process for making printing blocks directly by photographic methods using a coating of potassium dichromate and gelatine on steel plates. He called this process photoglyphic engraving. A combination of Bain's apparatus and Fox Talbot's technique could possibly have led to an earlier realisation of the suggestion advanced by Amstutz in 1893. He employed the properties of a dichromated gelatin film for his method of facsimile transmission. In this, a reversed negative was imprinted upon a glass sheet which had been coated with a gelatin emulsion containing potassium dichromate so that the lightest parts of the negative (corresponding to the shadows of the original) passed more light to the gelatin, making it hard, and *vice versa*. A simple washing in water then dissolved the softer parts of the emulsion so as to leave an image in relief in which the thickness of the film at each point was dependent on the original intensity of illumination.

Amstutz applied the film to a rotating cylinder and scanned its surface with a stylus, the movements of which caused, by a suitable mechanism, the electric current to vary in the transmission line. This use of a stylus to scan a relief surface had been put forward by Edison in 1881 [25] when he presented his autographic telegraph to the Exposition Internationale d'Electricité de Paris. Edison's apparatus was probably based on his phonograph of 1877. In operation, the sender of a message wrote his dispatch with a hard pencil so as to emboss the sheet of soft paper, which was then scanned by the agency of the stylus and cylinder combination.

The progress of still picture transmission proceeded in a rather erratic manner. Following Bain's and Bakewell's pioneering ideas on the subject in the 1840–50 decade, advances in the techniques of scanning and synchronisation occurred that enabled several practical schemes to be implemented in the 1865–75 period, only to lapse into disuse after a lack of commercial success. Then, in the first few years of the 20th century, renewed interest in the subject emerged which led to equipments being developed by Ritchie (1901), Korn (1902), Carbonelle (1907), Berjenneau (1907), Belin (1907), Thorne-Baker (1907), Semat (1909) and others (Fig. 8.7). Professor Korn, in particular carried out much work in this field and this led to the successful reception in October 1907 of a photograph in the French office of the weekly paper *L'Illustration* which had been sent from Berlin (Fig. 8.8). A Paris-London service was inaugurated on 7th November 1907 when a picture of King Edward VII was sent electrically to the London office of the *Daily Mirror*. The service was extended to Manchester but was not an unqualified success for the two systems of telegraphy then in use, Morse and Baudot, gave marked induction effects on the lines and at times it was possible for experts to read messages on some of the photographs which were transmitted.

The manager of the *Daily Mirror* photo-telegraphic department during the early years of this venture was Thorne-Baker [22]. He had studied with Korn in Paris in 1907 and on his return had developed an apparatus which he called a telectrograph. In a later version of this apparatus both the transmitting and receiving instruments were essentially alike and each comprised a brass cylinder driven by a clockwork motor and a fine pitch lead screw to traverse the transmitting or recording stylus along the cylinder. The picture to be transmitted was printed photographically as a negative onto a sheet of copper foil by the gum/bichromate process, commonly employed in the production of halftone printing blocks, but instead of using a cross-ruled screen, a screen ruled with parallel lines in one direction only was employed. The object or picture was photographed by placing the lined screen in front of the copper foil. As a consequence, the highlights consisted of wide black lines with very narrow spaces between them while those parts of the picture that were to appear black in the finished image consisted of wide white lines separated by thin black spaces.

The prepared copper foil was wrapped round the transmitter cylinder so that the lines of the image were parallel with its axis and the steel point, moved longitudinally by the lead screw, traversed the rotating surface of the cylinder. An intermittent current was consequently produced in an electrical circuit as the point moved over the copper foil, the hardened gum/bichromate emulsion acting as an insulator. This current could then be passed by a land line to the receiving apparatus, or made to operate the keying system of a continuous wave transmitter.

In the receiver, the copper foil was replaced by a sheet of moist paper treated with starch and potassium iodide. The platinum receiving stylus moved

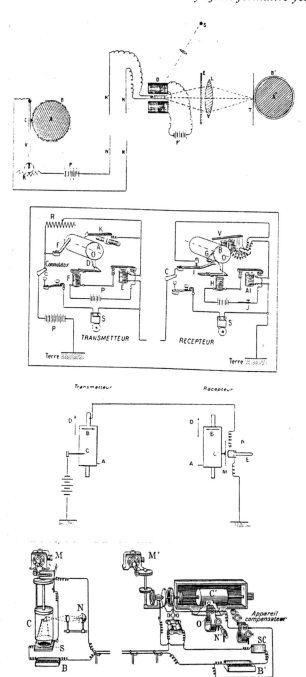

Figure 8.7   Schematic diagrams of the systems of Semat, Carbonelle, Korn and Belin.
Source: Boyer, op. cit.

*Figure 8.8   An example of an image transmitted, in 1906, by wireless using the Korn
system.*
Source: The Illustrated London News Picture Library.

synchronously with the steel transmitting stylus by the use of pendulums and,
when the current flowed through the moist paper, a blue stain was produced
by the iodide which was formed. The subsequent image changed colour to a
brownish tint as the paper dried but 'was reasonably permanent if not exposed
to bright light'. 'Very pleasing results, the novelty of which will doubtless appeal
to a considerable section of the public could certainly be produced in this
way.' The pictures measured about 5 in × 4 in (127mm x 102mm) and the time
occupied in the transmission was about six minutes [22].

   The telectrograph was first employed by the *Daily Mirror* organisation in
July 1909. Subsequently, many hundreds of photographs were sent on the Paris–
London and London–Manchester routes but, like its predecessors, the period
of utilisation of the telectroscope was short; the very high costs of hiring the
necessary telephone lines led to a termination of the service.

In 1928 the British Broadcasting Corporation transmitted wireless pictures for reception by the general public based on an adaptation of the telectrograph system. This is described in chapter 18.

Both Korn and Thorne-Baker wrote books on phototelegraphy: that of A. Korn and B. Glatzel [21], *Handbuch der phototelegraphie und telautographie* of 1911 remained the standard work on the subject for many years.

The development of facsimile transmission and the discovery of the photoconductive property of selenium led to a glut of schemes for 'distant vision' or 'seeing by electricity' (or in modern terminology television). Surprisingly, perhaps, although the first ever demonstration of rudimentary television was given in January 1926 the earliest 'schemes' for sending moving images date from c. 1879. These are discussed in the next chapter.

References

1 BAIN, A.: 'Certain improvements in producing and regulating currents and improvements in electric time pieces and in electric printing and signal telegraphs', British patent 9745, 27th November 1843

2 BURNS, R.W.: 'Alexander Bain, a most ingenious and meritorious inventor', *ESEJ*, 1993, pp. 85–93

3 BAIN, A., and WRIGHT, Lt. T.: 'Application of electricity to control railway engines and carriages, mark time, give signals, and print intelligence at distant places', British patent 9204, 21st December 1841

4 ANON.: a report, *The Times*, April 1844

5 HIGHTON, E.: 'The electric telegraph, its history and progress' (John Weale, London, 1852)

6 SCHAFFNER, T.P.: 'The telegraph manual: a complete history and description of semaphoric, electric and magnetic telegraphs of Europe, Asia, Africa and America, ancient and modern' (Pudney and Russell, New York, 1859)

7 RANGER, R.H.: 'Transmission and reception of photoradiograms', *Proc. IRE*, 1926, **14**, pp. 161–180

8 DAVY, E.: 'Telegraphs', British patent 7719, 4th July 1838

9 FAHIE, J.J.: 'A history of the electric telegraph to the year 1837' (E. and H.M. Spon, London, 1884)

10 RYDER, J.D., and FINK, D.G.: 'Engineers and electrons; a century of electrical progress' (IEEE Press, New York, 1984), chapter 9, pp. 149–171

11 BAKEWELL, F.C.: 'Electric telegraphs', British patent 12 352, 2nd June 1849

12 BAIN, A.: Letter to *The Times*, 17th November 1850

13 BURNS, R.W.: 'The electric telegraph and the development of picture telegraphy' in 'History of electrical engineering', The IEE conference publication 1988, pp. 80–84

14 Ref. 5, p. 168

15 ANON.: Report in *The Times*, 14th November 1850 (See also 19th November 1850, and 29th November 1850)

16 BAKEWELL, A.: Letter to *The Times*, 26th July 1894

17 CASELLI, G., and NEWTON, A.V.: 'Electric telegraphs', British patent 125 232, 10th November 1855

18 DILLON, T.A.: 'Transmitting messages and printed matter, etc., by electrical cables', British patent 1347, 4th April 1879

19 ANON.: Report in *The Times*, 22nd February 1862

20 BOYER, J.M.J.: 'La transmission télégraphiques des images et des photographies' (Paris, 1864)

21 KORN, A., and GLATZEL, B.: 'Handbuch der Phototelegraphie und Telautographie' (Nemnich, Leipzig, 1911)

22 THORNE-BAKER, T.: 'The telegraphic transmission of photographs', (Constable, London, 1910)

23 SMITH, W.: Letter to Latimer Clark, *J. STE*, 1873, **2**, p. 31

24 BECQUREL, E.: 'Recherches sur les effets de la radiation chimique de la lumiere solaire, au moyen des courants electrique', *Compte Rendu*, 1839, **9**, pp. 145–149

25 ISRAEL, P.: 'Edison. A life of invention' (John Wiley, New York, 1998), chapter 7, pp. 105–118

Chapter 9

Distant vision (c. 1880–1908)

The 1873 discovery, by Willoughby Smith, of the photoconductive property of selenium, is important historically, not so much for any practical value that selenium might have had in the field of optical telephony, (following Bell's and Tainter's work on photophones), but for the glut of schemes and proposals made for distant vision systems in the following years.

The earliest published accounts of schemes for seeing by electricity (whether still pictures or moving pictures) that included some details of the equipment which might be used were those of Senlecq [1] (1879), Ayrton and Perry [2] (1880), Carey [3] (1880), Sawyer [4] (1880) and Le Blanc [5] (1880). Others who contributed views or suggestions were W.L. [6] (1882), de Paiva [7] (1878), Perosino [8] (1879), Redmond [9] (1879) and Middleton [10] (1880). A. de Paiva was the first person to write a brochure on seeing by electricity: 'La telecopie electrique, basee sur l'emploi du selenium'. In addition, scanty announcements were made in the technical press of proposals by a Dr H.E. Licks [11] of Pennsylvania (1880) and by Connelly and MacTighe [12] of Pittsburg (1880).

Some of these workers claimed that they had given thought to the problem of distant vision for a number of years before they felt confident to commit their ideas in writing. Ayrton and Perry's plan was suggested to them in 1877 and more immediately by a cartoon in the magazine *Punch* (Fig. 9.1); Sawyer [4] wrote that the principles 'and even apparatus for rendering visible objects at distance through a single wire' were described in the fall of 1877 to a Mr James G. Smith, a former superintendent of the Atlantic and Pacific Telegraph Company; Senlecq claimed that his apparatus was invented in the early part of 1877 and some writers have stated that Carey's first idea was mooted in 1875 although no published evidence seems to exist to substantiate this statement.

Both the initial proposals of Carey and Senlecq were for the reproduction of still pictures rather than of moving scenes, and so not surprisingly the known techniques of picture telegraphy were adopted to some extent.

In one of his receiving instruments Carey suggested using chemically prepared paper while Senlecq's first receiver was to employ a tracing point of black

EDISON'S TELEPHONOSCOPE (TRANSMITS LIGHT AS WELL AS SOUND)

(Every evening, before going to bed, Pater- and Materfamilias set up an electric camera-obscura over their bedroom mantel-piece, and gladden their eyes with the sight of their Children at the Antipodes, and converse gaily with them through the wire.)

Paterfamilias (in Wilton Place). "BEATRICE, COME CLOSER, I WANT TO WHISPER."   Beatrice (from Ceylon). "YES, PAPA DEAR."

Paterfamilias. "WHO IS THAT CHARMING YOUNG LADY PLAYING ON CHARLIE'S SIDE?"

Beatrice. "SHE'S JUST COME OVER FROM ENGLAND, PAPA. I'LL INTRODUCE YOU TO HER AS SOON AS THE GAME'S OVER?"

Figure 9.1   Punch's Almanac for 1879 was the first magazine to illustrate, by means of a cartoon, a possible future application of 'seeing by electricity'.

Source: *Punch's Almanac*, 1879.

lead or pencil for 'drawing very finely'. In addition, his transmitter was to consist of an ordinary camera obscura containing at the focus an unpolished glass and any system of autographic telegraphic transmission' [1].

Willoughby Smith's discovery enabled an opticoelectrical transducer to be devised for a distant vision transmitter, but no progress had taken place on the design of a realistic electro-optical transducer for incorporation into a receiver. The only known chemical/physical properties that had been utilised in picture telegraphic systems at that time were the marking of chemically treated paper by an electric current and the marking of a sheet of plain paper by a pencil-electromagnet arrangement. Consequently, when the early workers appreciated that such reception methods would not suffice for moving picture reproduction, they had to consider what effects could be employed. Unfortunately, there were very few that were suitable for the purpose

The carbon arc had been spectacularly demonstrated by Sir Humphrey Davy at the Royal Institution lectures in 1808 and was later used in lighthouses and elsewhere for lighting. However, in the absence of electronic amplifiers, this form of illumination was of no importance to the inventors of distant vision in the 1870s – although many years later, in 1931, Baird demonstrated the feasibility of incorporating a modulated arc as the light source in his low definition television system.

Nevertheless, the 1870–80 decade was one of considerable activity in electrical science and on 18th December 1878 Joseph Swan showed an incandescent carbon filament lamp in operation at a meeting of the Newcastle Chemical Society. Ten months later, on 19th October 1879, Edison's lamp with a carbonised sewing thread filament was successfully exhibited. These events may have influenced Senlecq and Carey for both of these inventors advocated utilising incandescent platinum filaments in their receiving instruments. Sawyer thought that a spark produced by two platinum wires connected to an induction coil would solve the problem. The observation that a platinum wire glowed when an electric current passed through it was first noted by Staite and Moleyns in 1859 and, though this property was never subsequently the basis for a source of lighting in a television receiver, Baird did construct a honeycomb mosaic of lamps for a large screen televisor which he demonstrated to a cinema audience in 1930.

Ayrton and Perry made an original contribution in their 1880 paper to *Nature* when they put forward the notion that the then recently discovered Kerr effect [13] could feature in a 'seeing by electricity' receiver. (Interestingly, the Kerr effect would be employed in such receivers until 1936.) Their proposal for showing an image was not to vary the brightness of a lamp directly but to control its output flux indirectly by means of a variable shutter operated by the received current. In this way, a very high luminosity lamp could be utilised with a disc attached to a modified galvanometer. Here, again, developments that had occurred earlier during the advancement of electric telegraphy could be incorporated in the design of the shutter mechanism.

One of the fundamental questions that had to be answered by Ayrton, Perry,

Carey, Senlecq, Sawyer and others concerned the methods to be adopted in transmitting to the receiver the varying electrical signals from the selenium cell or cells. Two basic schemes were possible. First, in those transmitting systems based on a mosaic of small cells, each individual cell could be connected by a conductor to a separate receiving element of a corresponding receiver mosaic (Fig. 9.2). In this manner the problem of synchronisation would vanish, but the construction of the necessary transmission line would pose immense difficulties. Carey, and Ayrton and Perry, advanced such a solution. It is possible that they were influenced either by the working of the human eye and optic nerve or by the early developments that had taken place in electric telegraphy. As previously noted, during the latter half of the 18th century and the first quarter of the 19th century several inventors, C.M. (1753), Lesage (1774), Linget (1782), Reisner (1794) and Soemmering (1809) had described inventions for signalling in which the number of conductors between the sending and receiving equipments was equal to the number of letters of the alphabet and the numbers of sending and receiving elements were each made equal to this number. Ayrton and Perry's and Carey's distant vision ideas were thus analogous to these telegraphic systems. The second possibility was to employ some form of scanning procedure at the transmitter and receiver, as had been adopted in the picture telegraphy apparatuses of Bain, Bakewell and Caselli. Senlecq was the first person to propose this method and later much thought was given to scanning by Le Blanc, Nipkow, Weiller, Rosing and others.

As an illustration of the first type, Ayrton and Perry stated that their transmitter would be based on a mosaic of small, separate squares of selenium, each piece being connected to a corresponding receiver element. These elements would be shutters, of the magnetic needle type, which would be controlled by the transmitted currents so that by their movements they would open or close apertures through which light would pass to illuminate the back of small squares of frosted glass.

A more promising arrangement, said Ayrton and Perry, was suggested by Professor Kerr's experiments. Each receiver square would be made of silvered soft iron and would form the end of a core around which would be placed a coil. The surface formed from a multiplicity of these squares would be illuminated by a beam of light, polarised by reflection from glass, and the reflected beam would be viewed after having passed through an analyser. Consequently, the light flux received by the eye from each square would depend upon the rotation of the plane of polarisation of the light beam, produced by the iron-core coil unit.

Ayrton and Perry's proposal to utilise Dr Kerr's discovery of the rotation of the plane of polarised light reflected from the pole of a magnet was received rather sceptically and scornfully by a Mr J.E.H. Gordon [14], who mentioned, in a letter to *Nature*, published on 29th April 1880, some experiments that he had carried out in this field. However, Ayrton and Perry were not rebuffed by Gordon's views and reiterated their belief in the feasibility of their second plan in a further letter [15] to *Nature*: 'We still have no doubt that with a certain proper arrangement of the apparatus not only the effects observed by Dr Kerr

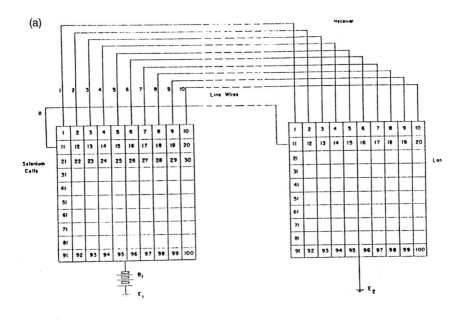

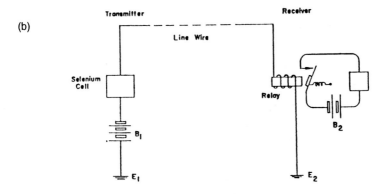

Figure 9.2   *(a) The use of a mosaic of light-sensitive cells at the transmitter, and a mosaic of light emissive elements at the receiver, eliminated the need for synchronising means between the transmitter and receiver. (b) In one form of (a), the transmitter elements could be selenium cells, and the receiver elements might be lamps operated, via relays, by the transmitted currents.*
Source: *Journal of the SMPE*, 1954, **63**.

but others of the Faraday polarisation of light effects might be practically made use of'. They then compared the advances made in a related field to lend weight to their arguments. 'For it must be remembered that the actual electric currents now used to transmit articulate speech are only one forty-millionth per cent as strong as those necessary to work even a delicate telegraph relay, whereas it required several Grove's cells to show in a decided way the old experiment of the sound emitted by an iron bar on being magnetised.'

In the same year (1880) Bell and Tainter successfully demonstrated their photophone. The history of the photophone is quite short and of no appreciable significance in the development of electrical engineering, but, at the time at which they conducted their experiments, the photophone was the first successful application of the selenium cell. The value of the photophone in the history of distant vision stems from this fact and the likely encouragement it gave to the early workers struggling to obtain an image of a distant object by electrical means. The photophone was a relatively simple device which could be constructed and developed by inventors and possibly lead to advances in the design of the all important optico-electric transducer. A number of patents for selenium cells were sealed following Bell's disclosure of the photophone, and this may also have led to the interest in distant vision shown in 1880. It is pertinent to note that Senlecq, Aryton and Perry, Middleton, Redmond, Le Blanc and others made their suggestions contemporaneously.

Since many of these suggestions were hopelessly ill-conceived and naive there was a need for some experimental evidence which would show the feasibility or otherwise of the ideas being advanced. Shelford Bidwell and Ayrton and Perry in February and March 1881, respectively, gave demonstrations of the photoconductive property of selenium as it might be applied in a distant vision system. Very sensibly, these three experimentalists reduced their apparatuses to the simplest forms possible and succeeded in showing the desired effects. At the same time, the enormity of the problems to be overcome was realistically highlighted with the probable consequence that future systems tended to be much more pragmatic than the conjectures of 1880.

Ayrton's and Perry's demonstration [16] was made using 'elements of their multi-element, multi-conductor scheme, and Shelford Bidwell restricted his practical tests [17] to showing that a selenium cell could be employed in a form of picture telegraphy apparatus [18]. This was exhibited at a meeting of the Physical Society on 26th February 1881 and was described in *Nature* on the 10th February 1881.

In a development of the latter equipment, a picture not more than 2 inches (5.08 cm) square was projected by a lens upon the side of a small rectangular box containing a selenium cell. The box was completely closed except for a small pinhole and was capable of moving up and down, through a distance of two inches, and at the same time laterally through a distance of 1/64 inch so that the pinhole passed successively over every point of the focused image. The receiver was slightly modified from Bakewell's form and enabled an image to be produced on a piece of paper which had been soaked in a solution of potassium cyanide (Fig. 9.3).

With this apparatus Shelford Bidwell transmitted 'simple designs in black and white, painted upon glass, and projected by a magic lantern. The image of a butterfly with well-defined marks upon its wings, and a rude drawing, in broad

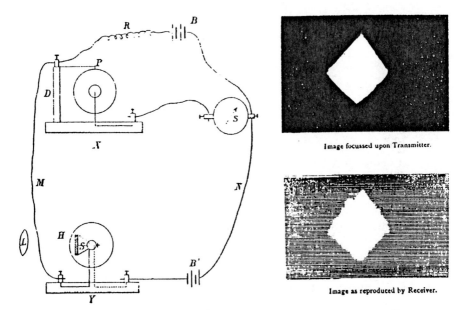

Figure 9.3   *Shelford Bidwell's transmitter, Y, and receiver, X, of 1881. The scanning cylinders were coupled to ensure synchronous operation. At the transmitter simple geometrical designs cut out of tinfoil were projected by a magic lantern onto the cylinder. A small aperture in it enabled some light flux to be incident on a selenium cell, S. The receiver's cylinder was covered with a sheet of paper soaked in a solution of potassium iodide. Simple designs could be reconstituted in a series of closely spaced lines.*

Source: *Nature*, 1881, **23**.

lines, of a human face' were among the objects which he 'most successfully' reproduced [17].

This was a start in seeing by electricity utilising selenium cells, although it was not television. Nevertheless, the equipment manifestly illustrated the limitations of such cells for this purpose; limitations that were not overcome until the mid-1920s. Bidwell wrote:

Slow rotation is essential, in order both that the decomposition may be properly effected and that the selenium may have time to change its resistance. The photophone shows that some alteration takes place almost instantaneously with a variation of the light, but for the greater part of the change a very appreciable period of time is required.

This rather ominous statement was to be confirmed again and again in the years ahead. For television systems, a minimum scanning rate of 10 frames per second was necessary whereas for picture telegraphy single frame scanning – which might take 20 minutes for a single scan – was appropriate. The extent of the difficulty facing inventors was obviously revealed by Bidwell's work.

Ayrton and Perry gave a demonstration to the Physical Society of their first idea for a distant vision probably to vindicate the proposal they had made. They desired to show the successful reproduction on a receiving screen of every change of illumination on one square of the sending screen. Their shutter was an elliptical, blackened aluminium disc (suspended in a blackened tube) of a special type of galvanometer and made an angle of 45° with the tube's axis. In this state no light transmission was possible but when the disc was deflected through 45° all the light passed through and formed an image of the square on the screen. Attached to the shutter was a small magnet, making an angle of 67.5° with it, and the two items were suspended by a silk fibre about one-twentieth of an inch in length. These particular angles were selected so that, first, all variations in intensity of illumination could be produced with a small motion of the shutter and, second, the magnet should always be in its most sensitive position in the coil through which passed the picture currents.

No accounts exist showing whether Ayrton and Perry extended their experiment to include multielement arrays of squares. However, they did suggest a method of putting, say, 30 or 40 selenium cells on a revolving arm, and hence disposing of a large number of cells, in order to transmit a complete picture which would also obviate the difficulty arising from abnormal variations of selenium. They were very much aware of the need to reduce the number of wires between the transmitting and receiving stations and mentioned that in practice a telegraph engineer would avail himself of the principles of multiplex telegraphy.

Ayrton's and Perry's experiment hardly advanced the new art: it was carried out with one element only under static conditions and did not illustrate what would happen, particularly at the receiver, under rapidly varying light states. Moreover, it failed to show the action that would result when a large number of magnetic elements, say 10,000, were in close proximity to each other at the receiver.

Of the two scientists, Perry was certainly aware of the advantages of distant vision. Later he wrote (concerning 'the people of one hundred years hence'): 'They will probably speak to one another at a distance without any artificial connection between them. . . . They will probably be able to see one another's actions at great distances, just as if they were close together.' These remarks, made before the pioneering work of Hertz on electromagnetic waves was widely known, were quite prophetic.

The first and only, so far as is known, systematic practical examination of the subject undertaken in the 19th century was that carried out by L.B. Atkinson. In 1882 he was a student at King's College, University of London, and performed his experiments under the direction of Professor Grylls Adams who during the period 1875–1877, had been engaged on a study of the properties of selenium and was acquainted with the technique of fabricating cells. Atkinson was thus well-placed to conduct an inquiry into the possible devices and ideas that had been advanced by others.

The accounts that exist of Atkinson's contribution are those that he related in a letter to the *Electrical Review* in 1889 [19] and during the discussion in 1924 of a paper by Campbell Swinton on 'The possibilities of a television'. Atkinson's apparatus was exhibited [20] before the Television Society (UK) on 5th March 1929 and is now preserved in the Science Museum, London.

Atkinson found from his experiments and the experiments of others that selenium was unsuitable as an opticoelectrical transducer 'on account of its great variation in sensitiveness with varying battery powers, and even in course of time by molecular changes, while its high resistance made the currents available very small' [19]. High resistance cells were desirable from the point of view of sensitivity but, since electronic amplifiers had not yet been invented, these small currents were required to actuate directly, or via an electro-mechanical relay, the electro-optical transducer at the receiver. Relays could be used in picture telegraphy apparatus but their speed of working was too slow for television purposes.

Atkinson also noted:

Coming to the receiver, I am aware that more sensitive or louder speaking magnetic telephones are made now [1889] than when I tried my experiments, but I was unable to produce, by means of the magnetic telephone any variation in a gas flame, with the movement of the plate, produceable by a variation in current, such as a selenium transmitter, or even a microphone (acoustic) transmitter will give off [19].

Another method he tried but 'which was quite hopeless, because it needed currents and forces which were in-applicable' [19], was rotation of the plane of polarisation of light reflected from the face of a magnet – the method suggested by Ayrton and Perry.

Atkinson did not limit his work to use of selenium cells, he also tried a transmitter comprising a mixture of carbon and sulphur [19].

In this case the luminous variation was first transformed into a heat variation, and owing to the sensitiveness of this mixture to heat, as affecting its resistance, into an electrical variation. Again, however, the variation was neither powerful enough nor rapid enough. Similar experiments with thin metal films, after the manner of Professor Langley's bolometer, were successful. Coming to the receiver, and abandoning the magnetic telephone, I tried the motograph or chalk-cylinder telephone of Edison, and in this case the gas flame movement could be obtained with the variation that could be produced in a microphone (acoustic) transmitter. . . . But here the defects of the gas flame receiver are apparent. It is not sufficiently dead-beat, the position at the gas flame depends not on the position of the telephone membrane but on the rate of change of its position. . . .

Of the other magnetic forms I tried an electromagnet with a light, stiff armature, carrying a thin plate, pressing against a slightly, convex plate, so as to give a series of Newton's rings. Viewed by monochromatic light (sodium flame), the centre spot is black or light by a variation in position of half a wavelength of yellow light. The chief defect of this form is that the illumination is scarcely intense enough to sufficiently impress the retina for the purpose in view. I also tried forms in which a small mirror was strung in a tightly stretched wire, so as to have a very small time of oscillation, the movement of the magnetic armature being highly magnified by several different devices. They did not prove, however to be insufficiently dead-beat.

Atkinson's highly interesting letter ended with two suggestions which he thought would be worthy of investigation. First, the utilisation of electrochemical transmitters and, second, the use of a Geissler tube, 'more especially a stratified tube, in the receiver'.

Atkinson, subsequent to 1882, never returned to the subject of his first researches except to mention, in 1920 in his presidential address [21] to the Institution of Electrical Engineers, the need for further research on, first, some action of light that would act instantaneously and enable a current variation to be set up and, second, some mass-less method of illuminating or darkening a surface, which could be varied by an electric current.

In 1881 the editor of the *Electrician* mused on the possibility of distant vision in a report [22] on the 1881 Electrical Exposition, Paris:

The telephotograph of Mr Shelford Bidwell even gives us the hope of being able, sooner or later, to see by telegraph, and behold our distant friends through the wire darkly, in spite of the earth's curvature and the impenetrability of matter. With a telephone in one hand and a telephote in the other an absent lover will be able to whisper sweet nothings in the ear of his betrothed, and watch the bewitching expression on her face the while, though leagues of land and sea divide their sympathetic persons. (See Fig. 9.4.)

For centuries the portrayal and the putative portrayal of illusions and images had attracted the attention of magicians, charlatans and pseudoscientists. There seemed to be a popular demand for visual displays and exhibitions of the unexpected as part of the social fabric of living. Even great writers were not immune from referring to magical illusory effects. An interesting example occurs in Sir Walter Scott's *My Aunt Margaret's mirror* which was published as one of the Waverly novels [23] in 1825. In this story the heroine, in an endeavour to

Figure 9.4　A 19th century impression of two-way television in the year 2000 AD.
Source: National Archives, Washington.

locate an unfaithful husband, consults a physician who has a reputation as a conjuror. She is led into a room containing a very tall and broad mirror, and then:

Suddenly the surface assumed a new and singular appearance. It no longer simply reflected the objects placed before it, but, as if it had self-contained scenery of its own, objects began to appear within it, at first in a disorderly, indistinct and miscellaneous manner, like a form arranging itself out of chaos; at length, in distinct and definite shape and symmetry. It was thus that, after some shifting of light and darkness over the face of the wonderful glass, a long perspective of arches and columns began to arrange itself on its sides, and a vaulted roof on the upper part of it; till, after many oscillations, the whole vision gained a fixed and stationary appearance, representing the interior of a foreign church.

An analogous scene occurs in the third act of George Bernard Shaw's play [24] *Back to Methuselah* which is set in the year 2170 AD.

The most remarkable examples of prophecy in this field were given by A. Robida in an engrossing, and prescient, book entitled *The XXth century, the conquest of the regions of the air* published in 1884. In this Robida foretells (for the year 1945) how the 'telephonoscope' will impact on people's lives [25]:

Among the sublime inventions of the XXth century and the thousand marvels of an age rich in magnificent discoveries, the telephonoscope may be considered as one of the most surprising, one of those inventions which bring the fame of scientists nearer to the stars.

The old electric telegraph, that childish application of electricity was dethroned by the telephone and then by the telephonoscope, which is the supreme and final development of the telephone. The old time telegraph allowed us to understand a correspondent at a distance, but the telephonoscope allows us both to see and hear him at the same time.

The invention of the telephonoscope was received with the greatest delight. The apparatus was attached to the instruments of all telephone subscribers who desired it, on payment of a supplementary charge. Dramatic art found in the telephonoscope an opportunity for immense prosperity. The theatrical performances transmitted by telephone became all the rage when it was also possible to see the performers as well as hear them.

Theatres had thus, besides the ordinary number of spectators in the building, a number of listeners-in and spectators in their own homes connected to the theatre with the wire of the telephonoscope. Here was a fresh source of gain and box-office receipts. No limit to profits now, 'no house full' limits to a theatre! When a show enjoys a big success, in addition to three or four thousand spectators in the theatre, fifty thousand spectators are in their homes, and fifty thousand at least in other countries of the world.

The Universal Company of Theatrical Telephonoscopy, founded in 1945, has now 600,000 subscribers scattered in all countries of the world. This company centralises the wires and pays the dues to the theatres for the reception of their performances and programmes.

The apparatus consists in a simple plate of crystal let into a wall of the apartment or placed like a mirror over any piece of furniture. The subscriber without disturbing himself, sits down before the mirror or plate, chooses his theatre; switches on the communication and enjoys the show.

The telephonoscope as the word indicates, allows us both to see and hear. Dialogue and music are transmitted as with ordinary telephone lines, but at the same time the stage, with all illumination, its decor, its actors appear on the glass screen as clear as anything seen in direct vision. The performance can be witnessed with both ears and eyes. The illusion is complete and absolute. It is like being in the box at the opera or theatre.

Of the cartoons included in Robida's book all those shown in Fig. 9.5 have their modern counterparts. Distance learning (cf. the Open University), instantaneous images of war, reports from foreign correspondents, televised opera and video conferencing are now everyday matters of fact. However, more than 40 years were to elapse before Robida's inspirations could be implemented. Many developments in pure science and technology relating to electricity, thermionic and secondary electron emission, the physics of phosphors, electromagnetic wave propagation, electronics and radio generally were necessary before the first flickering, crude televised images could be reproduced. But in one branch of television, that of analysing and synthesising an image by mechanical scanners, progress was made in the 1880s. Indeed, contemporaneously with the publication of Robida's book, a German inventor had patented a scanner which, for a few years in the late 1920s and early 1930s, was widely used in low definition television systems.

The 1880–1900 era was characterised by a new realism among the proponents of distant vision schemes. No longer, in the main, were hopelessly ill-conceived and naive systems advanced. Multi-conductor, non-scanning suggestions were abandoned, only to be raised for a brief moment by Shelford Bidwell in 1908. The inventors and scientists of this period had a clear conception of the principles of scanning. During the formative years, in the 1920s, of low definition television using mechanical scanning, the various scanners used were generally those advanced by Nipkow (1884), Weiller (1889), Brillouin (1891) and Szczepanik (1897). And the first mention of line and frame scanning dates from 1880.

Le Blanc's article [5], in *La Lumière Electrique*, (1880), contains drawings that unambiguously show two vibrating arms, each carrying a small mirror, for the execution, by line and frame scanning, of picture analysis and synthesis; although these would not have led to linear frame scanning as he stated but rather to Lissajous figure scanning. As usual with most of the early schemes no details were given by Le Blanc but the idea was adopted in 1926 by Rtcheouloff [26], who confirmed the mode of scanning.

In a letter to the *English Mechanic and World of Science* (1882), signed W.L. [27] the author (whose identity, William Lucas, was revealed in 1936) proposed another method of scanning using a 'pair of ordinary achromatic prisms, one placed vertically and the other horizontally ... each capable of turning about an axis.' Again, no details were given – 'for the simple reason that I have not, as yet, fully worked them out. The motions are somewhat involved and the fact that they must be synchronous and have a high speed still further complicates the mechanism by which they must be produced'. Several workers

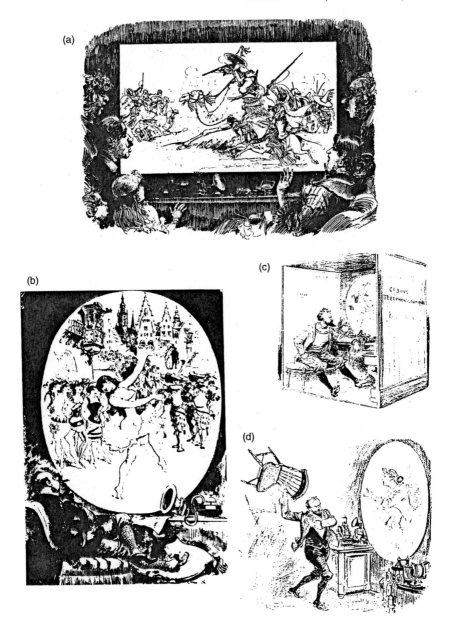

Figure 9.5   *In his book* The XXth century, the conquest of the regions of the air, *published in Paris in 1884, A. Robida correctly forecast some applications of 'seeing by electricity' (television). (a) Shows a family watching a televised report from a battlefield. (b) Illustrates a televised opera production being presented in a gentleman's drawing room. (c and d) Display aspects of two-way television.*

Source: A. Robida, *The XXth century, the conquest of the regions of the air* (Paris, 1884).

found various types of rotating prisms attractive for scanning and subsequently patents were taken out for their use by Jenkins, Zworykin, Vorobieff, Westinghouse and de Vet.

W.L.'s contribution was not confined to scanning, for he advanced a novel idea to employ a pair of Nicol prisms to modulate the light from a fixed source according to the variations of luminosity on the sending end photocell. This method depended on the fact that, if a beam of light is projected through a pair of Nicol prisms onto a screen, the intensity of the luminous spot so produced varies from a maximum to a minimum as one prism is turned through an angle of 90°. W.L. thought the angular displacement of one of the prisms could be varied by a mechanism actuated by an electromagnet which was itself to be operated by the modulated electric currents from the photocell. No record exists that such a device was ever worked in a successful demonstration of television, but the idea is interesting as it gives an indication of the ways by which inventors were trying to solve the problem of converting a variable electric current to a variable light flux. Also, W.L.'s proposal may have inspired the suggestion, first made by Nipkow two years later, to interpose a Kerr cell between two crossed Nicol prisms in order to achieve the same effect.

Of all the early scanning systems, that conceived by Paul Nipkow, who at the time was a student of sciences in Berlin, was to achieve the greatest fame. In his German patent [28] dated 6th January 1884, which A. Abramson has designated as the 'master television patent', the inventor delineated his television system. It was based on scanners that were to be employed by individual inventors and large manufacturing organisations alike, in many countries, until the 1930s. The success of Nipkow's method was undoubtedly due to its simplicity and, at a later stage of development of television, to its adoption by Baird for many of his experiments.

Figure 9.6 depicts Nipkow's scheme. The picture analysing and synthesising scanning discs were pierced by 24 apertures arranged at equal angular displacements along a spiral line. Each of the apertures had the shape of a picture element and allowed a light flux corresponding to the brightness of a picture element to be incident onto the selenium cell, S, via the condenser lens. With this mode of scanning a given aperture scanned a given line, and after one rotation of the disc the whole projected image had been explored. The rapid flyback of the scanning beam after each line and each frame was carried out automatically.

At the receiving end the image signal currents were fed to the light modulator PGA via a single line wire. The principle of operation of this modulator was founded on the magneto-optical Faraday effect [29] (discovered by Michael Faraday in 1845). The magnetic field established by the current flowing in the coil, G (situated between the crossed Nicol prisms P and A in the glass tube), rotated the plane of polarisation of the light, (emitted from the source), propagating through the slab of silica borate of lead glass, G. Consequently, the brightness of the image element seen behind the analyser, A, varied as the current varied. The complete image was synthesised by a combination of the receiver scanning disc and the phenomenon of persistence of vision.

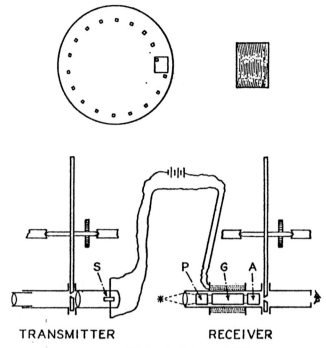

TRANSMITTER   RECEIVER

Nipkow's Patent of 1884.
S. Selenium cell.
P. Polarising prism.
G. Flint glass.
A. Analysing prism.

*Figure 9.6 Nipkow's 1884 proposal. His apertured scanning disc was used in television
systems until c. 1938.*
Source: German patent no. 30 105, 6th January 1884.

Initially, Nipkow wished to achieve synchronism of the two scanning discs by means of a clockwork mechanism, but later he recommended the use of two phonic wheels, of the type described by La Cour in 1878, for this purpose. Their speeds were to be controlled by using tuning fork oscillators.

In common with most of the early distant vision workers, Nipkow made no attempt to put his ideas into practice. At the time, 1884, their realisation was bound to fail, if only for the fact that the light modulator would have needed about 10 W of control power.

An appreciation of this truth may have induced Nipkow to propose an alternative receiver light modulating device. In October 1885, in an article [30] on 'Der Telephotograph und das elektrische Telescop', he expounded his concept of a light controlling element based on a telephone receiver. In essence the telephone acts in the inverse mode compared to the transmitter of the photophone.

Another type of light modulator was suggested to Henry Sutton [31] by the experiments [13] of Dr John Kerr. In 1875, Kerr, a mathematics lecturer of the Free Church Training College, Glasgow, outlined the experiments that he had carried out on electrostatic birefringence of optical media. Using plate glass as the medium Kerr noted the optical effect took 'a certain time to reach its full intensity'; but with liquids, such as carbon bisulphide, the 'electrostatic force and the birefringence power increase together; they also vanish simultaneously'.

Sutton, in 1890, was the first person to put forward the Kerr effect as a partial solution to the problem of transmitting optical images by the aid of telegraph wires. He also invented a new name for this subject and advocated calling it 'telephany', and the electro-optic instrument the 'telephane' [32], from the analogy between the transmission of luminous information by electricity and the transmission of acoustic information by the telephone. He clearly understood the nature of the problem, for he wrote: 'We have to take an optical image, seen on a surface, translate it into a line of consecutive varying electrical currents, and by means of these produce an effect as a surface, having the characteristics of the original image' [32]. Sutton thought his proposal offered a fair approximation to the solution of a very difficult problem. Actually, much use was made of the Kerr effect by Karolus in the 1920s.

Another system that was given much attention by later innovators was one advanced by Lazare Weiller [33,34] in 1889. His invention, according to his paper, was stimulated by the experiments of Lissajous, who had investigated the vibratory motion of bodies by optical methods. In several of these experiments Lissajous had attached small mirrors to the tines of tuning forks and had observed his now famous figures delineated on an opaque screen by reflecting a beam of light from the two mirrors in cascade.

Weiller generalised Lissajous's arrangement and proposed the use of a drum fitted with a number of tangential mirrors, each successive mirror being oriented through a small angle so that, as the drum rotated, the area of the image was scanned in a series of lines and projected onto a selenium cell. At the receiver an identical mirror drum, synchronised to the transmitter drum, reflected light from a 'telephone à gaz', to which the signal currents were applied, and created a raster on an opaque screen (Fig. 9.7). The telephone à gaz was an ingenious solution to the problem of establishing a variable light source. It combined the telephone and the capsule of Konig, and, as noted previously, was utilised by Giltay and Ruhmer in their work on photophones, although it was never incorporated into a practical distant vision receiver.

The mirror drum was to be an essential element in mechanically scanned television systems for many years. It featured in television schemes in the US, in the UK, and in Germany. By 1932 television equipment based on the Weiller drum scanner was commercially available for the home constructor, and by 1934 Fernseh AG of Germany was manufacturing mirror drums and mirror screws to a precision of ten seconds of arc in the positioning of the mirrors.

The drum was used by Rosing, Baird, EMI, Karolus and others. It was an alternative to the Nipkow disc for a considerable period even though it was not

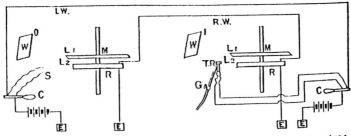

L.W., line wire ; R.W, regulator wire ; L_1, L_2, lenses ; M, mirror apparatus ;
R, regulators ; O, object ; I, image ; S, selenium transmitter ; C, commutators ; G, gas ; T.R, telephone receiver ; E, earth

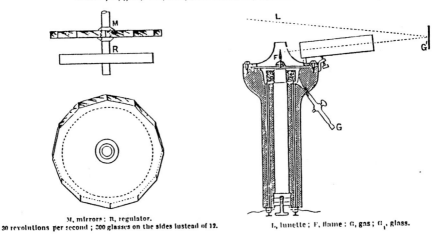

M, mirrors : R, regulator.
30 revolutions per second ; 200 glasses on the sides instead of 12.

L, lunette ; F, flame : G, gas ; G_1, glass.

Figure 9.7   *Weiller's mirror drum system, 21889. The mirror drum scanner was used by Baird, Alexanderson, Karolus and others in the 1920s and 1930s. When the number of lines per picture is less than 36, the mirror drum is more efficient than the Nipkow disc. The schematic diagram of the system, the mirror scanner, and telephone type receiver are illustrated.*
Source: *Le Génie Civil*, 1889, **XV**.

until c. 1930 that the relative efficiency of the two scanners was theoretically evaluated – the early papers on television are notably lacking in quantitative details and analysis. Nevertheless, it can be shown that when the number of lines per image is greater than 36 the aperture disc of Nipkow is more efficient than Weiller's mirror drum. For indirect scanning it can be deduced that the aperture disc and the mirror drum have practically the same efficiency when the number of lines is less than 85: for a larger number the mirror drum has an efficiency that decreases rapidly as the number increases.

Weiller made it clear in his paper that he had not fabricated his phoroscope, even though L.B. Atkinson had used mirror drums for scanning in 1882. (Atkinson's work was not published at that time and so Weiller is generally regarded as the inventor of this type of scanner.)

On Weiller's invention, in 1889, Jules Verne wrote an article [35], published in the *New York Forum*, in which he imagined a scene supposed to happen in the year 2889 AD. A machine that he called the phonotelephotograph featured in the imagery.

Francis Benett awoke that morning in a bad humour. His wife had been in France for eight days now, and he felt rather lonely. It may seem incredible, but in all the ten years of their married life, this was the first time that Mrs Edith Benett, professional beauty, had been absent so long from her husband.

As soon as he was aroused and fully awake, Francis Benett started to operate his phonotelephotograph machine, the wires of which ended in his own house in the Champs-Elysee quarter.

The telephone completed with the telephotographer another conquest of our civilisation! If the transmission of words by means of electric currents is a very old idea, it is only recently that we have been able to transmit images. A precious discovery for which Francis Benett is not the only one to bless the inventor when he sees a picture of Mrs Benett reproduced in the telephotographic mirror, despite the enormous distance between them.

A delightful sight! A little tired after the dance and the theatre, Mrs Benett is still in bed. Although it is nearly mid-day there in France, she is asleep with her pretty head sunk in the lace of the pillow.

Jules Verne's pessimism ('it is only recently', c. 2889 AD) was obviously extreme though the prospect of seeing by electricity in 1889 was not bright.

A pragmatic view of the need for visual images was given in the same year by E.H. Hall, Jr, the innovative vice-president of American Bell's long distance subsidiary, AT&T [36]. He opined that both sounds and sights were important in dialogues:

We all appreciate the advantage of speaking face to face above all other methods of communicating thought because ideas are then conveyed in three ways – by the words, by the tone and [by] the expression and the manner of the speaker. Until Bell invented the telephone only the first method was available – first by letter and later by the rapid letter – the telegram. It is perhaps fair to say that we obtain our impressions from all three of those things in equal proportions. I would not set bounds to the possibilities of inventive genius. Some day we may see as well as hear our distant friends when we communicate with them by the telephone.

Hall's prospect of the future was implemented by his company in 1930.

The last decade of the 19th century produced only a few proposals for a distant vision system. Inventors were possibly disenchanted with the subject following the complete lack of progress since the discovery of the photo-conductive effect in 1873. Many ideas had been put forward but despite the competence of some of the originators the implementation of a practical scheme seemed remote.

For a M. Brillouin the possibility of obtaining a successful solution seemed hopeless. Brillouin's paper [37] in the *Revue General des Sciences* (1891) was one of the first to contain an elementary analysis of the relation between picture definition and corresponding speed of signalling.

In carrying out his analysis Brillouin, who was the Maitre de Conferences de Physique a l'Ecole Normale Superieure, assumed that the transmitted image should be viewed at a distance of 11.8 to 15.7 in (30 to 40 cm) and that the width of each scanning line should be not more than one twentieth of a millimetre. He then correctly deduced that an image 1.6 × 1.6 in (4 cm × 4 cm) needed 640,000 dots to define it; all of which had to be illuminated in one tenth of a second. Consequently, the transmission apparatus would have to respond to changes in less than 1/ 5,000,000 of a second, in round numbers, Brillouin noted. This was too small for Brillouin, and so he modified his objective. He subsequently dealt with the problem of sending photographic images to a distant place by means of electric currents: this, of course, was the domain of picture telegraphy rather than television. His paper is important because it gives a contemporary view of the limitations of several previous notions.

Thus on Weiller's, Nipkow's and Sutton's schemes he made the following points:

1. The capsule and flame telephone (Weiller) 'cannot light up brightly enough, moreover it cannot guarantee that the illumination of the object and the flame are in proportion; it is insensitive and is difficult to regulate as a result of the small amplitude of the vibrating membrane'.
2. Of the polished membrane telephone (Nipkow), he wrote: 'it seems highly questionable that the membrane remains sufficiently plane in equilibrium not to be constantly misaligned; moreover, the changes in the curvature produced by the passage of current only result in variations of illumination between a maximum value and a minimum of more than zero; the image will always be saturated in light'.
3. The rotation of the plane of polarisation of light (Sutton)

   has in its favour an extremely fast reaction time; but it needs quite a strong electric current for any noticeable rotation, and, as with the cross[ed] Nichols, the brightness is a minimum or zero, a rotation of a given angle proportional to the intensity of the current results only in an increase in brightness proportional to the square of the intensity of the current, and as a result also proportional to the square of the illumination of the object being reproduced, which changes its character entirely.

In addition to these limitations, Brillouin felt that Nipkow's disc did not permit fineness or brilliance of reproduction and Weiller's 360 mirror cylinder was impractical to construct with precision. He did not state whether he had arrived at his conclusions by practical tests but, as he was a scientist of some standing, it is likely his comments were soundly based [38]. Indeed, they confirmed the negative results obtained a few years earlier by Atkinson.

At the turn of the century the principles that characterised the capture and reproduction of both static and moving images were well known. The science and the practice of photography were soundly based and photographs were a feature of everyday life. Cinematography had been established in the 1890s and was progressing rapidly. The Lumière Brothers had become the first persons to give a public exhibition of moving pictures – at the Grand Café in Paris on 28th December 1895, at which an admission fee had been charged – and soon afterwards, on 20th February 1896, their films had been shown at the Royal Polytechnic Institution in London. Also in 1886, the first film production unit in the world had been founded by George Méliès. His films – he produced about 4,000 – were to be considered classics of cinematography. In one of these, the famous *Journey to the moon*, trick photography was used for the first time. By the end of the century many different cameras and projectors, using a variety of film gauges, were on the market and cinemas had opened in numerous large towns in several countries.

With the growth of photography, cinematography and facsimile transmission it was to be expected that seeing by electricity would continue to attract attention. The subject had a wide appeal and during the first decade of the new century inventors in Germany, France, Belgium, Russia, the United Kingdom, Denmark, Sweden, the US and Norway advanced notions for its practical realisation (see Table 9.1). Not all the proponents of the divers schemes tried to reduce their ideas to practice and of the persons listed in Table 9.1 only Lux, Rosing, Dieckmann and Glage, Rignoux and Fournier, and Rhumer engaged in laboratory work. Their efforts have been dealt with in the author's book *Television, an international history of the formative years*.

The word 'television' was coined at the beginning of the century when Constantin Perskyi read a paper titled 'Television' at the International Electricity Congress, on 25th August 1900. Eight years later the British Patent Office introduced a new subject title – television – in its system of patent classification.

Despite the lack of really effective progress and achievement during the 35 year period following Willoughby Smith's announcement there were some people who thought television was almost at hand.

According to a telegram [39] from the Paris correspondent of *The Times*, dated 28th April 1908, the problem of distant electric vision was engaging the attention of a M. Armengaud, who 'firmly believed that within a year as a consequence of the advance already made by his apparatus, we shall be watching one another across hundreds of miles apart'.

This was too much for Shelford Bidwell and in a letter [40] to the editor of *Nature* he opined: 'It may be doubted whether those who are bold enough to attempt any such feat adequately realise the difficulties which confront them.' In particular, Shelford Bidwell emphasised the impracticability of such proposals because of the formidable nature of synchronising, by mechanical means, the receiver to the transmitter of a high definition system.

He calculated that a received image 1.9 × 1.9 in (50 mm by 50 mm) would need 150,000 elements if its definition were to be as satisfactory as that presented to the eye by a good quality photograph, and that necessarily the number of synchronised operations to be undertaken by the transmitting and receiving equipments would be 1,500,000 per second for a frame rate of ten per second. For a quality comparable with that of a coarse half-tone newspaper picture the number of elements would be about 16,000 and the video frequency would be

160,000 Hz for the same frame rate. Even this would be 'widely impracticable, apart from other hardly less serious obstacles which would be encountered', said Bidwell [40]. He mentioned the number of operations might be greatly diminished by employing an oscillating or rotating arm carrying a row of sensitive cells, but commented: 'for a coarse grained picture 50 mm square 120 of these would require 120 line wires, and [this] would also introduce a new series of troubles' [40].

Although the problem was superficially incapable of solution on the lines indicated, Shelford Bidwell felt that there was no reason beyond that of expense why vision should not be electrically extended over long distances. His plan was essentially one put forward by Carey in 1880, which was based on the structure of the eye. 'The essential condition is that every unit area of the transmitter screen should be in permanent and independent connection with the corresponding unit of the receiving screen' [40].

In this way, thought Bidwell, the difficulties due to synchronisation could be resolved without any grave complexity apart from that arising from the multiplication of components. He estimated the cost of such a scheme, assuming the transmitting and receiving stations to be 99.4 miles (160 km) apart, the received picture to be 1.9 in (50 mm) square and the length of a picture element or pixel to be 1/6 mm.

Of each of the elementary working parts – selenium cells, luminosity controlling devices, projection lenses for the receiver and conducting wires – there would be 90,000. The selenium cells would be fixed on a surface about 2.6 m square, upon which the picture would be projected by an achromatic lens (not necessarily of high quality) of 1.0 m aperture. The receiving apparatus would occupy a space of about 130 m³, and the cable connecting the stations would have a diameter of 20 cm to 25 cm.

Bidwell's estimate for such a design was £1,250,000. By an application of the three colour principle he thought it would be possible to present the picture in natural colours. The cost would then be £3,750,000.

Clearly, television was not going to be developed along the lines put forward by Bidwell and others. On the other hand, it seemed inconceivable that mechanical methods could be found which would allow 160,000 operations per second to be performed. The solution to this dilemma was first given by A.A. Campbell Swinton in a letter to *Nature* which was a comment on Bidwell's letter of 4th June 1908. He suggested the use of two cathode-ray tubes in which the beams of electrons would be synchronously deflected by the varying fields of two electromagnets, placed at right angles to each other, and energised by two alternating currents of widely differing frequencies, so that the moving extremities of the two beams would be swept synchronously over the whole of the image surfaces within the 0.1 s necessary to take advantage of visual persistence. He observed:

Indeed, so far as the receiving apparatus is concerned, the moving cathode beam has only to be arranged to impinge on a sufficiently fluorescent screen, and given suitable variations in its intensity, to obtain the desired result.

Table 9.1 Summary of the principal proposals made during the first decade of the 20th century for 'seeing by electricity'

Date	Name	Transmitter scanner	Receiver scanner	Transmission link	Synchronisation means	Opticoelectrical transducer	Electrooptical transducer	Remarks
1902	O. von Bronk Germany	mirror drum	commutator	signal wires	not stated	three selenium cells and associated red, green and blue filters	an array of Geissler tubes and associated red, green and blue filters	an early proposal for colour television
1902	M.J.H. Coblyn France	rotating cylinder having helicoidal slots	same as transmitter scanner	signal wires	few details given	selenium cell	lamp and light valve of variable shutter type based on Blondel oscillograph	not patented
1903	A. Nisco Belgium	a form of commutator – a steel blade passes over a static ebonite cylinder through the surface of which copper pins project slightly	slotted cylinder	signal wires	not stated	selenium mosaic, the wires from which connect with the transscanner	spark gap	an impractical scheme
1904	W. von Jaworski and A. Frankenstein Germany	mirror wheel with rotating sequential colour disc	same as transmitter scanner	signal wires	not stated	selenium cell	spark light source; spark gap modulated by vibrating wire connected to telephone diaphragm	use of rotating discs, having sequential red, blue and green filters, at transmitter and receiver for sequential colour television
1906	G.P.E. Rignoux France	two vibrating mirrors, scanning orthogonally, mounted on two tuning forks (500 Hz and 10 Hz)	same as transmitter scanner	two wires for the signal and four wires for synchronisation	receiver scanners connected directly to transmitter scanner	selenium cell	modulated arc	
1906	F. Lux Germany	non-simultaneous system	non-simultaneous system	multiwire	not necessary – simultaneous transmission	array of selenium cells each controlling the amplitude of an oscillator	array of vibrating resonant reed light shutters and light source	the n transmitter signals at $f_1, f_2, \ldots$ are received and applied to the array of light shutters each tuned to one of the frequencies $f_1, f_2, \ldots$
1907	N. Rosing Russia	two mirror drums scanning in orthogonal directions	cold cathode-ray tube, electron beam magnetically deflected	two signal wires and synchronising wires	deflection coils of receiver cathode-ray tube fed by currents generated at the transmitter	selenium cell, or photoelectric cell, or photovoltaic cell	electron beam, of c.r.t. modulated by video signals determines spot brightness	
1908	J. Adamian Germany	apertured disc	apertured disc	single wire	not stated	not stated	two Geissler tubes adapted to emit differently coloured lights	objective: to produce a picture having, to a certain extent, some natural colouring

Date	Inventor/Country	Transmitter scanner	Receiver scanner	Transmission medium	Synchronising	Photoelectric device	Light modulator/source	Remarks
1908	S. Bidwell, United Kingdom	non-simultaneous system	non-simultaneous system	multiconductor, 90,000 wires – diameter of cable 8" to 10"	not necessary – simultaneous transmission	array of 90,000 selenium cells, c. 8' square	array of 90,000 light valves	estimated cost: £1,250,000 (in 1908); size of receiver c. 4000 ft
1908	A.C. and L.S. Andersen, Denmark	apertured band or disc	apertured band or disc	single wire	electrically actuated clockwork	selenium cell with prism and rotating colour filters	lamp with moveable graded density filter and rotating colour filters	objective: use of rotating sequential filters at transmitter and receiver to enable colours of object to be reproduced
1906	M. Dieckmann and G. Glage, Germany	a rotating disc having 20 brushes arranged in a single spiral	electron beam in a cathode-ray tube	signal wires and synchronising wires	varying voltages from linear potentiometers	none; the 20 wire brushes of the transmitter contacted the object	fluorescent screen of receiver c.r.t.	not a television system – essentially a picture telegraphy system
1909	E. Ruhmer, Germany	none – simultaneous system	none – simultaneous system	multiconductor	not necessary – simultaneous transmission	5 × 5 array of selenium cells each connected to a sensitive relay; 25 signals transmitted; no half tones	5 × 5 array of light modulators (sensitive galvanometers) plus light source	system demonstrated on 26 June 1909: Ruhmer planned to construct a 10,000 element array (at a cost of Fr 6M) for the Brussels International Exposition
1909	G.E.P. Rignoux and A. Fournier, France	none – simultaneous system	none – simultaneous system	multiconductor	not necessary – simultaneous transmission	array of selenium cells (n)	n small galvanometers each with miniscule shutter	system demonstrated in 1911; no half tones reproduced
1910	G.E.P. Rignoux and A. Fournier, France	(i) array of n relays and commutator (ii) mirror drum	mirror drum	two wires	synchronous motors at transmitter and receiver	array of selenium cells (n)	lamp and light modulator based on Faraday effect	simple black and white figures reproduced; no half tones
1910	A. Ekstrom, Sweden	spiral scanning mirror	spiral scanning mirror	signal wires	not stated	single selenium cell	light source and light valve of Brillouin type	use of indirect scanning
1910	M. Schmierer, Germany	two commutators acting on row and column elements of array so that only one element at a given instant is connected in circuit	same as transmitter scanner	signal wires	not necessary – simultaneous transmission	array of light sensitive elements	array of light producing elements	an n-line television system would require n^2 elements for a 1:1 aspect ratio
1910	G.H. Hoglund, US	a pair of apertured discs: the discs rotate in opposite directions	same as transmitter scanner	signal wires	synchronous motors at transmitter and receiver	selenium cell	'speaking' arc	one disc has arcuate slots; the other has a 'plurality of slots extending in a stepped line outwardly from the centre of the disc'
1910	A. Sinding-Larsen, Norway	two vibrating mirrors mounted on two tuning forks (10 kHz and 100 Hz)	same as transmitter scanner	narrow tube with strongly reflective inner surface	transmitter and receiver driving elements coupled in series	not necessary since the light is transmitted directly from the transmitter	not necessary – the light is received directly from the transmitter	use of optic tube for tube for transmission (cf. modern optic fibre)

The real difficulties lie in devising an efficient transmitter which, under the influence of light and shade, shall sufficiently vary the transmitted electric current so as to produce the necessary alterations in the intensity of the cathode beam of the receiver, and further in making this transmitter sufficiently rapid in its action to respond to the 160,000 variations per second that are necessary as a minimum.

Possibly no photoelectric phenomenon at present known will provide what is required in this respect, but should something suitable be discovered, distant electric vision will, I think, come within the region of possibility.

Fortunately for the advancement of television important discoveries had been made by 1908 in the fields of photo-electric emission, cathode-ray tubes, the propagation and application of electromagnetic waves, and thermionic emission: and J.A. Fleming and L. de Forest had invented the diode and triode valves in 1904 and 1906 respectively. These discoveries and inventions would be greatly progressed during the next decade and lead to the first rudimentary demonstration of television a few years later. This work is considered in chapter 15.

Contemporaneously with the developments described in the present chapter, fundamental theoretical and experimental researches by Maxwell, Hertz and Lodge would greatly advance the physics of electromagnetic wave propagation and lead via Marconi, Jackson, de Forest, Fessenden and others to a transformation of the science and technology of communications. The work of the early radio physicists is the subject of the next chapter.

References

1 SENLECQ, C.: 'The telectroscope', *Les Mondes*, 1879, **48**, pp. 90–91. Also, *English Mechanic and World of Science*, 1879, (723), p. 509

2 AYRTON, W.E., and PERRY, J.: 'Seeing by electricity', *Nature*, 1880, p. 589

3 CAREY, G.R.: 'Seeing by electricity', *Sci. Am.*, 1880, **42**, p. 355. See also reports on 'The telectroscope', *Sci. Am.*, 1879, **40**; and 'Carey's diaphote', *English Mechanic and World of Science*, 1880

4 SAWYER, W.E.: 'Seeing by electricity', *Sci. Am.*, 1880, **42**, p. 373

5 LE BLANC, M.: 'Etude sur la transmission électrique des impressions lumineuses', *La Lumiere Electrique*, 1880, **11**, pp. 477–481

6 W.L.: 'The telectroscope, or seeing by electricity', *English Mechanic and World of Science*, 1882, **35**(891), pp. 151–152. Also, letter to *Nature*, 1936, **137**, p. 1076

7 PAIVA, A. de.: 'La telescopie électrique, basee sur l'emploi du selenium' (A.J. da Silva, Porto, 1878)

8 PERSINO, C.M.: 'Su d'un telegrafo ad un solo filo', *Atti Acad. Sci. di Torino I, cl. Sci. Fis. Math. Nat.*, 1879, **14**, p. 4a

9 REDMOND, D.D.: 'Seeing by electricity', a letter (15374), *English Mechanic and World of Science*, 1879, p. 540. See also BOLTON, H.E.: 'Seeing by electricity', a letter (15429), 1880, p. 235; GLEW, F.H.: 'Electric telescope', letter (15429), 1879, p. 586; MORSHEAD, W.: 'Electric telescope', letter (15430), 1879, p. 586

10 MIDDLETON, H.: 'Seeing by telegraph' a letter to the editor, *The Times*, 24th April 1880. See also 'Seeing by electricity', *English Mechanic and World of Science*, 1880, **31**, pp. 177–178

11 Reports in Design and Work, 1880, p. 283 and p. 437; and *The Times*, 24th April 1880

12 Reports in *La Lumiere Electrique*, 1880, **2**, p. 140

13 KERR, J.: 'A new relation between electricity and light: dielectrified media birefringent', *Phil. Mag.*, 1876, S. 4, **50**(332), pp. 337–348 and pp. 446–458

14 GORDON, J.E.H.: 'Seeing by electricity', a letter, *Nature*, 1880, p. 610

15 AYRTON, W.E., and PERRY, J.: 'Seeing by electricity', a letter, *Nature*, 1880, p. 31

16 Report on 'Seeing by electricity', *Nature*, 1881, **23**, pp. 423–424

17 BIDWELL, S.: 'Tele-photography', *Nature*, 1881, **23**, pp. 344–346

18 BIDWELL, S.: 'On telegraphic photography', Report of the British Association for 1881, transactions of Section G, pp. 777–778

19 ATKINSON, L.B.: 'Seeing to a distance by electricity', *The Telegraphic Journal and Electrical Review*, 1889, p. 683

20 Report on 'Television. A brief account of recent events', *Electr. Rev.*, 1929, p. 494

21 ATKINSON, L.B.: 'Inaugural address', *J. Inst. Electr. Eng.*, 1920, **59**, pp. 15–16

22 Editorial: 'The electrical exhibition at Paris', *Electrician*, 1881, pp. 40–41

23 SCOTT, W.: 'My Aunt Margaret's Mirror' (Black, Edinburgh, 1871), pp. 378–379

24 SHAW, G.B.: 'Back to Methuselah. A metabiological pentateuch' (Constable and Company, London, 1924)

25 ROBIDA, A.: 'The XXth century, the conquest of the regions of the air' (Paris, 1884)

26 RTCHEOULOFF, B.: 'Improvements in and relating to television and telephotography', British patent 287 643, 24th December 1926

27 W.L.: 'The telectroscope, or seeing by electricity', a letter (19941), *English Mechanic and World of Science*, 1882

28 NIPKOW, P.: 'Elektrische teleskop', German patent 30 105, 6th January 1884

29 FARADAY, M.: 'On the magnetisation of light', *Phil. Trans.*, 1846, p. 1

30 NIPKOW, P.: 'Der telephotograph und das elektrische teleskop', *Elektrotechische Zeitschrift*, 1885, pp. 419–425

31 SUTTON, H.: 'Telephotography', *The Telegraphic Journal and Electrical Review*, 1890, **27**, pp. 549–551

32 E.R.: 'Le probleme de la telephanie, d'apres M. Henri Sutton', *La Lumiere Electrique*, 1890, **38**, pp. 538–541

33 WEILLER, L.: 'Sur la vision a distance par l'electricite', *Le Genie Civil*, 1889, **XV**, pp. 570–573

34 H.W.: 'Sur la vision a distance par L Weiller', *La Lumiere Electrique*, 1889, **34**, pp. 334–336

35 See also: VERNE, J.: 'Le chateau des Carpathes' (Hamilton, London, 1963, Evans, I.O. (Ed.))

36 HALL, E.H.: Letter, *Electrical Review*, 1889, p. 9

37 BRILLOUIN, M.: 'La photographie des objets a tres grande distance par l'intermediaire du courant electrique', *Revue Generale des Sciences*, 1891, (2), pp. 33–38

38 BLONDIN, J.: 'Le telephote', *La Lumiere Electrique*, 1893, **43**, pp. 259–266

39 Paris correspondent: a telegram to *The Times*, 28th April 1908

40 BIDWELL, S.: 'Telegraphic photography and electric vision', a letter, *Nature*, 1908, pp. 105–106

Chapter 10

The early wireless pioneers

Every so often someone revolutionises a field of endeavour and thereby alters, irrevocably, our perception of the world. This requires a rare combination of insight, perseverance, talent and something more – a strange, magical touch, a kind of integrating faculty which transcends mere reason and defies analysis. The person who has this power and successfully applies it, we call a genius. Newton and Einstein were such men; and so too was Maxwell [1].

Maxwell was a brilliant polymath. After his death in 1879 his scientific papers were collected and edited by W.D. Niven, and published in two volumes by Cambridge University Press [2]. Among his papers, the following give some indication of the breadth of Maxwell's interests: 'Experiments on colour as perceived by the eye'; 'On the stability of motion of Saturns's rings'; 'On the dynamical theory of gases'; and 'On the dynamical theory of the electro-magnetic field'.

Maxwell is probably best known for his researches on electricity and magnetism which culminated in his great 'Treatise on electricity and magnetism', published in 1873 [3]. On this work Max Planck, who was awarded the 1918 Nobel prize in physics for his formulation of the quantum theory, wrote [4]:

it was his [Maxwell's] task to build and complete the classical theory of electromagnetism, and in doing so he achieved greatness unequalled. His name stands magnificently over the portal of classical physics, and we can say this of him: by his birth James Clerk Maxwell belongs to Edinburgh, by his personality he belongs to Cambridge, by his work he belongs to the whole world.

James Clerk Maxwell was born in Edinburgh on 13th June 1831, a few months before Faraday made his famed discovery of electromagnetic induction. Unlike Faraday, whose father was a Yorkshire blacksmith, Maxwell's father, John Clerk, an Edinburgh lawyer, came of landed gentry, with its lines extending back to the time of Mary, Queen of Scots. John Clerk adopted the surname Maxwell on inheriting the Galloway estate of Middlebie, which had passed into the family by a marriage with a Miss Maxwell. James's mother was Frances, the daughter of Robert Cay and his wife, of Charlton, Northumberland [5].

James's early childhood was spent in his father's country house of Glenlair, near Dalbeattie, but in 1841 James became a pupil at the Edinburgh Academy. He entered the University of Edinburgh in 1847 and attended classes on mathematics, natural philosophy (physics), chemistry and mental philosophy. Previously, Maxwell had shown some precociousness when at the age of 15 he had communicated a paper 'On the description of oval curves' to the Royal Society of Edinburgh. Further papers were submitted in 1849 and 1850. In this year he left Edinburgh for Cambridge, becoming an undergraduate at Peterhouse College; and then, in December of that year, an undergraduate at Trinity College. Maxwell graduated in 1854 as second wrangler, was elected a Fellow of Trinity the following year, and was appointed to the lecturing staff.

He returned to Scotland in 1856 following his acceptance of the position of Professor of natural philosophy in Marischal College, Aberdeen. Two years later he married Katherine Mary Dewar, daughter of the principal of the college. When, in 1860 this college amalgamated with King's College to form the University of Aberdeen, Maxwell vacated his chair and soon afterwards became Professor of natural philosophy in King's College, London. He resigned in 1865 and retired to private life at Glenlair but, in 1871, was persuaded to apply for the new chair of experimental physics, University of Cambridge. He was elected without opposition.

Maxwell's genius for original work was manifested in 1855 when his paper [6] on 'Faraday's lines of force' was read before the Cambridge Philosophical Society. Previously he had read Faraday's 'Experimental researches', and had entered into correspondence with Faraday who had conceived the notion of fields and lines of force, and had suggested that electric and magnetic forces did not act directly on distant bodies but rather were mediated by the conditions – the electric and magnetic fields – that prevailed in the intervening space. Another correspondent was Professor Sir William Thomson – later Lord Kelvin [7].

In 1845 Thomson, a theoretical physicist, wishing to acquire some experience in experimental physics, but being unable to do so in Great Britain, spent four months in Paris at H.V. Regnault's laboratory. While there, his interest in the theory of electricity was encouraged by J. Liouville [8]. Thomson set out to reconcile the apparent contradiction between the ideas of Faraday and those of Poisson and Coulomb. This work led to an important series of papers in which Thomson derived mathematical expressions for many of Faraday's ideas, and established the foundations for a mathematical theory of electricity. The complete formulation of a coherent electromagnetic theory, however, he left to Maxwell. He set himself the task of not attempting 'to establish any physical theory of a science in which [he had] hardly a single experiment, but to show how by a strict application of the ideas and methods of Faraday, the connection of the very different order of phenomena which he [had] discovered, may be placed before the mathematical mind' [8].

In his 1855 paper, Maxwell set out the basic equations of electromagnetic theory, which satisfied the known laws and work of Coulomb, Ampère and Faraday. He realised that the circuital theorem of Ampère, while appropriate to closed conductive circuits, could not be applied to circuits that included a capacitive element, but was unable to postulate a necessary change or addition to the theorem. Six years later, while contemplating a most ingenious and elaborate conceptual model (based on molecular vortices and intervening particles [9]) of the electromagnetic field, Maxwell arrived at a solution to his difficulty. He assumed that space comprised a material medium, the ether, and that the electric and magnetic fields in space were to be considered as mechanical states of this medium. Maxwell conceptualised that for a coherent description of the electromagnetic field equations he had to assume that the medium possessed elastic properties and deduced that an extra term, that he called the displacement current, should be added to the conduction current term in Ampère's theorem. The new equation became one of the four fundamental equations of the electromagnetic field, the set of which are now known as Maxwell's equations [10]. (For completeness the equations are: curl $E = -\partial B/\partial t$, curl $H = J + \partial D/\partial t$, div $D = \rho$, div $B = 0$, where the symbols have their usual meanings.)

The assumption of an elastic medium suggested the notion that it would support the propagation of an electromagnetic wave: a notion that Maxwell easily confirmed by deriving the wave equation from his fundamental equations. The solution to the wave equation was an electromagnetic wave, in which the electric and magnetic field vectors were mutually orthogonal and transverse to the direction of propagation, travelling at a velocity that (using the then known permittivity and permeability parameters) was within 1 per cent of the velocity of light. On this finding Maxwell wrote [11]:

The velocity of transverse undulations in our hypothetical medium, calculated from the electromagnetic experiments of Messrs Kohlrausch and Weber, agrees so exactly with the velocity of light calculated from the optical experiments of M. Fizeau, that we can scarcely avoid the inference that light consists in the transverse undulations of the same medium which is the cause of electric and magnetic phenomena.

Maxwell's conclusions were presented in a paper 'A dynamical theory of the electro-magnetic field' which was read to the Royal Society on 8th December 1864. In the introductory paragraphs he stated:

The theory I propose may . . . be called a theory of the *Electro-magnetic Field* because it has to do with the space in the neighbourhood of the electric or magnetic bodies, and it may be called a *Dynamical Theory* [12] because it assumes that in the space there is matter in motion, by which the observed electro-magnetic phenomena are produced [13].

Maxwell's theory was of monumental importance. R. Feynman, a Nobel prizewinner in physics, has said that in the far future 'the most significant event of the 19th century will be judged as Maxwell's discovery of the laws of electrodynamics. The American Civil War will pale into provincial insignificance in comparison with this important scientific event of the same decade'. P.G. Tait, a contemporary of Maxwell, and a noted physicist, described the formulation of Maxwell's equations as 'one of the most splendid monuments ever raised by the genius of a single individual'; and A. Einstein, of relativity fame, opined: 'To few men in the world has such an experience been vouchsafed' [14].

Curiously, 22 years would elapse between the publication of Maxwell's 1864 paper and the first experiments, brilliantly conducted by Hertz, which demonstrated conclusively the existence of electromagnetic waves.

Heinrich Rudolf Hertz was born on 22nd February 1857 in the Hanseatic city of Hamburg [15]. He was the eldest of the three sons and daughter of Dr G.F. Hertz, a prominent lawyer who subsequently became an appeal court judge and, later, a senator of the city, and his wife – the daughter of a physician, Dr J. Pfefferkorn of Frankfurt-am-Main. It was a prosperous and cultured environment which influenced Heinrich's early formation.

Hertz showed outstanding promise from a young age; his memory was excellent, he had a great capacity for learning – particularly mathematics and languages in which he excelled, and he was skilful in using tools of all kinds. Until he was 15 years of age Hertz attended a private school which, two years afterwards, he described in the following words:

All of us, or at least the better scholars amongst us, were unusually fond of Dr Lange's school, despite the hard work and the great strictness. For we were ruled strictly; detentions, impositions, bad marks for lack of neatness or ill-behaviour rained down upon us; but what particularly sweetened the strictness and profusion of work for us was, in my opinion, the lively spirit of competition that was kept alert in us and the conscientiousness of the teachers who never let merit go unrewarded nor error unpunished [16].

After leaving Dr Lange's school Hertz spent two years studying at home before entering, in 1874, the Johanneum Gymnasium in Hamburg. A year later he was awarded a certificate, which qualified him to gain entry to a university, and was placed first in his year.

During this period Hertz had not determined on the course of his future career. He had gained much fluency in mathematics and the classics, but also he excelled in working with his hands. He spent much of his spare time in his workshop/laboratory, which he had set up in the basement of his parent's house, where he could engage in undertaking experiments in chemistry and physics and in constructing mechanical contrivances. His thoughts, at the age of 18, on the possibility of his becoming a scientist were:

I intend, if I succeed in passing the matriculation examination, to go to Frankfurt-am-Main and work for a year under a Prussian architect, as I would be ultimately required to do for the state licensing examination for professional engineers. Only if I were to prove unsuited for this profession or if my interest in the natural sciences were to increase further, would I devote myself to pure science [16].

Following his period at the Hamburg Gymnasium, Hertz spent a year as an apprentice in the city engineer's office at Frankfurt, and then in April 1876, at the age of 19, he enrolled on an engineering course at the Dresden Technische Hochschule. He left on 30th September 1876 for his mandatory year of military service – with the First Railway Guards Regiment in Berlin. By now Hertz had resolved to follow the natural sciences and in October 1877 he became a student at the University of Munich. Here he sought the advice of the professor of physics, Dr P.G. von Jolly, on which courses he should study, and chose advanced mathematics and mechanics, experimental physics, and experimental chemistry.

A further move was made in October 1878 since in Germany at that time it was customary for a student to progress from one university to another to gain status and promotion. Hertz decided to transfer to the University of Berlin where he soon attracted the attention of Professor Hermann von Helmholtz (1821–1894), one of Germany's great men of science.

Helmholtz, from 1847, had been interested in the spark discharge from a Leyden jar and had proposed, from purely theoretical principles based on the conservation of energy, that the spark discharge was not a uni-directional flow of electricity but was of a high frequency oscillatory nature. Moreover, experimenters had noted that sometimes these discharges had an influence at a distance away from them, but their efforts to explain the observed effects were unsuccessful. However, with the publication of Maxwell's papers, which predicted that oscillatory disturbances would result in the propagation of electromagnetic waves, a way forward seemed to exist. Helmholtz was the first person on the Continent to study Maxwell's papers and to appreciate their importance in future physical investigations. He wanted to reconcile Maxwell's theory of electromagnetism with the theory, based mostly on Newtonian mechanics, of F. Neumann and W.E. Weber.

Helmholtz later wrote [17]:

This plentiful crop of hypotheses had become very unmanageable, and in dealing with them it was necessary to go through complicated calculations. . . . So at that time the domain of electromagnetics had become a pathless wilderness. With the object of clearing up this confusion I had set myself the task of surveying the region of electromagnetics, and of working out the distinctive consequences of the various theories, in order, wherever that was possible, to decide between them by suitable experiments.

To resolve the confusion the Prussian Academy of Sciences (Berlin) in 1879 decided [18], following a proposal from Helmholtz, to offer a prize – often referred to as the Berlin prize. The Academy poses the following question for the 1882 prize:

The theory of elecrodynamics which was brought forth by Faraday and was mathematically executed by Mr J.C. Maxwell presupposed that the formulation and disappearance of the dielectric polarisation in insulating media – as well as in space – is a process that has the same electrodynamic effects as an electrical current and that this process, just like a current, can be excited by electrodynamically induced forces. According to that theory, the intensity of the mentioned current would have to be assumed equal to the intensity of the current that charges the contact surfaces of the conductor. The Academy demands that decisive experimental proof be supplied either

for or against the existence of electrodynamic effects of forming or disappearing dielectric polarization in the intensity as assumed by Maxwell, or for or against the excitation of dielectric polarization in insulating media by magnetically or electrodynamically induced electromotive forces.

Helmholtz considered that Hertz would be the most likely person to succeed in the investigation and promised the young student the assistance of the university if he decided to accept the challenge. Hertz was a sound choice. As later events were to show he had the intellectual ability to understand Maxwell's theory and also the necessary practical skill to undertake an experimental study. But as Hertz stated [19]:

I reflected on the problem, and considered what results might be expected under favourable conditions by using the oscillations of Leyden jars or of open induction coils. The conclusion at which I arrived was certainly not what I had wished for; it appeared that any decided effect could scarcely be hoped for, but only an action lying just within the limits of observation. I therefore gave up the idea of working at the problem; nor am I aware that it has been attacked by anyone else.

Hertz's misgivings about the project probably stemmed from the fact that no means appeared to exist for either the generation of electromagnetic waves or their detection. Instead Hertz undertook an analytical study of the induced currents in a rotating metal sphere in a magnetic field and obtained his doctorate in 1880, with summa cum laude, at the age of 23 [20].

After graduation he remained as an assistant to Helmholtz for three years and then moved to the University of Kiel as an instructor in theoretical physics. The post was not entirely to Hertz's liking for he had no access to any experimental facilities. Nevertheless, in 1884, he published a significant paper 'On the relations between Maxwell's fundamental electromagnetic equations and the fundamental equations of the opposing electromagnetics'. He was now coming to the conclusion that 'if the choice rests only between the usual system of electromagnetics and Maxwell's, the latter is certainly to be preferred' [21].

On 29th March 1885 Hertz obtained a post as a professor at the Technische Hochschule, Karlsruhe. He had been offered promotion at Kiel but did not wish to follow theoretical physics to the exclusion of experimental physics. At Karlsruhe he had his own department, including a well-equipped laboratory, workshop, and some staff [22]. He married Elizabeth Doll, the daughter of Dr Max Doll, the lecturer in geometry, on 31st July 1886, and commenced his famous series of investigations on electromagnetic waves later that year, aided by his assistant Julius Ammons [23].

Among the apparatus in his laboratory was a pair of 'Knochenhauer spirals'. These were flat, air-cored, spiral coils of copper wire with the ends of the wires terminated in small, closely spaced, spherical electrodes. While experimenting with these Hertz noted that when he connected a battery across the electrodes of one of the coils and then opened the circuit a spark occurred. This was not surprising since the coil possessed inductance; however, Hertz was intrigued to discover that a small spark occurred simultaneously across the other coil. Further investigation showed that charged Leyden jars could be used in place of the battery, and even the discharge of a small induction coil sufficed for the same purpose. He wrote:

In altering the conditions I came upon the phenomenon of side-sparks which formed the starting point of [my] research.

At first I thought the electrical disturbances would be too turbulent and irregular to be of any further use; but when I discovered the existence of a neutral point in the middle of a side-conductor, and indications therefore of a clear and orderly phenomenon, I felt convinced that the problem of the Berlin Academy was now capable of solution [24].

The experiments with the Knochenhauer coils were crucial to the subsequent success of Hertz's work on electromagnetic waves for he had discovered a means to detect the radiation from a primary spark gap. His detector now became a single turn, circular or square coil (i.e. a loop), in which the ends of the wire loop were arranged to form an adjustable air-gap – the loop and the air-gap being co-planar and thereby sensitive to the plane of polarisation of the received electromagnetic waves. At resonance the loop is electrically $\lambda/2$ in length with the potential difference between any two points on the loop being greatest across the two terminals.

For his first high frequency generator Hertz abandoned the Knochenhauer spiral and replaced it with a distributed resonant circuit. In a letter [16] dated 5th December 1886 to von Helmholtz he described his new invention.

I have succeeded in demonstrating quite visibly the induction effect of one open rectilinear current on another rectilinear current, and I may hope that the way I have now found will enable me to solve one or other of the questions connected with this phenomenon.

I produced the inducing rectilinear current in the following manner: A thick, straight copper wire, 3 m long, [was] attached at the ends to two spheres, 30 cm in diameter, or two conductors of similar capacity. There is a break in the midpoint of the wire for a spark gap of [0.75] cm between two small brass spheres. Across them the crackling sparks of a large induction machine [were] allowed to pass, whereupon electric oscillations characteristic of the rectilinear circuit [were] excited (which to be sure, was scarcely to be anticipated), and these oscillations now [exercised] a relatively strong inductive effect on the environs. I obtained sparks in a simple square circuit 75 cm on a side, consisting of a thick copper wire and containing only a short spark gap, even at a distance of two metres from the induction path. (Fig. 10.1a.)

Unfortunately Hertz never explained the thinking which led him to replace the Knochenhauer spiral by a rectilinear circuit arrangement. His invention – now known as a half-wave antenna or half-wave dipole – was a major step forward and is used throughout the world.

Additional experiments, in which each of the two parallel sides of the square loop detector were varied in length from 4 in to 98 in (10 cm to 250 cm), enabled Hertz to tune his detector to the frequency of the transmitted electromagnetic waves, and to show that his detector was capable of responding to variations in the strength of the received electric field of 10 to 1 – corresponding to a change of gap from 0.12 in to 0.01 in (3.0 mm to 0.3 mm) [16] (Fig. 10.1b).

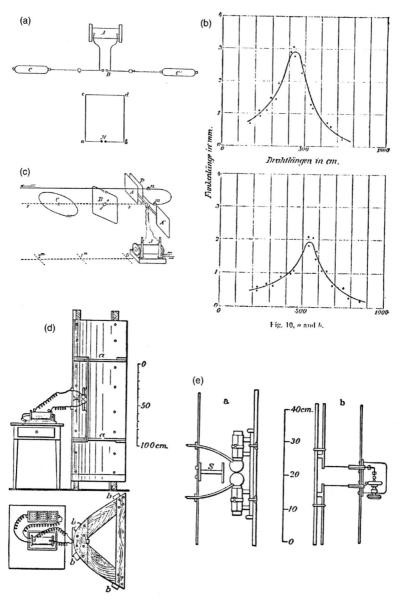

Figure 10.1　*(a) Schematic diagrams of the basic transmitter and detector of electromagnetic waves used by Hertz. (b) By varying the lengths of the sides of the detector Hertz obtained the resonance curve shown in the figure. (c) The layout of Hertz's experiment to determine the velocity of propagation of electromagnetic waves. (d) Hertz's half-wave dipole and reflector transmitting antenna. (e) Hertz's receiving antenna.*

Source: H. Hertz, *Electric waves* (MacMillan, New York, 1893).

While engaged on these experiments with sparks Hertz noted that the behaviour of the detector spark was erratic. Whenever it was exposed directly to the transmitter spark the maximum length of the detector spark was always increased. The effect was particularly troublesome when Hertz was determining the resonance characteristic of his circuit and needed to measure the maximum length of the spark gap which would just break down. Careful observations showed that other light sources produced the same result in varying degrees, and that a light shield placed between the two spark gaps was sufficient to ensure the proper functioning of his detector.

Being the scientist that he was, Hertz realised he was observing a new phenomenon and temporarily abandoned his electromagnetic wave experiments to carry out an extensive inquiry. The results and those of his other investigation were presented in two papers: 'On the effect of ultraviolet light upon the electrical discharge' [25] (published in July 1887); and 'On the action of rectilinear electric oscillations upon a neighboring circuit' [26] (published in March 1888). In the first of these Hertz concluded: 'I confine myself at present to communicating the results obtained, without attempting any theory respecting the manner in which the observed phenomenon are brought about.' Soon, these observations would be taken up by, among others, Elster and Geitel, and subsequently lead to the science of photo-electricity (see Chapter 14).

Hertz now felt he had the apparatus and experimental techniques to pursue the Berlin Prize problem proposed eight years earlier by Helmholtz. The problem was to confirm or disprove Maxwell's equations, based on three assumptions, namely that [27]:

1. the variation of dielectric polarization of a dielectric produced a magnetic field which was equivalent to that produced by a conduction current,
2. a magnetic field travelling through a dielectric would produce polarization,
3. air and empty space behaved in the same way as other dielectrics, and that, as Maxwell had predicted, electromagnetic waves travelled with a finite velocity.

For the award of the prize the Academy required a candidate to consider just one of the first two assumptions. Hertz experimentally confirmed the first of these and sent a manuscript to Helmholtz requesting him to present it to the Prussian Academy of Science [28]. Whether Hertz received the prize seems doubtful, since the time limit had expired in 1882, but it is known that no one else had entered the contest.

Hertz now 'felt that in the third hypothesis [was] contained the gist and the special significance of Faraday's and Maxwell's view, and that it would thus be a more worthy goal for [him] to aim at'. In his paper 'On the finite velocity of propagation of electromagnetic actions' [29] Hertz describes his first experiment to measure the velocity of propagation of electromagnetic waves. Fig. 10.1c illustrates his apparatus. In this, the transmission line (of the single-wire-over-groundplane form) is capacitively coupled, via plate P, to the c. 6 m oscillator/antenna (of the dipole type with square plates AA' for loading). The wire was positioned 1.5 m above the floor and horizontal polarisation was used.

With the line open-circuited at the far end, reflection of the incident waves occurred and vector addition of the incident and reflected waves produced standing waves. These were detected, and from measurements of the distance between the maxima and minima of the pattern, Hertz determined the wavelength as 5.6 m. He estimated the capacitance and inductance of the dipole resonator and calculated the velocity as 2.10^8 m/s. (Professor Sir William Thomson had shown in 1853 that the resonant frequency of a circuit is a function of its inductance and capacitance.) However Hertz's calculation contained an error and after this was pointed out to him, by H. Poincare [30], the velocity was determined to be $2.8.10^8$ m/s, a value within 7 per cent of the accepted value – a most outstanding achievement.

Further investigations were undertaken of the phenomenon of the skin effect, which Maxwell had explicitly mentioned in his electromagnetic theory, of guided waves travelling on a coaxial transmission line, of electromagnetic waves propagating in air and being reflected from a conducting surface, of the fields radiated from a dipole antenna, of the generation of 60 cm radiation, and of the similarity of the characteristics of electromagnetic radiation and light waves [31].

For the latter inquiry Hertz placed the c. 60 cm wavelength half-wave dipole, (which was connected to an induction coil) in the focal line of a parabolic cylinder reflector. The reflector was fabricated from a sheet of zinc 2 m × 2 m bent over a wooden frame to the necessary curvature to give a focal length of 12.5 cm and an aperture of 2 m × 1.2 m (Fig. 10.1d). A similar reflector, together with a dipole, (comprising two straight pieces of wire, each 50cm long and 5 mm in diameter), was employed as the receiving antenna, Fig. 10.1e. From the centre terminals of the dipole a short length of parallel transmission line was connected to Hertz's adjustable spark gap detector. Both the transmitting and receiving antennas were mounted on castors. Also for these experiments Hertz purchased 800 kg of asphalt, 450 kg of tar, and 100 kg of sulphur.

With his focused-beam apparatus Hertz was able to demonstrate the reflection, refraction and polarisation of electromagnetic waves; and to state in a lecture, delivered on 20th September 1889 at a meeting, held in Heidelberg, of the Natural Scientists [32]:

All these experiments in themselves are very simple, but they lead to conclusions of the highest importance. They are fatal to any and every theory which assumes that electric force acts across space independently of time. They mark a brilliant victory for Maxwell's theory. No longer does this connect together natural phenomena far removed from each other. Even those who used to feel that this conception as to the nature of light had but a faint air of probability now find a difficulty in resisting it. In this sense we have reached our goal.

Hertz possibly felt the need to make this point since, as he later told Lodge, he had 'difficulty in getting his ideas accepted in Germany, where the professors . . . did not understand Maxwell' [33].

Hertz did not profit financially from his outstanding research: he was a pure classical physicist for whom the thought of patenting his methods and applications was quite foreign to his thinking and outlook. His interests in physics were broad: though he is esteemed for his endeavours on electromagnetic waves he also carried out inquiries in mechanics, instrumentation, friction, magnetics, meteorology, electricity, cathode rays and electrical discharges in gases.

Alas, he did not live to see the technological advancements that his fundamental work, and that of Lodge, engendered, for he died tragically on 1st January 1894, at the age of just 36 years. For some time prior to his death Hertz had been troubled by 'an extremely old and therefore stubborn' tooth abscess. An operation was performed on 23rd September 1893 but septicaemia from the infection of his jaw proved fatal. A few months afterwards Lodge demonstrated, for the first time anywhere, the basic principles of wireless telegraphy.

In his last letter dated 9th December 1893 to his parents Hertz wrote: 'If anything should really befall me, you are not to mourn, rather you must be proud a little and consider that I am among the especially elect destined to live for only a short while and yet to live enough [34].

Subsequently his papers were collected and published in Leipzig in 1892 [35].

Hertz was aware of Lodge's earlier experiments and commented: 'Inasmuch as he entirely accepted Maxwell's views, and eagerly strove to verify them, there can scarcely be any doubt that if I had not anticipated him he would also have succeeded in observing waves in air and thus also proving the propagation with time of electric force.'

Sir Oliver Lodge's work on radio propagation came about in a somewhat indirect way. During the summer of 1872 electric storms had been such a frequent occurrence that on 6th August William Preece, the Divisional Engineer, Post Office Telegraphs, wrote [36] to *The Times* on the subject of lightning conductors and recommended that, because of the damage caused by recent lightning strikes, each householder should clear the soot from his chimneys – since it was an 'admirable conductor to lead the lightning into our very rooms'. He suggested 'everyone [could], if he [chose], at the expense of a few shillings, render his house absolutely safe with a perfect system of lightning conductors'.

His letter initiated a lively exchange of views [37] and some controversy on the practical construction of these conductors and on the area of protection they afforded. One correspondent quoted Sir William Thomson: 'If I urge our manufacturers to put up conductors, they reply it is cheaper to insure'. He requested Preece to give some guidance. Another writer [38] said the cost was several pounds not several shillings: 'As I am advised, half the lightning conductors now put up are useless and a good proportion of the other half dangerous'.

Such was the state of knowledge of the phenomenon of lightning, that one Past President of the Society of Telegraph Engineers, Latimer Clark, observed [39]:

A person reclining on a sofa or bed at a distance from all the walls of the room would scarcely suffer injury, even in a house struck by lightning, but a most absolute security [was] obtained by lying on an iron or brass bedstead of the form known as the Arabian bedstead in which the head is surmounted by an iron erection supporting the curtains. . . . [This bedstead formed] the most complete lightning conductor which could be devised.

A week later, Preece [40] replied to some mild criticism and mentioned his 20 years experience in an industry in which, he claimed, all telegraphic apparatuses were protected. What he had said was not a 'mere sciolism', since he did not pretend to be an authority on atmospheric electricity, rather, his views had been gained from 'the best of all teachers – experience'. Later, in 1879, at the British Association Meeting in Sheffield, Preece admitted that much telegraph apparatus had not been protected because 'the remedy [was] worse than the disease', although the Post Office had tried a number of devices. The type which then was being brought into general service was the high voltage by-pass. This comprised two metal plates, one of which was earthed, separated by 0.002 in (0.051 mm) of paraffin wax.

The following year, protection from lightning was the theme of a specially convened 'Lightning Rod Conference'. Recommendations were made in a report submitted to the British Association in 1881, but the signatories accepted that they knew little about electric storms. It was evident that the subject should be investigated by a competent scientist. Oliver Lodge, Professor of Experimental Physics at the newly established University College of Liverpool, was selected to undertake the inquiry [41]. Subsequently, his experimental researches on the discharge of Leyden jars (capacitors) led to confirmation of the existence of electromagnetic waves as postulated by Maxwell. Lodge's work was undertaken contemporaneously with that of Hertz. Hertz's first paper 'On the action of rectilinear electric oscillations upon a neighbouring circuit' was published in March 1888; while Lodge's paper [42] 'On the theory of lightning conductors' – which dealt with electromagnetic waves travelling along a transmission line – was published in July 1888. Moreover their endeavours were complementary; Hertz, at first, investigated the properties of unguided electromagnetic waves whereas Lodge, initially, pursued his inquiries using guided electromagnetic waves. During his work he devised a sensitive detector of electromagnetic radiation, which he called the coherer [43], and established the importance of tuning, or syntony as he called the condition when one or more circuits are adjusted to be resonant at the frequency of an applied signal [44]. By 1894 Lodge's knowledge and experience of electromagnetic waves was unique and probably influenced many later scientists and inventors, including Marconi. Indeed, one writer has opined [45]: 'without Lodge there would have been no Marconi. Not only did Lodge prepare the ground over at least a decade in which the seeds of Marconi's work were to germinate but, by providing independently a number of essential elements, he made possible the development of Marconi's ideas into a practical system of wireless communication.'

Oliver Joseph Lodge [46] was born on 12th June 1851 at Penkhull, Stoke on Trent, the first son of Oliver Lodge (the third), a cashier on the North Staffs

Railway at Stoke, and his wife Grace Heath. He was self-motivated and, several years after the birth of his eldest son, he realised that there were better prospects than railway work to be sought by participating in the prosperity of the Potteries and the Five Towns. He acquired an agency from B. Fayle & Co. to supply the numerous pottery firms in the district with china clay and other pottery materials, and soon established a profitable business. This allowed his growing family to enjoy an increasingly affluent style of living, though the children – eight boys and one girl – were raised with very little domestic help. Their mother must have been highly industrious for, apart from raising a large family, she assisted her husband in the business and engaged in pursuing her hobby of wet plate photography.

Lodge's school days were for him the most miserable days of his life. Relief came when he finished schooling at the age of 14 and joined his father in his business. This association continued until he was 22 years of age. An important influence on his early development was his London Aunt Anne who aroused in him a love of the sciences, a love the young Lodge eagerly embraced and developed by his attendances at various educational courses in scientific subjects provided locally at the Wedgwood Institute, at the Mechanics Institute, and at the Athenaeum. These courses had been instituted early in the 1860s, by the Science and Art Department, South Kensington, in a number of industrial towns throughout the Midlands. Lodge's enthusiastic participation in the classes were a welcome change from the drudgery of his employment. During one course Lodge was invited to assist in the preparation of the demonstrations of one of the lecturers – a Mr J. Angel, who had to travel from Manchester to present his lectures and so had little time to organise the associated practical illustrations. In this way Lodge learned a great deal of practical chemistry. Later Lodge widened his knowledge by enrolling on courses on heat, light, sound, electricity and mathematics. His assiduity was rewarded when in the ensuing examinations he took first place in each of the eight subjects in which papers were set. Lodge's industry had a most welcome outcome. Parental pressure to follow his father in the family business was relaxed, somewhat unwillingly, and Lodge was permitted to enrol on a government assisted course for teachers at South Kensington. His studies gained for him, in 1871, matriculation to the University of London for a BSc degree. Courses in chemistry, physics and mathematics were followed at the Royal College of Science, King's College, and University College, and in 1875 he was awarded his degree. Lodge obtained his DSc by examination in 1877. He wrote [47] in his autobiography: '[It] was the easiest examination I ever went in for; for one only had to take one subject. I chose electricty, and though one was supposed to take it at a high grade, I had by that time read most of Clerk Maxwell, and knew at least as much about it as the examiners – which in time they found out. So I got through without any difficulty.' This study would stand him in good stead when, in the 1880s, he began his investigation of electromagnetic waves.

Shortly after his DSc examination Lodge married Mary Marshall, the daughter of some family friends in Newcastle-under-Lyme. At this time he was working as a research assistant/demonstrator to Professor Carey Foster, the professor of physics at University College, London, for a remuneration of just £50 per annum. Fortunately, he was able to augment this sum by lecturing in physics, and later in chemistry, to the ladies at Bedford College, and by marking innumerable examination papers [48].

In 1881 Lodge was appointed as the first Professor of Physics at University College, Liverpool. No laboratory facilities or apparatus existed and the only building the new college possessed was an abandoned lunatic asylum. Nevertheless Lodge's post demanded the existence of a teaching laboratory and, as few, if any, good models existed in the United Kingdom, Lodge set off on a continental tour to see for himself what could be achieved and to purchase items of apparatus. In Berlin he met Heinrich Hertz, who was working as a demonstrator to Helmholtz, and in Chemnitz he was able to purchase some equipment from a Professor Weinhold. He was head of 'a sort of technical institute', but to augment his salary manufactured laboratory apparatus and fittings. To Lodge [49] the apparatus was excellent: 'not made to sell, but to use'. The bought items included several Leyden jars (capacitors), 'exceptionally well made, with no chains, wooden lids, and other gimcrack arrangements, such as were usual in this country [the UK].' With his new laboratory resources Lodge returned to Liverpool to engage in teaching by chalk and talk until his laboratory – a former padded cell of the asylum – was ready for use. Lodge's extensive teaching duties and those of administration in establishing ab initio a new department effectively meant that he had little time for research work during the early years of the 1880s. Nevertheless Lodge maintained his interest in Maxwell's theory and the possibility of producing electromagnetic radiation.

He corresponded with G.F. Fitzgerald, a senior Fellow of Trinity College, Dublin whom he had met in 1878, at the British Association Meeting in Dublin. Fitzgerald became Lodge's 'special friend'.

At the 1883 B.A. Meeting Fitzgerald presented two papers. In the first[50], on 'The energy lost by radiation from alternating electric currents', he derived an expression for the power of the radiation in terms of the period of oscillation, and showed how much more effective the radiation would be if short waves were used rather than those having a longer wavelength. The second paper[51] proposed 'A method of producing electromagnetic disturbances of comparatively short wavelengths' and, as the summary of the paper in the annual volume of the British Association mentioned, 'This [was] by utilising the alternating currents produced when an accumulator [was] discharged through a small resistance. It would be possible to produce waves of as little as ten metres wavelength, or even less'.

In his autobiography Lodge observed [47]:

These papers were the sequel to a previous paper [52] of his and to many discussions with me on the possibility of producing Maxwell's waves artificially. A title only appears on the previous volume [of the British Association] of 1880. It runs: 'On the possibility of originating wave-disturbances in the ether by electromagnetic forces'. This was an extremely general mathematical paper, and its original title had been 'On the *im*possibility . . .' for it was a doubtful point whether the waves could be generated directly by electrical means.

Lodge afterwards felt he could have had the honour of discovering electromagnetic waves but for the delay caused by Fitzgerald's doubts about the possibility of such waves.

Fitzgerald's 1883 suggested method of generating electromagnetic radiation was subsequently generally adopted. As Lodge later (1931) said [47]: 'The fact is that it is ridiculously easy to produce the waves. . . . The trouble was to detect them.'

Towards the end of 1887 Lodge was invited to present two public lectures, in tribute to Dr R. Mann who had been an enthusiastic advocate of lightning rods in South Africa, on the subject of lightning to members of the Society for the Encouragement of the Arts, Manufactures and Commerce. His first lecture, delivered on 10th March 1888, was entitled 'The protection of buildings from lightning' and was mainly a description of the phenomena of lightning and a consideration of the current views on the construction of lightning conductors. He drew the attention of his audience to the similarities of a lightning stroke and the discharge of a Leyden jar, and made the assumption that the two discharges were oscillatory in nature. Joseph Priestly had established the oscillatory character of a Leyden jar discharge in 1770 and by 1888 the subject had been investigated by several physicists. Given this assumption it follows that both the resistance and the reactance of the lightning conductor are of importance. Lodge stressed the importance of minimising the self inductance of the conductor and thereby its reactance [53].

Actually, Lodge's assumption was incorrect; a lightning discharge is not oscillatory but consists of a series of independent unidirectional transients. However, the inductance (L) of a lightning conductor will generate a Ldi/dt voltage which will greatly exceed the iR voltage due to the resistance (R) of the circuit, and so Lodge's work was valid [54].

In his second lecture, given a week later, Lodge demonstrated a number of experiments with his Chemnitz Leyden jars and showed that the discharges were determined 'far more by inertia [self inductance] than by conductivity'. These experiments (see Fig. 10.2) had commenced in February 1888 with the collaboration of Lodge's demonstrator, A.P. Chattock and assistant E. Robinson. They employed a Voss electrostatic machine and a range of capacitors, including sliding-tube types (90 pF to 500 pF), Leyden jars (240 pF to 6.3 nF), and 'double thickness window glass condensers' (0.028 μF) [55]. All these were compared with a well-constructed standard, guard-ring, capacitor using a ballistic galvanometer. In addition five loops of wire, each c. 32.8 yards (30 m) long, and varying from 1 SWG copper (0.025 Ω to 27 SWG iron (33 Ω) were suspended on silk threads in the laboratory [56].

In his 'alternative path' experiment (Fig. 10.2a), the spark gap B was shunted by a stout copper wire, 40 in (101.6 cm) in length, having a resistance of just 0.025 Ω. However, when the Leyden jars were discharged the path of the

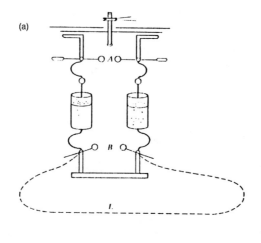

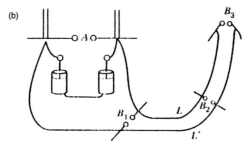

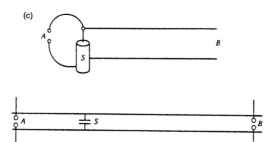

Figure 10.2 (a) The 'alternative path' experiment. The discharge takes place at B, and not via the loop, with a spark gap of c. 40 mm. The experiment showed the importance of self inductance when oscillatory discharges are considered. (b) The 'recoil experiment'. A discharge at A produces longer discharges at B_1, B_2, and B_3, the longest being at B_3. 'Plainly what is happening is this: the discharge at "A" sets up electrical oscillations . . . and the long spark at the far end of the wires is due to the recoil or kick at the reflection of the wave' (Lodge, O.J.: 'Experiments on the discharge of the Leyden jars', Proc. Roy. Soc., 1888, 50, pp. 2–39.). (c) The 'recoil kick' experiment: 'A', primary spark gap; 'B', secondary spark gap; 'S', capacitor.
Source: Author's collection.

discharge was across the gap B and not via the wire, showing that the inductive reactance of the wire offered a greater impedance than the spark gap, and that the discharge was oscillatory. Essentially the resonant circuit formed by the wire (of self inductance L) and the Leyden jars (of capacitance C) was shock excited by the discharge that took place at the spark gap A.

An interesting aspect of this experiment was the production of a more intense spark at B than at A, the terminals of the Voss electrostatic generator which Lodge employed. He investigated this apparent anomaly with the 'recoil kick' experiment (Fig. 10.2b) and demonstrated the physical existence of electric waves propagating along the wires, and being reflected by a short circuit. Both experiments illustrated the importance of resonance (or syntony, as Lodge called it).

In a footnote to his printed paper Lodge mentioned [55]: 'Since the delivery of the lecture, a great number of quantitative observations on these lines have been made. Evidence of electromagnetic waves 30 yards (27.4 m) long has been obtained. I expect to get them still shorter.' (His wire, in his first experiment, was 95 ft (28.95 m) in length and the measured resonant frequency was c. 5 MHz.)

By his endeavours Lodge had succeeded in generating, detecting and measuring the wavelength of electromagnetic waves guided by a metallic structure.

As a consequence of his investigations, Lodge said the 'lightning rod' signatories (of 1881) were little less than criminals'; to which Preece remarked: 'Well . . . I am a criminal and every signatory to that Conference is a criminal: and why? Because he believes in himself and not in Oliver Lodge' [40]. This was a beginning of a minor feud between the experimental physicist and the experienced engineer (Fig. 10.3). The consequential importance of this is not concerned with the subject of lightning conductors per se; rather, the dispute affords an explanation for Preece's action in ignoring the later brilliant researches of Lodge in favour of patronising Marconi.

Lodge, in a paper 'On the theory of lightning conductors' published in the *Philosophical Magazine* in August 1888, referred again to the recoil experiment (Fig. 10.2c) and wrote [42]:

The jar discharges at A in the ordinary way and simultaneously a longer spark is observed to pass at B at the far end of the two long leads . . . .. The theory of the effect seems to be that oscillations occur in the A circuit according to the equation: $T = 2\pi\sqrt{(LS)}$, where L is the inductance of the A circuit and S is the capacity of the jar. These oscillations disturb the surrounding medium and send out radiations, of the precise nature of light, whose wavelength is obtained by multiplying the above period by the velocity of propagation.

On the optimal conditions for producing the 'recoil kick' Lodge stated: 'The best effect should be observed when each wire is half a wavelength, or some multiple of half a wavelength, long. The natural period of oscillation in the wires will then agree with the oscillation period of the discharging circuit, and the two will vibrate in unison like a string or a column of air resounding to the reed' [42].

Figure 10.3   The 'battle' between engineering (Preece) and experiment (Lodge).
Source: *Electrical Plant*, December 1888.

Lodge's paper was dated 7th July 1888. Before it was published, Lodge learnt of Hertz's work on electromagnetic waves and forwarded a postscript from Cortina where he was holidaying. The physicist warmly welcomed the news, and wrote:

Since writing the above, I have seen in the current July number of *Wiedemann's Annalen* an article by Dr Hertz, wherein he establishes the existence and measures the length of aether waves excited by coil discharges; converting them into stationary waves, not by reflexion of pulses transmitted along a wire and reflected at its free end, as I have done, but by reflexion of waves in free space at the surface of a conducting wall . . ... The whole subject of electrical radiation seems to be working itself out splendidly.

Had Lodge's paper been published in an earlier issue of the *Philosophical Magazine*, as he had anticipated, his claim to simultaneous discovery with Hertz of electromagnetic waves would have been greatly strengthened.

The relationship between Hertz's work and that of Lodge was demonstrated when Lodge presented a popular discourse on 'The discharge of a Leyden jar' at a Friday meeting of the Royal Institution on 8th March 1889 [57]. He repeated and extended the experiments that he had described the previous year and particularly stressed the oscillatory nature of the discharge, and explained the phenomenon of resonance. He referred to vibrating mechanical systems and the influence of mechanical friction in controlling the rate of decay of the vibrations, and mentioned that in an electrical system resistance was analogous to mechanical friction but

there [was] another cause [of decay], and that a most exciting one. The vibrations of a reed [were] damped, partly indeed by friction and imperfect elasticity but partly by the energy transferred to the surrounding medium and consumed by the production of sound. It [was] the formation and propagation of sound which largely [damped] the vibrations of any musical instrument. So it [was] with electricity. The oscillatory discharge of a Leyden jar [disturbed] the medium surrounding it, carv[ed] it into waves which [travelled] away from it into space . . .

The second cause, then, which damps out the oscillations in a discharge circuit [was] radiation.

A further paper 'On electrical radiation and its concentration by lenses', which Lodge read to the Physical Society on 11th May 1889, described experiments that he and an assistant, Dr Howard, had undertaken to progress Hertz's work. By utilising plano-convex cylindrical lenses of mineral pitch Lodge showed that Hertzian waves of approximately one metre wavelength could be focused in an analogous way to light waves. Additionally, his studies disclosed that a half wave linear oscillator was a more efficient radiator of waves than the circuits which had been employed previously.

Strangely, neither Hertz nor Lodge saw any practical use for their discoveries, although at the 1888 British Association meeting Lodge prophesied [58] that 'the now recognised fact that light is an electrical oscillation must have before long a profound practical import'. No patents were sought by either experimentalist. Both had shown imagination in devising brilliant series of experiments to show, for the first time ever, the existence of unguided (free-space) and guided electromagnetic waves – in confirmation of Maxwell's theory – but both seemed to lack the imagination necessary to relate their work to a pragmatic advantage, e.g. signalling. Hertz and Lodge were pure scientists, and the tradition and ethos of the natural sciences precluded patenting embodiments of ideas and scientific findings for commercial gain. It is noteworthy that the greatest experimentalist of the 19th century, Michael Faraday, never applied for a patent. And yet, by 1888, as noted previously, several inventors, including Edison, had demonstrated free-space systems of telegraphy based on electric induction.

Albert Einstein once remarked: 'Imagination is more important than knowledge'. One person who displayed such a quality and prophesied about the future of telegraphy and telephony was Sir William Crookes. In a lengthy article [59] 'On some possibilities of electricity', published in the 1st February 1892 issue of the *Fortnightly Review*, he expounded his thoughts on the 'new and astonishing world' that was opening up as a consequence of the work of Hertz and Lodge; a world with 'an almost infinite range of ethereal vibrations or electric rays, from wavelengths of thousands of miles down to a few feet'. Crookes observed it was difficult to conceive that this world 'should contain no possibilities for transmitting or receiving intelligence'.

Rays of light will not pierce through a wall, nor, as we know only too well, through a London fog. But the electrical vibrations of a yard or more in wavelength . . . will easily pierce such mediums, which to them will be transparent. Here, then, is revealed the bewildering possibility of telegraphy without wires, posts, cables, or any of our present costly appliances. Granted a few reasonable postulates, the whole thing comes well within the realms of possible fulfillment. At the present time experimentalists are able to generate electrical waves of any desired wavelength from a few feet upwards, and to keep up a succession of such waves radiating into space in all directions . . .. Also an experimentalist at a distance can receive some, if not all, of these rays on a properly constituted instrument, and by concerted signals messages in the Morse code can thus pass from one operator to another.

Crookes was well aware of the need for tuning if confidentiality of message transmission were required.

I assume here that the progress of discovery would give instruments capable of adjustment by turning a screw or altering the length of a wire, so as to become receptive of wavelengths of any pre-concerted length . . .. Considering that there would be the whole range of waves to choose from, varying from a few feet to several thousand miles, there would be sufficient secrecy; for curiosity the most inveterate would surely recoil from the task of passing in review all the millions of possible wavelengths on the remote chance of ultimately hitting on the particular wavelength employed by his friends whose correspondence he wished to tap. By 'coding' the message even this remote chance of surreptitious straying could be obviated.

Crookes stressed that this was no mere dream of a visionary philosopher. 'All the requisite skills', he said, 'needed to bring it within the grasp of daily life [were] well within the possibilities of discovery, and [were] so reasonable and so clearly in the path of researches which [were] now being actively prosecuted in every capital of Europe that we [might] any day expect to hear that they have emerged from realms of speculation into those of sober fact.'

Lodge remained unmoved by this prospect. He was a scientist and it was the responsibility of scientists to investigate the phenomena of nature and to deduce the principles that characterised the phenomena: the development of any commercial aspects of such enquiries were best undertaken by others. For Lodge, the details of implementing practically a system of telegraphy based on Hertzian waves 'could safely be left to those who had charge of the Government monopoly of telegraphs, especially as their eminent Head [Preece] was known to be interested in this kind of subject' [60]. Four years would pass before Crookes's scientific prophecies would begin to be realised. Then, curiously, in 1896, Preece wholeheartedly welcomed an unknown, 21 year old, foreign, amateur inventor with an interest in wireless into his circle of influence and gave him much encouragement, and the patronage of the Post Office. The brilliant university trained British physicist, who had investigated and demonstrated the scientific principles of wireless was ignored.

Following Hertz's death, on 1st January 1894, Lodge was invited to deliver a memorial lecture [61] at the Royal Institution on 'The work of Hertz'. Experiments on the reflection, refraction, diffraction, interference and polarisation of electromagnetic waves were shown, and the importance of syntony was stressed.

Shortly after the lecture, Alexander Muirhead, a well-known manufacturer of telegraphic apparatus, was so stimulated by the prospects which seemed to be evolving that he 'went to Lodge with the suggestion that messages could be sent by the use of waves' [62]. Lord Rayleigh, too, was greatly impressed by this possibility and told Lodge: 'Well, now you can go ahead; there is your life['s] work!' [63].

In retrospect, the lecture probably had a considerable impact on future developments. The text of Lodge's presentation was given wide circulation by its publication, first, in serial form in several consecutive issues of *The Electrician* [64]; and, second, (with additional material) as a small book [65] entitled *The work of Hertz and some of his successors*. Furthermore, Lodge's detailed description of his coherer, a device he had discovered during his lightning rod experiments and which later became the first, practical, commercial detector in wireless telegraph receivers, enabled others to fabricate easily this 'most astonishingly sensitive detector of Hertzian waves'.

These writings of Lodge, and also those of Hertz, provided the essential information to other experimentalists to enable them to replicate and, possibly, to advance further the investigation of radio wave propagation. The publications were circulated widely and were cited by J.C. Bose [66] of India (in 1895), by A.S. Popov [67] of Russia (in 1896), and by Jackson [68] of the United Kingdom (in 1899).

Later that month, on 13th June, at the annual 'Ladies Conversazione' at the Royal Society, Lodge once again displayed his detector, and succeeded in demonstrating wireless telegraphy. For his purpose Lodge replaced the headphones, which he had used in previous tests, by a sensitive mirror galvanometer – lent by Muirhead – so that his audience could observe the effects which he wished to demonstrate. A paragraph in the august journal *Nature* described his equipment as follows [69]:

Professor Oliver Lodge, FRS, exhibited a compact and sensitive detector for electric radiation, and a spherical radiator of short Hertz waves. The apparatus consisted of a small copper cylinder containing a piece of zinc and sponge, forming a battery, a coil and suspended needle-mirror, forming a galvanometer, and a ball contact or 'coherer', or else a tube of filings, in circuit with the other two. Electric surgings in the air, or in a scrap of wire pegged into the lid, increased the conductance of the circuit. A light tap on the cylinder reduced it again. A handy lamp and scale enabled the deflexion of the needle to be seen. The surgings could be excited by giving sparks to an insulated sphere not far off, especially if the knobs supplying the sparks were well polished.

Lodge presented a similar lecture [70] 'On experiments illustrating Clerk Maxwell's theory of light' at a meeting of the British Association for the

Advancement of Science in Oxford on 14th August 1894, but on this occasion his apparatus included a Kelvin marine galvanometer and a Morse inker provided by Muirhead. With these Lodge was able to show his audience the transmission (through two stone walls) of signals representing letters of the alphabet by means of electromagnetic wave propagation.

The transmitter, installed in the Clarendon Laboratory, comprised an induction coil, having a Morse key in the primary circuit, and a Hertzian radiator; the receiver in the lecture theatre of the Oxford Museum situated 65.6 yards (60 m) away, consisted of his coherer, the marine galvanometer and a cell. With these 'he was able to transmit a dot or a dash signal and by suitable combinations to send any letter of the alphabet in the Morse code and consequently intelligible messages . . .. It is, therefore, unquestionable that on this occasion Lodge exhibited electric wave telegraphy over a short distance' [71]. It was the first demonstration anywhere of the principle of wireless telegraphy.

Lodge's demonstrations 'excited great interest' in the lecture theatre, which was 'crowded to overflowing' with an enthusiastic audience that included Lord Rayleigh, A. Muirhead, J.A. Fleming, S.P. Thompson, A. Trotter (the editor of the *Electrician*), and other important figures. According to *Nature*, the audience 'repeatedly showed its appreciation of Professor Lodge's beautiful experiments' [72].

Remarkably, there is no evidence to suggest that Lodge in 1894 had the slightest interest in the commercial development of wireless telegraphy, or in obtaining patent protection for his system. He ignored Lord Rayleigh's advice and the suggestion of Muirhead and almost immediately after the BA meeting went to the south of France for an extended vacation. There, he engaged in a series of lengthy experiments investigating the powers of a famous medium who was later exposed as a charlatan [73,74].

Lodge afterwards (1902), following the success achieved by Marconi, realised he had been unwise and candidly acknowledged he had been stupid and blind.

Signalling was easily carried on from a distance through walls and other obstacles . . . [but] stupidly enough no attempt was made to apply any but the feeblest power so as to test how far the disturbance could really be detected . . .

The idea of replacing a galvanometer . . . by a relay working an ordinary sounder or Morse was an obvious one, but so far as he [Lodge] was concerned he did not realise that there would be any particular advantage in thus with difficulty telegraphing across space instead of with ease by the highly developed and simple telegraphic and telephonic methods rendered possible by the use of a connecting wire. In this non-perception of the practical uses of wireless telegraphy he [Lodge] undoubtedly erred. But others were not so blind. [Ref. 44, chapter 6]

Among these 'others' were an enterprising officer of the British Royal Navy, Captain H.B. Jackson, and a young Italian amateur wireless enthusiast, G. Marconi. They would use the principles of electromagnetic wave propagation to evolve practical wireless schemes which would have immense significance and application in future communication systems. The civilised world would be changed for ever. Their endeavours form the basis of the next chapter.

Note 1

Some pre-Hertzian wave observations [75]

Prior to Hertz's and Lodge's experiments on unguided and guided electromagnetic waves, several scientific workers had observed some unknown phenomenon for which there appeared to be no explanation. By way of illustration, in 1842 Professor Joseph Henry found that the discharge of a Leyden jar could magnetise needles situated in the basement of his house, 30 ft (9.14 m) below his workroom, even though two 14 in (35.56 cm) thick floors intervened. The same effect could be produced by lightning strokes seven or eight miles distant. And in 1875, Professor Elihu Thomson, while experimenting on the first floor of the Central High School in Philadelphia, noticed that small sparks could be seen between a sharp pencil point and a brass door-knob whenever a circuit (which comprised a sparking coil, having one end earthed and the other end connected to a large insulated metal can) was operated. A Dr Mahlon Loomis, of Washington D.C., in 1866, discovered that one kite (A) of two kites (A and B), each held aloft by c. 600 ft (182.8 m) of thin copper wire and grounded through a galvanometer to earth, could influence the galvanometer of the other kite (B) whenever the circuit of A was opened.

Of the pre-1886 experiments on electromagnetic waves, those of Professor David E. Hughes are well documented. He was born in London on 16th May 1830 but received his early education in Kentucky, Virginia when his family emigrated to the USA in 1837. Subsequently he became a professor of music at the University of Kentucky. Later, after a brief period spent in Paris, he returned to London and began experimenting at his house at 40 Langham Street (just c. 300 m from Broadcasting House) from c. 1878.

Hughes was an inventor of some merit: when he died, in London on 22nd January 1900, he left a fortune of c. £400,000 – most of it being bequeathed to hospitals in the capital.

During some of his investigations Hughes experienced difficulty in balancing an induction balance on which he was working, but succeeded in tracing the cause of the problem to a loose connection in his circuit. Adventitiously at this time Hughes was also experimenting with his 'loose contact' microphone. He observed that when the faulty contact in his bridge arrangement was made and broken a noise signal was produced in an earpiece connected in series with a battery and his microphone.

To further his study of this effect Hughes constructed a crude spark generator, consisting of an induction coil driven by a battery and a clockwork contact breaker, to provide a repeatable cause of the effect. In some experiments a metal fireguard was attached to the transmitter to act as a radiator. His receiver/

detector comprised a glass jar, fitted with a turned boxwood lid, into which a steel needle lightly touched a small lump of coke, together with a battery and an earpiece. He also joined two short wires to his detector thereby providing an antenna. With this equipment Hughes demonstrated that sounds could be heard in his earpiece whenever the generator/transmitter was operative. He has written (in 1890):

After trying unsuccessfully all distances allowed in my residence in Portland Street, my usual method was to put the transmitter in operation and walk up and down Great Portland Street with the receiver in my hand, with the telephone to my ear. The sounds seemed to slightly increase for a distance of 60 yard [54.8 m] and then gradually diminish until at 500 yards [457 m] I could no longer with certainty hear the transmitted signals [75].

On the 20th February 1880 Hughes demonstrated his experiments to Mr Spottiswoode, the President of the Royal Society, and several Fellows of the Royal Society. The meeting did not have a favourable outcome for Hughes, although his experiments worked well. In one of his notebooks he recorded:

Mr Spottiswoode, . . . Prof. Stokes and Prof. Huxley, visited me today at half-past 3 pm and remained until quarter to 6.00 pm, in order to witness my experiments with the Extra Current Thermopile. The experiments were quite successful, and at first they were astonished at the results, but at 5.00 pm Prof. Stokes commenced maintaining that the results were not due to conduction but to induction, and that results then were not so remarkable, as he could imagine rapid changes of electric tension by induction. Although I showed several experiments which pointed conclusively to its being conduction, he would not listen, but rather pooh-poohed all the results from that moment.

This unpleasant discussion was then kept up by him, the others following his suit, until they hardly paid any attention to the experiments, even to the one working through the gas pipe in Portland Street to Langham Place on the roof. They did not sincerely compliment me at the end on results, seeming all to be very much displeased, because I would not give at once my Thermopile to [the] Royal Society so that others could make their results . . . . They left very coldly and with none of the enthusiasm with which they commenced the experiments. I am sorry at these results of so much labour, but cannot help it. (20th February 1880) [75]

Hughes was so discouraged by his visitors' reactions to his work he refused to write a paper on the subject. Nevertheless he continued his experiments for 'some years, in the hope of arriving at a perfect scientific demonstration of the existence of aerial electric waves'. However, he lacked the training in physics that Hertz had received and failed to progress his own achievement. Sir Oliver Lodge, in an article published in *Popular Wireless* September 1923 summed-up Hughes' work when he wrote:

He was a man who 'thought with his fingers', and who worked with the simplest home-made apparatus – made of match-boxes and bits of wood and metal, stuck together with cobbler's-wax and sealing-wax. Such a man constantly working is sure to come upon phenomena inexplicable by orthodox science. And orthodox science is usually too ready to turn up its nose at phenomena which it does not understand, and so thinks it simplest not to believe in.

References

1 TOLSTOY, J.: 'James Clerk Maxwell' (Canongate, Edinburgh, 1981), p. 1
2 NIVEN, W.D. (ed.): 'James Clerk Maxwell, Scientific Papers', 2 vols (Cambridge University Press, 1890)
3 MAXWELL, J.C.: 'Treatise on electricity and magnetism' (Clarendon Press, Oxford, 1873)
4 BORDEAU, S.P.: 'Volts to Hertz, the rise of electricity' (Burgess Pub. Co., Minneapolis, 1982), p. 206
5 GLAZEBROOK, R.T.: entry on 'Maxwell, J.C. (1831–1879)' in the 'Dictionary of National Biography' (Oxford University Press, London, 1949)
6 MAXWELL, J.C.: 'On Faraday's lines of force', *Trans. of the Cambridge Philosophical Society*, 1856, **10**, pp. 27–83
7 GARRATT, G.R.M.: 'The early history of radio from Faraday to Marconi' (IEE, London, 1994), p. 21
8 BURCHFIELD, J.D.: entry on 'Kelvin, William Thompson, Baron', 'Dictionary of National Biography' (Oxford University Press, London, 1949), pp. 354–355
9 SIEGEL, D.M.: 'Innovation in Maxwell's electromagnetic theory' (Cambridge University Press, 1991), ch. 3, pp. 56–84
10 Ref. 9, pp. 85–119
11 MAXWELL, J.C.: 'On physical lines of force, part III: The theory of molecular vortices applied to statical electricity', *Phil. Mag.* (4th series), 1862, **23**, p. 22
12 MAXWELL, J.C.: 'A dynamical theory of the electromagnetic field', *Phil. Trans. Roy. Soc.*, 1865, **155**, pp. 459–512
13 Quoted in Ref. 4, p. 197
14 Quoted in Ref. 1, p. 2
15 SUSSKIND, C.: 'Heinrich Hertz: a short life', pp. 802–805; and KRAUSS, J.D.: 'Heinrich Hertz – theorist and experimentalist', pp. 824–829, IEE Trans. on Microwave theory and techniques, 1988, **36**(5)
16 HERTZ, J.: 'Heinrich Hertz, memoirs, letters and diaries', revised edition prepared by M. Hertz and C. Susskind, (Eds) (San Francisco Press, Inc., and Physik Verlag, GmbH, 1977)
17 Quoted in ref. 4, p. 223
18 Monthly report of the Prussian Academy of Sciences in Berlin, July 1879, pp. 519, 528–529
19 HERTZ, H.: 'Electric waves, being researches on the propagation of electric action with finite velocity through space' (MacMillan, New York, 1893, and Dover, New York, 1962), trans. D.E. Jones
20 HERTZ, H.: 'On induction in rotating spheres'; in 'Miscellaneous Papers' (trans. D.E. Jones and G.A. Schott), (MacMillan, New York, 1896), ch. 2
21 Ibid, ch. 17
22 LEHMANN, O.: 'History of the Technical Institute of the Technische Hochschule' (Karlsruhe, 1892)
23 CICHOU, D.J., and WIESBECK, W.: 'The Heinrich Hertz wireless experiments at Karlsruhe in the view of modern communications', The IEE Conference Publication, No. 411 on '100 years of radio', 1995, pp. 1–6

24 Ref. 19, ch. p.2
25 HERTZ, H.: 'On the effect of ultraviolet light upon the electrical discharge', *Annalen der Physik*, 1887, **31**, p. 983
26 HERTZ, H.: 'On the action of rectilinear electric oscillations upon a neighbouring circuit', *Annalen der Physik*, 1888, **34**, p. 155
27 Ref. 4, p. 230
28 Ref. 16, pp. 233–234
29 HERTZ, H.: 'On the finite velocity of propagation of electromagnetic actions', *Annalen der Physik*, 1888, **34**
30 POINCARE, H.: letter to Hertz, Sept/Oct 1890, The Deutsches Museum, no. 3001
31 BRYANT, J.H.: 'Heinrich Hertz' (IEEE Press, New York, 1988), p. 50
32 Ref. 16, ch. 20
33 LODGE, O.: 'Advancing science' (Harcourt Brace, New York, 1931), p. 111
34 Quoted in ref. 23, p. 2
35 HERTZ, H.: 'Untersuchungen, uber die Ausbreitung der elektrischen Kraft' (Johann Ambrosins Barth, Leipzig, 1892)
36 PREECE, W.H.: letter to *The Times*, 6th August 1872
37 See letters to *The Times* for 8th, 10th, 12th, 13th, 14th, and 16th August 1872
38 Letter to *The Times*, 8th August 1872
39 CLARK, L.: letter to *The Times*, 13th August 1872
40 PREECE, W.H.: letter to *The Times*, 13th August 1872
41 BAKER, E.C.: 'Sir William Preece, F.R.S.. Victorian Engineer Extraordinary' (Hutchinson, London, 1976), pp. 293–308
42 LODGE, O.J.: 'On the theory of lightning conductors', *Phil. Mag.*, 1888, **26**, (5th series), pp. 217–230
43 PHILLIPS, V.J.: 'Early radio wave detectors' (Peter Peregrinus Ltd, London, 1980), ch. 3, pp. 18–64
44 ROWLANDS, P. and WILSON, P.J. (Eds.): 'Oliver Lodge and the invention of radio' (PD Publications, Liverpool, 1994), p. 80
45 Ref. 7, p. 51
46 WILSON, P.J.: 'Oliver Lodge: a sketch of his life', ch. 1, pp. 1–6, of Ref. 44
47 LODGE, O.J.: 'Past years – An Autobiography' (Hodder and Stroughton, London, 1931)
48 Ref. 46, p. 3
49 Ref. 47, pp. 153–154
50 FITZGERALD, G.F.: 'On the energy lost by radiation from alternating electric currents', *Sci. Trans. Royal Dublin Society*, 1883, **III**, pp. 57–60
51 FITZGERALD, G.F.: 'A method of producing electromagnetic disturbances of comparatively short wavelengths', Trans. of Section A, Report of the British Association, 1883, **LIII**, pp. 404–405
52 FITZGERALD, G.F.: 'On the possibility of originating wave disturbances in the ether by means of electric forces', *Sci. Trans. of the Royal Dublin Society*, 1877–83, **I**, Series 2, pp. 133–134, 173–176, 325–326
53 LODGE, O.J.: 'Lightning conductors and lightning guards' (Whittaker, London, 1892)
54 GOLDE, R.H.: 'Lightning protection' (Arnold, London, 1973)

55 LODGE, O.J.: Experiments on the discharge of Leyden jars', *Proc. Roy. Soc.*, 1888, **50**, pp. 2–39

56 ROWLANDS, P.: 'Oliver Lodge and the Liverpool Physical Society' (Liverpool University Press, 1990)

57 LODGE, O.J.: 'Easy lecture experiment in electric resonance', *Nature*, 1890, **41**, p. 368

58 ANON.: a report, *The Times*, 8th September 1888

59 CROOKES, W.: 'On some possibilities of electricity', in the *Fortnightly Review*, 1st February 1892, pp. 173–181

60 LODGE, O.J.: 'Signalling without wires' (Van Nostrand, New York, 1902)

61 LODGE, O.J.: 'The work of Hertz', *Nature*, 1894, **50**, pp. 133–139

62 MUIRHEAD, M.F.: 'Alexander Muirhead' (Published privately, Oxford, 1926), p. 39

63 Quoted in ref. 33, p. 122

64 LODGE, O.J.: 'The work of Hertz and some of his successors' *The Electrician*, 1894, **XXXIII**, pp. 153–155, 186–190, and 204–205

65 LODGE, O.J.: 'The work of Hertz and his successors' (Electrician Printing and Publishing Co., London, 1894)

66 BOSE, J.C.: 'On the determination of the indices of refraction of various substances for the electric ray', *Proc. of the Roy. Soc.*, 1895, **LIX**, pp. 160–167

67 POPOV, A.S.: 'Zh Russ fiz-Khim Obschestra', 1896, **XXVIII**, pp. 1–14

68 'Statement of Captain Jackson's Claims as Regards the Invention of Wireless Telegraphy', appended to a letter from Captain F.T. Hamilton to the C-in-C, Devonport, 28th January 1899, ADM 116/523, PRO

69 ANON.: a report, *Nature*, 1894, **50**, pp. 182–183

70 LODGE, O.J.: 'On experiments illustrating Clerk Maxwell's theory of light', a lecture to a joint meeting of Sections A (Physical Sciences) and I (Physiology), British Association for the Advancement of Science, Oxford, 14th August 1894

71 FLEMING, J.A.: 'Guglielmo Marconi and the development of radio communications', *J. Roy. Soc. Arts*, 1937, **86**, No. 4436, pp. 42–63

72 ANON.: a report, *Nature*, 1894, **50**, p. 463

73 Ref. 44, p. 83

74 JOLLY, W.P.: 'Sir Oliver Lodge. Psychical researcher and scientist' (Constable, London, 1974)

75 FAHIE, J.J.: 'A history of wireless telegraphy' (Blackwood, Edinburgh, 1899), pp. 289–295

Chapter 11

Early experimental wireless
telegraphy (1895–1898)

During the 1870s a new weapon was developed which would transform the tactics of sea warfare. It was a weapon that, in two world wars, caused immense destruction of both merchant shipping and naval vessels, and loss of life. The weapon was the torpedo. From 1914 to 1918 U-boats sank 4,837 ships totalling 11,135,000 tons, (11,300,000,000 kg) and from 1939 to 1945 U-boats destroyed 2,775 British, Allied and neutral merchant ships of c. 14,500,000 tons. (14,700,000,000 kg) [1]. The battle against the Atlantic U-boats during World War II, (known as the Battle of the Atlantic), led Churchill, the British war-time prime minister, to write after the war [2]:

The Battle of the Atlantic was the dominating factor all through the war. Never for one moment could we forget that everything happening elsewhere, on land, at sea, or in the air, depended ultimately on its outcome, and amid all other cares, we viewed its changing fortunes day by day with hope or apprehension.

The battle caused Churchill more anxiety than any other battle. Grand Admiral Doenitz, the commander of the German U-boat fleet, had a simple strategy for German victory. In 1940 he said: 'I will show that the U-boats alone [with their torpedoes] can win this war . . . nothing is impossible to us' [3].

The invention of the torpedo, and the consequential evolution of a vessel – the torpedo boat – to launch it, necessitated the development of a non-visible communication system which could be used at night, or in daylight when ships and boats were obscured by smoke, to enable offensive torpedo attacks to be coordinated, and to allow defensive actions to be taken. Captain H.B. Jackson, of the Royal Navy was the first naval officer to engage in this development.

The first notion for a locomotive torpedo appears to have occurred to an Austrian naval officer named Captain Lupis, but it took the form of a small vessel containing within itself some propulsion means by which it could move over the surface of the water, its course being guided by ropes or lines from a ship or from the shore. The fore part of the vessel was to hold an explosive charge, to be fired automatically when it came into contact with the side of a hostile boat or ship. However the Austrian authorities felt that the system of guidance was impractical and that the methods of obtaining motive power, by clockwork or steam power, were objectionable [4].

They sought the assistance of a Mr Whitehead who at that time was director of an engineering establishment at Fiume. Whitehead devoted himself to solving the problem of designing a torpedo which would be able to travel beneath the surface of the water and be independent of the launching facility for its guidance.

In 1870 the British Admiralty purchased two of Whitehead's torpedoes and in trials one of them blew a large hole in an old corvette. The range of these early torpedoes was limited to 730 m and the speed to six knots, and it was evident that by day quick-firing guns should be able to sink an attacking vessel before it could get within torpedo range. It was clear that the torpedo would be most effective when employed at night against ships at anchor in open roadsteads.

Night action seemed to be a hazard which even the strongest fleet might be unable to avoid and so received serious consideration. In 1873 the Admiralty set up the Torpedo Committee to investigate the whole question [5]. Following a series of trials that extended over a six month period, the Committee concluded that 'any maritime nation failing to provide itself with submarine locomotive torpedoes would be neglecting a great source of power, both for offence and defence'. Upon this recommendation the Admiralty immediately purchased from Whitehead, for £15,000, the secret of the internal mechanism of his invention and the rights of manufacturing it. Soon, the torpedo was further developed and by 1876 its speed was 18 knots (9.26 m/s), by 1884 24 knots (12.3 m/s), and by 1895 30 knots (15.4 m/s)

This activity led to the design and construction of special vessels, torpedo boats, capable of launching the new type of missile, and this in turn necessitated the introduction of torpedo boat destroyers, later known as destroyers. The torpedo boats were a potent new weapon: they were cheap to build, they were fast and highly manoeuvrable, they required just a very small crew to operate them, and they carried a projectile capable of sinking the mightiest capital ship (Fig. 11.1).

The importance attached to the prospective use in war of the Whitehead torpedo may be illustrated by the fact that at the end of 1890 the number of torpedo boats built or laid down in Great Britain was 206, and in France 210; while other nations followed with numbers proportionate to their means. Forty torpedo boat destroyers were in construction for the British navy towards the close of the year 1896 (the year Marconi came to England), and in 1897 the French navy announced that the number of torpedo boats and torpedo boat destroyers would be increased by 175.

The small size of torpedo boats made them difficult targets to observe at night. Torpedo nets could be used by ships at anchor and until c. 1880 they were the principal means of defence against night-time torpedo attacks.

Figure 11.1   Torpedo boat No. 2 (1886). Torpedo boats were small, fast, highly
manoeuvrable, and cheap to construct, but were capable of sinking the
mightiest capital ship.
Source: The Imperial War Museum, Q21974.

In 1873 the Admiralty Torpedo Committee considered the various defensive measures available and recommended the use of illuminants to detect any hostile craft [6]. Fortunately Gramme's 'electromagnetic induction machine' was in a reasonably developed state at this date and so the utilisation of arc lights became a practical proposition. Mr Wilde's 'electric light' (of 11,000 candle power) was tried in HMS *Comet*, a corvette, in 1874. In trials it was found that although the arc light did not meet the Committee's requirement to floodlight the entire area around a ship to beyond torpedo range it was far superior to other illuminants. 'The value of the light is decided and considerable', the Committee reported, and in the following year an improved 'electric light' was fitted in the battleship *Minotaur*. During the ensuing years from 1882 searchlights, as the 'electric lights' came to be called, were fitted to all British naval ships (Fig. 11.2). The searchlight was not a British prerogative and other nations developed and fitted it concurrently.

The serious threat posed by torpedo boats, and the need to train seamen in the handling and operation of torpedoes and searchlights, caused the Admiralty to establish a torpedo training school. In 1884 HMS *Defiance*, a laid-up 91-gun ship of 5,270 tons (5,350,000 kg) that had been launched in 1861, was recommissioned at Devonport as the Admiralty's Torpedo Training School. This establishment augmented the torpedo school for officers at HMS *Vernon*.

One of the officers who completed the coveted torpedo course at HMS

Figure 11.2   HMS Dreadnought. *Five searchlights can be clearly seen – by the aft tripod
mast, by the aft funnel and on the wing of the bridge, (1907).*
Source: The Imperial War Museum, Q38713.

Vernon in 1883 was Lieutenant Henry Bradwardine Jackson, a person of whom
one historian has written [7]:

Henry Jackson (1855–1929) is among Britain's greatest unknown heroes of the 20th
century. A true Yorkshireman, he was the first Englishman (and the second person
anywhere in the world) to develop a practical radio telegraph. He was, moreover, First
Sea Lord at the time of the greatest surface-ship battle in the Royal Navy's history. His
achievements as an applied scientist were recognised by election to Fellowship of the
Royal Society: his achievements as a naval officer raised him to the rank of Admiral of
the Fleet.

Jackson joined the Royal Navy as a cadet in 1868, at the age of 13 years,
and was posted to HMS *Britannia,* the cadet's training ship at Dartmouth.
Here, he received an excellent technical education and in 1871 obtained First
Class passes for both seamanship and academic studies. Leaving HMS *Britannia*
as a midshipman, Jackson spent three years at sea and, at the end of his training,
passed the Lieutenant's examination. He served on HMS *Active* during the Zulu
War and, after further advanced technical training at the Royal Naval College,
Greenwich, was posted to HMS *Vernon,* Portsmouth, for the torpedo course.
HMS *Vernon* was then the Admiralty's establishment for electrical engineering
development and research and was considered to be an excellent posting for
young ambitious officers.

On completing the course, Jackson was given command of the Royal
Navy's latest vessel, HM *Torpedo Boat No. 81.* From his training and experience

Jackson would have been aware of the potency of boats of this type for night actions, and of the need to communicate between torpedo boats during attacks to ensure properly coordinated manoeuvres. Indeed, in 1887 Lieutenant Jackson had been in command [8] of HMS *Vesuvius*, the tender to the HMS *Vernon* torpedo school, and so as an ambitious young naval officer would have been well aware of the difficulties and limitations of signalling at night between units of a flotilla of torpedo boats. It is likely he read the Annual Report of the Torpedo School for 1887 which included 'Appendix A – Extract from "Report on Mining and Torpedo Operations" carried out by the Mediterranean Squadron at Livathi Bay, near Argostoli, Cephalonia, April 1887' [9]. During the operations eight torpedo boats, in two divisions, carried out a mock attack on a squadron of battleships at anchor in the bay. The ships were defended by two lines of guard boats positioned across the bay, and the means employed for the detection of the 'attacking' force comprised both fixed and moveable searchlight beams, and look-outs on the shore.

One report of the exercise noted that the torpedo boats in the first division were 'seen as they crossed the beam [of a searchlight], but [were] obscured by the smoke of blanks when fire was opened' [9]. Further 'Remarks by Captain of "Vernon"' stressed that 'every effort should be made to obtain some satisfactory plan of identifying friends and distinguishing them from foes' [10]. Twelve months later the Annual Report observed [11]:

No perfectly satisfactory method of distinguishing between friendly and hostile torpedo boats appears yet to have been devised, and no ship could in war allow a torpedo boat to approach on a dark night. It would appear very difficult for a torpedo boat to identify an enemy without exposing herself to destruction.

Interestingly, at the commencement (1939) of the Second World War the Royal Air Force had no satisfactory method of detecting friend from foe, and several British aircraft were shot down by the RAF's fighters as a consequence.

In 1890 Jackson married his cousin Alice Burbury, the daughter of Samuel Burbury, FRS, a barrister and mathematician. He had published a treatise, in 1889, based on contemporary developments in Maxwell's theory of electromagnetic wave propagation. The possibility exists that Jackson knew of this work and that it stimulated him to consider the likely applications of the theory. He has recorded [12]:

my idea of utilising Hertzian waves for signalling from torpedo craft originated in about 1891, and I should have started experimenting in this direction earlier than I did (in 1895), had I had the time and instruments at my disposal and had I heard of the coherer principle sooner, the first time I heard of it being through reading some of Dr Bose's experiments in 1895. This important detail was all that was required, in my mind, to obtain signals by Hertzian waves from a distant vessel under way.

In 1895 Jackson was appointed to command HMS *Defiance*. Here, he had access to some limited facilities for experimental investigations and began, in December 1895, to construct his first apparatus for an inquiry into the uses of Hertzian waves. Moreover, he had acquired copies of Hertz's [13] and

Lodge's [14] books and was familiar with some of the published papers on the subject.

Jackson's earliest equipment was modelled on Bose's apparatus and was not very satisfactory [15]. However, by March 1896 he was using a more powerful induction coil in his transmitter, and had fabricated a coherer which comprised a glass tube filled with metal filings and an electric bell trembler. By the summer of this year he had discarded the glass and pitch lenses he used for focusing the radiation from his transmitter because he had found that elevated wires (antennas) at the transmitting and receiving ends of his system gave improved results. With this apparatus he demonstrated, on 20th August 1896, the transmission and reception of Morse code signals, and by the end of the month he had achieved a range [16] of 54.6 yards (50 m), the maximum working distance available on HMS *Defiance*. Also, at the end of the month he met the young Italian inventor named G. Marconi.

A. Righi, Professor of Physics at the University of Bologna, was another person who was aware of the work of Hertz and Lodge. He had written an obituary of Hertz for *Il Nuovo Cimento* in April 1894, and had followed this with a paper that described experiments which illustrated Hertz's work [17]. Righi's 4-ball spark-gap transmitter was similar to that of Lodge's transmitter of 1890 (although Lodge was not named), but his receiver used a Geissler tube as a detector rather than a coherer. Significantly, Righi soon afterwards replaced this detector by a coherer.

Among those who read Righi's obituary was Guglielmo Marconi, who was holidaying at Biellese in the Italian Alps. He subsequently said that, as a result of reading the article, the notion of wireless telegraphy using Hertzian waves suddenly came to him: he would now devote his life to implementing a system of wireless telegraphy.

Marconi's decision was extraordinary. He had not previously followed an apprenticeship or a formal, structured course in engineering, and he did not possess a recognised qualification in this discipline. Yet, undaunted by his technical limitations, he proceeded to investigate his chosen subject with much zeal and determination. He made an outstanding contribution to the development and commercial success of wireless telegraphy and radio communications. Honours, including prestigious orders, honorary degrees, honorary fellowships of learned societies, and the Nobel Prize in physics, were heaped on him [18]. At his funeral on 21st July 1937 the Italian radio services observed a five-minute silence [19]: in Great Britain, operators and engineers throughout the Post Offices of the land maintained a two minutes' silence from 18:00 hours, the hour of the funeral. (See Marconi's obituary [20].)

Guglielmo Marconi was born on 25th April 1874 at no. 7 Via delle Asse in Bologna. He was the second son of Giuseppe Marconi, a prosperous landowner, and his wife, Annie Jameson, a member of the well-known whisky family of Ireland [21].

Guglielmo's early childhood was privileged and was spent mostly in the family's country residence, the Villa Griffone, near Bologna, but when he was three years of age he was taken to England. After three years Guglielmo returned to Italy, was educated, initially, by a private tutor at the Villa Grifone, and then at a school in Florence. According to one biographer, W.P. Jolly, Guglielmo 'did not relish' the private tuition, and 'loathed almost everything about the ["inhospitable"] school': and it seems 'the school did not take to Guglielmo' [21].

In 1887 Guglielmo enrolled at the Leghorn Technical Institute and began to receive instruction in physics – including electricity – and chemistry. These subjects, particularly electricity, excited him so much that he persuaded his mother to arrange some extra private lessons for him from a Professor V. Rosa. The scientific studies led to Guglielmo experimenting at home, much to the displeasure of his father. At the end of seven years the young Marconi sat the matriculation examination of the University of Bologna but failed and was refused admission.

At this stage Annie Marconi used her influence, and her charms, to further the career of her favourite son. One of her neighbours was Professor A. Righi, of the University of Bologna. He was a most capable experimental physicist and a fine lecturer, and his reputation throughout Europe ensured that his classes were filled with students from many countries. Annie decided to meet the distinguished professor and tell him of her son's interests, aspirations, and difficulties, and to seek his support.

Righi, with many demands on his time, cannot have been enthused by the prospect of becoming involved with an amateur young experimenter who had failed to matriculate. However, Annie's persuasive appeal, which must have been considerable, brought success: her son would be allowed to use certain university facilities which were under Righi's control. Guglielmo was permitted to set up experiments in Righi's laboratory, possibly to borrow material for further experiments at home, and to have the use of the university's library [22].

After reading the obituary of Hertz, during the summer of 1894, Marconi returned home to commence his life's work on wireless telegraphy and radio communications. Many years later he observed [23]:

The idea [of wireless telegraphy] obsessed me more and more and, in these mountains of Biellese, I worked it out in imagination. I did not attempt any experiments until we returned to the Villa Grifone in the autumn, but then two large rooms at the top of the house were set aside for me by my mother. And there I began experiments in earnest.

My chief trouble was that the idea was so elementary, so simple in logic, that it seemed difficult for me to believe that no one else had thought of putting it into practice. Surely, I argued, there must be much more mature scientists than myself who had followed the same line of thought and arrived at an almost similar conclusion.

He told Righi of his aspirations but Righi, who had been researching the subject of Hertzian waves for several years and had written an account of Hertz's work, does not seem to have been enamoured of the prospect of Marconi making important advances in this field when his grasp of the necessary fundamentals was still somewhat insecure.

Nevertheless, the budding Italian inventor began by assembling the type of apparatus that was being employed by most, if not all, the other Hertzian wave researchers of the day. In a letter to Lodge, dated 18th June 1897, Righi wrote [24]:

Perhaps you have seen an interview with me, published in a Bologna journal, à propos of the so called invention of young Mr Marconi.

'I know this young man, who is very intelligent, although but little instructed in physics. I have advised him to pursue a regular university course. I shall be very curious to know about his apparatus, but I suspect it much resembles what he rigged up here with my oscillator and Lodge's coherer.'

By the end of 1894 Marconi had succeeded in achieving the typical ranges of contemporary workers, that is, ranges of the order of 9.8 yards (9 m), the distance across his attic workroom. Slowly the transmission distance was extended, but not dramatically so, during the spring and summer months of 1895. Then, one day, when he was experimenting in the grounds of the Villa Grifone and was using apparatus that had large metal bodies attached to the terminals of the spark gap, and similar plates connected to the terminals of the coherer, success was achieved. The range increased substantially. 'By chance [Marconi had] held one of the metal slabs at a considerable height above the ground and [had] set the other on the earth. With this arrangement the signals became so strong that they permitted [Marconi] to increase the sending distance to a kilometre [25].' Marconi initially had intended to use the plates to increase the wavelength of operation – by increasing the capacitance of the radiating system. This serendipitous discovery of earthing one end of each of the transmitting and receiving antennas and holding the other ends aloft (see Fig. 11.3) was the breakthrough which now enabled Marconi to progress rapidly. Whereas, with one set-up of tin boxes, of 12 in (30 cm) side positioned successively at heights of 6 ft (2 m), 13 ft (4 m), and 26 ft (8 m), gave effective ranges of just 98 ft (30 m), 328 ft (100 m) and 1,312 ft (400 m) respectively, another set-up using tin boxes of 39 in (100 cm) side placed 26 ft (8 m) above the earth gave a range of 7,874 ft (2,400 m) [26].

Towards the end of the 1895 summer another chance discovery aided Marconi [27].

I [Marconi] was sending waves through the air and getting signals at a distance of a mile, or thereabouts, when I discovered that the wave which went to my receiver through the air was affecting another receiver which I had set up on the other side of the hill. In other words, the waves were going through or over the hill. It is my belief that they went through, but I do not wish to state it as a fact.

Motivated by his progress and aware that financial support and commercial development would be needed to further his work Marconi sought the advice of his parents. Giuseppe Marconi consulted friends in the district and learnt that Dr Gardini, the family physician, knew the Italian Ambassador in London, General Ferrero, and was willing to request him to approach the Italian Government on Guglielmo's behalf [28]. In addition, Annie Marconi wrote to

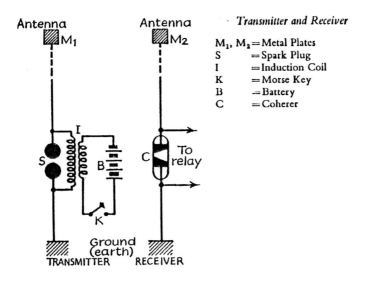

Transmitter and Receiver

M_1, M_2 = Metal Plates
S   = Spark Plug
I   = Induction Coil
K   = Morse Key
B   = Battery
C   = Coherer

Marconi's Improved Coherer (actual size)

A, B, = Evacuated glass tube
TT   = Platinum terminal wires
PP   = Silver bevelled plugs
S   = Side tube for exhaustion

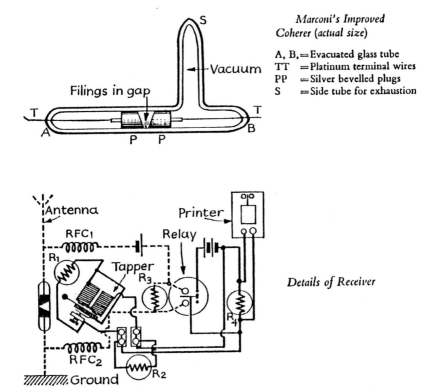

Details of Receiver

Figure 11.3   Marconi's apparatus for wireless telegraphy, 1894–1896.
Source: W.J. Baker, *A history of the Marconi Company* (Methuen, London, 1970).

her relatives in England in the expectation that they would be able to use their influences. Her eldest sister's son, Colonel H. Jameson-Davis, who was a professional engineer with an office at 12 Mark Lane in the City of London, had good contacts in both the scientific and the financial circles of the capital and was prepared to assist the young inventor. Other relatives offered accommodation if Annie and Guglielmo were to travel to London. Naturally, as Guglielmo was keen, in the first instance, to offer his invention to the Italian Government, General Ferrero was asked to make this position known to the Ministry of Posts and Telegraphs, in Rome, in his correspondence with them. Much to Marconi's bitter disappointment, the Italian Government saw no immediate future for his invention but did advise him to safeguard Italian interests when it was patented.

In February 1896 Annie and Guglielmo arrived in London. After a short stay with friends they moved to a private hotel at 71 Hereford Road, Bayswater, London found for them by Jameson-Davis. A few months later they moved to 21 Burlington Road, St Stephen's Square, and again, early in 1897, to 67 Talbot Road, Westbourne Park [29].

Soon after their arrival Marconi filed an application for a patent to protect his invention. It was filed at the Patent Office on 5th March 1896 and had the title 'Improvements in telegraphy and in apparatus therefor'; he was just 21 years of age at that time. The patent [30] (no. 5028) was subsequently abandoned – possibly it was inadequately drafted – and no details have survived.

Colonel Jameson-Davis was the contact that led to A.A. Campbell Swinton, electrical engineer, being given a private demonstration of Marconi's equipment in the Bayswater rooms. Swinton, on 30th March, wrote to W.H. Preece, of the Post Office [31]:

I am taking the liberty of sending to you with this note a young Italian of the name of Marconi, who has come over to this country with the idea of getting taken up a new system of telegraphing without wires, at which he has been working. It appears to be based upon the use of Hertzian waves, and Oliver Lodge's coherer, but from what he tells me he appears to have got considerably beyond what I believe other parties have done in this line. [Fig. 11.4]

It has occurred to me that you might possibly be kind enough to see him and hear what he has to say and I also think what he has done will very likely be of interest to you.

For reasons unknown, Marconi did not meet Preece until July. Meantime, in May, Marconi had written [32] to the War Office and declared that he 'had discovered electrical devices which enable[d] [him] to guide or steer a self propelled boat or torpedo from the shore or from a vessel without any person being on board the said boat or torpedo'. Since the coastal defence of the United Kingdom was the responsibility of the army, Marconi presumably felt that his 'invention' would be of interest to them. However, the officer, Major C. Penrose, deputed to investigate Marconi's system was interested more in the wireless telegraphy aspects of the invention than in the method of torpedo control [33].

Figure 11.4   Marconi with his apparatus.
Source: The Marconi Company.

At a further meeting, arranged by the War Office's Torpedo Committee and held on 31st August, Penrose introduced Marconi to Major G.A. Carr, Instructor in Electricity at the School of Military Engineering, Chatham, and to Captain H.B. Jackson – he had been promoted in June 1896 – who attended as the Royal Navy's representative

Jackson [15] soon learnt that 'the principles on which Signor Marconi's apparatus were constructed were similar to those employed by "Defiance", but [were] more fully developed, and the instruments themselves were much more sensitive'. Jackson told Marconi that he had been working from December 1895 on what was 'evidently the same thing which rather upset him, till I told him that I had no idea of patenting it'.

The improved sensitivity of Marconi's detector was not surprising for Marconi had been able to work full time on his inquiries whereas Jackson had had the detailed duties and responsibilities of his command to limit the time he could spend on his project.

Both Marconi and Jackson used coherers of similar dimensions but dissimilar mixtures; namely, fillings of nickel and silver amalgam, and tin and iron filings respectively. The improved sensitivity of Marconi's coherer stemmed from the very extended programme of fabrication and testing that he had initiated. Jackson has recorded that Marconi investigated the cohering properties of c. 500 different mixtures before he determined on his coherer's design [34].

After the War Office meeting, Penrose and Carr wrote a report on 'Signor Marconi's inventions' and opined [35]: 'this application of the invention is capable of considerable further development without the apparatus becoming too cumbersome for military purposes. It is understood that the GPO [General Post Office] are experimenting with the subject and it appears therefore unnecessary for the War Office to conduct experiments at present'.

Preece met Marconi, one morning in July 1896, in his office in the General Post Office's West building. The meeting has been described [36] by P.R. Mullis, who at that time was a young telegraph maintenance worker:

[Marconi] had two large bags. After handshakes and while the Chief cleaned his gold-rimmed spectacles Marconi placed on the table the contents of the bags: they included a number of brass knobs, a large sparking coil, and a small glass tube from each end of which extruded a rod joined to a disc fitted in the tube. The gap between the two discs was filled with metal filings. The Chief seemed particularly interested in that piece of apparatus. I obtained a Morse key, batteries, and wire. We joined up two circuits, fitting rods ending in knobs to the coil. We put the second circuit, containing the glass tube, on another table. When this had been completed the Chief – who was the kindest man I have ever met – pulled out his gold hunter. He said very quietly, 'It has gone twelve now. Take this young man over to the refreshment bar and see that he gets a good dinner on my account, and come back here again by two o'clock.'. . . .

Back in Mr Preece's office Marconi depressed the key in the sparking coil circuit whereupon an electric bell in the coherer circuit rang. Marconi tapped the coherer tube and the bell stopped ringing. He had to repeat that action each time he caused the bell to ring. I knew by the Chief's quiet manner and smile that something unusual had been effected. . . .

For the rest of the week there was further experimenting . . . Marconi would say, 'We will try this', 'We will try that' . . . The Chief arranged for Marconi's apparatus to be greatly improved in the Mechanics' Shop for demonstration to the Admiralty.

Another demonstration [37], arranged by Preece, was given on 27th July, before members of the GPO's administrative staff. Signals were successfully transmitted from the roof of one of the GPO's buildings at St Martins le Grand to a receiver on the roof of the Savings Bank in Knightrider Street, situated c. 295 yards (270 m) distant, and were recorded on the paper tape of the inker. The apparatus included a parabolic reflector of sheet copper made in the Mechanics' Shop, and an improvement that had been added in the Mechanics' Shop, which comprised an electric bell, wired in parallel with the inker, so that the coherer was 'de-cohered' by the bell's gong after each signal was received. At the demonstration Marconi met George Kemp (1857–1933), a Post Office technician, and retired petty officer from the Royal Navy's torpedo school, who would become a devoted admirer, and loyal assistant of Marconi and give him much valuable practical support extending over a period of c. 37 years.

The superior sensitivity of Marconi's detector was demonstrated during a major trial – the first in the UK – of his equipment, undertaken on Salisbury Plain, one week following the War Office conference. Again a spark-gap and coherer were employed and the radiation was focused by parabolic copper reflectors, 'somewhat like a searchlight projector', but during the series of tests the reflectors were replaced by antenna wires raised to heights of 100 ft (30.5 m) to 150 ft (45.7 m) above the ground. Preece later claimed that Marconi had not understood 'the proper use and function of vertical wires' before the 'Post Office made them for him', even though wire antennas had been used by several of the early wireless telegraph pioneers [36].

Ranges of c. 2.4 miles (4 km) were obtained in varying weather conditions. These transmission distances were not sufficient to impress Preece who, rather disappointedly, wrote [38]: 'The report is interesting but not encouraging. I think our plan of using electromagnetic waves is cheaper and more practical. . . . It is amusing to find the War Dept waking up to a system of signalling without wires which we have been working at for nearly ten years!'. (Here, Preece was referring to his inductive telegraphy system.) Still, Marconi had shown a considerable improvement in his equipment's capability, compared to the London demonstrations, and although Preece had presumably anticipated a much greater transmission range, there was no reason why additional development should not be expected. Certainly, Preece did not reject Marconi's invention, as evidenced by first, his address to the British Association, in September, when, for the first time, he introduced the assembled delegates to Marconi's system (Fig. 11.5); and second, his public lecture at the Toynbee Hall, East London on 12th December 1896.

At the Salisbury trials Captain Jackson attended as the Admiralty's

Figure 11.5   The first Marconi transmitter.
Source: The Marconi Company.

representative. He immediately saw the potential of the new system and reported to the Commander-in-Chief at Devonport that 'for military purposes, as an auxiliary signal for fog, and transmitting secret intelligence, its adoption would be almost invaluable'. Again, the apparatus 'would be invaluable for friendly torpedo boats to signal their approach' to units of the fleet. Using the new communication means an Admiral in his flagship would be enabled to maintain contact with all the ships under his command even 'when the ships were not visible, or even aware of their proximity to each other'. The application to collision avoidance at sea was obvious [39].

For these reasons Jackson considered the system would be 'worthy of a trial'. He submitted that

the apparatus [should] be first fixed up on the two torpedo boats attached to this school [HMS *Defiance*] for preliminary trial. . . . It should then be fitted to ships in commission for trial and report under service conditions. Three or more receivers . . . should be included in the order (with one transmitter) to ascertain that they behave similarly under all conditions, and these could be eventually distributed amongst several ships of the squadron, the flagship alone having the transmitter [39].

He [Jackson] recommended the use of 'all round lenses' rather than reflectors for naval service, to permit a given message being signalled to the whole fleet.

He also informed the Commander-in-Chief, Devonport: 'It may be of interest to state that the energy consumed by this apparatus to transmit signals (2 miles) at Salisbury was 13 watts, that for working the masthead flashing lamp being about 260 watts' [39] The apparatus in its current form was not, of course, ruggedised for ship-borne use; nevertheless Jackson considered 'that a design might well be prepared and made suitable for the roughest usage' [39].

During the autumn of 1896 Jackson continued his experiments at Devonport (Fig. 11.6). He increased the length of his receiver's antenna to 8 ft (2.4 m) and succeeded in working his apparatus to 300 yd (274 m). Further improvements to his transmitter and an increase in length of the receiving antenna to 70 ft (21.3 m) led by early March 1897 to a signalling range of 1,200 yd [40] (1,097 m).

An examination of the literature – both private letters and reports, and published papers – of the second half of the 1891–1900 decade shows Captain Jackson to have been a most honourable and considerate naval officer and gentleman. He never felt aggrieved because Marconi's apparatus was more sensitive than his own, or because his own position as an experimenter in the new field of wireless telegraphy would most likely be surpassed by an amateur whose age and standing were much junior to those of himself. Indeed, Jackson went out of his way to be helpful to Marconi. After the Salisbury trials Jackson set out in a letter [41] to the inventor the general specification for a ship-borne WT set:

I think personally that your apparatus is worth a trial, and would be of use to the service, if the signals can be made over 3 miles, without reflectors: all round lenses would be permissible.

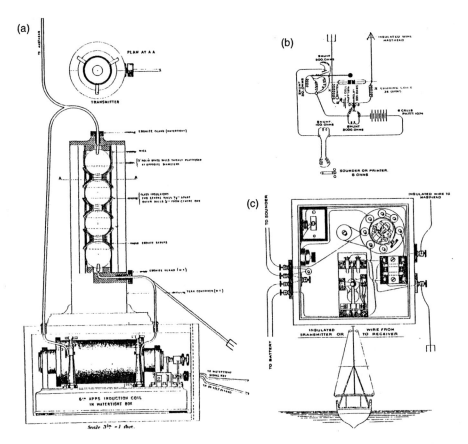

Figure 11.6 *(a) Plan of the apparatus used by Captain Jackson for telegraphing without wires between HMS* Scourge *and HMS* Defiance. *(b) Sketch of the receiver circuit. (c) Box containing receiving apparatus.*
Source: The Admiralty Library.

The size of the transformer would not be of importance but I would state roughly 4 cubic feet [0.11 m³], and a weight of 2 cwt [101 kg] should not, if possible, be exceeded. The power available would be a continuous pressure of 80 volts, of which 5 horsepower (3730 W) would practically not be much felt and be always available. All parts of the apparatus would have to be protected from wet and capable of standing rough usage, and heavy shocks from the firing of guns.

Three months after the September 1896 Salisbury Plain tests, Preece, during his lecture [42] at Toynbee Hall 'indulged his sense of the dramatic by telling his audience that he had the greatest pleasure that day in telling Mr Marconi that the Post Office had decided to spare no expense in experimenting with the apparatus.' Although the phrase 'spare no expense' was probably rhetorical, since no formal Treasury approval had been sought for any expenditure for the experiments, nevertheless Preece was in a position to direct staff and materials under his control to the furtherance of Marconi's endeavours.

Preece's decision to use his influence in a Government Department to give encouragement and patronage to a young, foreign, amateur inventor was extraordinary. It is unique in the annals of British electrical engineering history. At that time, it seemed to some scientists and engineers, including Lodge's friends, Fitzgerald and Heaviside, that Preece's judgement was strange [43], given

1. the apparent absence of any substantial degree of originality embodied in Marconi's system; and
2. the brilliant series of experiments which had been conducted, from c. 1888, by the British professional, university physicist, Lodge, on electromagnetic wave propagation which had culminated in the demonstration of wireless telegraphy in 1894.

Indeed, the suggestion has been made that Preece was being vindictive, in supporting Marconi rather than Lodge, because of the dispute that had arisen between them over the lightning conductors controversy [44]. Lodge, himself, seems to have accepted Preece's patronage of Marconi with good grace and stoicism. In a letter dated 16th October 1896, Lodge wrote [45]:

Dear Mr Preece
   I thank you for your friendly courtesy in sending me Kempe's report on Marconi's trials. Some of my friends seemed to feel aggrieved, but I told them that if I had come asking for facilities for large scale experiment you would doubtless have given it me 2 years ago, and that it was well for a man who took the trouble to work the thing out in detail to have facilities given him. (I did not realise that it was much of a desideratum)
   At the same time there is nothing new in what Marconi attempts to do, it could have been done any time these 2 years in a laboratory way, and I thought it better to demonstrate practically to the section [presumably the British Association section] that I too could work a Morse instrument by electric waves and get dots and dashes without[?] any notice. Ld Kelvin and others saw it working on the Wednesday morning, but you were busy in the divided section. . . . I have no doubt that Marconi uses a similar plan to mine viz a coherer, shaken by [a] mechanism, (as I showed at Oxford) back to zero, and actuating a relay whenever waves fall upon it, said waves being of course collected by a synchronised resonator attached to it . . .
   Yours faithfully
   Oliver J. Lodge

Notwithstanding his reasonable attitude Lodge does seem to have been aggrieved by Marconi's success. In another letter to Preece, dated 29th May 1897, that is after some further experiments by Marconi, Lodge noted [46]:

My dear Preece
   The papers seem to treat the Marconi method as all new.
   Of course you know better, and so long as any[?] scientific conferences are well informed it matters but little what the public says.
   The stress of business may however have caused you to forget some of the details published by me in 1894. I used brass filings in vacuo then too. It could all have been done

3 years ago had I known that it was regarded as a commercially important desideratum. I had the automatic tapping-back and everything, see enclosed pamphlet

I see you are lecturing on the matter next Friday.

I was in town yesterday, and nearly called to see you, but hardly thought it likely that I should find you without an appointment. Besides, as you said in your letter last October, you are not likely to forget my share in the work, or the fact that Marconi must have got all his initial ideas from my little book.

Yours very truly
Oliver J. Lodge

(Lodge's letters are not carefully punctuated.)

Lodge, Preece and Righi, were essentially correct when they inferred that Marconi's system was based on work that had been undertaken by Lodge and Righi. However, what Marconi brought to the United Kingdom was a practical knowledge of the means by which the range of signalling could be increased. Neither Lodge nor Righi had carried out any investigations on determining the factors that would lead to their apparatus having a commercial viability, but Marconi fully appreciated that, if the Righi oscillator–Lodge coherer combination were to have a commercial application, the distance between the transmitter and receiver of a communication link would have to be substantially increased. He set out to discover how this could be achieved. His empiric experimental work on antennas – particularly his attempts to determine the relationship between range and heights of the antennas – was the pivotal factor which enabled the laboratory apparatuses of Lodge and Righi to be transformed into commercial systems.

Furthermore, Marconi had the vision, like Crookes in 1892, and Jackson in 1891, but unlike Lodge in 1894, to see the practical applications of wireless telegraphy. Preece, too, was aware of the likely uses of the new communication system. In a lecture given in June 1897 at the Royal Institution he said [47]: 'There are a great many practical points connected with this system that require to be thrashed out in a practical manner before it can be placed on the market, but enough has been done to prove its value and to show that for shipping and lighthouse purposes it will be a great and valuable acquisition'.

Preece's opinion was presumably predicated on the unfavourable prospect of either the conduction or induction methods of electric telegraphy being successful for long range signalling. These methods, as noted previously, had attracted the attentions of many workers from the 1840s until the 1890s, but in all the trials that had been executed the maximum range of message transmission had not substantially exceeded the length of the wire used at the transmitting/receiving sites. Preece, himself, had initiated researches on both methods from 1882 and knew well the limitations of these forms of signalling. Indeed, when Lodge was investigating the properties of Herztian waves in 1894, Preece was directing the first experiments of wireless telephony using a non-Herztian wave method.

In February 1894 trials were conducted across Loch Ness to determine the laws governing the transmission of Morse signals by Preece's electromagnetic method of wireless telegraphy. Two parallel wires, properly earthed, were laid, one on each side of the loch, and subsequently systematically shortened to determine the minimum length necessary to record satisfactory signals. During the trials Mr (later Sir) John Gavey (1842–1923), Assistant Engineer-in-Chief, decided to compare telephonic and telegraphic signalling [48]. The tests showed that it was possible to exchange speech across the loch at an average distance of 1.3 mile (2.1 km) between the parallel wires when the length of the wires themselves was reduced to four miles on each side of the water'. Clearly, this type of communication system did not augur well for the furtherance of long distance telephony across open water.

Marconi's early work has to be set against the times when it was undertaken. In the 1890s no theories of either antennas or electromagnetic wave propagation existed. Marconi could not have consulted the research findings of others in these fields for there were none, even though several experimenters, including Elihu Thomson, Thomas Edison, Amos Dolbear and David Hughes had observed the efficacy of vertical radiators – but not in association with wireless waves. Marconi had to determine the optimum conditions for propagation by practical tests. His modus operandi, given his background, had to be of an empiric nature. He carried out numerous tests so that when he came to the United Kingdom he was able, in September 1896, to demonstrate wireless telegraphy over a greater distance than had been achieved by Preece during his 1894 wireless telephony experiments at Loch Ness. Not surprisingly, Preece, who had a responsibility, as the Engineer-in-Chief of the General Post Office, to keep abreast of all developments in the field of communications and to inform the Postmaster General of these developments, was anxious to explore the advantages and disadvantages of any new method of signalling vis-à-vis the existing methods, namely line telegraphy and line telephony. Marconi was an enthusiastic experimenter who had the determination and imagination, if not the fundamental knowledge, to ensure the success of the Lodge method, if it could be realistically achieved. Preece's encouragement of Marconi stemmed from the pragmatic outlook of the young experimenter, which contrasted with the scientific attitude of the university physicist.

Aware that his policy of patronage might be thought to be curious, if not vindictive, towards Lodge, Preece in a lecture to a large audience at the Royal Institution presented on 4th June 1897 stated [49]:

In July last Mr Marconi brought to England a new plan. Mr Marconi utilises electric or Hertzian waves of very high frequency. He has invented a new relay which for sensitiveness and delicacy exceeds all known electrical apparatus. The peculiarity of Mr Marconi's system is that, apart from the ordinary connecting wire of the apparatus, conductors of very moderate length only are needed, and even these can be dispensed with if reflectors are used. . . .

Excellent signals have been transmitted between Penarth and Brean Down, near Weston-Super-Mare, across the Bristol Channel, a distance of nearly 9 miles (14.5 km). On Salisbury Plain Mr Marconi covered a distance of four miles (6.4 km).

On the point reported by some commentators that Marconi had contributed nothing new, Preece said [50]:

He has not discovered any new rays; his receiver is based on Branly's coherer. Columbus did not invent the egg, but he showed how to make it stand on its end, and Marconi has produced from known means a new electric eye more delicate than any known electrical instrument, and a new system of telegraphy that will reach places hitherto inaccessible. . . . Enough has been done to prove and show that for shipping and light-house purposes it will be a great and valuable acquisition. (See Fig. 11.7.)

Marconi's first patent was published on 2nd July 1897. It included 19 claims, the first of which was: 'The method of transmitting signals by means of electrical impulses to a receiver having a sensitive tube or other sensitive form of imperfect contact capable of being restored with certainty and regularity to its normal condition substantially as described' [51]. As a US historian, H.G.J. Aitken, who has examined the patent in detail, has pointed out [52]: 'It would have been difficult indeed for Marconi to prove that he had literally been the first to "invent" what this claim, or in fact most of the other claims, described. He was the first to claim these methods, these pieces of equipment, these circuits, as *property*, and under British patent law that was what counted'. Elsewhere, in the United States of America and Germany the patent laws attached more weight to priority of discovery and so, given Lodge's lectures and demonstrations in 1894, the practical consequence of Marconi's first patent was nullified.

None of the diagrams reproduced in the patent contain tuned or tunable circuit elements, although the patent does contain the statement that the receiving antenna should be of a size 'as to be preferably tuned with the length of wave of the radiation emitted from the transmitting instruments'. No reference is made to the problem of interference, and hence the desirability of there being tuned circuits to minimise this problem is not considered. All of these facts were too much for the *Electrician*, the leading electrical journal of the day, and in a trenchant editorial the editor summarised Marconi's patent, matched its claims against Lodge's demonstration in 1894, and declaimed:

Dr Lodge published enough three years ago to enable the most simple minded 'practician' to compound a system of practical telegraphy without deviating a single hair's-breadth from Lodgian methods. . . . It is reputed to be easy enough for a clever lawyer to drive a coach and four through an Act of Parliament. If this patent be upheld in the courts of law it will be seen that it is equally easy for an eminent patent-counsel to compile a valid patent from the publicly described and exhibited products of another man's brain. No longer is it necessary to devise even so much as 'a novel combination of old instrumentalities', and the saying 'ex nihilo nihil fit' [out of nothing nothing comes] evidently was not intended to apply to English patents at the end of the nineteenth century [53].

From his study Aitken concluded [54]: 'There was nothing in Marconi's patent that was new in concept, with the single possible exception of the vertical antenna used for transmission'.

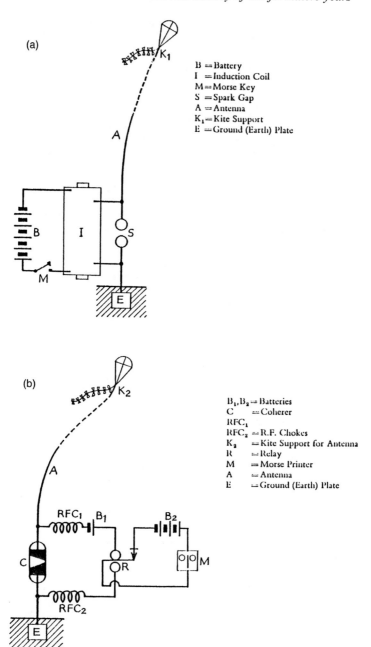

(a)

B = Battery
I = Induction Coil
M = Morse Key
S = Spark Gap
A = Antenna
K_1 = Kite Support
E = Ground (Earth) Plate

(b)

B_1, B_2 = Batteries
C = Coherer
RFC_1
RFC_2 = R.F. Chokes
K_2 = Kite Support for Antenna
R = Relay
M = Morse Printer
A = Antenna
E = Ground (Earth) Plate

Figure 11.7   (a) Marconi's apparatus for wireless telegraphy, 1897; the transmitting system. (b) Marconi's apparatus for wireless telegraphy, 1897; the receiving system.

Source: J.A. Fleming, *The principles of electric wave telegraphy* (Longmans, Green, London, 1908).

The work of Marconi in the UK, in 1896–1897, had an important effect on Lodge and the progress of wireless communications. He stirred himself from the state of lethargy into which he had fallen and applied his intellect to the problem of syntony. The importance of tuning in wireless communications systems, so that different wireless transmissions could be distinguished by a receiver, had not been lost on Lodge. In his lecture on 'The work of Hertz', given on 1st June 1894 [55], Lodge had stressed the need for syntony or resonance tuning for message discrimination. Again, in a letter [56] to S.P. Thompson, dated 14th April 1897, Lodge observed: 'Marconi is nothing but coherer and relay[.] I am first going to publish or patent a plan of precise tuning between send[er] and receiver'. Lodge's patent no. 11,575 on 'Improvements in syntonised telegraphy without line wires' was filed on 10th May 1897 before Marconi's first patent [57], no. 12,039 which had been submitted on 2nd June 1896, had been published and its technical details made known. A further Lodge patent [58] with a similar title, no. 29,069, was registered on 10th December 1897; and two Lodge patents [59,60] relating to improvements in coherers, nos. 16,405 and 18,644, were also filed in 1897.

During 1897 and 1898 Marconi worked to improve the range of his equipment and to demonstrate its usefulness for signalling across open waters. He had in mind ship-to-ship, and shore-to-ship communications – applications for which no modern technology existed. Moreover, there was the possibility that wireless communications would prove to be more reliable and cheaper than line communications for signalling between a shore and a lighthouse or lightship.

He continued his trials [61] on Salisbury Plain from 15th March to 25th March 1897. Prior to this month Marconi had observed that the distance at which signals could be received 'increased almost exactly in direct proportion with the square of the height' of his antenna's metallic body from earth, for heights up to about 30 ft (9.1 m). He wished to determine whether this law applied for heights of 100 ft to 150 ft (30.5 m to 45.7 m). For his experiments Marconi had the assistance of H.R. Kempe, and a sergeant and two sappers from the Royal Engineers.

Using kites and balloons, flown in very variable weather conditions, Marconi found that 'good signals' could be obtained at 4.3 miles (6.9 km), and 'indications of signals' at 6.8 miles (10.9 km), from the transmitter. 'One of the most important facts which I [Marconi] have learned from these experiments is that signals can be transmitted to considerable distances in all directions, without employing bulky conductors or capacities in the air, it being apparently sufficient to have a wire leading to the top of a pole or mast to as great a height as possible' [61].

The poor weather prevented Marconi from corroborating the law, (range = function of height^2), though he was of the opinion that it still held.

Captain H.B. Jackson, RN witnessed the tests on the 24th: 'He was very pleased of the results' wrote Marconi to Preece [62]. A week later Jackson informed the C-in-C, Devonport, of the outcome of the trials and of his own work following these. Antenna wires had been fitted to the masts of HMS *Defiance* and a marked improvement had been obtained.

One of the questions that Jackson had to consider was whether a design, for a W/T set, could 'be prepared and made suitable for the roughest usage' aboard naval vessels. This was an issue of prime importance since the main applications of wireless telegraphy in the late 1890s seemed to be those associated with ship-borne use. Consequently, Jackson implemented and tested suitable apparatus, and on 19th May 1897 told Marconi: 'I tried my apparatus in heavy rain this morning, the transmitter being entirely exposed to the weather and all the braiding of the wires soaking wet, yet my results were very good and the signals were apparently quite unaffected by the wet wires' [63].

Jackson subsequently organised a series of trials, involving torpedo boats attached to HMS *Defiance*. The C-in-C was formally invited to observe the trial, held on 20th May 1897, when, with the transmitter housed in the gunboat HMS *Scourge* and the receiver on board HMS *Defiance*, the transmission of signals was successfully demonstrated over distances of up to c. 3.1 miles (5km) [64].

Additionally, Jackson established that the system for use on board ship did not affect the *Scourge's* compass, 'though within 10 feet (3.05 m) of it', and did not cause premature firing of the guns or torpedoes 'at the ready'. He found 'the various parts of the apparatus are not more delicate than other electrical instruments in the service, and . . . require no more attention than them, when once adjusted' [65].

These sea trials boded well for the adoption of wireless telegraphy on board naval vessels, but one problem remained – a problem that Jackson had identified in his September 1896 report to the Commander-in-Chief, Devonport – namely, the problem of tuning. When a ship transmitted a signal all receivers in the fleet would, except under certain conditions, 'record every signal made by every ship (unless switched off), thus rendering it impossible for two ships to signal at once, or all ships to answer a general signal together' [66].

Jackson's direct involvement with wireless telegraphy ended when he was posted, on 1st November 1897, as the Naval Attache to the British Embassy in Paris, to report on the submarines then being constructed for the French Navy. His position as Captain of HMS *Defiance* was taken by Captain F.T. Hamilton, RN who continued Jackson's work.

Following the Salisbury Plain experiments, Marconi, having confirmed that his apparatus could signal over ground for distances of several miles, now wished to determine the characteristics of his system when propagation occurred over water. He, or possibly Preece, selected the same sites, Lavernock Point and Flat Holme island, that Preece had used for some of his induction telegraphy experiments – presumably to enable a comparison to be made between the two forms of telegraphy [67].

The distances from Lavernock Point, near Cardiff, to Flat Holm island in the Bristol Channel and from Lavernock Point to Brean Down, near Weston-Super-Mare, are 3.3 and 8.7 miles (5.3 km and 14 km) respectively. The trials, in May, effectively simulated the situation that would prevail when signalling between a land station and an offshore lighthouse or lightship. According to Preece, 'excellent signals' were received at Lavernock Point from Flat Holme island and from Brean Down across the Bristol Channel.

Among the guests invited to observe Marconi's experiments were representatives of the Royal Commission on Lightship Communications, and Professor A. Slaby, of the Technische Hochschule, Charlottenburg, Berlin (see Chapter 12).

Meanwhile, a proposition had been put to Marconi that would lead to a discontinuance of the friendly and helpful relationship he had enjoyed with Preece. On 10th April Marconi wrote to the Engineer-in-Chief of the GPO and said:

I am in a difficulty. Those gentlemen (Mr Jameson, Davis [sic] and others) which desired to form a company for acquiring the rights of my invention, and to which I had notified that I could not deal with them, or give them any definite answer until the experiments I am carrying out with your assistance are concluded, have notified to me through my solicitors that they want to know without much delay whether I intend to accept their offer or not [68].

The terms of the offer would give Marconi £15,000 in cash and half the shares in the proposed company which would have a working capital of £25,000. This was a serious proposal that required his very careful consideration. During his experiments Marconi, with the support initially of his father, and later of the GPO, had carried the burden of financing his experimental investigations but obviously this state of affairs could not continue indefinitely. There had to be additional financial resources to allow the system to be developed – and be protected by patents – to the stage where it would be a worthwhile, marketable product. Only then could revenues be anticipated.

In his letter Marconi wrote: 'Mr Jameson and his friends think that the Company would make money through constructing apparatus for use on board ships for the purpose of enabling the ships to be warned in fogs, when in the proximity of rocks or dangerous shallows, or for preventing collisions.' He stressed that he had never sought the offer or given encouragement to the promoters. However, Jameson Davis, his mother's cousin, had been helpful in introducing him to Preece, and, other than selling Marconi's system to the General Post Office – which presumably would restrict its applications to the United Kingdom – it is difficult to posit how Marconi could have advanced his system without the formation of a company.

Preece's immediate response to this letter does not appear to be available, but his reply to a further letter from Marconi would be blunt and unambiguous.

By July 1897 reports of Marconi's successes had reached the Italian Government: Marconi was summoned to return to Italy. He carried out demonstrations at Spezia and showed that wireless telegraphy was possible between the San Bartolemeo shipyard, Spezia, and the cruiser *San Martino* 12 miles (19.3 km) distant. This was followed by another successful demonstration (in August), seen by the King and Queen of Italy, which led to the adoption, at a later date, of the Marconi system by the Italian Navy [69].

While in Italy, Marconi sent another letter to Preece telling him that, on the advice of all his legal advisors, he had entered into a contract with an English company and that he held half the shares. Many motives had induced him to accept the company's offer. First, he said, there was the need to construct his apparatus 'in a more practical form' and to proceed with further 'extensive experiments'; second, he opined that 'the business [was] too large for himself alone, as all the governments in Europe want experiments carried out'; and third, he felt the expenses associated with patent applications were 'too much for [him], especially as [he had to] patent some further improvements'. 'This added to the great uncertainty of the life of the patents, and the vigorous opposition made to [him] by Lodge in England[,] Tesla in America[,] and others in Europe [had] induced [him] to take this step' [70].

Marconi ended his letter by expressing the hope that Preece would continue his benevolence towards him, and by stating that all Preece's great kindness would never be forgotten by him 'in all [his] life'.

Preece's reply, dated 6th August 1897, was stern. 'I am very sorry to get your letter. You have taken a step that I fear is very inimical to your personal interests. I regret to say that I must stop all experiments and all action until I learn the conditions that are to determine the relations between your company and the Government Departments who have encouraged and helped you so much' [71].

The Wireless Telegraph and Signal Company Limited was established on 20th July 1897 with the objective of developing commercially the Marconi system. It had a nominal capital of £100,000, comprising 100,000 £1 shares. Marconi was handsomely rewarded for his efforts and received 60,000 shares, the balance of 40,000 being available for public subscription. In addition out of the proceeds from the sale of these shares Marconi was awarded £15,000 for his patents, but had to pay the formation expenses. He was just 23 years of age [72].

£15,000 was a very considerable sum of money in 1897 as evidenced by the costs of goods and services. A labourer earned £1 per week, a 2-bedroom house could be bought for £300, rent for a working class family was 35p per week, coal was £1.15 per ton, income tax averaged 3.5p in the pound during the decade, and a gallon of beer and an ounce of tobacco were 5p and 2p respectively [73].

Jameson Davis was the company's first Managing Director, and H.W. Allen was appointed the first Secretary: they operated from offices at 28 Mark Lane in the City of London. Here demonstrations of the Marconi system were given to interested parties.

On returning to England Marconi continued his experiments for the GPO and in October established communications between Salisbury and Bath, a distance of 34 miles (54.7 km) [72].

Marconi's vision for wireless telegraphy was to communicate with (and eventually between) ships at sea. As the principal shareholder in the new company he had a responsibility to the other shareholders to secure the company's financial well-being, which would enable dividends to be paid from profits earned. Profits could be realised from the sale or hire of wireless telegraph apparatus, from licence fees from the users of the system, and from royalties obtained from the use, by other manufacturers, of the Marconi patents.

As part of his plan to further his company's objective Marconi installed coastal stations at the Needles Hotel, Alum Bay, in the Isle of Wight and, initially, at Madeira House, South Cliff, Bournemouth, and later, in September 1898, at the Haven Hotel, Poole [74]. Masts – 120 ft (36.6 m) high were erected at both stations to support the antenna conductors (usually stranded conductors of 7/20 copper wire insulated with rubber and tape). A ten inch (25.4 mm) induction coil worked by a battery of 100 Obach cells, 'M' size, was utilised at each station, and signals were transmitted to receivers placed on board the steamers *Solent* and *May Flower*, which operated daily between Alum Bay and Bournemouth and Swanage (18.5 miles – 29.8 km – distant). Soon telegraphic contact (at four words per minute) could be maintained with the ships for the whole of their voyages, even during inclement weather. A demonstration was given to the GPO on 23rd December 1897.

In January 1898 an adventitious event gave Marconi much good publicity. During the winter a heavy snow storm had for a time isolated Bournemouth telegraphically from London [75]. This caused concern to the newspaper reporters who had gathered in Bournemouth, to cover the death of W.E. Gladstone, but who could not send their messages to the capital. Marconi heard of their predicament and was able to transmit by wireless telegraphy their copy to his Alum Bay station, from whence the messages were forwarded to London by the standard telegraph lines. Much goodwill and publicity were thereby engendered.

By now responsibility for the introduction of wireless telegraphy into the Royal Navy had been transferred from the Torpedo Schools to the Admiralty's Signal Committee, an 'estimable body', according to Admiral Sir Reginald Bacon, 'who knew the colour of every flag and all about the Morse code, but to whom electricity was a sealed science'. The committee's ignorance on wireless communications was short-lived. On 7th May 1898, Commander Evan-Thomas, the Secretary of the Signal Committee, together with Admiral Lord Charles Beresford, the formidable Commander in Chief, Portsmouth, visited Marconi at the Alum Bay station. In addition, since Evan-Thomas was not technically qualified in electrical matters, a technical expert, Lieutenant Hornby of the Torpedo School, HMS *Vernon*, accompanied the party. Demonstrations were given of signals being transmitted to and from Bournemouth.

In his report to the Signal Committee, Commander Evan-Thomas [76] commented approvingly on the demonstrations but unerringly exposed a weakness of the system.

With the apparatus in its present stage, it seems possible, given instruments of the same, or nearly the same tune [i.e. wavelength], for an enemy to spoil your communications by simply transmitting perpetual signs, these would be shown on all receivers in distance and convert all communications into a jumble. For instance, if a blockaded port has a transmitter, the communications of the blockaders (when you had once got their tune) could always be thrown out if in distance.

Again, cruisers watching a channel to which this system might be particularly useful, would find their communications thrown out when most needed, by the presence of hostile ships also using the system.

These examples, however, would show rather a reason for having the system in the service than not, if only for the purpose of incommoding our possible enemies.

During the same month Marconi, at the request of Lloyds, successfully transmitted signals between Ballycastle, Ireland and Rathlin Island, 7.5 miles (12.1 km) distant, on which there was a lighthouse [74].

Following this test, Marconi was commissioned by the *Daily Express* of Dublin to report 'from the high seas the results and incidents of the Kingston Regatta' [77]. A steamer, the *Flying Huntress*, was chartered to follow the racing yachts and the necessary equipment placed on board. In the grounds of the harbourmaster a 110 ft (33.5 m) pole to carry the antenna wire was erected, and a telephone link between the Kingston land transmitting station and the *Daily Express* office in Dublin was established. Seven hundred messages were signalled wirelessly, over ranges varying from 10 to 25 miles (16 to 40 km), to give the newspaper the latest news – well in advance of its competitors – on the progress and result of each race. The scoop brought Marconi not only a desirable financial gain but, more importantly, much publicity in the world's newspapers.

Immediately after the finish of the regatta, he was requested to install his system on the Royal yacht *Osborne,* where the Prince of Wales (the future King Edward VII) was convalescing, and in Osborne House, Isle of Wight, where Queen Victoria was in residence. The Prince had suffered an injury and the Queen wished to be kept informed of her son's well-being but intervening hills made visual signalling impossible [78].

On 4th August 1898 the first medical bulletin ever transmitted by wireless was sent from the Royal yacht, which was lying off Cowes, to Osborne House, a distance of 1.75 miles (3 km):

From Dr Fripp to Sir James Reid

H.R.H. the Prince of Wales has passed another excellent night, and is in very good spirits and health. The knee is most satisfactory [74].

During the next 16 days another 149 messages were exchanged, including some when the Royal yacht was cruising off the Needles, a distance of 7 miles (11.3 km), and when Osborne House was completely screened from the yacht by hills.

An amusing anecdote has been related about this trial. The 100 ft (30.5 m) pole that supported the receiving antenna wire was situated at Ladywood Cottage in the grounds of Osborne House. One day the young Marconi while walking through the grounds on his way to inspect the apparatus was stopped by a gardener and told to go via a longer route, since Her Majesty was in the grounds and would object to his intrusion of her privacy. Marconi refused to deviate from his path and the issue was reported to the Queen. She was 'not amused', and is supposed to have commented: 'Get another electrician', unaware that this was not possible.

The above mentioned demonstrations of some of the possible applications of wireless telegraphy gave Marconi a great deal of free publicity for himself and the The Wireless Telegraph and Signal Company, but in addition the various trials exposed a weakness in the organisation of the company. Until November 1898 Marconi had been the only technical authority in the company: his investigations and demonstrations had been undertaken with the support of just one technician, G.S. Kemp, and a few assistants/helpers, including W.W. Bradfield. Marconi's apparatus towards the end of 1898 was still, in its essential features, that which he had used two years previously, although improvements had been incorporated by his empirical method of 'trying this and trying that'. He was aware of the urgent need to keep ahead of his competitors and to achieve this objective required staff who had a greater intellectual and technical competence than his own. A physicist was needed who could progress the experimental aspects of wireless telegraphy.

In November 1898 Dr J. Erskine Murray, assistant professor of physics at Heriot-Watt College, Edinburgh resigned his position and became Marconi's principal experimental assistant [79]. From this date Marconi began to surround himself with a group of very able engineers and scientists [80]. Moreover Marconi fully appreciated the worth of specialists and later employed them for many years. He sought out some of the most promising university scientists who were interested in wireless and appointed them as consultants. Professor A. Fleming and Professor M. Pupin [81] were two of the notable consultants of the WT&S Company. Marconi was never hesitant in engaging staff who were his superiors in academic attainments. His policy certainly brought benefits to the company. One of his most important early demonstrations of wireless telegraphy, the transatlantic venture of 1901, owed much to Fleming's work. It was Fleming who designed the transmitters, each of which incorporated an ingenious circuit having two spark-gaps operating in cascade at different frequencies.

Erskine-Murray began work for Marconi at the Haven Hotel, Poole; there was much to be done. One of the ground floor rooms in the hotel was adapted as the main laboratory, and various huts in the hotel grounds were utilised for further experimental work. A visitor to the Haven station in 1899 mentioned that in the main laboratory

he found two of the earliest employees, the brothers Cave, at one table making coherers; at another, P.W. Paget was winding receiver chokes and at a third Marconi himself was busy fitting V-gap plugs into an experimental coherer. Outside, along the foreshore, Dr Erskine Murray was conducting parabolic mirror reflector tests, using centimetric [sic] wavelengths [82].

Life seems to have been congenial for Marconi's small staff. At meal times they, Marconi, his mother, his brother Alfonso (when present), Mrs Erskine and any visitors all shared a common table. Frequently, after the evening meal, musical entertainment was provided by a trio comprising Marconi (piano), Alfonso (violin) and Murray (cello) [83].

Many demonstrations were given to individuals and organisations who could further the aspirations of the young inventor. Some of these had been noted in the Postmaster General's published report for the year ending 31st March 1898: 'A series of trials was undertaken with the Marconi apparatus with special reference to its adaptability for lighthouse, lightship, and other communications: but although signals were successfully transmitted a certain distance, no [commercially] practical results have yet been achieved' [36].

Here was a potentially valuable market. The Corporation of Trinity House had long wanted some reliable means of communicating with their off-shore lighthouses and lightships but none existed. Two systems had been tried, namely, the use of a cable, and the use of inductive telegraphy, but both had been found wanting. The cables were short-lived because of abrasion caused by their rubbing on rocks; and the inductive system was not satisfactory since it required two parallel conductors – one on the shore and the other on the lighthouse island – each of which had to have a length comparable to the distance between shore and lighthouse.

And so, with the unsatisfactory history of previous trials much in their minds, the Elder Brethren of Trinity House offered WT&S the opportunity of demonstrating to them the utility of Marconi's system between the South Foreland lighthouse and one of the following lightships, viz., the *Gull*, the *South Goodwin*, and the *East Goodwin*. The latter vessel was selected – it was moored c. 12 miles (19.3 km) from South Foreland – and Kemp was delegated to undertake the necessary installation, in foul weather, in December 1898.

Numerous two-way exchanges between ship and shore were conducted, during which officers of Trinity House were shown the apparatus being worked by the crew of the lightship themselves. On one occasion during a strong gale in January 1899 part of the bulwarks of the *East Goodwin* was carried away by very heavy seas. A report of the damage sustained was promptly telegraphed to South Foreland and thence to the superintendent of Trinity House. Two months later, on 17th March, when the *Elbe* went ashore on the Goodwin Sands in dense fog, the Ramsgate lifeboat was called out, by the same means, to render assistance. And on 28th April, when the SS *R. F. Matthews*, in dense fog, collided with the lightship and caused severe damage to its stem, the wireless installation was again used to seek the aid of a lifeboat [84]. These incidents highlighted the enormous potential importance of wireless telegraphy for safety at sea, if the range of working could be increased dramatically.

The application of maritime wireless telegraphy is considered in Chapter 13, but before this can be discussed certain developments in Russia, Germany and the United States of America must be considered.

References

1 BURNS, R.W.: 'Impact of technology on the defeat of the U-boat, September 1939–May 1943', *IEE Proc.-Sci. Meas. Technol.*, (5) 1994, **141**, pp. 343–355

2 CHURCHILL, W.S.: 'Closing the ring' (Cassell, London, 1954), p. 20
3 Quoted in ANON.: 'The battle of the Atlantic' (HMSO, London, 1946), p. 5
4 ROUTLEDGE, R.: 'Discoveries and inventions of the nineteenth century' (George Routledge and Sons, 1903), pp. 227–248
5 HEZLET, A.: 'The electron and sea power' (Peter Davies, London, 1975), pp. 1–16
6 BURNS, R.W.: 'The background to the development of early radar, some naval questions', pp. 1–28, in BURNS, R.W. (Ed.): 'Radar development to 1945' (Peter Peregrinus Ltd, London, 1988)
7 POCOCK, R.F.: 'Admiral of the Fleet Sir Henry Jackson (1855–1929)', Proc. of the 24th The IEE Weekend Meeting on the *History of Electrical Engineering*, 1996
8 The Navy List, 1887, PRO
9 ANON.: Annual Report of the Torpedo School (HMS Vernon) for 1887, Appendix A – 'Extract from "Report on mining and torpedo operations"', Naval Library, MOD
10 Ref. 9, p. 149
11 Annual Report of the Torpedo School (HMS Vernon) for 1888, p. 8
12 JACKSON, H.B.: communication to Admiral Sir John Fisher, 28th November 1900, ADM 116/570, PRO
13 HERTZ, H.: 'Electric waves' (MacMillan, London, 1893)
14 LODGE, O.J.: 'The work of Hertz and his successors' (Electrician Printing and Publishing Co., London, 1894)
15 'Statement of Captain Jackson's claims as regards the invention of wireless telegraphy', appended in a letter from Captain F.T. Hamilton to the Commander-in-Chief, Devonport, 28th January 1899, ADM 116/523
16 'Tabular statement of dates of working out important points of "Defiance's" system', ADM 116/523
17 RIGHI, A.: 'Su alcuno disposizioni speriementali per lo dimonstrazlione e lo studio delle ondulazioni di Hertz', *Il Nuovo Cimento*, 1894, **XXXV**, pp. 12–17
18 JACOT, B.L. and COLLIER, D.M.B.: 'Marconi – Master of Space' (Hutchinson, London, undated), pp. 279–281
19 BAKER, W.J.: 'A history of the Marconi Company' (Methuen, London, 1970), p. 295
20 FLEMING, J.A.: obituary, *J. of the Roy. Soc. of Arts*, 1937, **26**, pp. 57–62
21 JOLLY, W.P.: 'Marconi' (Constable, London, 1972)
22 Ibid, p. 16
23 Ref. 18, p. 24
24 RIGHI, A.: extract from a letter to Professor O.J. Lodge, 18th June 1897, The IEE Archives
25 Quoted in ref. 21, p. 27
26 MARCONI, G.: 'Wireless telegraphy', *J. IEE*, 1899, **28**, pp. 273–297
27 Interview, *McClure's Magazine*, March 1897
28 Ref. 21, p. 32
29 GARRATT, G.R.M.: 'The early history of radio from Faraday to Marconi' (Institution of Electrical Engineers, London, 1994), p. 78

30 MARCONI, G.: 'Improvements in telegraphy and in apparatus therefor', patent application filed on 5th March 1896, entered as no. 5028 in the Official Journal of the Patent Office

31 CAMPBELL SWINTON, A.A.: letter to W.H. Preece, 30th March 1896, ENG 23109, PRO

32 MARCONI, G.: letter to the Secretary of State for War, 20th May 1896, WO 32/989, file 84/M/3975, PRO

33 PENROSE, C.: memorandum to the Inspector of Submarine Defence, 20th June 1896, WO 32/989, file 84/M/3975

34 JACKSON, H.B.: letter to the Commander-in-Chief, Devonport, 16th September 1896, ADM 116/534

35 PENROSE, C., and CARR, G.: 'Signor Marconi's inventions', WO 32/989, file 84/M/3975, PRO

36 BAKER, E.C.: 'Sir William Preece, FRS' (Hutchinson, London, 1976), pp. 266–280

37 KEMP, G.G.: 'Diary of wireless experiments at GPO and Salisbury, 1896 and 1897', Marconi file HIS 64. Also E.C. Baker, 'Sir William Preece, FRS', p. 267

38 PREECE, W.H.: note appended to the report of the Salisbury Plain trials, 20th September 1896, E 23109/1899, file 2, Post Office Archives

39 JACKSON, H.B.: letter to the Commander-in-Chief, Devonport, 15th September 1896, ADM 116/523

40 JACKSON, H.B.: letter to the Commander-in-Chief, Devonport, 31st March 1897, ADM 116/523

41 JACKSON, H.B.: letter to G. Marconi, 15th September 1896, Marconi file HIS 64

42 ANON.: a report, *Westminster Gazette*, 14th December 1896

43 ROWLANDS, P., and WILSON, P.J. (Eds.): 'Radio begins in 1894', pp. 75–114 in 'Oliver Lodge and the invention of radio' (PDS Publications, Liverpool, 1994)

44 BURNS, R.W.: 'Discovery and invention. Lodge and the birth of radio communications', *IEE Review*, 1994, pp. 131–133

45 LODGE, O.J.: letter to W.H. Preece, 16th October 1890, The IEE Archives

46 LODGE, O.J.: letter to W.H. Preece, 29th May 1897, The IEE Archives

47 PREECE, W.H.: 'Signalling through space without wires', abstract of a Friday Evening Discourse delivered before the Royal Institution, 4th June 1897, *The Electrician*, 11th June 1897, pp. 216–218

48 FAHIE, J.J.: 'A history of wireless telegraphy' (Blackwood, Edinburgh, 1899), p. 152

49 PREECE, W.H.: 'Signalling through space without wires', *Proc. Roy. Inst.*, 1897, **XV(ii)**, pp. 467–470

50 Ref. 36, p. 249

51 MARCONI, G.: Report to W.H. Preece, 31st March 1897, The IEE Archives

52 AITKEN, H.G.J.: 'Syntony and spark – the origins of radio' (John Wiley, New York, 1976), p. 205

53 ANON.: editorial, *Electrician*, 17th September 1897, pp. 686–687

54 Ref. 52, p. 208

55 ANON.: reports, *Nature*, 1894, **50** pp. 133–139, and 160–161

56 LODGE, O.J.: a letter to S.P. Thompson, 16th March 1897. See also S.P. Thompson, a letter to O.J. Lodge, 12th April 1897, and O.J. Lodge to S.P. Thomson, 14th March 1897, University College, University of London archives

57 MARCONI, G.: 'Improvements in transmitting electrical impulses and signals and in apparatus therefor', British patent no. 12,039, 2nd June 1896

58 LODGE, O.J., and MUIRHEAD, A.: 'Improvements in syntonic telegraphy', British patent no. 29,069, 10th December 1897

59 LODGE, O.J., and MUIRHEAD, A.: 'Improvements relating to electric telegraphy', British patent no. 16,405, 10th July 1897

60 LODGE, O.J., and MUIRHEAD, A.: 'Improvements relating to electric telegraphy', British patent no. 18,644, 11th August 1897

61 MARCONI, G.: report to W.H. Preece, 31st March 1897, The IEE archives

62 Ibid, p. 6

63 JACKSON, H.B.: letter to G., Marconi, 19th May 1897, HIS 64, Marconi archives

64 POCOCK, R., and GARRATT, G.R.M.: 'The origin of marine radio' (HMSO, London, 1972), p. 11

65 JACKSON, H.B.: report to the Commander-in-Chief, Devonport, 22nd May 1897, ADM 116/523, PRO

66 JACKSON, H.B.: report to the Commander-in-Chief, Devonport, 16th September 1896, ADM 116/523, PRO

67 Ref. 29, pp. 80–86

68 MARCONI, G.: letter to W.H. Preece, 10th April 1897, The IEE archives

69 Ref. 21, pp. 39–41

70 MARCONI, G.: letter to W.H. Preece, 27th July 1897, HIS 43, Marconi archives

71 PREECE, W.H.: letter to G. Marconi, 6th August 1897, HIS 43, Marconi archives

72 Ref. 19, p. 35

73 PRIESTLY, H.: 'The what it cost the day before yesterday book, from 1850 to the present day' (Kenneth Mason, Hampshire, 1979)

74 MARCONI, G.: 'Wireless telegraphy', *Proc. IEE*, 1899, **XXVIII**, pp. 273–297, and discussion pp. 300–316

75 Ref. 19, pp. 36–37

76 EVAN-THOMAS, Commander.: report to the President of the Signal Committee, 10th May 1898, ADM 116/523

77 Ref. 21, pp. 53–54

78 Ref. 21, pp. 54–56

79 Ref. 19, p. 40

80 MacLAURIN, W.R.: 'Innovation and invention in the radio industry' (MacMillan, New York, 1949), p. 49

81 ANON.: report, *Electrical World*, 6th June 1903, p. 961

82 Ref. 19, pp. 42–43

83 Ref. 21, p. 99

84 POCOCK, R.: 'The early British radio industry' (Manchester University Press, 1988), p. 139

Chapter 12

Other wireless developments

Since 1945, in the former USSR and now in Russia, Aleksandr Stepanovich Popov (1859–1905) has been claimed to be the 'inventor of radio'. In that year, the 50th anniversary of his lecture on his work, the Soviet authorities made resolute efforts to establish the title for Popov.

C. Susskind, historian, of the University of California, who has conducted an exhaustive inquiry [1] into the validity of the claim has written:

An elaborate celebration was held [in 1945] in Moscow, characterised by what one foreign observer, the Australian botanist Sir Eric Ashby, has described as 'shrill festivities'. Several books were published about Popov, all in Russian. The most elaborate [was] a two-volume history of radio comprising 1) an historical account of the work of Popov's predecessors and 2) a collection of papers, letters, and documents pertaining to the question of his priority. A similar (but less complete) collection was published simultaneously. There was a fictionalised popular biography and an account written from the Navy's viewpoint. A.I. Berg and M.I. Radovsky co-authored two books on Popov in 1945 and 1948, and the Soviet Academy of Sciences published a Popov lecture by the editor-in-chief of the Great Soviet Encyclopaedia in 1948. There [was] also a booklet of reminiscences by Popov's assistant, a later book of recollections by several contemporaries, and the . . . biobibliography. May 7, the date of Popov's 1895 lecture according to the new-style calendar, was declared 'Radio Day', to be celebrated annually.

On 25th April (7th May new style) 1895, Popov, a physics instructor at the Imperial Torpedo School in Kronstadt, Russia read a paper [2] 'On the relation of metallic powders to electric oscillations' at a meeting of the Russian Physico-Chemical Society. The minutes included a one paragraph summary of the meeting.

Popov had seen an account of Lodge's 1894 lecture and had decided to experiment with a Branly filings type coherer tube [3]. Shortly afterwards, in July 1895, while studying the phenomena of atmospheric electricity at the Institute of Forestry, St Petersburg, he utilised his version of the coherer, together with a recording cylinder, to detect and record distant atmospheric electrical discharges. During observations, one end of the coherer was connected to a lightning conductor and the other end joined to earth. In use the lightning conductor, acting as an antenna, received electromagnetic waves, generated by the discharges, which were detected by the coherer. The consequent change of the coherer's resistance closed the circuit (see Fig. 12.1 [4]) of a telegraphic relay, the functioning of which then actuated an electric bell. Popov's apparatus was so disposed that the hammer of the bell struck the coherer tube causing the iron particles to de-cohere [5]. A pen recorder was joined in parallel with the electric bell. Popov has stated that his apparatus was operational from July 1895 to January 1896 and worked well as a lightning recorder.

His first paper [6] on this investigation, 'Apparatus for the detection and recording of electrical oscillations', was published in January 1896. It ended

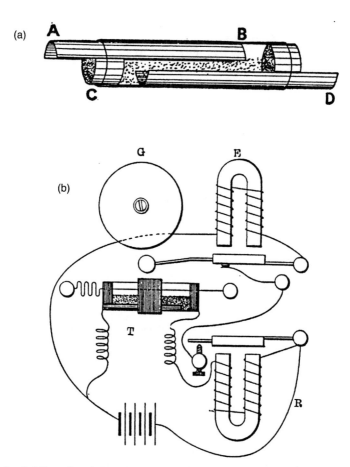

Figure 12.1   (a) Popov's coherer. (b) Popov's electromagnetic tapper for tapping back
the metallic filings tube to wave sensitiveness. T, coherer; R, relay; E,
electromagnetic tapper; G, gong.

Source: J.A. Fleming, *The principles of electric wave telegraphy* (Longmans, Green, London, 1908).

with the statement: 'In conclusion I may express the hope that my apparatus, when further perfected, may be used for the transmission of signals over a distance with the help of rapid electric oscillations, *as soon as a source of such oscillations possessing sufficient energy will be discovered*'. (Author's italics.)

Two months later, on 12th March 1896 (24th March new style), Popov gave a demonstration before the Physico-Chemical Society. Unfortunately, no extended record of the event exists: the minutes [7] of the meeting note simply: 'A.S. Popov shows instruments for the lecture demonstration of the experiments of Hertz. A description of their design is already in the Zh. R.F.-Kh. Obshchestva'.

In September 1896, while working as the electrical engineer in charge of the power plant at the annual fair in Nizhni Novgorod, Popov read a newspaper account of Marconi's demonstration and was much surprised. He returned to Kronstadt, resumed his experiments, and in January 1897 wrote [8] to a local newspaper and expressed an opinion that Marconi's receiver – details of which had not yet been published – was probably quite similar to his.

In the same year Popov replied [9] to a letter from Eugene Ducretet, a Paris instrument maker, who presumably had written to Popov requesting details of his work in relation to that of Marconi, and stated: 'I have no other printed papers at my disposal that can prove my participation in the practical solution of the problem of wireless telegraphy, other than the article already known to you. Nevertheless, I regard that article as sufficient to prove the identity of the constituent parts and their disposition in my apparatus with those in the *receiving station* of Mr Marconi.' (Author's italics.)

Until the turn of the century Popov had not published any account of his work that required a *wireless transmitter*, though he continued to be involved in the development of wireless telegraphy. In 1900 he set up a workshop at Kronstadt to repair and manufacture WT equipment but, according to one Soviet writer [10], 'the workshop had neither adequate equipment nor sufficient personnel to supply the Russian Navy with radio stations'. Indeed, at the outbreak of the Russo-Japanese War of 1904–1905, the Russian Navy was obliged to supply its ships with German manufactured wireless telegraphy transmitters and receivers.

From his enquiries, Susskind has summarised the evidence for the title of 'inventor of radio' to be accorded to Popov [11]:

The argument turns principally on whether Popov used his instrument merely to register lightning flashes and their man-made equivalents or whether he actually did publish, before mid-1896, a description of his use of his instrument for the transmission of intelligence.

The record shows that he did not. There is no mention of radiotelegraphic experiments in the minutes of the 1895 meeting, nor in the terse minutes of the 1896 demonstration. There is no mention of radiotelegraphic experiments in the 15-page paper that appeared in January 1896, nor in the artfully updated excerpts that appeared in *The Electrician* in 1897, nor yet in the translation that the *Electrical Review* printed in 1900.

The assertion that Popov had demonstrated a basic signalling system was first made in 1926 when V.S. Gabel, an official in the Soviet Bureau of Weights and Measures, sent a note [12] to the British magazine *Wireless World* about the forthcoming commemoration of the 30th anniversary of Popov's lecture on 7th May 1895. The note was published and stated simply that Popov had demonstrated 'the first successful transmission ... of a communication by means of electromagnetic waves. ... As each signal in Morse code was received, the President of the Society wrote the corresponding letter on the blackboard, and the enthusiasm of the audience was very great when the words "Heinrich Hertz" were eventually spelt out'. Curiously, neither this far-reaching experiment nor the resulting enthusiasm were mentioned in the one paragraph summary of the meeting.

Subsequently, the editor of *Wireless World* requested Gabel to provide additional information. He in turn wrote to three of Popov's associates – O.D. Khvolson, V.K. Lebedinsky, and V.V. Skbecyn – who recalled the meeting. Other than a change of date to 24th March 1896 and not 7th May 1895, no contemporary written evidence was provided [11].

Histories based on the spoken word should always be treated with some circumspection. People's memories are not always infallible; a person's recollection may be coloured by what he/she has been told or has read; it is possible that an oral account may be unintentionally distorted because a person stresses the importance of his/her contribution, real or imagined, in the saga of events; and it is possible that nationalistic zeal may inhibit the impartiality that should attend historical research.

Meantime, during the formative period of Marconi's system in the late 1890s, Lodge had been provoked to renew his activity in electromagnetic wave propagation. He had been granted an important patent [13] on syntony in 1897, and in the same year had collaborated with A. Muirhead [14] on further experiments and patent applications [15,16,17]. By the end of 1897 Lodge, and Lodge and Muirhead had two and three patents respectively. Marconi at this time had just two patents [18,19]. Some secrecy, for commercial reasons, now surrounded Lodge's and Muirhead's work. It is known that some transmissions took place between the Victoria tower and the tower of Lewis's shop in Liverpool, and others between the University of Liverpool and Lodge's house in Grove Park, a distance of c. 1.5 miles (2.4 km) [20].

These experimental investigations seem to have been conducted at a more leisurely pace than those of Marconi, since it was not until 1901, five years after the founding of the Wireless Telegraph and Signal Company, that Lodge and Muirhead felt sufficient progress had been made to form the Lodge-Muirhead Syndicate [21]. Their system (Fig. 12.2) for sometime known as the 'Loghead' system, was installed in 1903 on the Eastern Extension Telegraph Company's cable ships, *Restorer* and *Patrol*. During the same year the syndicate applied to the Post Office for licences for shore stations at Dover, Isle of Wight, Lizard and Fastnet – presumably to establish a marine service. The application was refused, though alternative unspecified non-Channel sites were offered.

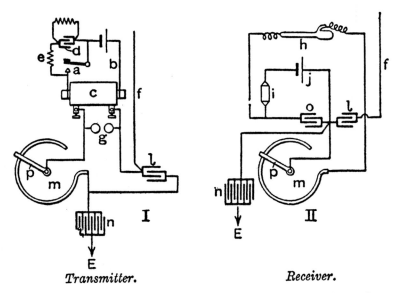

Transmitter.   *Receiver.*

Figure 12.2   Diagram of connections of the Lodge–Muirhead wireless telegraph transmitter and receiver.
Source: J.A. Fleming, op. cit.

In 1905 the Midland Railway Company installed Lodge–Muirhead equipment on the Midland Railway's ships operating from Heysham to the Isle of Man, and a shore station was set up at Heysham. The railway company also experimented with the system on trains.

The Lodge–Muirhead Syndicate's system was not widely adopted although it was competitive and based on sound scientific principles. It might have created a difficulty for the Marconi system, but the non-availability of licences severely restricted the business potential of the syndicate. As a consequence it was limited to the manufacture and sale of apparatus. One system was purchased by the Indian Government for communications between Burma and the Andaman Islands, a distance of c. 300 miles (483 km). Other installations were: 1) at shore stations in Singapore and Hong Kong; 2) for the government link connecting Trinidad and Tobago; 3) for the African Direct Telegraph Company in Lagos; and 4) for the syndicate's own experimental stations at Downe and its factory at Elmer's End, Kent [21].

Lodge's patents on syntony were, of course, of very considerable importance to any company that aspired to establish wide-ranging wireless telegraphy. They constrained Marconi's freedom of action, notwithstanding his '7777' patent of 1900, and some accommodation between Lodge and Marconi became necessary. The Lodge–Muirhead patents were acquired by the Marconi company, the syndicate ceased to exist, and Lodge became a scientific advisor to the company – but was never consulted. Of some note, when questions of priority of invention are discussed, is the ruling in 1943 (sic) of the United States Supreme Court that the only valid patent of the three held by the Marconi organisation in the field of tuning was that of Lodge [22].

As noted previously, Professor Adolf Slaby was a witness at the demonstration given by Marconi in 1897. Afterwards, he recorded [23] how the occasion changed his perspective on wireless telegraphy.

In January, 1897, when the news of Marconi's first successes ran through the newspapers, I myself was earnestly occupied with similar problems. I had not been able to telegraph more than one hundred meters through the air. It was at once clear to me that Marconi must have added something else – something new – to what was already known, whereby he had been able to attain lengths measured in kilometres. Quickly making up my mind I travelled to England. . . . Mr Preece . . . in the most courteous and hospitable way, permitted me to take part in [Marconi's demonstrations]; and in truth what I there saw was something quite new. Marconi had made a discovery.

Slaby's presence did not please Marconi. Slaby was a rival who in association with Count von Arco subsequently devised variants of Marconi's scheme. Moreover, Slaby's indirect approach to Preece was somewhat devious rather than being plainly honest. In March Preece received a letter [24] from a Mr Gisbert Kapp:

My friend, Privy Councillor Slaby, whom you will perhaps remember from the Francfort days, is the private scientific advisor of the Emperor. Any new invention or discovery interests the Emperor and he always asks Slaby to explain it him. Lately the Emperor has read up your and Marconi's experiments . . . and he wants Slaby to report on this invention. . . . I shall be much obliged if you will kindly answer the following questions

1. Is there anything in Marconi's invention?
2. If yes, could you arrange for Slaby and myself to see the apparatus and witness experiments if we come over to London towards the end of next week?

As the Emperor is in a hurry to get Slaby's report, will you kindly let me have a line by return? Please treat this letter as confidential and *say nothing to Marconi about the Emperor*.' (Author's italics.)

By this machination Slaby became familiar with the 'secret' of Marconi's success, namely, the use of an elevated antenna and good earth connection, and subsequently became an ardent worker in the field of wireless telegraphy. He soon realised, as had Lodge and Marconi before him, that a transformer was necessary to match the impedance of the coherer circuit of a WT receiver to that of the associated antenna. Slaby used a novel solution. Instead of a wound transformer, he employed a transmission line having a length equal to a quarter of the wavelength of the received radiation. Such a line transformer has the property that an impedance, Z, connected across one end of the line is transformed to an impedance of A/Z, (where A is a constant characteristic of the line), at the other end of the line. Thus, a low impedance is transformed to a high impedance and vice versa. The arrangement is shown in Fig. 12.3, taken from the German patent, no. 130,723 dated 16th October 1900, of Slaby and his

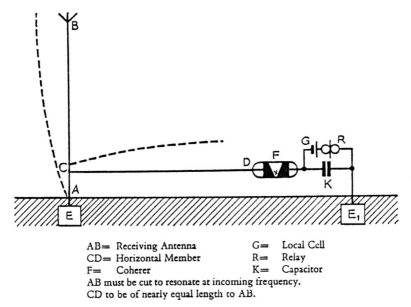

AB= Receiving Antenna   G= Local Cell
CD= Horizontal Member   R= Relay
F= Coherer   K= Capacitor
AB must be cut to resonate at incoming frequency.
CD to be of nearly equal length to AB.

Figure 12.3 Slaby-Arco receiving antenna.
Source: J.A. Fleming, op. cit.

associate Count von Arco. Various arrangements [25] of the basic scheme were devised and patented, including those that enabled duplex, or simultaneous, wireless telegraphy to be implemented.

On 22nd December 1900, Slaby, in the presence of the Emperor of Germany, demonstrated the duplex method during a lecture on 'Syntonic and multiple spark telegraphy' given in the conference room of the Allgemeine Elektricitats Gesellschaft (General Electric Company) of Berlin. Signals were simultaneously transmitted from the transmitting station, at the Technische Hochschule, Charlottenburg, to the company, (a range of 2.4 miles – 4 km), and to a cable manufactory at Oberschoneweide, (a distance of 8.7 miles – 14 km). The transmission wavelengths were c. 700 yards (640 m) and 262 yards (240 m) [26].

Professor Ferdinand Braun (1850–1918), of the University of Strasburg, also devoted much effort to evolving practical syntonic wireless telegraph systems. Since, in 1909, he shared the Nobel Prize in physics with Marconi for his practical contributions to wireless telegraphy, a few words on his background are apposite [27]. He was born in Fulda, Germany, and studied at the University of Marburg and the University of Berlin from which he received in 1872 his doctorate for a thesis on the vibrations of elastic rods and strings. Two years later Braun published the results of his researches on the conduction of current through metal sulphides. He found that these crystals conducted electric currents in one direction only – a property that was of considerable significance in the early years of the 20th century when attempts were being made to improve the detectors of electromagnetic waves.

Subsequently Braun held various positions at Wurzburg, Leipzig, Marburg, Karlsruhe, Tubingen – where he founded the Physical Institute – and Strasbourg. Following a brief period from 1880–1883 at Strasbourg, he returned there permanently in 1895 and was appointed Professor of Physics and Director of the Physical Institute. A visit to the United States to give evidence in a litigation case involving radio broadcasting led to his detention when the United States entered the 1914–1918 war. He died in a Brooklyn hospital on 20th June 1918.

Braun is possibly best known for his cathode ray oscilloscope tube – the Braun tube – which dates from 1897. Of the many investigators who had investigated the properties of cathode rays and had devised cathode ray tubes to study the properties of these rays, the most eminent was J.J. Thomson. He determined the ratio of charge-to-mass of the cathode rays (electrons) using both electric and magnetic fields to deflect the rays. However the practical form of the cathode ray tube, as an instrument suitable for application in a general physics laboratory was due to Braun [28].

Braun's work on wireless telegraphy seems to date from after his appointment at Strasbourg for his first patent is dated 14th October 1898. He experimented with his methods, during the summers of 1899 and 1900, and succeeded in sending signals between Cuxhaven and Heligoland, a distance of 39 miles (63 km) [29]. This work led to an association with the firm of Siemens and Halske of Berlin.

Prior to 1903 the patents, methods and systems of Slaby, von Arco, and the Allgemeine Elektricitats Gesellschaft, and those of Braun and Siemens and Halske, were in competition and had been in conflict in the courts. But on 30th May 1903 a Berlin court awarded a favourable decision to Braun and Siemens and Halske. However, the Kaiser ordered an end to the rivalry and accordingly the two systems were amalgamated in 1903 into a new company called the Gesellschaft fur Drahtlose Telegraphie. It became a formidable rival to the Marconi Company, to the present day, and marketed a system known as the Telefunken system [30].

Shore stations were constructed and important ships of the German Navy were equipped with Telefunken apparatus. Then in 1905 a subsidiary company was established in the United States and a powerful station was erected in New York City.

The growth and spread of the German system did not, of course, give pleasure to the Marconi International and Marine Communication Company which aspired to establish a monopolistic position world-wide. The company adopted an aggressive policy to protect and extend its interests and this caused some contention with rival systems. In an endeavour to resolve certain issues by diplomatic means, which had not been amenable to commercial bargaining, the German Government in 1903, (and also in 1906), called an International Conference on the subject of wireless telegraphy. The results of these conferences are discussed in the following chapter, suffice to mention here that Telefunken's position was greatly enhanced as a consequence.

In the United States of America the principal contributors to the progression of wireless telegraphy were de Forest and Fessenden. Their work was outstanding from an engineering viewpoint rather than from the viewpoint of physics. Neither person sought to emulate Maxwell, or Hertz or Lodge in determining and understanding the fundamental physical principles which would advance electromagnetic wave signalling. Instead, they sought to incorporate improvements and to implement new ideas which would enhance and extend the basic systems of wireless telegraphy and wireless telephony.

Reginald Aubrey Fessenden (1866–1932) was the first important American inventor to experiment with wireless [31]. His abilities were not confined to engineering, for he was able to work with equal facility in chemical-, electrical-, metallurgical- and mechanical-engineering. In the wireless field he devised an electrolytic detector, invented the heterodyne method of reception, advocated and used high frequency alternators in continuous wave wireless systems, invented a wireless telephone system and transmitted the first voice signals to be sent over a long distance. He obtained 230 US patents [32]. Among his admirers, Elihu Thomson, a noted engineer and inventor, is reputed to have described Fessenden as 'the greatest wireless inventor of the age – greater than Marconi' [31].

Fessenden was the son of the Reverend E.J. Fessenden and his wife, and was born, on 6th October 1866, in East Bolton, Quebec, Canada where his father had charge of a small parish. Five years later, in 1871, the family moved to Fergus, Ontario, and then in 1875 to Niagara Falls, Ontario. The young Fessenden attended the De Veaux Military Academy, Niagara Falls but after one year moved to Trinity College School, Port Hope, Ontario. He won several prizes and was described by his headmaster as one of the best pupils he had taught [33].

At the tender age of just 16 years Fessenden accepted a mathematics post at Bishop's College, Lennoxville, Quebec. Here his interest in science was stimulated by his reading the periodicals *Nature* and *Scientific American*. Four years later, at the age of 20, he was appointed to the principalship of Whitney Institute, Bermuda. This position must not have been to Fessenden's liking for after only one year he obtained employment at Thomas Edison's laboratory in East Orange, New Jersey. A further move was made when Fessenden was laid off in 1890. He went to work for Westinghouse and made several significant contributions in electric lamp manufacturing technology, and developed a silicon steel for transformers and electric machines [33].

In 1892 he accepted the chair of electrical engineering at Purdue University and, though he remained there for just one year, he was responsible for setting up the Department of Electrical Engineering. (By today's standards the appointment seems extraordinary given that Fessenden did not hold a degree in *any* subject.) One year afterwards, another move was undertaken, this time to the University of Pittsburgh where he accepted a post similar to that at Purdue. It appears that Fessenden must have carried out some good work while at Westinghouse, from 1890 to 1892, and as a consequence George

Westinghouse was anxious for him to return to Pittsburg. A substantial honorarium paid by him was a provocative factor in Fessenden's decision to make the transfer [33].

The positions he held at Purdue and Pittsburg entitled Fessenden to call himself Professor Fessenden, a title that he retained after he left academic life in 1900 to join the US Weather Bureau. The Bureau was seeking a means of electrically transmitting weather forecasts. Several wireless telegraph stations were erected and tested successfully.

During his short period with the Bureau Fessenden succeeded in giving a demonstration, on 23rd December 1900, of the wireless transmission of voice signals using spark apparatus: the system used two 50 ft (15.2 m) masts positioned one mile (1.6 km) apart. Necessarily, the reproduction of the signals must have been grossly distorted and almost unintelligible, but the demonstration so impressed the Weather Bureau that further work on a larger scale was planned and partially implemented. However, Fessenden's choleric personality was not conducive to good relations with some members of the Bureau and he resigned, in August 1902, following a quarrel over patent rights [34].

In September he secured the financial support of two Pittsburg millionaires, T.H. Given and H. Walker, and together they formed the National Electric Signalling Company (NESCO). Three stations were constructed, initially, at Washington, Brooklyn, and Jersey City, and, later experimental stations were built at Machrihanish, Scotland and Brant Rock, Massachusetts, for trans-Atlantic communications [35].

The spark system of wireless telegraphy has some basic disadvantages. The waves set up by the spark discharge are a series of highly damped oscillations, which means that the energy of the spark is not concentrated in a single frequency wave but is spread out on either side of the nominal frequency component determined by the constants of the transmitter circuit. Furthermore, since the bursts of radiation are separated by pauses, the average power transmitted is very much less than the peak power. As a result sharp tuning is not possible at the reception point and interference is caused to neighbouring stations [36].

From the outset Fessenden's objective was the accomplishment of good wireless telephony rather than wireless telegraphy. For this purpose unquenched spark generators are quite useless since they cannot be modulated: moreover coherer detectors cannot be utilised because, being bistable devices, they cannot be employed with continuous signals. Ideally, Fessenden had to engineer a system based on the generation of a continuous high frequency signal, a modulator, and a detector capable of responding to the modulation envelope of the received signal.

He sought a device that would follow the complex signal produced by the voice and, initially, attempted to fabricate a hot-wire 'barretter' [37] (Fig. 12.4); a device that consisted essentially of a very short length of extremely fine wire, contained in a closed vessel. In operation the received antenna current heated the wire and changed its resistance, and thereby varied the current flowing in

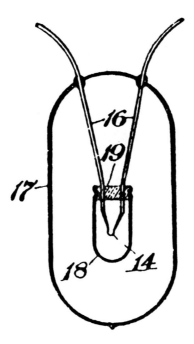

Figure 12.4   Fessenden's fine wire barretter.
Source: V.J. Phillips, *Early radio wave detectors* (The IEE, London, 1980).

the circuit that comprised the barretter, an earpiece of a telephone head-set, and a battery.

The fine wire was generally produced by the Wollaston method [38]. In this process a thin piece of platinum wire was first coated with silver to improve its strength, and then the composite wire was drawn through a series of dies until it was as fine as possible. Next, the silver was removed by dipping the wire in a bath of nitric acid to leave an extremely fine wire of platinum having a diameter of c. 0.01 mm.

One day, while Fessenden was removing the silver coating from a platinum wire, he noticed that it was responding to the signals received from an automatic sender. An investigation showed that the wire had broken. Serendipitously, Fessenden had discovered that a Wollaston processed wire dipping into a 20 per cent solution of nitric acid was far more sensitive and reliable than any other type known to him: he had invented the electrolytic detector, though he continued to call it a barretter. Patents were applied for in 1903 and 1904. (This type of detector was also independently developed by Schloemilch [39] of Germany.) According to S.M. Kintner, one of Fessenden's associates and later the Director of Research at Westinghouse, the electrolytic detector was used as the standard of sensitivity until it was replaced by the diode in c. 1913. They were used by the United States Navy from 1908–1913, by the United Fruit company, and by the United Wireless company until it was prohibited from doing so by a court injunction.

Fessenden's detector was, of course, useless for the reception of unmodulated continuous waves since all that would be heard would be the 'clicks' when the Morse key opened and closed. What was needed was a detection system that would give an audible tone when an unmodulated carrier wave was received. Fessenden's solution was to generate a local carrier signal at the receiver, and combine this signal with the received carrier wave to produce an audible signal having a frequency equal to the difference between the frequencies of the local and received signals [40,41,42].

The heterodyne system was subjected to US Navy tests in 1910 and, in further tests between the Fessenden equipped station at Arlington and the cruiser USS *Salem*, in 1913, the superiority of the method was demonstrated [43]. However, general adoption of the system was much delayed by the cumbrous apparatus which had to be used to generate the local carrier signal. Fessenden utilised an arc generator but this was noisy, difficult to adjust and costly. Subsequently, the method of heterodyne reception – the word is derived from the Greek roots *hetero* meaning difference, and *dyne* meaning force – has proven to be of outstanding utility: it has been the basis of radio receiver design for many years.

The constraints imposed by the existing technology of the early years of the 20th century, and the non-availability of equipment that could be suitably adapted, led Fessenden to consider synchronous, rotary, quenched spark gap, transmitters. With a quenched spark gap the oscillations in the primary circuit decay to zero after a few cycles: the oscillations in the coupled antenna circuit also decay but with a smaller decrement since the damping of the circuit can be made less than that of the primary circuit. A synchronous spark gap phased to fire on both the positive and the negative peaks of the 3-phase waveforms, from a 125 Hz generator, would produce an output having a frequency of 750 Hz [36]. The output would not be a perfect continuous wave signal but it would be an approximation to it; and the signal would be characterised by a musical tone, when heard with headphones, which would enable it to be read easily in the presence of atmospherics and interference from other transmitters.

Fessenden's Brant Rock station, completed on 28th December 1905, employed equipment of this type. The rotary gap was 6 feet (1.8 m) in diameter at the stator and 5 feet (1.5 m) in diameter at the rotor, and was coupled to a 125 Hz, 3-phase, 35 kVA alternator (a.c. generator). Fifty poles were mounted on the rotor and four on the stator [36].

On 10th January 1906 Fessenden succeeded in sending and receiving two-way wireless communications [44] across the Atlantic Ocean, between Brant Rock and Machrihanish, Scotland. Antenna masts, 420 ft (128 m) high were employed at each end of the link and the signals were transmitted using one of three frequencies. With this link Fessenden made a study of the propagation characteristics between the two places, and carefully recorded the signal strengths at various times of day and night and as a function of the month of the year.

Regrettably the study came to an end when a gale destroyed the Machrihanish antenna mast on 5th December 1905: the tower was never rebuilt.

Although impressive in an engineering context the station did not satisfy Fessenden's desire for a truly continuous wave station. He probably knew that Tesla in 1890 had designed, and had had constructed, a high frequency alternator, with 384 poles, which generated a 10 kHz output; this would be much more satisfactory than the rotary spark gap. Fessenden's first high frequency alternator was designed by Steinmetz, the 'presiding' genius at General Electric, and was manufactured by GE in 1903, but its output was insufficient for Fessenden's application. He pressed GE to design machines of higher and higher power outputs [45].

Fessenden seems to have had a very single-minded personality; he did not suffer fools, had great confidence in his own ideas and his engineering prowess, and was somewhat averse to the opinions of others when they did not coincide with his own. All of this could make for difficult relationships with those who were trying to assist him. Some of these traits became evident in his negotiations for another high frequency alternator from GE. Fessenden wanted a machine without iron but Dr E.F.W. Alexanderson, a very able and well educated and trained engineer with much experience (who later became Chief Engineer of RCA) was opposed to this. He has written [46]:

I submitted a design of an alternator without iron but reiterated my opinion that iron [was] preferable [to wood]. . . . I did the work contrary to my own opinion, hoping to ultimately persuade Fessenden to go back to my early designs. On October 16 1905 I reported to Fessenden on the test of one of his models and pointed out the great difficulties and suggested the abandonment of that line of designs and the adoption of the one which is a compromise between his design and my early one.

Subsequently, the manufactured Fessenden-Alexanderson machine had a fixed armature, and a revolving field magnet with 360 teeth. At a speed of 139/s, an output of 65 V, 300 W, 50 kHz was obtained [47].

Meantime, Fessenden was continuing his work on wireless telephony. With his latest alternator he gave a demonstration of the new technology on 11th December 1906. Signals were sent from Brant Rock to Plymouth, Massachusettes, a distance of 11 miles (17.7 km); and in July 1907 speech was transmitted between Brant Rock and Jamaica, Long Island, a separation of 180 miles (290 km) [44]. The station at Brant Rock was modulated by a carbon microphone connected in series with the antenna feeder wire. Another demonstration was given on Christmas Eve, 1906, when Fessenden played Handel's Largo on the violin. This was the first ever broadcast of music: it was heard by wireless operators on board several US Navy and United Fruit Company ships, equipped with Fessenden's receivers, in the Atlantic Ocean. The broadcast was repeated on New Year's Eve [48].

By early 1907 the exploits and successes of Fessenden in the area of wireless telephony led the Bell System to consider whether the time had now arrived when it should take a commercial interest in acquiring some patents in this field.

A report to President Fish of the Bell System mentioned: 'I feel that there is such a reasonable probability of wireless telegraphy and telephony being of commercial value to our company that I would advise taking steps to associate ourselves with Mr Fessenden if some satisfactory arrangement can be made' [49].

Negotiations between the principals were opened and were close to completion when the 'financial panic of 1907 hit the country'. The management of the Bell System was changed, the Chief Engineer and President were replaced and all expenditures were critically examined. Another, more realistic and critical, appraisal of the Fessenden patents was undertaken. On 8th July 1907 T.D. Lockwood, the patent counsel of the company, submitted a 26 page report on the position to President Vail. Lockwood noted that Fessenden's patent no. 706,747 was the first patent issued for voice transmission by wireless but observed:

Nevertheless, the field of commercial operation in sight is relatively so small; there are so many systems competing for it; and there are apparently so many ways of performing every function, that . . . it is difficult to see where the business can ever be successful or profitable unless the several systems can unite, and be operated and managed as a single concern. Even under these conditions, it is hard to see from whence sufficient business can be derived for many years to come [50].

Bell System decided not to buy. There was a conviction that wireless telephony 'cannot and will not reach any practical realisation within the term of years yet remaining to Fessenden's fundamental patents' [50]. President Vail expressed the attitude of the new management when, in response to a large English investment house, he wrote [50]: 'As to the "wireless": I can only refer you to the success of the wireless telegraph and the [negligible] inroad made by it upon the general telegraphic situation as compared with the promises and prophecies. The difficulties of the wireless telegraph are as nothing compared with the difficulties in the way of the wireless telephone.' (It should be mentioned that at this time, 1907, although the audion valve had been invented by de Forest in 1906, the valve was not an engineering circuit element and its uses as an amplifier and oscillator were not known. For wireless telephony the valve would be an essential component.)

The result of the company's examination must have been a grave disappointment to Fessenden. For several years he had urged Given and Walker to manufacture and sell apparatus to selected customers, including the US Navy, but the policy of Given and Walker was to develop a patent structure for a complete system of wireless communication and then sell.

This policy was changed in 1908 when permission was given by Fessenden's backers to sell by contract to the United States Navy and the United Fruit Company. By then quarrels were becoming frequent: Fessenden's difficult temperament and disagreements with his sponsors about policy, and Given's and Walker's lack of managerial skill and tact, came to a head at the end of December 1910. Fessenden was dismissed on 8th January 1911.

Fessenden immediately engaged in litigation for breach of contract, won his case in the lower court and was awarded damages of $400,000. NES went into receivership in 1912 but continued its development work with a reduced workforce. Eventually, following further litigation between NES and the Marconi organisation over infringement of each other's patents, NES received c. $300,000 in royalties from the Marconi organisation and paid to them c. $30,000. The patents of NES were acquired by purchase by Westinghouse. Later, as a consequence of the 1921 Westinghouse–RCA agreement, the patents became RCA property [51].

Actually, Fessenden's work with high frequency generators would have come to an end a few years after 1912, even if he had not been dismissed, for during the period when he was developing his c.w. transmitter system, inventions had been patented that would revolutionise trans-continental, and trans-oceanic telegraph and telephone communication. The inventions were those of the diode valve (1904), due to A.J. Fleming, and the triode valve (1906) due to L. de Forest. Of these devices, the triode became of exceptional utility, and must be considered to be one of the great inventions – probably the greatest in electrical engineering – of the 20th century. Without it amplifiers, oscillators, pulse generators, modulators and many other types of electronic circuit would not be possible. Television, radar, stereophonic broadcasting, fast computers, satellite communications and numerous applications in industry would not have been realised.

Lee de Forest was born on 26th August 1873 [52]. His father, H.S. de Forest, DD, of Huguenot background, was the minister of the Congregational church at Council Bluffs in Iowa where Lee was born. His mother, A.M. Robbins, was the daughter of the Congregational minister of the town of Muscatine in Iowa, where Lee spent the first six years of his life.

Following later schooling at the Mt Hermon School in Massachusetts, de Forest was admitted to the three year mechanical engineering course of the Sheffield Scientific School at Yale University, where his father had been awarded his Doctor of Divinity degree. Lee graduated in 1896 and then spent a further three years working for a PhD degree, which he obtained in 1899 for a thesis on 'Reflection of Hertzian waves from the ends of parallel wires'. At that time his knowledge of the theoretical and experimental properties of electromagnetic waves must have been comparable to that of Lodge, Righi, Marconi, Slaby and a few others. De Forest determined to pursue his research work and sought a position where this could be undertaken.

His first post was with the Western Electric Company, Chicago. Here, after two months spent – somewhat uselessly – in the dynamo department, he was promoted to work in the telephone laboratory, 'goal of my dreams'. De Forest, in his diary, described his work there: 'I have begun a systematic search through *Science Abstracts, Wiedemann's Annalen*, etc, for some hint or suggestion of an idea for a new form of detector for wireless signals. I had built a Branly coherer at Yale and used it. Marconi's coherer, and tapping back, did not appeal to me. It was too slow and complicated' [53].

Soon, his experiments on his new wireless 'responder' were occupying more and more of his time, to the impatience of his supervisor at Western Electric.

On 8th April 1900 de Forest recorded in his diary [54]: 'At last I have the opportunity to do experimental work in wireless telegraphy. This came as the result of my having written to Professor Johnson of Milwaukee, President of the newly organised American Wireless Telegraph Company. Not long after, he came to see me in Chicago and asked me to join his concern.'

De Forest accepted the invitation and stayed for three months, from 1st May 1900, but was 'fired' when he refused to assign his responder invention to the company. And so another position had to be urgently obtained by the impecunious engineer. He was appointed the assistant editor of the *Western Electrician* at a salary of $10 per week, to translate papers on electricity, but continued to devote his free time to experimenting with his 'electrolytic anti-coherer'. A friend from his Western Electric days, E. Smythe, became his financier. 'Smythe was comparatively rich, earning $30 a week. Naturally, our budget for experimental work was very limited' [55].

Again, his employment was short-lived.

I have begun to hazard my job with the *Western Electrician* by working half-time in the laboratory of [the] Armour Institute, teaching two nights weekly at [the] Lewis Institute. I am risking mediocrity and weak contentment for a chance of great success . . .

Soon the experiments became so engrossing that it was impractical for me to continue to work even half time for the *Western Electrician*. So once more I crossed the Rubicon, burned my bridges, and with only the amount of $5 paid by [the] Lewis Institute per week, and an equal amount advanced by Smythe, determined to continue my life as an inventor [55].

With an investment of $1,000, obtained from a business acquaintance, de Forest established the Wireless Telegraph Company of America, rented a small machine shop in Jersey City, and constructed equipment for reporting the International Yacht Races. The venture was not a success and de Forest was of necessity compelled to seek further finance. After many rebuffs, he met, in the autumn of 1901, A. White, a stock promoter. He suggested the formation of a public company, and in 1901 the De Forest Wireless Telegraph Company (NJ) was incorporated with an authorised capital of $3,000,000. Numerous additional companies were formed subsequently.

Initially, some successes were achieved. De Forest received an order from the US War Department to install his receiving apparatus on one of the Army's tugboats, and to erect two land stations for the US Signal Corps. When these were found to be satisfactory he was commissioned, by the US Navy, to construct two land stations, one in Washington and the other at Annapolis which previously had purchased its wireless sets from Slaby–Arco [56]. Then, in 1904, de Forest's company received a contract, from the United Fruit Company, to erect a radio link between Costa Rica and Panama.

An adventitious factor, commencement of the Russian–Japanese war on 6th February 1904, aided de Forest's cause. He sent two of his most experienced staff to China where they set-up a wireless station on a 150 ft (45.7 m) cliff near Wei-Hai-Wei. Fortunately, a correspondent for *The Times*, on his way to report the war, had with him a de Forest wireless set. This was installed on a tug and, from the war zone, messages were transmitted to Wei Hai Wei from whence they were sent to *The Times* in London. The result was a scoop for the newspaper and much publicity for the inventor [57].

A further US Navy order, to build and equip five transmitting and receiving stations along the Gulf of Mexico, was awarded in 1905.

Eventually more than 90 stations were erected by the company. White had ambitious aspirations for expansion, based on de Forest's engineering prowess, and planned to set up a network of stations that would rival those of Western Union and Postal. However, static interference severely limited the working of the stations, communications were unreliable and sometimes impossible, and many of the stations never sent a message. The company faced bankruptcy. White, by rather devious means, succeeded in selling a substantial amount of stock but finally, in 1907, the company collapsed, its assets being sold to the United Wireless Telegraph Company. De Forest was forced to resign. Wisely, he retained control of his patents pending on the triode.

The genesis of the triode valve, or audion as de Forest called it, begins in 1900 when he was working with Smythe on the responder. De Forest noted that the gas light in the laboratory dimmed when his spark telegraphy apparatus was in operation, and that the light returned to its full brilliance when the apparatus was switched off. This was a new phenomenon for him and he considered whether the effect could be the basis of a new type of detector of wireless waves. His diode type detector (Fig. 12.5) seems to have worked reasonably well for W.H. Eccles described it as 'having properties of a most felicitous and commodious character' [58]. Then de Forest fabricated a glass vessel containing two separate electrodes and partially filled it with a gas. His idea was to heat the electrodes, electrically, and expose the tube to radiation. By a series of experiments he transformed his Bunsen burner arrangement into a device that was very similar to Fleming's diode. The further transformation of the diode structure to that of a triode has been narrated by de Forest [59].

Although I now had proof that I was on the right track, I was still not satisfied. My diode detector permitted part of the high frequency energy to pass to earth through the telephone and battery circuit instead of concentrating it upon the ions between the plate band and the filament. To overcome this imperfection and to improve still further the sensitivity of the detector, I wrapped a piece of a tin foil around the outside of the cylindrical-shaped glass tube and connected this third electrode to the antenna or to one terminal of the high frequency tuner. I then realised that the efficiency could be still further enhanced if this third electrode were introduced within the tube. I have therefore had McCandless construct another audion – as I now for the first time began to call it. This new device contained two plates with a filament located midway between them. This detector showed distinct improvement over its predecessors.

It now occurred to me that the third, or control, electrode could be located more efficiently between the plate and the filament. Obviously, this third electrode so located

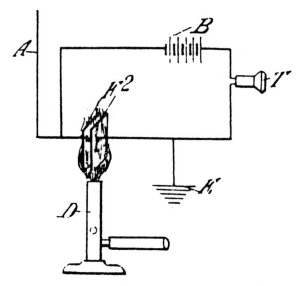

Figure 12.5 De Forest diode-type detector using the ionisation in a gas flame.
Source: V.J. Phillips, op. cit.

should not be a solid plate. Consequently, I supplied McCandless with a small plate of platinum, perforated much better than anything preceding it, but in order to simplify and cheapen the construction I decided that the interposed third electrode would be better in the form of a grid, a simple piece of a wire bent back and forth, located as close to the filament as possible.

De Forest has averred that he knew nothing of Fleming's diode invention, though Fleming had described his new detector at a lecture given before the Royal Society in February 1905, and had taken out a British patent on 16th November 1904 [60].

According to one account [60], the third electrode was added to the tube on 31st December 1906. Investigation showed that by varying the potential difference between the grid and the heated electrode, or cathode, the flow of electrons between this electrode and the other electrode, the anode, could be altered. An application for a patent was made on 29th January 1907, and for several years afterwards de Forest fabricated audions. Due to inadequate manufacturing facilities these tubes were non-uniform in performance – owing to the presence of residual gas that de Forest believed was essential to the working of the audion – and were less satisfactory than some other detecting devices such as the electrolytic, magnetic, and crystal types. Commercial users of wireless receivers found the audion superior to these detectors, but de Forest's triodes needed such constant attention and such frequent adjustment of the anode potential and filament current that they preferred the earlier forms of detector. Consequently, the early audions had few applications. Surprisingly, de Forest undertook few scientific experiments with the valve in the five-year period after its invention.

On 24th October 1914 the Marconi Wireless Telegraph Company of America – the parent company of which owned the rights to the use of the Fleming patent – brought an action against the de Forest Radio Telephone Company in the Federal District Court of New York seeking to restrain the company from manufacturing and selling the tube (or valve) [61]. Eventually the lengthy action between the two antagonists was decided by the District Court on 20th September 1916. The Court affirmed that the application by Fleming of the Edison effect to wireless detection was patentable, but on de Forest's addition of the grid the Court concluded:

De Forest had long been proceeding on a theory different from that of Fleming. Having read Fleming's article he began to experiment with the incandescent lamp. He probably doubted its efficiency at first but within a very short space of time – perhaps a week, perhaps a month – he changed his mind, and, discovering that Fleming was right, wrote his solicitor, after he had filed his application for [patent] No. 824,637, that the new receiver is the best yet. Thereafter he used the language of the incandescent lamp. . . . De Forest in his three-electrode audion has undoubtedly made a contribution of great value to the art . . . but on the other hand Fleming's invention was likewise a contribution of value, and is to be treated liberally and not defeated either by unconfirmed theory or by association in apparatus, where later developments have taught how other useful adjuncts can be employed [62].

The result was that de Forest's invention had infringed that of Fleming and so could not be manufactured without the consent of the Marconi company.

This was not the end of the matter, however: rather it was the beginning of a long and bitter controversy which was only resolved in 1943 when the US Supreme Court ruled in de Forest's favour. By this time, of course, the exceptional utility of the triode valve, to radio communications, telecommunications, radar, television and electronics generally had been well established, and de Forest's patent had long expired. Fleming eventually recognised that there was an essential difference between the two inventions. He wrote [63]: 'Sad to say, it did not occur to me to place the metal plate [the anode] and the zig-zag wire [the grid] in the same bulb, and to use an electron charge, positive or negative, on the wire to control the electron current to the plate. Lee de Forest, who had been following my work very closely, appreciated the advantage to be gained'.

As an aside, de Forest's invention led to an amusing anecdote. In March 1912 he was arrested together with some directors of his company and charged with 'using the [US] mail to defraud, by selling to the public in a company incorporated for $2,000,000[,] whose only assets were the de Forest patents directed chiefly to *a queer little bulb* like an incandescent lamp which he called an audion and which device had been *proven to be worthless* and was not even a good lamp'. (Author's italics.) (After a trial lasting six weeks, de Forest was acquitted but two of his associates were found guilty. The inventor was given a stern homily by the Federal judge and told to get 'a common garden variety of job and to stick to it'.) [Ref. 34, p. 84]

Interestingly, de Forest despite his doctoral training, did not understand the operation of his device, which had a tendency to become unstable. Clarification came when Dr H.D. Arnold, a young physicist with AT&T, suspected, following a demonstration by de Forest to AT&T in 1912, that the instability of the audion was caused by gas ionisation, and that improved evacuation of the gas from the tube would negate the defect [64]. Dr I. Langmuir, a chemist with General Electric, also, at about this date, was led to a similar view.

My active interest in thermionic currents began in connection with some experiments on electrical discharges occurring within tungsten lamps. . . . As a result of this work, we became firmly convinced that the electron emission from heated metals was a true property of the metals themselves and was not, as has so often been thought, a secondary effect, due to the presence of gas [65].

He developed methods for producing high vacuum, or hard, valves. In 1914, E.H. Armstrong also confirmed that the presence of gas in a valve was not fundamental for its performance. Subsequently, both AT&T and GE initiated programmes of research and development to improve the working of de Forest's tube.

Even so, the full understanding of the operation of the triode valve was not delineated until 1938 [66]. Necessarily then, de Forest's appreciation of the modus operandi of his tube must have been very limited.

Actually, de Forest was not a physicist who was concerned with extending the laws and principles of physics; he did not have the outlook of a pure physicist such as Maxwell or Hertz or Lodge. De Forest was primarily an inventor who sought patents for his work rather than a scientist who published papers on fundamental physics. He read, extensively, the scientific literature with the objective of finding new principles which would enable him to invent some device or method. His limitation was a lack of persistence in perfecting, or attempting to perfect, a particular invention: he did not have the stamina of, for example, Marconi or Baird in relation to his work. Nonetheless, de Forest was a highly creative, energetic and enterprising individual; he held 216 US patents.

He would sweep down on a problem with a hungry rush and his imagination had an astonishing faculty for leaping difficulties. If the quarry snagged or proved elusive, however, he had to hop to something else. When necessity did compel him to work at something without respite, his nerves rebelled. 'The jumpies' de Forest called these attacks [67].

Following the invention of the audion in 1906–07, de Forest next began to apply his energies to wireless telephony. The 'de Forest Radio Telephone Company' was formed in 1907, with a capitalisation of $2,000,000 [68], and in the Spring of 1907 messages were sent between a Lackawanna ferry and its terminals at Hoboken and Manhattan. The US Navy became the first customer for the system, which was based on an arc transmitter and either a crystal or vacuum tube detector, and ordered 27 sets for use in fleet manoeuvres [69] in the Pacific. In tests, communications over distances of c. 20 miles (32 km) were viable. According to one report a wireless operator named Meneratti developed the practice of broadcasting daily to the fleet using phonograph records.

Ever enterprising, and aware of the benefits of good and spectacular demonstrations, de Forest in the summer of 1908 travelled to Europe, with several assistants, in the hope that his system of radio telephony would solicit sales. While in Paris he obtained permission for his antennae to be suspended from the Eiffel tower. During the subsequent tests phonograph records were played by a Pathe talking machine which modulated the carrier signal. Some receptions of the tests were received as far away as Marseilles [70].

Although these demonstrations gave an inkling of what would eventually become commonplace, de Forest was keen to transmit 'live' music. His plans suffered a setback in 1908 when a fire destroyed his records and laboratory: it took a year before he was able to re-engage in laboratory work.

Early in January 1910 he concluded arrangements with the Metropolitan Opera Company in New York to broadcast one of its performances. His biographer has described the events associated with the first programme. It

was one of the Metropolitan's double bills, 'Cavalleria Rusticana' and 'Pagliacci' with Caruso appearing as Turridou in the former. A little half kilowatt telephone transmitter through which the chief arias were to pass was installed in a vacant room at the top of the Opera House. . . . Listeners-in were stationed at the Park Avenue laboratory, at the Metropolitan Life Building and at the Newark plant, supplemented by many curious engineers and amateurs and a specially invited audience at one of the hotels in the Times Square district. In spite of the crude arrangements . . . Caruso's voice went out over the ether on that date memorable in the history of radio, January [13], 1910, and was heard by perhaps 50 listeners. Wireless operators on ships in New York harbour and nearby waters, and at the Brooklyn Navy Yard, and a group of newspaper men who had gathered with great interest in the factory in Newark, all were lavish in their praise of the reception [70].

Rather more brutal opinions were published in the city's newspapers. *The New York Times* reported: 'At the receiving station in Mr Turner's office the homeless song waves were kept from finding themselves by constant interruptions. . . . Signor Caruso and Mme Destinn, it was learned later, sang finely at the Opera House. Mr Turner said that sometimes he could catch the ecstasy . . . but the reporters could hear only a ticking.'

The 'ticking', sometimes referred to as a 'tick-ta-ta-tick' was caused by the transmitting apparatus. Unlike Fessenden – who early in his work sought to negate the objectionable signal distortion heard at the receiver by utilising continuous waves – de Forest, ever impatient for quick results, employed spark discharge equipment. But this system was inherently incapable (for fundamental technical reasons) of reproducing satisfactory music or speech reception. As a consequence sales of the system did not flourish, the position of his company became dire, and finally it went into bankruptcy in 1911, following an unsuccessful merger with North American Wireless Corporation. de Forest now went to Palo Alto to work for the Federal Telegraph Company, at a salary of $300 per month. He later described his time with the FTC as among the 'happiest and most useful years' of his life.

His four 1911 patents [71–74] – one on a 'System for amplifying feeble electric currents', one on 'Space telegraphy', and two on 'Space telephony' – give some indication of his thoughts in 1911. A few months later de Forest, while working for the Federal Telegraph Company, undertook some experiments with the audion, based on its amplifying property. He obtained a patent on 'A method of and apparatus for amplifying and reproducing sounds', in January 1914. His amplifying system had the distinguishing feature that a circuit arrangement was employed which permitted the use of two or more amplifiers connected up in cascade and the employment of a single lighting battery. He connected the output of one tube to the input of another, and then took the output of the second tube and applied it to the input of a third tube, and showed that the final output was much greater than could be obtained from one tube. This cascade arrangement of tubes, to form a multi-stage amplifier, clearly had potential as a telephone repeater for voice signals on long lines.

Since the AT&T company was known to be interested in trans-continental communications, de Forest on 30th October 1912 approached the company, via a friend – J. Stone, a consulting engineer in New York – with a view to effecting a financial deal, either an outright sale of, or a licensing of, the patents of his audion amplifier. J.J. Carty, AT&T's Chief Engineer, invited de Forest to demonstrate his amplifier, and in October 1912 the system was tested. Success was once more achieved by the inventor, but the resulting financial negotiations in 1913, when AT&T agreed to purchase the telephone repeater rights in the audion for $50,000 [75], left de Forest feeling he had been cheated. He had optimistically expected these rights to be worth c. $500,000. Regrettably, his impecunious state and his trial for fraud severely limited his freedom of action in such matters. Afterwards, the company paid $90,000 for the radio rights to the triode.

The position in England regarding de Forest's audion patent was rather different to that prevailing in the US because, although the inventor had taken out a British patent in 1908, he allowed it to lapse on account of nonpayment of the renewal fees due in January 1911. Hence, the audion was freely available in England and the Fleming patent, of 16th November 1904, became a master patent. The Fleming patent was held by the Marconi company and so further advancement in valves took place largely in that company's laboratories. The two principal workers were H.J. Round and C.S. Franklin. Because of their company's interests, they directed their attention towards the solution of practical problems: hence the use of valves developed faster in England than in the US.

Still, undaunted, de Forest returned to Palo Alto and continued his study of the audion. He found that the tube, when properly configured, could be used not only as a detector, and as an amplifier, but also as an oscillator. This was a finding that, in a few years, would be of immense significance in communications generally. It would lead to the abandonment of the Marconi spark apparatus, the Poulsen arc and the Alexanderson alternator, which were expensive, cumbrous and noisy, as generators of electromagnetic waves. Indeed, such was the importance of the triode valve as an oscillator that a four party interference action was initiated in the United States involving Langmuir, Meissner, Armstrong and de Forest. De Forest eventually, in 1924, won the interference case and was awarded the covering patents.

The use of the valve oscillator in the heterodyne reception of signals was soon implemented by de Forest. Though Fessenden had devised this method of detection his local oscillator was the unsatisfactory arc apparatus. The small arcs supplied by the Fessenden company to the US Navy, which by this time had completely adopted continuous wave principles, were unreliable and subject to frequent breakdowns. Now, by means of the audion, a compact local oscillator could be produced: de Forest called his circuit the 'ultraudion'. It was installed in Federal Telegraph Company stations and was sold to the US Navy in considerable numbers.

An oscillator is essentially an amplifier, of appropriate gain and phase characteristics, in which its output is fed back, via a feedback network, to the input to sustain oscillations. If the gain of the amplifier is reduced to a value below that at which oscillations take place, then it is found that the amplifier gain with feedback is greater than the amplifier gain without feedback. Both de Forest and Armstrong claimed priority for this very important principle, and so, once again, the issue had to be resolved by the courts. The litigation became, possibly, the most controversial in radio history. The detailed arguments advanced by the two parties cannot be considered here, suffice to state that in 1928 the Supreme Court of the United States reversed the decisions of the lower courts and awarded priority of invention to de Forest. Six years later, when the case was the subject for further legal debate, a similar verdict was reached.

By this date de Forest had long given up his activities in radio and had turned his attentions to 'talking movies'. In 1917 he sold his feedback, telephone, and vacuum tube inventions to AT&T for $250,000, and from that time ceased to advance radio technology [76].

De Forest was an outstanding inventor. His 216 US patents display a breadth of imagination and pragmatic ability which classifies him as one of the truly great radio inventors of the 20th century. His experimental endeavours in the field of the three-electrode valve and its concomitant applications in detector, amplifier and oscillator circuits were brilliantly conceived and executed. Though he titled his autobiography [77] 'Father of radio', he can also be considered to be the 'father of electronics'.

Note 1 [38]

Detectors

An extensive literature exists on the history, principles of operation, construction and characteristics of the means by which electromagnetic signals can be detected. V.J. Phillips, in his exhaustive study of these devices, *Early radio wave detectors*, has classified them as: spark-gap detectors; coherers; electrolytic detectors; magnetic detectors; thin-film and capillary detectors; tickers, tone-wheels and heterodynes; and miscellaneous detectors. Fortunately, for the historian, 'very few indeed progressed beyond the laboratory stage to see active service in practical communication systems' [78]. 'Most of the detectors', Phillips says, 'were, by their nature, only able to provide a simple yes/no indication of the presence of a received signal. A few were able to give a quantitative measure of the strength of that signal, but most of these were unable to operate with sufficient speed to follow the dots and dashes of a telegraphy signal' [38].

An indication of the worth of the many types of detector during the formative period of wireless telegraphy may be obtained from standard text books of the period. In the 1911 edition of the US Navy's *Manual of wireless telegraphy for the use of naval electricians* the author, S.S. Robison, wrote: 'There are but two types of detector now in general use, crystal or rectifying detectors, and the electrolytic. Coherers and microphones are practically obsolete and comparatively few of the audion or valve detectors have been installed'. Two years later, in the 1913 edition, he amended this statement to: 'Only one type of detector is now in use, the crystal. The electrolytic is used as a standard of comparison. Coherers and microphones are practically obsolete and comparatively few magnetic and audion or valve detectors have been installed'. Finally in the 1915 edition Robison observed: 'The detectors now in general use are the crystal or rectifying detector and the audion'.

In the British *Admiralty handbook of wireless telegraphy* for 1920 the writer mentions: 'Crystal detectors are being replaced by valve detectors which are more stable, easier to adjust and generally more satisfactory'. Interestingly, during the formative period of microwave engineering, it was found that silicon (crystal) detectors were preferable to valve detectors. In the first week of April 1940 the first real comparison on a pulse [radar] system receiving reflected echoes was made between a crystal mixer and a mixer using a concentric diode. An appreciable improvement was observed with the crystal mixer. From this time the device steadily improved its lead over competitive mixer arrangements, many of which, including various diode designs, autodyne triode circuits and velocity modulation tubes, were tried.

An analysis of the advantages and disadvantages of the various types of detector was made by S.M. Powell in 1911. He summarised his findings in a table, reproduced in Table 12.1.

A comparison of the sensitivities of some detectors was attempted by P. Edeman in 1915/16. He used as his standard for comparison the energy required to enable a detector to indicate reliably the arrival of a Morse dot. His results are given in Table 12.2.

Table 12.1 Comparison of various types of detectors

	Maxima	Coherer	Magnetic	Electrolytic	Thermo-electric	Rectifier	Walter's Ta-Hg	Valve
Automatic decoherence	5	0	5	5	5	5	5	5
Adjustability	5	0	3	3	5	5	5	0
Tuning	5	3	4	5	4	4	4	3
Quick acting	5	5	4.5	5	4.5	5	5	5
Sensitivity to								
(a) weak signals	5	4	5	5	4.5	2	2	4.5
(b) strong signals	3	3	2	3	3	2	3	3
Resistance	4	2	3	2	3.5	2.5	2.5	2.5
External effects								
(a) vibration	4	5	4	4	3	4	4	4
(b) field	4	4	3.5	4	4	4	4	3.5
Ease of construction and erection	2	1	1	2	2	2	2	0
Ease of procuring material	1	1	1	1	0.5	0.5	1	1
Total	43	28	36	39	39	36	37.5	31.5

Table 12.2 *Detector sensitivities*

Type of detector	Energy required (ergs per dot)
Electrolytic	0.00364–0.00040* 0.007§
Silicon	0.00043–0.00045*
Magnetic hysteresis	0.01§
Hot-wire barretter	0.08§
Carborundum	0.009–0.014*

(*according to Pickard; § according to Fessenden)
1 erg = 0.1 µJ

Note 2

The birth of the valve [78]

The science of thermionic emission developed side by side with that of photoelectric emission and the science of the conduction of electricity through gases. Great discoveries were made in these fields, during the last quarter of the 19th century and the first decade of the 20th century, which had important consequences for the furtherance of radio and television broadcasting.

The first systematic investigations in thermionics were carried out by Elster and Geitel during the years 1882 to 1889. They studied in detail the charge acquired by an insulated metal plate, mounted close to a metallic filament within a glass bulb, under different conditions of filament temperature and gas pressure. In the same period, at the Philadelphia Exhibition of 1884, Edison exhibited a discovery he had made while investigating incandescent lamps, and which is now known as the Edison effect. Neither Edison nor Sir William Preece, who subsequently performed some experiments on this effect, gave any explanation of the phenomenon, nor was any practical application made of it.

Later, in 1890, Professor J.A. Fleming showed that, when the negative leg of a heated carbon loop filament was surrounded by a cylinder of either a metal or an insulating material, the Edison effect disappeared. Other experiments of a similar nature demonstrated that the action was due to the passage of negative electricity from the incandescent filament to the cold electrode and corroborated the findings of Elster and Geitel.

Ensuing investigations by Thomson in 1899 indicated, in the case of a carbon filament glowing in hydrogen at very low pressure, that the negative electricity was given off by the filament in the form of free electrons. This conclusion applied, too, to the electric current emanating from a lime-covered platinum cathode.

The first application of the phenomena associated with thermionic emission was Fleming's 1904 device for rectifying alternating voltages. It consisted of a carbon filament incandescent lamp provided with a separate insulated electrode in the shape of a flat or cylindrical metal plate, or another carbon filament sealed into a bulb. Later, in 1908, Fleming noted that much improved results were obtained when the valve was constructed with a tungsten filament and an insulated coaxial cylindrical copper anode.

The next step in the evolution of the thermionic valve was contributed by de Forest in 1906 when he introduced a third electrode into the rectifying valve, sited between its filament and anode structures. The principle of grid control had been used previously by the German physicist Lenard for studying the motion and nature of the electrons liberated from a zinc cathode by ultraviolet light, but Lenard had not conceived of its use for the detection or amplification of wireless signals.

The development of large power valve transmitters was first shown to be within the realms of practical accomplishment when AT&T in August 1915 used such a transmitter to send speech signals from Arlington, Virginia, to Darien in the Panama Canal zone, 2,100 miles (3,380 km) distant. About one year later the same company communicated with Paris and employed a bank of 500 valves, each having a capacity of 15 W, in its Arlington transmitter.

The First World War gave an enormous impetus to the utilisation of valves in signalling systems and so stimulated advances in technique that by 1918 triodes could be manufactured to cover a wide power range and were suitable for both transmitting and receiving purposes; their theory and operation, over the frequency bands used at that time, were both thoroughly understood.

During the war, valve transmitters had been used principally for communications between ground and aircraft in flight over France and, in this work, the Marconi company played a most valuable role. The valve transmitter had quickly ousted the spark, the arc and the high frequency alternator types of transmitters, and the progress of continuous wave radio communications had moved forward rapidly to the stage where commercial sound broadcasting could be seriously contemplated shortly after the cessation of hostilities in 1918. By 1921 the Marconi company had a valve transmitter with a rated output of 100 kW installed at Caernarvon, North Wales. And if sound signals could be propagated by radio, then surely vision signals too could be transmitted. This is considered in Chapter 15, but first the applications of wireless telegraph to maritime signalling must be discussed.

References

1 SUSSKIND, C.: 'Popov and the beginnings of radiotelegraphy', *Proc. IRE*, 1962, **50**, pp. 2036–2047
2 POPOV, A.S.: 'On the relation between metal powders and electric

oscillations', *Zh. Russ. Fiz.-Khim. Obshchestva* (Physics, Part 1), 1895, **27**, pp. 259–260

3 FLEMING, J.A.: 'The principles of electric wave telegraphy' (Longmans, Green, London, 1908), p. 361

4 Ibid, p. 363

5 POPOV, A.S.: letter, *The Electrician*, 1897, **40**, p. 235

6 POPOV, A.S.: 'Apparatus for the detection and recording of electrical oscillations', *Zh. Russ. Fiz.-Khim. Obshchestva* (Physics, Part 1), 1896, **28**, pp. 1–14

7 ANON.: minutes of a meeting, *Zh. Russ. Fiz.-Khim. Obshchestva* (Physics, Part 1), 1896, **28**, pp. 121–124

8 POPOV, A.S.: letter, *Kotlin*, no. 5, 8th January 1897, p. 2

9 GEOGIEVSKY, N.N.: quoted in *Elektrechestvo*, no. 4, 1925, pp. 214–215

10 VVEDENSKY, B.A. (Ed.).: 'Great Soviet Encyclopaedia' (Gosdarstvennoe Nauchnoe Izdatelstvo 'BSE', Moscow, 1955), **34**, p. 159

11 Ref. 1, p. 2043

12 GABEL, V.S.: note, *Wireless World and Radio Review*, 1925, **16**, p. 410

13 LODGE, O.J.: Improvements in syntonised telegraphy without wires, British patent no. 11,575, 10th May 1897

14 ROWLANDS, P., and WILSON, P.J. (Eds.): 'Oliver Lodge and the invention of radio' (PD Publications, Liverpool, 1994), p. 81

15 LODGE, O.J., and MUIRHEAD, A.: 'Improvements relating to electric telegraphy', British patent no. 16,405, 10th July 1897

16 LODGE, O.J., and MUIRHEAD, A.: 'Improvements relating to electric telegraphy', British patent no. 18,644, 11th August 1897

17 LODGE, O.J., and MUIRHEAD, A.: 'Improvements in syntonic telegraphy', British patent no. 29,069, 8th December 1897

18 MARCONI, G.: 'Improvements in transmitting electrical impulses and signals and in apparatus therefor', British patent no. 12,039, 2nd June 1896

19 MARCONI, G.: 'Improvements in apparatus employed in wireless telegraphy', British patent no. 29,306, 10th December 1897

20 Ref. 14, pp. 98–100

21 Ref. 14, pp. 181–183

22 Ref. 14, p. 199

23 SLABY, A.: 'The new telegraphy', *The Century Magazine'*, 1898, **55**, p. 867

24 KAPP, G.: letter to W.H. Preece, 19th March 1897, The IEE archives

25 Ref. 3, pp. 473–477

26 SLABY, A.: 'Abgestimmte und Mehrfache Funkentelegraphie', *Elektrotechnische Zeitschrift*, 1901. And: *The Electrician*, 1901, **46**, p. 475

27 SHARLIN, H.I.: article on Ferdinand Braun, 'Dictionary of Scientific Biography' (Charles Scribner's, New York, 1970) pp. 427–428

28 BRAUN, F.: 'Ueber ein Verfahren zur Demonstration zum Studium des Zeitlichen Verlaufes Variabler Strome', *Annalen der Physik*, 1897, **60**, p. 552

29 Ref. 3, p. 499

30 Ref. 3, p. 500–501
31 SUSSKIND, C.: article on R.A. Fessenden, 'Dictionary of Scientific Biography' (Charles Scribner's, New York, 1970) p. 601
32 KRAEUTER, D.W.: 'Radio and television pioneers: a patent bibliography' (Scarecrow Press, Metuchen, 1992)
33 BELROSE, J.S. : 'Fessenden and the early history of radio science', *Proc. of the Radio Club of America*, November 1993, pp. 6–23
34 MACLAURIN, W.R.: 'Invention and innovation in the radio industry' (MacMillan, New York, 1949), p. 59
35 Ibid, p. 64
36 Ref. 33, p. 12
37 Ref. 33, p. 14
38 PHILLIPS, V.J.: 'Early radio wave detectors' (Peter Peregrinus Ltd, London, 1980), p. 38
39 Ibid, p. 74
40 FESSENDEN, R.A.: 'Wireless telegraphy', US patent no. 706,740, 12th August 1902
41 FESSENDEN, R.A.: 'Electric signalling apparatus', US patent no. 1,050,441, 14th January 1913
42 FESSENDEN, R.A.: 'Signalling', US patent no. 1,050, 728, 14th January 1913
43 HOGAN, J.V.L.: 'The heterodyne receiving system and notes on the recent Arlington-Salem tests', *Proc. IRE*, 1(3), 1913, pp. 75–97
44 Ref. 34, p. 64; and ref. 33, p. 17
45 Ref. 34, p. 59
46 Quoted in 'History of radio to 1926', by G. Archer (American Historical Co., New York, 1938), p. 85
47 RUHMER, E.: 'Wireless telephony', trans. from the German by J. Erskine Murray, with an appendix on R.A. Fessenden by the translator (Crosby Lockwood, London, 1904)
48 Ref. 33, p. 18
49 FAGEN, M.D.: 'A history of engineering and science in the Bell System. The early years (1875–1925)' (Bell Telephone Laboratories, 1975), pp. 362–363
50 Ref. 34, p. 66
51 Ref. 34, pp. 67–69
52 SUSSKIND, C.: article on L. de Forest, 'Dictionary of Scientific Biography' (Charles Scribner's, New York, 1970), pp. 6–7
53 Quoted in Ref. 34, p. 71
54 Quoted in Ref. 34, p. 72
55 Quoted in Ref. 34, p. 73
56 CARNEAL, G.: 'The reflection of short Hertzian waves from the ends of parallel wires' (Yale University, 1899), p. 125
57 KINGSFORD, P.W.: 'Electrical engineers and workers' (Edward Arnold, London, 1969), p. 201
58 ECCLES, W.H.: a note, *Electrician*, 1908, **60**, p. 588
59 Ref. 57, pp. 201–202
60 ARCHER, G.L.: 'History of radio to 1926' (Arno Press, New York, 1971), p. 60

61 Ibid, p. 115
62 Ibid, p. 135
63 Ref. 57, pp. 202–203
64 Ref. 49, pp. 260–262
65 LANGMUIR, I.: 'The pure electron discharge', *Proc. IRE*, 1915, **3**(3), pp. 266–269
66 BURNS, R.W.: 'The evolution of modern British electronics, 1935–1945', *Transactions of the Newcomen Society* (to be published)
67 LUBELL, S.: 'Magnificent failure', *Saturday Evening Post*, 24th January 1942, p. 41
68 Ref. 56, p. 198
69 Ref. 67, p. 38
70 Ref. 60, pp. 97–99
71 FOREST, De L.: 'System for amplifying feeble electric currents', US patent no. 995,126, 13th June 1911
72 FOREST, De L.: 'Space telegraphy', US patent no. 995,339, 13th June 1911
73 FOREST, De L.: 'Space telegraphy', US patent no. 1,006,635, 24th October 1911
74 FOREST, De L.: 'Space telephony', US patent no. 1,006,636, 24th October 1911
75 Ref. 34, p. 77
76 Ref. 60, p. 135
77 FOREST, De L.: 'Father of radio' (Chicago, 1950)
78 BURNS, R.W.: 'Television, an international history of the formative years' (IEE, London, 1998)

Chapter 13

Maritime wireless telegraphy

At the end of 1898 Marconi's numerous investigations and demonstrations, and the experiments of Jackson, had clearly shown that:

1. a wireless telegraph system could operate over land and water, and could work particularly well between a shore station and ships;
2. the range of working was much greater than could be achieved by either a conductive or inductive electric telegraph system;
3. wireless telegraphy could be used at night as well as during the day; and was unaffected by bad weather, such as rain, fog or storms – provided the insulation of the antenna wire was not degraded;
4. the interposition of hills and trees did not prevent communication between a sender and a receiver;
5. the apparatus could be used by any telegraphist after a short induction period;
6. the system that comprised the transmitter and receiver was not costly, though the installation of a high mast and associated antenna might be expensive;
7. the dimensions of the various units, with the exception of the antenna and mast, were such that the units could easily be accommodated on board ships and large yachts; and
8. wireless telegraphy had many applications.

Two immediate problems remained to be solved before the new communication system became commercially worthwhile. First, the range of working had to be greatly increased; and, second, means had to be implemented that would permit signals to be sent to a particular recipient rather than to anyone who possessed the basic apparatus.

The solution of the range problem was central to the viability of the new communication method for maritime use, and to the financial success of the Wireless Signal & Telegraph Company. Longer transmission distances were being slowly, but steadily, achieved and Marconi was optimistic that with further effort the problem would be solved. Then, given the demonstrated advantages of wireless telegraphy for message transmission from shore-to-lighthouse and from shore-to-lightship, for ship-to-ship signalling, for safety at sea, for fleet manoeuvres, for identifying friend from foe during naval exercises and naval engagements, for alerting shipping companies of the positions of their ships and unknown hazards, and for general commercial and public intelligence communications, the prospects of good sales, or hire, of equipment could be anticipated. These prospects seemed to be imminent: there was a need to establish a manufacturing base.

In December 1898, one month after Dr Erskine Murray's appointment as Marconi's principal experimental assistant, WS&T acquired a building in Hall Street, Chelmsford, Essex [1]. The premises that had been a silk factory and later a furniture warehouse now became the first wireless factory in the world. Dr Erskine Murray left the Haven experimental station in July 1899 and became responsible for the technical work at the factory.

Meantime, experimental work [2], under the direction of Captain F.T. Hamilton [3], had continued on board HMS *Defiance* throughout 1898. He had been particularly keen to test the Royal Navy's apparatus under adverse sea conditions and to ensure that its operation would not influence the safety of any naval ship on which it was installed. This was a matter of some importance since the Admiralty's Signal Committee had expressed its fears about the integrity of ships' magazines when electromagnetic waves were used. Captain Jackson [4] had given his view (see Chapter 11) but such was the committee's concern that it asked several distinguished scientists for their opinions. Lord Kelvin [5] agreed with Jackson: 'I believe there would be absolutely no danger from the use of this system on board a ship of war of any class. The risk of accident by the explosion of detonators or of ammunition of any kind for guns or torpedoes will not, in my opinion, be increased in the slightest degree by the introduction and use of Marconi's apparatus for wireless telegraphy'. Preece also agreed [6]. The way was now open for the Admiralty to consider fitting wireless equipment, either that of Jackson or that of Marconi, or a combination of these, generally to its fleet.

Accordingly, the Admiralty commenced some negotiations with WS&T for the purchase of apparatus, initially to equip two ships of the navy [7]. The company readily responded to the enquiry and agreed to the sale, subject to the payment of an annual royalty for the use of Marconi's patents. At the end of March 1899, WS&T wrote: 'The amount of this royalty . . . I shall inform you of when our board has decided what it ought to be'.

Exclusive of royalty, the price quoted for the equipment was: (1) induction coil attachments and transmitting key, £4.27.5; (2) induction coil, £38.57.5; (3) 100 M size Obach cells, £22.0; (4) complete receiver in box, £13.80; (5) inkwriter, £15.0; (6) bell, £17.5 [8].

The question of the royalty fee was considered by the WS&T Board on 17th April 1899, and a figure of £100 per ship per annum was agreed. This charge was felt by the Director of Naval Ordnance to be very high and so he enquired whether the company was exempt from the Government's telegraph monopoly. The Admiralty's solicitor was consulted and he opined that the monopoly held only within the United Kingdom and consequently did not apply to the use of wireless telegraphy at sea. Further referral of the royalty issue to the General Post Office elicited the response that the GPO was awaiting Treasury permission to commence negotiations with the Wireless Signal and Telegraph Company and hence no advice could be offered. All of this meant that the Admiralty could not arrive at a conclusion as to whether the royalty charge would be reasonable or not. Accordingly it deferred its reply to WS&T's quotation until the GPO–WS&T position was resolved.

At this time, the Admiralty was anxious to equip two of its ships with the Marconi apparatus because it wished to assess the wireless telegraph system during the Fleet manoeuvres which had been planned for the summer of 1899. Fortunately, Captain Jackson's tour of duty at the British Embassy in Paris was coming to a close and, as he had established a good rapport with Marconi himself and was familiar with his apparatus, and wireless telegraphy generally, he appeared to be the obvious choice to take charge of the trials.

Obviously, the Wireless Signal and Telegraph Company could not adopt a negative attitude to the Admiralty's need for two sets of apparatus. The company had been in existence for just two years and was anxious to seek lucrative governmental and commercial markets for its products. The potential markets, embracing both mercantile and naval fleets world-wide, were vast; good publicity was an indispensable condition for future sales; and reasonable and agreeable relations with customers were necessary conditions for commercial success. WS&T agreed to lend the Admiralty the apparatus it needed for the manoeuvres without cost, other than the cost of carriage for the sets and the expenses of the company's technicians who would install and maintain the equipment [9]. Furthermore WS&T agreed to defer the general question of payment to a later date.

Of the two sets, one was installed on the cruiser HMS *Juno*, commanded by Captain Jackson during the summer exercises, and the other was put aboard the battleship HMS *Alexandria*, the flagship of Vice-Admiral Sir Compton Domville. Marconi was invited to observe the manoeuvres and sailed with the *Juno*. Another ship, the cruiser HMS *Europa*, was equipped with a third set, at the Admiralty's request [10].

The objective of the exercise was to determine the optimum conditions for deploying a large force of cruisers in association with a fleet of battleships, and to evaluate the desirability of including torpedo boats and destroyers in such a large battle group. For this purpose the warships were divided into two fleets: 'A' Fleet, under the command of Admiral Sir Harry Rawson, which comprised 8 battleships and 19 cruisers; and 'B' Fleet under the control of Vice-Admiral Sir Compton Domville, which consisted of 10 battleships and 20 cruisers. Of the two fleets, only the 'B' Fleet possessed wireless equipment [11].

The strategic plan of the exercise assumed that a convoy of merchant ships, represented by HMS *Calliope* and HMS *Curacoa*, had sailed, with a single cruiser as escort, from Nova Scotia en route to Milford Haven, and was waiting at a rendezvous which was known only to the Admiral commanding 'B' Fleet. His task was to arrive, in strength, at the rendezvous before the 'enemy', represented by 'A' Fleet, could do so, and then to escort the convoy safely to Milford Haven.

The exercise was an outstanding success for 'B' Fleet. Jackson's ship HMS *Juno* was the first to sight the convoy and immediately communicated by wireless telegraphy to HMS *Europa*, more than 59 miles (95 km) away, which then relayed the signal to HMS *Alexandria*. The total distance between *Juno* and *Alexandria* was 93.2 miles (150 km) [12]. This was vastly greater than could be accomplished by semaphores, flags or lamps. Moreover, the propagation time for the message was instantaneous whereas a fast naval vessel would have taken three or more hours, depending on the sea state, to have conveyed the intelligence to the commander of 'B' Fleet [13].

The transmission distance of 59 miles (95 km) was the longest that had been achieved by the summer of 1899 and was attributed by Marconi to his use of resonance and a 'jigger'. Essentially, this was a transformer that coupled, and matched, the receiving antenna to the coherer circuit.

When a vertical wire antenna is set up with its lower end connected to an earth plate, or to the earth itself, the fundamental mode of the voltage and current distribution established along the wire is such that a node of potential and an anti-node of current exist at the lower, or earthed end, of the wire. At the top, or open end, of the antenna a current node and a voltage anti-node occur. If now a coherer is inserted in series with the antenna near its lower end, the voltage developed across the coherer, when an electromagnetic wave is received, is much smaller than that developed if the coherer could be inserted in the antenna wire near its top end. Since the latter position would be impractical, and since the coherer is a voltage operated device, means must be taken to increase the potential difference across the coherer to achieve good sensitivity. Such a means comprises a matching radio frequency transformer, or 'jigger' as Marconi called the transformer, (Fig. 13.1).

All of this was known to Lodge and, in his patent [14] no. 11,575 dated 10th May 1897 on 'Improvements in syntonised telegraphy without line wires', which is concerned with resonant systems, he clearly showed the use of such a transformer. However, Lodge did not give any details of its practical construction. Marconi on the other hand wound and tested several hundred transformers, using different ratios of turns, varying wire diameters and diverse couplings. He patented three of the most promising designs. According to Baker [15] (who does not give any supporting evidence), Marconi's experiments with jiggers began 'well before Lodge's patent of 1897', and were conducted between March and December 1897.

During the naval exercises of 1899 when a signalling distance of 60 miles (96.5 km) was achieved, Marconi determined that a range of just 7 miles (11.2 km) was possible when the 'jigger' was removed, thereby giving proof to the importance of matching and tuning [16].

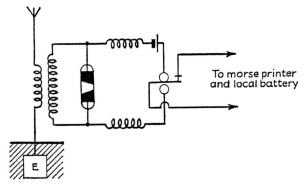

Marconi's H.F. Transformer or 'Jigger', 1897

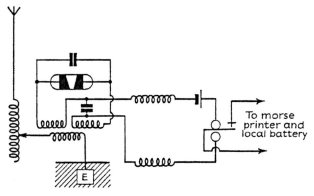

Marconi's H.F. Transformer with Capacitors, 1898. (Pat. No.
25,186, 19 December.) (The capacitor across the coherer and that across
the split in the secondary winding constitutes a form of pre-set tuning.)

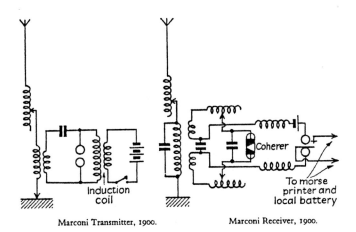

Marconi Transmitter, 1900. Marconi Receiver, 1900.

Figure 13.1   *Examples of the 'jigger' circuits used by Marconi in 1897, 1898 and 1900.*
Source: Author's collection.

Captain Jackson's report to Sir Compton Domville also ascribed an increase in signalling speed to the device. Jackson's enthusiasm for Marconi's new system led him to recommend its installation in the flagships and the scouting cruisers of the Channel and the Mediterranean Squadrons, in the Commander-in-Chief's offices of the home ports and Malta, in the torpedo school ships at home, and in HMS *Vulcan*, the torpedo depot ship in the Mediterranean. Furthermore, he proposed: (1) that two trained signalmen should be deputed to operate each installation; (2) that courses of instruction should be established in *Vulcan* and in the torpedo schools; and (3) that sailors who successfully completed the wireless operator's course should be given a small pay rise [17].

Domville strongly endorsed Jackson's suggestions and in a covering note to the Admiralty he wrote [18]:

I entirely concur with the whole of Captain Jackson's remarks, and, speaking from my own experience of its use during the last month, I consider, from the Admiral's point of view, that this system of wireless telegraphy is absolutely invaluable. It can be trusted, especially at night or in a fog, when no other system is perfectly reliable.

I would submit that the proposal to fit certain ships with it be carried into effect as soon as possible.

The Second Sea Lord at the Admiralty, Rear Admiral Douglas, agreed and proposed that a further eight sets should be ordered immediately from WS&T, in anticipation of the general adoption of wireless telegraphy in the British navy, and utilised for training suitable seamen of the Channel and Mediterranean Squadrons [19].

This general enthusiasm for Marconi's system was somewhat dampened by the terms of the contract which WS&T wished to impose on any purchaser of its apparatus. The £100 royalty fee per installation per annum was still considered to be excessive [20]. Moreover, the General Post Office had not yet informed the Admiralty of the conclusions of their negotiations with WS&T for the right to use the Marconi patents. On enquiries being made the Admiralty, on 4th September 1899, was apprised [21] that: 'the Company have made a proposal, but it is of such a nature that the Postmaster-General does not think that Her Majesty's Government could possibly entertain it. The whole matter is, however, under . . . consideration, and it is possible that after further negotiations the Company may be disposed to modify their terms.'

Still, the Admiralty persevered but on 13th October 1899 the Director of Naval Contracts minuted [22]: 'The terms proposed by the Company are preposterous, and they refuse to sell instruments exclusive of royalty, leaving the amount to be settled afterwards.'

Events, involving the General Post Office, the Treasury, the Admiralty and the Wireless Signal & Telegraph Company, now dragged on for several months but, on 16th March 1900, the Admiralty had to concede defeat [23]:

My Lords [Commissioners of the Admiralty], however, feel so strongly as to the absolute necessity of obtaining the free use of this invention for HM Navy that ... they see no alternative but to accept unconditionally the proposals of the Company.

I am therefore to request that the Lords Commissioners of the Treasury will give their sanction to the Lords Commissioners of the Admiralty entering at once into an agreement with the Marconi Company for the use of their system of Wireless Telegraphy by HM Navy on the payment of a Royalty of £100 per annum for each installation.

Treasury approval [24] was given on the 4th April but discussions within the Admiralty on points of detail deferred their notification of approval of the royalty payments to WS&T until the 8th May 1900 [25].

Subsequently, the company's quotation for the supply of 32 complete sets, at a cost of £196.14s.4d per installation [26], was accepted and delivery of the sets to the Admiralty commenced at the end of July 1900. The order was completed by November and the sets were allocated as follows [27]:

Channel Squadron	6 sets
Mediterranean Squadron	6 sets
Reserve Squadron	4 sets
Home Ports	4 sets
Torpedo Schools	4 sets
Signal stations	4 sets
Cruisers on manoeuvres	4 sets

Apart from these sets, the Admiralty had already acquired four Marconi sets which had been supplied, on 10th October 1899, to the Royal Engineers, in South Africa, on the commencement of hostilities with the Boers [26]. The Army sets had failed to meet the performance expected from them and were never taken to the front line. Actually, the sets were fault free (see Chapter 16).

These sets were installed in four ships of the Durban Squadron which were employed in the blockade of Lourenco Marques to prevent supplies from reaching the Boers in the Transvaal. One of the ships, HMS *Thetis*, was fitted with the apparatus on 17th March 1900 and became the first ship to be equipped with WT in a theatre of war [26].

Meantime, during the period of indecision when talks were being held to resolve the question of the royalty payment, the Admiralty decided to invoke the power retained for the Crown in the Patents, Designs and Trade Marks Act of 1883 [28]. This Act enabled the Government to manufacture and use a patented invention if it was felt to be necessary in the national interest, even if the holder of the patent withheld his/her permission. Accordingly, the Admiralty decided to manufacture sets of wireless transmitters and receivers, employing the facilities of HMS *Vernon* and those of sub-contractors. Captain Jackson had been appointed to HMS *Vernon* on 20th October 1899 and was, of course, very familiar with Marconi's system. He designed the equipment to replicate [29], as far as possible, Marconi's set so that operators could easily make the transition from one to the other, and so that duplication of spare parts could be avoided.

By 2nd December 1899 two sets had been completed and installed initially in HMS *Hector* and HMS *Jaseur*. Trials, which continued for approximately three months, showed that signals could be consistently received at 20 miles (32 km) range, and occasionally at 32 miles (51.5 km). The sets were inferior in performance to those of WS&T but they solved the problem of providing apparatus for the training of wireless operators [30].

At the end of 1900 the Royal Navy possessed 32 Marconi and 19 Jackson sets distributed as follows [31]:

Allocation	Marconi sets	Jackson sets
Channel Squadron	6	2
Mediterranean Squadron	4	3 (+1 spare)
Reserve Squadron	3	3
Training Squadron	4	–
China Squadron	3	3 (+1 spare)
Torpedo Schools		
HMS *Vernon*	2	2 (+1 spare)
HMS *Defiance*	2	2 (+1 spare)
Shore Stations	3	–
Shore Stations (awaiting Installation)	5	–

These installations made the British Fleet the most modern in the world, from a communications viewpoint. Elsewhere, at the turn of the century, the position varied from one of indecision to a slow implementation of a few sets on board naval ships [31].

1. In Italy, as a consequence of Marconi's visit and demonstrations in July 1897, the Italian Navy had set up shore stations, using his equipment, at Spezia, Leghorn, San Bartolemo and Varugiano; and had installed wireless apparatus on a few ships.
2. The Japanese Navy had purchased five Marconi sets and had fitted them to the *Fuji*, the *Yashima*, the *Asama*, the *Akashi* and the *Yaeyana*.
3. No French ships carried wireless telegraphy equipment though the French Navy had undertaken extensive exercises following the establishment of the Dover to Calais WT link, and was veering towards the adoption of the Popov–Ducretet system.
4. The United States Navy was without WT and had deferred its policy on such matters pending the development of a suitable tuned system.
5. In Germany, Slaby had produced a workable system for the army and this was available to the navy when needed.
6. The Russian Navy was without WT in 1900, though, it is said, Popov, in the spring of 1896, had carried out some short range tests in Kronstadt harbour with his apparatus fitted in the cruisers *Russia* and *Africa*. In January 1900, when the battleship *Grand-Admiral Apraxin* ran aground on Suursaari

island in the Gulf of Finland, Popov established a WT link between the island and Kotka, the nearest mainland town, 31 miles (50 km) distant. This link remained operational until April 1900.

The 51 equipments of the Royal Navy gave it an invaluable amount of data on the characteristics of the Jackson and Marconi sets, under varying conditions. Analysis of 19 separate features showed that, for the Mediterranean Squadron, the desirable attributes were distributed as: Marconi, seven; Jackson, 10; equal, two. The corresponding figures for the Home Fleet were 10, 8 and 1 respectively [32].

By 31st December 1901 the Royal Navy had a total of 105 sets, of which there were at least 30 Marconi sets and at least 60 Jackson sets. In addition there were eight shore stations. Later, in 1909, high power 100 kW, 250 kHz spark stations were established at Cleethorpes, and Horsea, and at Gibralter in 1910, and then Rinella, Malta. Further stations, of 14 kW output, were completed in 1911 at the Admiralty, Aberdeen, Ipswich and Pembroke: the Horsea 100 kW arc transmitter began working in 1912. Finally, in 1915, arc transmitters were installed at Cleethorpes, Aberdeen, Ipswich and Pembroke.

Apart from its lucrative contracts with the British Admiralty in 1900, Marconi's Wireless Telegraph Company Ltd (the changed company name was registered on 14th March 1900 [33]) in that year received an order from an unexpected source – the Norddeutscher Lloyd shipping line of Bremen [34]. This order was for the installation of apparatus on board the liner *Kaiser Wilhelm der Grosse*. Permission was also given by the German Government for wireless equipment to be fitted to the Borkum Riff lightship and at the Borkum lighthouse. Soon, these installations were followed by others on the Belgian cross-Channel *Princess Clementine*, a state-owned railway steamer [35], and at a shore station at La Panne on the Belgian coast to enable messages to be sent to and be received from the steamer. In all these cases MWT sold rather than hired its apparatus to its customers.

In 1900 it was clear that the immediate prospect of financial returns from the sale or hire of wireless telegraph equipment was from maritime, and ship-to-shore, installations. As a consequence the board of MWT decided to form a subsidiary company – the Marconi International Marine Communication Company Ltd (MIMCC). Incorporated on 25th April 1900 the company opened offices in London and Brussels, and agencies in Paris and Rome, with the objective of furthering the maritime working of the Marconi system [36]. The fitting of WT sets into ships did not immediately render the sets useful. There had to be a network of coastal stations which could receive any signals being transmitted from the ships. Unfortunately, these stations were not being provided by governments.

In 1901 the policy of the company changed to counter this inhibitory effect on sales. The company now committed itself to the organisation of a public wireless telegraph service primarily for maritime purposes, but was not free to establish such a service ab initio; the British Telegraph Acts of 1868 and 1869 imposed limitations on what could be undertaken by a private company. The principal relevant features of the Acts were [36]:

1. the clauses in the Acts would permit a private company to send messages for its own use, but not for gain as a public telegraph service;
2. a private company could own and maintain a private service on behalf of some other organisation so long as no charge was made for the sending of telegrams.

These restrictions led to the following conclusion: a company could install its equipment and operators on board ships for the purposes of communicating with shore stations, and other similarly fitted ships on the high seas, provided no direct charge was made for the messages handled.

In 1901 MWT changed its policy to reflect this situation: apparatus would now be hired, rather than sold, to shipping companies; the company's wireless operators would be in charge of the ship-borne installations and be responsible for the transmission and reception of all traffic while at sea; and the company would establish its own shore stations for the handling of all messages to and from ships using the Marconi system. It was envisaged that the company's profits would come from the receipts of messages sent, particularly those from the trans-Atlantic liners. Consequently, the shore stations were sited, initially, to cover the approaches to the main British ports, but later the network of stations was extended to cover the approaches to New York harbour. The earliest (1901) coast stations were situated at Holyhead, Caister, and North Foreland in England, and Rosslare and Crookhaven in Ireland.

The *Lake Champlain* became the first British ship to be equipped [37], (in 1901) and was followed by several ships of the Cunard Line. Soon the majority of the trans-Atlantic liners – British, German and French – were using the Marconi system. (See Figs 13.2 and 13.3). For the ship owner, the early news of a ship's impending arrival in port, and the organisation of the necessary unloading and loading arrangements to achieve a rapid turn-around, was an incentive to the hiring of the wireless telegraph.

From an early stage in the pursuit of its policy, the Marconi International Marine Communication Company (MIMCC) declared that ships installed with its apparatus would not, except in an emergency, communicate with ships or shore stations employing the equipment of a competitor. The company averred that such a restriction would prevent chaos by maintaining the service under one authority, and claimed that if a contrary situation existed it would be difficult to litigate against infringements of its patents.

The creation of a monopoly position, while being very attractive and beneficial to any company, had the obvious disadvantage of denying commercial competition to a ship owner, with the consequential prospect of extortionate rates for the hiring of the company's equipment and operators. Once the company had set up a chain of coastal stations and had its system on board several ships, the ship owner would be faced with accepting the company's terms, or accepting a very limited service from a competitor.

(a)

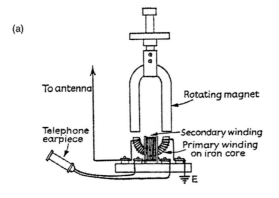

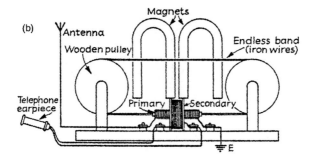

(b)

(c)

Figure 13.2   *(a), (b) In 1902 Marconi patented two forms of the Rutherford type of magnetic detector. Compared to the coherer, the detector was more sensitive, it produced clearer signals and enabled a higher speed of transmission and reception to be achieved. (c) An early coherer receiver unit, 1904.*

Sources: (a), (b) W.J. Baker, *A history of the Marconi Company* (Methuen, London, 1970); (c) The Marconi Company.

Figure 13.3   The radio room of RMS Olympic – *similar to that on RMS* Titanic.
Source: The Marconi Company.

The Marconi company's contract with the British Admiralty possibly stimulated Lloyds of London for, in 1901, it signed an agreement [38], to be operative for 14 years, with MIMCC whereby the company would install WT equipment at a number (initially ten) of coastal stations. Apart from insurance, Lloyds provided marine intelligence to shipping companies, and maintained a network of look-out stations throughout the world which enabled ships to pass messages by semaphore or lamp to the shore stations. Another agreement, between MIMCC and Trinity House, for the installation of WT on six lightships, was signed in 1905 [39].

These contracts seemed to augur well for the future of MIMCC but difficulties quickly arose. Lloyds were never in favour of the Marconi company endeavouring to establish a world monopoly [40], and were opposed to any restrictions on inter-communications. Lloyds were refused licences for stations in Jamaica, Ceylon, Barbados, St Helens, Perim, the Straits Settlement and Mauritius because, with Marconi apparatus installed in them, such stations could only communicate with ships equipped with Marconi sets and operators [41]. The difference in outlook between the two parties was not conducive to good relations and to the implementation of the agreement. Considerable litigation ensued and their agreement was varied in 1905. Outside the United Kingdom many countries held a similar view to Lloyds and refused to permit Marconi stations to be established on their coasts [42].

MIMCC's policy favoured competitors who were opposed to restricted inter-communications. The de Forest system, for example, was adopted by several shipping lines, including the Royal Mail Steam Packet for its ships on the Central and South American routes, and it appears that the growth of the de Forest system in America curtailed the expansion of the Marconi system. By 1907, despite the litigious action of MIMCC to protect its patents in the United States, and notwithstanding some important judgements given in their favour, approximately half the marine installations of wireless telegraphy were those of competitors.

Action to oppose a world monopoly of wireless telegraph services was also taken diplomatically. There seemed to be an opinion, and a fear, particularly in Germany, that the activities of MIMCC and Lloyds would create monopolistic wireless telegraph services which would parallel those of the very extensive, world-wide, British cable services. And so, when proposals were put forward for the amalgamation of the Slaby-Arco and Marconi interests and were rejected, the German Government took diplomatic steps to protect their national interest [43].

In 1903 the German Government organised, as stated previously, an International Conference on the subject of International Legislation of Wireless Telegraphy [44]. It was held at the Imperial Post Office in August 1903 and was attended by representatives of Great Britain, France, Germany, Italy, Austria-Hungary, Russia, Spain and the United States of America [45].

The German proposals were outlined by Herr Sydow, the Under Secretary of the Post Office. They were predicated on the basis that all wireless telegraph messages to and from ships should be accepted and transmitted by all systems of WT; that is a ship equipped with a German set should be able to send messages to, and receive signals from, a Marconi shore station. Such a proposal, while being very advantageous to the Marconi company's competitors, was commercially inequitable since the English company had had to bear the burden and great cost of setting-up the network of coastal stations. MIMCC had to be operated on sound business principles and not on generous altruistic grounds.

The outcome of the conference was that the majority of the delegates adopted several resolutions, which it was hoped would be a foundation for a future conference:

1. the coastal stations should be obliged to receive and transmit all telegrams from and to ships at sea, without respect to the system, in order to facilitate communication between the ships and stations. As far as possible, all necessary technical information as to their equipment should be published;
2. the coastal stations should give precedence to telegrams relating to ship-wrecks and appeals from ships for assistance;
3. the states having coastal stations should fix a tariff, with the consent of the WT company, for forwarding messages, which should be based on the tariff in force for ordinary telegrams plus a special WT charge;
4. the tariffs should be determined by the number of words of the messages;
5. the individual WT services should be regulated so that the stations produced the minimum disturbance possible.

A protocol acceding with these resolutions was signed by representatives of the above mentioned countries with the exception of Italy and Great Britain. The Italian Government's delegates could not sign because the Government had contracted with Marconi to use his system exclusively for 14 years and perforce could not exchange telegrams with stations equipped with other systems. The British delegates indicated that they would refer the recommendations of the conference to their Government [45].

One consequence of the German Government's stance on WT was the closure (following some diplomatic pressure) by the US Government in 1904, of MIMCC's station on the Nantucket lightship because it would not accept non-Marconi traffic [46].

The involvement of the United States Navy in wireless telegraphy dates from early 1902 when it purchased: two sets from the German companies of Slaby-Arco, and Braun Siemen's Halske; two sets from the French firms of Popoff Ducretet, and Rochefort; and two sets from the American manufacturers of De Forest, and Shoemaker. Later the navy ordered two sets from Lodge-Muirhead. Comparative trials of these equipments gave the following maximum ranges:

Slaby-Arco	62 miles (99.8 km)
De Forest	54 miles (86.9 km)
Lodge-Muirhead	33 miles (53.1 km)
Braun Siemen's Halske	25 miles (40.2 km)
Popoff Ducretet	24 miles (38.6 km)
Rochefort	13 miles (20.9 km)

By 1903 the US Navy had concluded that the Slaby-Arco set as used by the German Navy was the best and ordered 20 of them. A further 25 were ordered in September. Five shore stations on the Atlantic coast were constructed and during the summer naval exercises were held to determine the impact of wireless telegraphy on fleet tactics. The exercises were an outstanding success and the Commander-in-Chief (Atlantic) reported he was satisfied that wireless had come to stay; that it was capable of development and would be of great use in time of war and in peace. Immediately after these trials seven more shore stations were established on the eastern seaboard. By November 1904 there was a chain of 17 stations stretching along the Atlantic coast from Cape Elizabeth (Maine) in the north to Key West (Florida) in the south: there were three stations on the Pacific coast and four in the Caribbean region; and two existed in the Philippines. Many of the stations were equipped with 1.25 kW Slaby-Arco transmitters having a range of 74.6 miles (120 km).

Of the European countries, the position c. 1904 was as follows:

1. Germany had five naval wireless stations covering the North Sea coast and about 75 ships fitted with Telefunken apparatus: four stations were being planned to cover the Baltic Sea. In addition Telefunken had built a very large transmitting station at Nauen for the government. One of

the Imperial Navy stations – that at Norddeich near Emden had a range of 901 miles (1,450 km).

2. Italy had 19 government coastal stations for use by both commercial and naval ships. When opened in July 1904, the Coltano station near Pisa had the most powerful transmitter in the world. All utilised Marconi sets.

3. France had five stations on the Channel and Atlantic coasts and one in the Mediterranean region equipped with Ducretet Rochefort transmitters; and more were being planned for Corsica and North Africa.

4. Norway, Sweden, the Netherlands and Austria-Hungary had adopted the Telefunken system and had commenced to establish shore stations and to install sets in their ships.

In 1904 the British Government, anxious to prevent the unregulated growth of wireless telegraphy, enacted the Wireless Telegraphy Act. This provided that no person should establish or work a wireless telegraph apparatus in Great Britain or on a British ship except under licence from the Postmaster General. It was furthermore decreed that the Act would be in force until July 1906, unless Parliament determined otherwise. In February 1906 an Amendments Act was passed which extended the 1904 Act for a further six years.

Additionally, in 1904, an understanding was reached between the Government and MIMCC according to which:

1. the company assented to accept the terms of any international agreement entered into by the Government;

2. the Government granted licences to the company for all its existing stations (at Poldhu, Lizard, Rosslare, Crookhaven, Caister, Niton, Holyhead, North Foreland, Haven, Withernsea and for installations on board ships) and provided facilities whereby telegrams could be handed in at post offices for transmission to Atlantic liners [47];

3. the company would receive double payment, until 1911, for all messages exchanged with other systems if free inter-communication were established – the surcharge being paid by the Post Office.

Of the other stations that were initially licensed, there were the three Lloyd's stations at Malin Head, Inistrahull and Fastnet (all of which used Marconi sets), and the Eastern Telegraph Company's station at Porthcurno (which employed a Maskelyne-de Forest installation). To these were added during the next two years: the Clifden trans-Atlantic Marconi station, two more Marconi stations for Lloyds, and a station at Cullercoats which utilised apparatus of the De Forest Wireless Telegraph Syndicate. Applications from Telefunken, and from the owners of the Orling-Armstrong systems were refused; and the Lodge-Muirhead Syndicate was, as noted previously, denied licences for stations at Dover, Isle of Wight, Lizard and Fastnet although unspecified alternative sites not on the English Channel coast were offered.

In 1906 the German Government organised another International Convention on Wireless Telegraphy [48]. It was attended by representatives from

Argentina, Austria-Hungary, Belgium, Brazil, Bulgaria, Chile, Denmark, Egypt, France, Germany, Greece, Great Britain, Italy, Japan, Mexico, Monaco, Montenegro, the Netherlands, Norway, Persia, Portugal, Roumania, Russia, Siam, Spain, Sweden, the United States of America and Uruguay. The conference discussed, article by article, the text of a protocol outlining draft regulations for the international control of radiotelegraphy. Of these articles, Article 3 provided that coastal stations and ship stations should be compelled to exchange wireless telegrams regardless of the system used to transmit them. Other articles were concerned with the specification of the wavelengths for particular services, and with the establishment of an International Bureau for collecting and disseminating information. Following provisional agreement by the conference to the protocol, it was referred to the represented governments for ratification. The British Government ratified the protocol, by an Order in Council, with effect from 1st July 1908 [49].

One consequence of the convention was the agreement to replace the distress call CQD, which was the ordinary telegraphic call for 'all stations', by SOS. On British ships the priority of message handling was: (1) distress signals, (2) messages to the British Admiralty and Government departments, (3) navigation messages, (4) wireless service messages, and (5) ordinary correspondence.

One effect of the ratification was the entry of the General Post Office into the marine wireless telegraphy field. The first of the Post Office's wireless stations was erected in 1906 and was intended for internal communications only in the event of a failure of the electric telegraphy land lines; but in 1908 a station was opened at Bolt Head specifically to communicate with ships [48]. Then, on 30th September 1909, the Post Office assumed control of all the MIMCC and Lloyds coastal maritime communication stations [50]. (The Cullercoats station which was operated by the Amalgamated Radiotelegraphic Company – previously the de Forest Company – was excepted [49].) By this action all the difficulties that had arisen between MIMC and Lloyds now became part of the history of wireless telegraphy. Table 13.1 lists all the stations under the Post Office administration in 1910.

The terms of the takeover provided that:

1. MIMCC would surrender all their rights under the 1904 agreement with the Post Office; the Post Office would have the free use (i.e. without royalty payments), for 14 years, of the company's existing and future patents for communications from Great Britain to Ireland and outlying islands, between any two outlying islands, and, except for public telegrams, between any two stations on the mainland;
2. the price to be paid for the MIMCC stations would be £15,000, and for the Lloyds' stations the value of the plant.

The agreement was a bargain for the British Government and though, prima facie, it seemed to frustrate the aspirations of the Marconi organisation, the agreement made good business strategy. Troublesome concords with the Post Office and with Lloyds were now at an end and MIMCC had funds that con-

Table 13.1 List of Post Office transmitter stations

Station	Post Office control	Nominal power	Transmitter
Caister*	29th Sept. 1909	1.5 kW	Rotary converter and plain spark
North Foreland*	31st Dec. 1909	3.0 kW	Motor generator and plain spark
Niton*	29th Sept. 1909	0.25 kW	Coil and plain spark
Lizard*	29th Sept. 1909	0.25 kW	Coil and plain spark
Seaforth*	29th Sept. 1909	3.0 kW	Motor generator and plain spark
Bolt Head	11th Dec. 1908	3.0 kW	Motor generator and plain spark
Rosslare*	29th Sept. 1909	0.25 kW	Coil and plain spark
Tobermory	1st May 1907	1.5 kW	Rotary converter and plain spark
Lochboisdale	1st May 1907	1.5 kW	Rotary converter and plain spark
Crookhaven*	29th Sept. 1909	3.0 kW	Motor generator and plain spark
Malin Head	31st Dec. 1909	0.25 kW	Coil and plain spark
Hunstanton	1st Dec. 1906	0.25 kW	Motor generator and plain spark
Skegness	1st Dec. 1906	0.25 kW	Motor generator and plain spark
N. Ronaldshay	21st Oct. 1910	50 W	Coil and plain spark
Sanday	21st Oct. 1910	50 W	Coil and plain spark

*MIMCC stations

siderably relieved its financial position, since at about this time the company had 'no liquid capital'. 'It was doing much too big a business for the capital at its disposal' [51].

'Liquid capital' was certainly an important consideration for the Marconi organisation at the end of the first decade of the new century. The company was in competition with the Lodge-Muirhead Syndicate, the British Radio Telegraph and Telephone Company (which employed the Balsillie system), the Helsby Company and Siemens Brothers (which had the Telefunken system). All these companies depended on patents, and the holding of master patents was a matter of great commercial significance – *vide* the Marconi company's acquisition of the Lodge-Muirhead patents on syntony. Patent infringements provoked litigious actions but the high costs of such actions demanded a financial resource. In 1910, with their new capital, the Directors of the Marconi organisation 'resolved that it was time there should be an end to the infringements of the company's patents' [52]. The company took suit against BRT&T [53] in 1911, and against the Helsby Company in 1913. In both cases the defendants had sold apparatus for use on board ships, and in both cases the courts ruled that the companies had infringed the '7777' patent [54] of the Marconi company.

In 1911, Siemens Brothers, at that time a wholly owned subsidiary in the UK of Siemens and Halske – which had a half share in the Telefunken Company – stated that it was licensed to use the Telefunken patents and consequently was in a position to install radio installations. Telefunken had developed the quenched spark method (see Note 1) of generating electromagnetic waves, which could be fitted on board ships, and this was offered to customers with an indemnity against any patent litigation by the Marconi company.

Actually, the Telefunken apparatus infringed the Lodge patent – which Siemens admitted – and in January 1912 it applied for a licence to use the patent. The licence was allowed on a royalty basis of £10 per kilowatt per annum. Also, at the end of 1912 Siemens acknowledged the validity of Marconi's '7777' patent, and it became evident that some commercial contract between the two principal competing radio companies seemed necessary [55].

MIMCC's involvement with German shipping companies began – before the 1906 Convention – when agreements, extending to 1914 and 1917, were reached with the North German Lloyd and the Hamburg-America shipping lines respectively. The agreements seemed to indicate a promising future for MIMCC's European aspirations but unfortunately the German Government, after the ratification of the convention in 1907, exerted its influence on ship owners to persuade them to use Telefunken equipment and to replace their MIMCC equipment. The shipping companies acceded to this pressure and by 1911 86 German ships had been fitted with Telefunken gear but only 38 with MIMCC apparatus [56]. Both companies leased their equipments.

Fortunately for MIMCC, Telefunken's infringement of the '7777' patent gave it some commercial bargaining power. The Belgian Marconi Company, which controlled the Marconi interests in Germany, and Telefunken came to an amicable arrangement to end the competition over the selection of their systems by the German mercantile marine [57]. A new company known as 'Debeg' was formed to take control of the 124 German ship installations: it became the licensee, world-wide, of all MIMCC's and Telefunken's patents [58].

Apart from Telefunken, MIMCC's main overseas competitor was the United Wireless Telegraph Company of the United States of America, which had c. 500 ship installations and 70 land stations, and was pushing forward for business in Europe [59]. But again, according to MIMCC, it was infringing the '7777' patent and legal redress had to be obtained. Encouraged by the success of its litigation against the British Radio Telegraph and Telephone Company, the Marconi organisation instituted proceedings against the United Wireless Telegraph Company. The case ended in a verdict for the Marconi company. Unusually, by then UWTC was bankrupt – the President and first Vice-President, in 1911, had been arrested on a stock fraud charge and had been found guilty – and so the American Marconi Company was able to take over all UWTC's c. 500 ship installations and 70 shore stations.

When hostilities commenced in 1914, the assets of Siemens Brothers were seized though they continued to manage their c. 70 existing radio installations. At this time, the Marconi International Marine Communication Company was in an unassailable position. Its patent strength had been either tested in courts of law or admitted without litigation by competing companies. The company's principal competitor was now an enemy and so MIMCC had become a prime provider of marine installations. Of the 706 coast stations and 4,846 ship installations which existed at the end of March 1915, some 225 coast and 1,894 ship stations were those of MIMCC [60].

In 1914, The Safety of Life at Sea Convention, was signed in London following an international conference. The agreement specified the minimum radio equipment to be carried by vessels of various types, and made the provision of emergency apparatus on board ships obligatory. (Fig. 13.3 shows the radio room of RMS *Olympic*.)

Note 1

Transmitter development

The basic configurations of directly coupled and inductively coupled antenna – induction coil circuits are shown in Fig. 13.4. Analysis [61] shows that if the length of the antenna is adjusted so that its free resonant frequency is the same as that of the isolated primary capacitor circuit then when the two circuits are coupled together the arrangement can lead to the radiation of electromagnetic waves having wavelengths of λ_1 and λ_2. If the coupling is very weak, c. three per cent, a single wave is radiated.

The basic spark transmitter has several serious limitations. It is inefficient because: (1) power is not transmitted continuously – the duration of the damped spark train is usually a small fraction of the period of the interrupter; (2) the coupling must be weak to prevent two waves from being radiated; and (3) power is lost because the radiated electromagnetic radiation contains many harmonic components which cannot be received by a tuned receiving antenna. In addition the high harmonic content of a powerful transmitted signal can cause interference and prevent stations receiving signals on other wavelengths, a matter of some importance when important messages have to be sent to specific ships.

In 1906 M. Wien [62] discovered, during the course of some researches on electrical discharges between metal electrodes in close proximity, that tight coupling is possible, without the transmission of two waves, if the spark in the primary circuit is rapidly quenched. If this condition is satisfied the antenna radiates at its resonant frequency. Following this finding many investigators – Lowenstein, Massie, Fleming, Richardson, Koehler, Shaw, Lepel, Peuket, Lodge and Chambers, Chaffee, Marconi, Telefunken and others [63] – devised methods whereby the primary circuit spark can be quenched. Among the methods, the use of air blasts, magnetic fields, electrodes of relatively substantial mass and rotating electrodes may be mentioned. The essential requirement is the removal of the gaseous ionisation generated by the spark.

Figure 13.5 illustrates Telefunken's quenched spark gap [63], (patented 1908). Its modus operandi is based on the rapid conduction of the heat generated by the spark away from the sparking region. The spark gap comprises a number of very short gaps in series with each other. The electrodes are a pile of spaced copper electrodes, plated with silver (a good conductor of heat), and provided with heat radiating fins. Mica washers, about 0.008 in (0.2 mm) in thickness, separate the copper discs, the number of which is determined by the breakdown

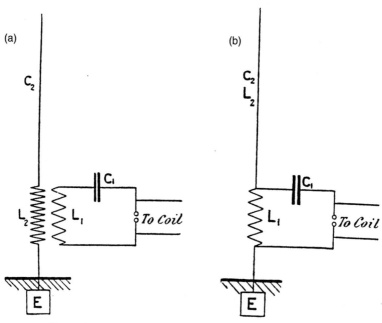

C₁, condenser; L₁, inductance; E, earth plate; C₂, antenna.

Figure 13.4   *(a) Antenna and capacitor circuit inductively coupled; (b) antenna and capacitor circuit directly coupled.*
Source: J.A. Fleming, *The principles of electric wave telegraphy* (Longmans, Green, London, 1908).

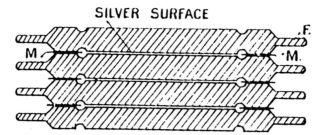

Figure 13.5   *The Telefunken quenched spark discharger in section.*
Source: H.M. Dowsett, *Wireless telephony and broadcasting* (Gresham Publishing Co., London, 1924).

voltage of the primary circuit capacitor. Sparking takes place between the exposed surfaces of the electrodes but is quenched because of the rapid heat loss from the gaps, and because the mutual repulsion of the ions in the ionised air between the discs forces the spark to the outer edges of the sparking surfaces where it becomes lengthened across the circular grooves. The Telefunken quenched spark gap worked efficiently.

In September 1907 Marconi patented his rotating disc discharger [64] (Fig. 13.6a). This consisted of an insulated metal disc, A, rotated at a 'very high speed', and two discs, C_1 and C_2, (called polar discs), also driven at high speed,

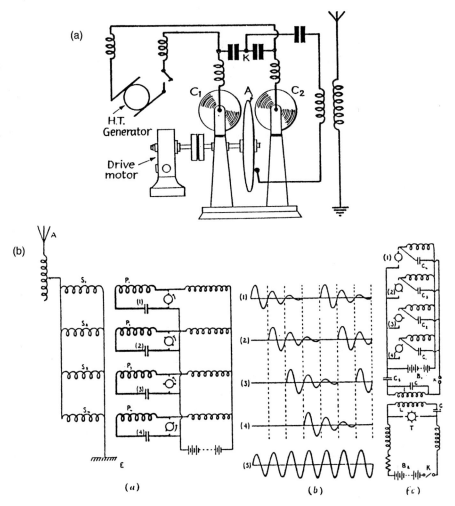

Figures 13.6 (a) Marconi's rotating disc discharger, 1907. (b) 'Timed spark' method of generating a continuous wave.
Sources: (a) W.J. Baker, op. cit. (b) H.M. Dowsett, op. cit.

placed 'very close' to A. When the rotating discs were connected as shown in Fig. 13.6a a discharge was created which Marconi described as 'neither an oscillatory spark nor an ordinary arc', but which generated continuous oscillations of up to 200 kHz. In a variation, a row of small studs was fixed around the periphery of A to interrupt the oscillations and allow the radiated wave to be detected using a crystal detector.

The culmination of the development of spark transmitters was probably reached in 1912 with the 'timed spark' method [65] of Marconi's Wireless Telegraph Company. Figure 13.6b shows how by synchronous timing the several outputs combine to produce a continuous wave. MWT manufactured transmitters of this type to provide up to 200 kW to the antenna, at a wavelength of 14,000 m (21,400 Hz), from a 300 kW generator.

The advantages of continuous wave working were, of course, known shortly after the inception of wireless telegraphy, but the difficulty of generating the waves in the pre-valve era seemed very great. The breakthrough came in 1900 when W. Duddell [66] demonstrated the 'musical arc'. He showed that if a series L-C circuit was connected in shunt across a direct current arc, (the electrodes being two solid carbon rods), the arc produced a sound the frequency of which was the frequency of oscillation of the L-C circuit. Duddell realised that if the frequency of operation could be increased the principal problem of wireless telephony would be solved. Two years later V. Poulsen [67] provided the solution.

His improvements consisted of the use of (1) a hydrogenous atmosphere, (2) a magnetic field transverse to the arc, and (3) a water-cooled copper anode and a rotating carbon cathode (see Fig. 13.7). Signals could be made by short-circuiting some of the turns of P, but other methods were also devised.

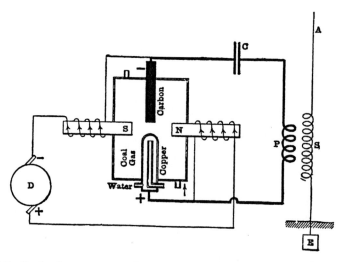

Figure 13.7   Poulsen's arrangement for producing undamped electric oscillations.
Source: J.A. Fleming, op. cit.

Many different types of arc generators were patented by workers in several countries (see Ref. 63), and they were manufactured up to 450 kW in size.

The advantages of continuous wave (c.w.) working, especially for wireless telephony, were so compelling that several attempts were made in the first decade of the 20th century to achieve a high frequency output from an electrical machine. With a conventional alternator (a.c. generator) the frequency (f) of the output is given by the product of the angular speed (N revs per second) and the number of pole pairs (p). Thus a two-pole machine generates an output of 50 Hz at 3,000 r.p.m.. For wireless telegraphy e.m. waves having frequencies of several tens of kilohertz are required and so N and p must be as large as is practicable. Two engineering problems result: (1) the difficulty of keeping the eddy current losses in the iron of the machine within reasonable bounds; and (2) the difficulty of accommodating a large number of poles in a machine driven at very high speeds.

In 1908 R. Goldschmidt devised an ingenious machine based on standard machine theory and external static components [68]. A stationary alternating magnetic field of frequency f (Hz) and amplitude H, produced by a coil S, is identical with the resultant field established by two rotating magnetic fields having angular velocities of f revolutions per second and −f revolutions per second, each having an amplitude of H/2. Consequently if another coil, R, is rotated at f_1 revolutions per second in the stationary alternating magnetic field, e.m.fs. having frequencies of $(f_1 - f)$ and $(f_1 + f)$ will be generated. If f_1 is equal to f the resultant e.m.f will have a frequency 2f. This e.m.f. will now induce in the stationary coil e.m.fs. of frequencies $(2f - f)$ and $(2f + f)$, i.e. f and 3f. This reasoning may be extended to show that e.m.fs. of f, 3f, 5f . . . are induced in S, and e.m.fs. of 0, 2f, 4f . . . are induced in R. For the currents produced by these e.m.fs. to be of appreciable magnitude it is essential that low impedance paths be provided for them. A series LC circuit has a low impedance at resonance and may be used to select the appropriate harmonic in S or R (see Fig. 13.8).

Of course, the above process cannot be continued indefinitely since losses are associated with the tuned circuits and the machine itself. These losses (particularly the magnetic losses) in practice limit the transmitted frequency to no more than four or five times the fundamental frequency. The leakage flux must be minimised and Goldschmidt designed his machines with an air-gap (between the rotor and stator) of just c. 0.8 mm (sic). The manufacturing problem of constructing such a machine having a 5080 kg mass rotating so that its peripheral speed was 200 m/s must have been formidable.

The first Goldschmidt generator (12 kW, 60 kHz) in the UK was installed at Slough in 1912. A year later a 100 kW machine was commissioned at Neustadt-am-Ruebenberger, Germany, and a 120 kW machine was set-up at Tuckerton, USA. The radio station at Laeken, Brussels was equipped with a 250 kW Goldschmidt generator.

Mention has been made, in Chapter 12, of the high frequency alternators designed by Steinmetz, Fessenden and Alexanderson. The Alexanderson alternator was especially important and effectively led to the establishment

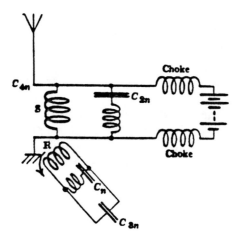

Figure 13.8   Schematic diagram of Goldschmidt's alternator showing the rotor (R) and stator (S) coils and the circuit elements which comprise the resonant circuits.
Source: IEE Conference Publication, No. 411.

of the Radio Corporation of America, (see Chapter 20). A total of 12 200 kW alternators were fabricated and sold around the world, but only one remains [69]. It is installed at the Grimeton station, near Varberg, Sweden and was demonstrated in 1995 at an IEE International Conference on '100 years of radio'. Alternators of this size were capable of generating outputs having frequencies of c. 30 kHz. Smaller machines were also designed to produce 2 kW at 100 kHz, and 5 kW at 200 kHz.

Apart from its applications to naval manoeuvres, maritime signalling and safety at sea, wireless telegraphy communications had important commercial uses. This is the topic of the following chapter.

Note 2

Marine signalling

Note 2 gives a brief chronological overview of the steps that led to an efficient means of signalling at sea, prior to the introduction of wireless telegraphy.

1337–51   Two flag signals were listed in the *Black Book of the Admiralty*; the first to notify all captains to board the Admiral's flagship for a meeting; the second to report the sighting of an enemy.

1596   Queen Elizabeth I's secretaries drew up 'the first regular sets of signals and orders to the commanders of the English fleet'.

1647   The Parliamentary Government introduced the first code of flags for use at sea. Unfortunately all details of this reform appear to have been lost.

1653	The Generals-at-Sea issued the first surviving *Code of Instructions* with accompanying signals. Five flags were used but the meaning of a given flag depended on which of five prominent positions – fore-, main-, mizzen-trucks, spanker gaff, and after flagstaff – it was hoisted.
1665	The Duke of York (later King James II, 1685–88) endeavoured to codify the system of naval flag signals.
1673	The Duke of York issued the first British Signal Book. The number of flags and hoisting positions were increased and their meanings were incorporated in various articles of instructions.
1689	The date of the earliest surviving manuscript Signal Book.
1705	The *Permanent Instructions* were issued. Twenty instructions were listed based on four coloured flags – red, white, blue and yellow – and the Union flag.
1715	The date of the first known printed Signal Book. It contains diagrams of a ship showing the appropriate coloured flag and hoist position for different signals.
1738	M. de la Bourdonnais advanced the notion of having flags to signify numbers. He employed ten variously coloured pennants to express the numerals from 0 to 9. Three sets of these pennants enabled numbers from 0 to 999 to be signalled.
1746	In the signal book of 1746, a flag is shown at the top of a page and the signals corresponding to the different hoist positions of the flag are given below. Sixteen flags are displayed in the sixteen pages of the book. A later manuscript Signal Book, published in 1762, has twenty-eight flags illustrated.
1763	S.F. de Bigot published his *Tactique navale ou traite des évolutions et des signaux*. The book was translated and published in England in 1767.
1776	Lord Howe issued the first *official* Signal Book for ships under his command.
1777	Admiral Kempenfelt, among others, began to develop and experiment on ways by which *any* signal, rather than one of a given set of signals, could be communicated by means of flags. He was familiar with the ideas of de la Bourdonnais, and possibly of de Bigot, and had read B. de Villehuet's work, *Le Manoeuvrier*, which outlined a 'freedom of speech' method. In Kempenfeldt's 1780 tabular or matrix scheme, the *N* flags in use were displayed along the top and down the left hand side of a table or matrix. The row and column elements of the matrix represented page numbers and articles on a page respectively. Each flag hoist showed two flags; the upper one signified the page, the lower one indicated the article on the page. Thus for sixteen flags there are 256 elements; and the element in the 10th row, 15th column of the matrix refers to page 10, article 15 of the Signal Book.
1782	Lord Howe was appointed Commander of the Channel Fleet. He was senior to Kempenfelt and was impressed by his thinking. Howe's *First Numerary Code* followed, shortly before Kempenfelt's untimely death.

1790	Lord Howe's *Second Numerary Code* was adopted for the Royal Navy under the title of *Signals Book for the Ships of War*. In this code nine flags were used. Consequently, a hoist of three flags could signify one of 999 messages; a hoist of four flags could connote one of 9,999 signals, and so on.
1797	The first American Signal Book using the numeracy system was produced.
1800	Sir Home Popham produced his *Telegraphic Signals or Marine Vocabulary*. It was issued officially in 1803. For the first time a Commander could signal any message by means of flags. Twenty-five flags represented 25 letters of the alphabet (I and J were accorded the same flag). However, to simplify the process of communication, Popham also wrote a dictionary of some 1,000 words commonly used in naval procedures, each word of which could be signalled by three flags. Nelson's famous signal before the Battle of Trafalgar, 'England expects that every man will do his duty', was communicated to his fleet by Popham's system. Nelson had originally used the word 'confides', but when his flag lieutenant, Pasco, pointed out that 'confides' was not in the dictionary and would have to be spelt out, whereas 'expects' was listed, Nelson readily agreed to the change.
1808–14	During the Peninsular War, the Royal Navy also used a system of five balls, one flag and one pennant for signalling. Their positions allowed from 1 to 999 messages to be transmitted according to a pre-arranged code.
1817	J. Macdonald's work, *A Treatise Explanatory of a New System of Naval, Military and Political Telegraphic Communication of General Application* was published. It included 77 pages of text, around 1,000 pages of a dictionary and references to 12 recommended signal systems.
1815	Popham's semaphoric telegraph system was approved for test by the Admiralty. By June 1816 the telegraph was working accurately. Two lines were planned – the first from London to Portsmouth and the second from London to Plymouth. Only the former was completed, in 1824. It operated until 1847. Popham recived £2,000 for his invention and signal books. Some admirals requested this type of telegraph for their shore stations and ships but were told to construct the apparatus themselves.
1816	C.W. Pasley invented a semaphore system which was later, in 1854, in operation during the Review of the Fleet at Spithead by Queen Victoria. From 1876 the system was officially approved for operational signalling. It was alphabetic in character and was based on the utilisation of 21 letters, with common signs for I and J, S and Z, Q and X, and U, V, and W.
1852	Captain R.W. Jenks's work on *The brachial telegraph: an original method of conversing and signalling on land and at sea by means of*

the human arms at any and all distances within furthest range of the telescope was published in North America. This work described an alphabetic system but the positions of the arms were not those of the later accepted flag system. The Royal Navy seems to have used arm and flag signalling from around 1879.

1862   Captain P. Colomb and Major Bolton independently proposed to adapt to lights the well known 'dot/dash' telegraph system as a means of telegraphing at night. They carried out 'conjointly a great many experiments' and found that 'From a station on the Isle of Wight to a steamer at sea we succeeded in getting quite as rapid communication as had ever been got by sea at two miles [3.2 km] at a distance of 30 miles [48 km]'. Unfortunately Colomb, in his paper, 'A sketch of the progress of sea telegraphy' (*J. Soc. Tel. Eng.* 1872, 1, pp. 57–70) does not give details of the equipment employed. Elsewhere in his paper he states that, at sea, an oil lantern was 'capable of displaying its light over an arc of 180° horizontally and to a distance not exceeding six miles in clear weather'. However, to obtain a greater range Colomb and Bolton designed a lamp known as the Chatham Light. 'The flashes, which are extremely brilliant and are visible at least twelve miles, are produced by blowing finely powdered magnesium diluted with a resinous substance into the flame of a spirit lamp.' The flashing light system was adopted by the Admiralty in 1867. The Royal Navy represented the dots and dashes as short and long exposures of light respectively; some nations used the converse of this arrangement. Lamp signalling was widely used for both day and night signalling during the Second World War.

1902   Various semaphore arrangements were tried out in the 19th century and by 1902 all Royal Navy ships and dockyards had masthead or wagon-mounted semaphores installed. They were retained on board ships for many years. The 1910 Encyclopaedia Britannica mentions that a record message of 350 words was signalled simultaneously to 21 ships at the rate of 17 words per minute, and that under good conditions of visibility a message could be read at 16 to 18 miles (26 to 29 km). The Royal Navy used semaphore arms with lengths of 9 to 12 ft (2.7 to 3.7 m).

References

1   BAKER, W.J.: 'A history of the Marconi Company' (Methuen, London, 1970), p. 44
2   POCOCK, R.F., and GARRATT, G.R.M.: 'The origins of maritime radio' (HMSO, London, 1972), p. 16
3   HAMILTON, F.T.: report to Commander-in-Chief, Devonport, 28th January 1899, ADM 116/523, PRO
4   POCOCK, R.F.: 'The early British radio industry' (Manchester University Press, 1988). p. 18

5 KELVIN, LORD.: letter to Sir Evan MacGregor, 10th October 1898, ADM 116/523, PRO

6 HAMILTON, F.T.: report to the Commander-in Chief, Devonport, 2nd November 1898, ADM 116/523, PRO

7 Ref. 1, p. 46

8 WIRELESS TELEGRAPH and SIGNAL Co.: letters to the Secretary of the Admiralty, 29th March 1899 and 13th April 1899, ADM 116/523, PRO

9 WIRELESS TELEGRAPH and SIGNAL Co.: letter to the Director of Navy Contracts, 3rd July 1898, ADM 116/523, PRO

10 ANON.: Annual Report of the Torpedo School for 1899, Naval Library, MOD

11 Ref. 2, pp. 24–25

12 Ref. 2, p. 29

13 Ref. 2. pp. 26–30

14 LODGE, O.J.: 'Improvements in syntonised telegraphy', British patent no. 11,575, dated 10th May 1897

15 Ref. 1, p. 54

16 MARCONI, G.: 'Wireless telegraphy', *Nature*, 1900, **61**, pp. 377–378

17 JACKSON, H.B.: report to Sir Compton Domville, the Admiral Commanding 'B' Fleet, 10th August 1899, ADM 116/523, PRO

18 DOMVILLE, C.: remarks to Captain Jackson's report of 10th August 1899, 11th August 1899, ADM 116/523, PRO

19 ANON.: Admiralty Minutes, 17th August 1899 to 21st August 1899, ADM 116/567, PRO

20 ANON.: Admiralty Minute, 23rd August 1899, ADM 116/567, PRO

21 GENERAL POST OFFICE.: letter to the Secretary of the Admiralty, 6th September 1899, ADM 116/567, PRO

22 ANON.: Admiralty Minute, ('Director of Navy Contracts'), 13th October 1899, ADM 116/567, PRO

23 Ref. 2, p. 38

24 TREASURY.: letter to Admiralty, 5th April 1900, ADM 116/567, PRO

25 ADMIRALTY.: letter to Marconi's Wireless Telegraph Co., 8th May 1900, ADM 116/567

26 Ref. 2, p. 40

27 Ref. 2, p. 39

28 Ref. 4, p. 153

29 Ref. 4, p. 155

30 Ref. 2, pp. 36–37

31 Ref. 2, p. 44

32 ANON.: Annual Report of the Torpedo School for 1901, Naval Library, MOD

33 Ref. 1, p. 52

34 Ref. 1, pp 57–58

35 Ref. 1, p. 60

36 Ref. 1, p. 59

37 Ref. 1, p. 86

38 Ref. 1, p. 87

39 Ref. 1, p. 114

40 INGLEFIELD, E.F.: evidence to House of Commons, 18th April 1907

41 Ibid.: question 2044, p. 146
42 SMITH, H.B.: evidence to House of Commons, question 298, p. 27
43 HALL, C.: evidence to House of Commons, 16th April 1907, question 1309, p. 96
44 ANON.: report, *Electrical Times*, 1902, p. 458
45 Ref. 1, p. 96
46 ANON.: report, *Electrical Times*, 1904, p. 177
47 STURMEY, S.G.: 'The economic development of radio' (Duckworth, London, 1958) p. 54
48 Ref. 1, pp. 115–116
49 Ref. 47, p. 57
50 ANON.: report, *Electrical Times*, 1909, p. 321
51 ISAACS, G.: evidence to House of Commons, 1913, question 4880, p. 253
52 ANON.: report, *Electrical Review*, 1911, p. 69
53 Ref. 1, p. 130
54 MARCONI, G.: British patent no. 7777, dated 26th April 1900
55 Marconi's Wireless Telegraph Co. versus British Radio Telegraph Co., reports of the Patent, Design and Trade Mark Committee, 1911
56 ANON.: report, *Electrical Review*, 1911, p. 226
57 ANON.: report, *Electrical Times*, 1911, p. 62
58 Ref. 47, p. 60
59 ANON.: report, *Electrical Review*, 1910, p. 393
60 ANON.: report, *Electrical Review*, 1915, p. 720
61 FLEMING, J.A.: 'The principles of electric wave telegraphy' (Longmans Green, London, 1908), pp. 564–573
62 DOWSETT, H.M.: 'Wireless telephony and broadcasting' (Gresham Publishing Co., London, 1924), p. 101
63 BLAKE, G.G.: 'History of radio telegraphy and telephony' (Radio Press, London, 1926), pp. 156–175
64 Ref. 1, pp. 117–119
65 Ref. 62, p. p. 105
66 Ref. 63, p. 176
67 Ibid, pp. 177–180
68 FRERIS, L.L.: 'A re-appraisal of the Goldschmidt generator', pp. 71–75; a paper in The IEE Conference Publication No. 411 on 'The International Conference on 100 years of radio'
69 WEEDON, K.: 'The Alexanderson alternator, a "near perfect" system of W/T transmission', pp. 69–70; a paper in The IEE Conference Publication No. 411 on 'The International Conference on 100 years of radio'

Chapter 14

Point-to-point communications

By 1900 Marconi's plans and pretensions for the development and application of his system of wireless telegraphy were extensive and all-embracing. As previously mentioned, he had demonstrated land station-to-land station, shore-to-lightship, shore-to-lighthouse, and ship-to-ship, communications and had shown their worth to officers of the Army and the Royal Navy, to civil servants of the General Post Office and to officials of Trinity House. The potential of wireless telegraphy for strategic and tactical Service operations, for saving lives at sea, for coastguard duties, for business and social intercourse was unquestioned and seemed to be limitless. Some of these uses – for example, ship-to-ship signalling – could not, of course, be provided by the well-developed cable systems.

Moreover, for those applications that could be implemented by the trans-oceanic cables the current rates for message transmission were considered to be excessive. Until December 1884 the rate was two shillings per word (10p). Thus a message of just 10 words cost more than a labourer earned in a week. Some reduction in the rate – to 1s 8d (8.3p) per word – took place when the Commercial Cable Company opened for service. And in June 1886 the 'pool' companies – which had been formed in 1869 to protect rates – reduced their rate to 6d (2.5p) per word in an attempt to break the Commercial Cable Company. However, when this company likewise reduced its rate and the 'pool' companies found that their profits were declining, all the companies agreed to increase their rate to 1s (5p) per word – a rate which remained unchanged until 1923 [1].

The changes in the rates caused consequential changes in the traffic handled by the cables. When the rate was reduced from 1s 8d to 6d per word the number of words transmitted increased by 140 per cent. The effect was not linear though, for the rise from 6d to 1s led to a decrease in traffic, for the last quarter of 1888, of 26 per cent compared to the same quarter of 1887. Because of these high rates the cables operated at just about half of their maximum capacity, transmitting between 25 million and 30 million words per annum [2].

By the turn of the century the world-wide cable network, while providing an important service for press agencies, diplomats and government officials, military personnel, and stockbrokers, was felt to be too costly for many businesses and the general public. But, if the new system of communications, wireless telegraphy, could function at a lower operating cost than any of the cable systems there was a prospect of a very substantial increase in message sending, and profits for the wireless companies. Prima facie, it appeared reasonable that a decrease in word rate would be achieved by wireless since the cost of installing a transmitter and receiver station was likely to be much lower than a corresponding cable installation. Trans-oceanic cable laying required specialist ships and crews, it was hazardous, repair work could be expensive and time consuming, and the cables themselves were costly. On the other hand, wireless stations were fairly cheap and easy to assemble, though the antenna masts and wiring required experienced riggers. A possible disadvantage of trans-oceanic wireless communications might be the longer time needed to send a message: signals would have to be passed by electric telegraph offices, and land-lines, to the wireless stations and delays could arise, especially if the cable companies gave priority to their own traffic.

These factors indicated that a wireless service could carry messages at a lower cost than a cable service but would not be so expeditious.

In June 1900, following his return from the United States, where he had signalled over a distance of more than 100 miles (161 km), Marconi decided to attempt the feat of sending wireless messages across the Atlantic ocean [3]. He was just 26 years of age. For his attempt the equipment would have to be of an altogether different order of size and complexity. Until 1900, Marconi's apparatus had utilised 10 in (25.4 cm) or 20 in (50.8 cm) induction coils, rated at probably 200 W to 300 W, and no difficulty had been found in making and breaking the 15 A to 20 A primary currents. Now, a much more powerful transmitter would have to be engineered, possibly rated at 10 kW. Marconi, himself, was unable to undertake the necessary engineering and so, in July 1900, asked Professor J.A. Fleming, of University College, University of London, to specify the nature of the electrical engineering plant to be used, and to design special parts of the apparatus for generating the electromagnetic waves (Fig. 14.1). 'This involved many experiments on a small scale before embarking on the construction of large and costly plant of an entirely new type' [4].

Meantime, a convenient transmitting site at Poldhu, near Mullion, on the coast of Cornwall, was leased in August 1900, and the construction of the various buildings, by the Marconi Wireless Telegraph Company, began in October 1900. Fleming has provided a chronology of the subsequent events, as follows [4].

November 1900   commencement of the erection of the machinery specified by Fleming;

December 1900   building work far advanced;

January 1901   Fleming at Poldhu for experimental work 'to ascertain how far it [the electric plant] would be efficient for the purpose in view';

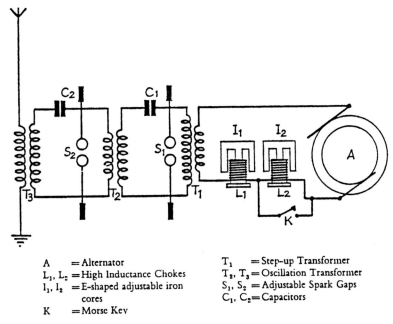

A   = Alternator
L_1, L_2 = High Inductance Chokes
l_1, l_2 = E-shaped adjustable iron
       cores
K   = Morse Key

T_1 = Step–up Transformer
T_2, T_3 = Oscillation Transformer
S_1, S_2 = Adjustable Spark Gaps
C_1, C_2= Capacitors

Figure 14.1 Simplified circuit diagram of the Poldhu transmitter.
Source: W.J. Baker: *A history of the Marconi Company* (Methuen, London, 1970).

Easter 1901 Fleming at Poldhu to conduct experiments, using a short temporary antenna, of propagation between Poldhu and the Lizard, a distance of 6 miles 'to show that the work was being conducted on the right lines';

Next four months 1901 'much work was done by Mr Marconi and [Fleming] together, in modifying and perfecting the wave generating arrangements, and numerous telegraphic tests were conducted during the period by Mr Marconi between Poldhu, in Cornwall, and Crookhaven, in the south of Ireland, and Niton, in the Isle of Wight';

18th September 1901 a storm wrecked a number of masts;

end of November 1901 'sufficient restoration of the aerial [antenna] was made . . . to enable Mr Marconi to contemplate making an experiment across the Atlantic';

27th November 1901 Marconi sailed for Newfoundland, on the SS *Sardinia*, taking with him his assistants Kemp and Paget, and a number of balloons and kites;

c. 5th December 1901 arrival of Marconi at St John's. Arrangements made 'for sending up a balloon and an attached aerial wire';

9th December 1901 Marconi cabled his assistants at Poldhu to commence the transmission of the letter 'S' in Morse code, from 3.00 p.m. to 6.00 p.m. each day;

11th December 1901   signal transmission begins; balloon made first ascent, carrying with it the aerial wire, but it broke away and was lost;

12th December 1901   kite, with aerial wire attached, flown at 400 ft (122 m). Suddenly, at 12.30 p.m., Newfoundland time, Marconi handed his earpiece to Kemp and said: 'Can you hear anything, Mr Kemp?' Kemp affirmed that he could hear the Morse code 'S' signals, and Marconi recorded in a pocket notebook:

'Sigs. at 12.30, 1.10 and 2.20'

13th December 1901   Marconi confirmed the reception;

14th December 1901   cable sent to Major Flood Page, of the Marconi Wireless Telegraph Company: 'St. John's, Newfoundland, Saturday, December, 14, 1901. Signals are being received. Weather makes continuous tests very difficult. One balloon carried away yesterday.'

Fleming [4] has estimated that in these experiments the transmitter power was not more than 10 kW or 12 kW. The sending antenna consisted of 50 bare stranded copper wires, 7/20 gauge, suspended from a triatic stay, strained between two masts 160 ft (48.7 m) in height and 200 ft (61 m) apart, the wires being arranged in a fan shape and connected together at the base by a bar. A single wire joined to the bar was then taken into the transmitter building for connection to the transmitter. Marconi's epic experiment was conducted during very squally weather. At times 'the kite was rising and falling irregularly' [5], and the consequent variation in the capacitance of the antenna made the use of any form of syntonic apparatus impossible. Marconi had to abandon his new tuned receiver for an older, untuned model.

No independent confirmations of the reception of the 'S' signals were made and as a consequence some scepticism was cast, not only by the lay Press, but also in some technical journals, on whether Marconi and Kemp had really heard the signals (Paget was ill in bed at that time). The *Electrical Review* (20th December 1901) wrote [6] that this 'latest feat of Marconi is so sensational that we are inclined for the present to think that his enthusiasm has got the better of his scientific caution', and suggested that the signals had come from an American station or were a practical joke.

This doubt was not unreasonable when the conditions of the experiment were pondered. The receiving system comprised an inefficient antenna coupled to an untuned receiver that possessed no form of amplification whatsoever, and a signal detector that was less sensitive than the crystal detectors of several years later. Moreover, the transmission (on the 12th) took place at the worst possible time for electromagnetic propagation across the Atlantic ocean, for the whole of the signal path was in daylight, and the Poldhu transmitter was experimental in design.

Some modern estimates, notably those of Ratcliffe [7] (1974), of the power, frequency and propagation conditions that prevailed on the 12th December 1901, have tended to support the doubt that surrounded these early experiments. However, in 1995, at an international conference on '100 years of radio',

MacKeand and Cross read a paper [8] on 'Wide-band high frequency signals from Poldhu?', which was a summary of a comprehensive study they had undertaken of all the important parameters that influenced the reception of the signals, and concluded: 'We therefore argue that in December 1901, Marconi is likely to have received high frequency wide band signals, spurious components of the spark transmitter output, propagated across the Atlantic by sky waves near the maximum usable frequency.'

Following Marconi's trial, the Anglo-American Telegraph Company somewhat peremptorily notified Marconi that they held a monopoly of all telegraphic communications in Newfoundland, and ordered him to cease his operations. The world's Press noted this incident with indignation, and the Newfoundlanders were furious at the crassness of the Anglo-American Telegraph Company, but soon generous offers of assistance were forthcoming from the Canadian and United States Governments [9]. In particular, following a very successful visit to Ottawa and meetings with the Governor General, the Canadian Premier and Government Officials, Marconi was offered a free site, and £16,000 towards the cost of establishing a new station [10], on condition that the rate for wireless telegrams across the Atlantic should not exceed 10 cents per word, against the 25 cents per word then being charged for cable telegrams [11]. Glace Bay at Cape Breton Island was ultimately chosen to be the site of the new station (Fig. 14.2).

The Canadian Government's generous offer was not wholly altruistic. With the further development of wireless, the government saw a way to breach the monopoly stronghold of the co-operating cable companies and thereby appease the commercial users and potential users of the trans-Atlantic cables who had been complaining about the excessively high cable rates [12].

Marconi returned to England in February 1902 and made arrangements for a permanent station and antenna installation to be erected at Poldhu. Four wooden lattice masts, each c. 210 ft (64 m) high, spaced at the corners of an imaginary square of side 200 ft (61 m), were constructed. They carried insulated rope triatic stays from which the 400 copper wires of the antenna were suspended in a conical arrangement. Similar antennas and station buildings were erected at Cape Cod in Massachusetts, USA, and at Glace Bay, Nova Scotia [13] (Fig. 14.3).

In February 1902 Marconi travelled back to Canada on the SS *Philadelphia*, and, using an insulated antenna wire 197 ft (60 m) high fixed to the ship's masts, conducted reception tests of signals sent from Poldhu: they were printed on Morse tape. Readable messages were obtained as far west as 1550 miles (2484 km) from Cornwall, and indications of signals were obtained at up to 2100 miles (3379 km). Of especial interest to those on board was the observation that during the day signals ceased entirely at 700 miles (1126 km), but at night they were detectable at up to 1550 miles [13] (2494 km).

Further voyages, made by the Italian warship *Carlo Alberto* (which had been put at the disposal of Marconi by the Italian Government), to the Baltic (July 1902) to the Mediterranean (August 1902), and to Nova Scotia (October 1902),

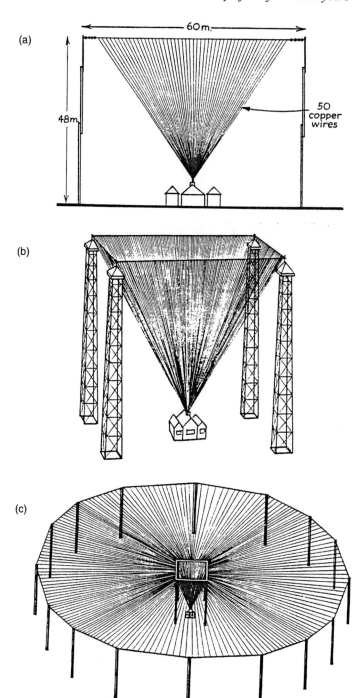

Figure 14.2   Diagrams of the antennas utilised at Poldhu (a), and at Glace Bay (b) and (c).
Sources: Author's collection.

Figure 14.3   View of the first station erected in Canada in Glace Bay, 1902, from which the first wireless transatlantic messages were sent.
Source: The Marconi Company.

all confirmed that signals could be received at distances of many hundreds of miles (Fig. 14.4). There was no longer any cause for doubt on this matter as there had been in December 1901 [13].

On 21st December 1902 the structures at the Glace Bay and Cape Cod stations were sufficiently well advanced to enable Marconi to send the following message to England from Nova Scotia:

I beg to inform you that I have established wireless telegraphic communication between Cape Breton, Canada, and Poldhu, in Cornwall, England, with complete success. Inauguratory messages, including one from the Governor General of Canada to King Edward VII, have already been transmitted (December 21) and forwarded to the Kings of England and Italy [14].

Less than one month later, on 19th January 1903, a wireless message was sent across the Atlantic, from Cape Cod (Fig. 14.5) to Poldhu, from Mr Roosevelt, the President of the United States of America, to King Edward VII [15].

His Majesty King Edward the seventh (by Marconi's transatlantic wireless telegraph)
In taking advantage of the wonderful triumph of scientific research and ingenuity which has been achieved in perfecting the system of wireless telegraphy, I extend on behalf of the American people the most cordial greetings and good wishes to you and all the people of the British Empire.
Theodore Roosevelt, White House, Washington

These successes appeared to indicate that all was now ready for a commercial news service across the Atlantic, and a press service for *The Times* was started in

Figure 14.4　Wireless telegraphy at sea: the Italian warship Carlo Alberto *fitted for communication with the station at Poldhu, Cornwall.*
Source: The Illustrated London News Picture Library.

Figure 14.5　The antenna at Cape Cod, USA, erected by Marconi's Wireless Telegraph Company.
Source: The Marconi Company.

1903. It was discontinued after a few days because the Glace Bay antenna came down due to a silver thaw [16]. Actually, further engineering effort was required before a reliable service could be inaugurated. In March 1904 the transmitter power of the Glace Bay station was doubled to 150 kW [17], and a new station was constructed at Clifden [18], on the west coast of Ireland, to replace the Poldhu station which was on a constricted site and did not allow a larger and more powerful station to be built.

The Glace Bay–Clifden wireless link was opened for a limited public service in October 1907, and an unrestricted service was operative from February 1908. Apart from interruptions due to a fire at Glace Bay and the transfer of the Clifden station to Caernavon during the Irish Uprising, the Marconi company's link provided an efficient wireless service between Canada and Great Britain until 1928. The cost of sending messages was calculated at the rate of 7.5d (3.125p) per word, a considerable reduction on the cable rate of 1s (5p) per word [19].

Of course, the successful trans-oceanic transmission of wireless messages was not the end of the Marconi organisation's long distance communications problem. There was the problem of dispatching the messages received at Glace Bay to their inland destinations, and the cable companies were not willing to be magnanimous to a competitor and accept messages from Great Britain for onward inland transmission across the continent. A person in Ottawa who wished to send a message 'via Marconi' to a person in Leeds, would have to take the message to a Marconi office where a clerk would send an inland telegram to Glace Bay, and similarly at the British end of the link. The handling of the messages gave rise to extra costs and made the service expensive – particularly so because the Marconi company had to pay and charge its customers the ordinary inland telegram rates, whereas the cable companies were given preferentially lower rates on inland telegrams. This unsatisfactory position was not resolved until early 1912. Agreements were reached with the Western Union and the Great Northern Telegraph companies that permitted 'Marconigrams' to be accepted at telegraph offices throughout the USA and Canada at the prevailing, and preferential, rates which applied to ordinary inland telegrams [20].

In the United Kingdom the distribution of 'Marconigrams' was aided by the Wireless Telegraphy Act of 1904. Following an accord between the Post Office and the Marconi International Marine Communication Company, the Post Office consented to collect messages for transmission by wireless to ships for a handling charge of 0.5d (c.0.2p) per word. Later, on 1st May 1912, this service was extended to cover the sending of wireless telegrams to the USA and Canada [21].

Towards the end of the first decade of the 20th century the Marconi organisation had amassed a wealth of experience – both commercial and engineering – of long distance wireless telegraphy. The time now seemed opportune for an extension of its trans-Atlantic service to other parts of the world, and, as the British Empire was an empire 'on which the sun never sets', it was logical for the British based MIMCC to seek an expansion of its services within the

Empire. Indeed, the question of Imperial communications had been the subject of two critical discussions during 1909; first, at the Imperial Press Conference, and second, at a conference held in Melbourne where it was agreed that the issue should be referred to the Imperial Defence Conference to be held in 1911.

In March 1910 MIMCC submitted to HM Government a plan for an 'all-red' radio communications network throughout the Empire ('wireless' was replaced by 'radio' at the 1903 German Conference). Probably, the news, early in 1910, that MIMCC's main rival, the Telefunken company, had obtained a contract to install its equipment in two stations in Australia – one in Sydney and the other in Freemantle – precipitated their action.

According to the plan MIMCC would construct 18 radio stations on British soil at key positions in the Empire [22], viz.:

England	Bombay	New Zealand
Alexandria	Colombo	St Helena
Aden	Singapore	Sierre Leone
Mombasa	Hong Kong	Bathurst
Natal	North Australia	British Guiana
South Africa (Cape	Sydney, Australia	West Indies
Colony or the Transvaal)		

and would operate the stations for 20 years, after which the British Government would have the option to purchase the stations.

The many advantages of such a radio network were outlined by the company [23]:

1. The system of communications would be British owned, and, in the event of war, would be in the complete control of the Government.
2. The danger of the German Company being able to arrange for a similar system on foreign territory would be minimised, as we have reason to believe that foreign countries will give preference to a British system, and will be indisposed to assist a competitive line.
3. The existing rates would be cut in half and in some cases still further reduced, resulting in a much greater volume of traffic.
4. Cheap means of communication results in greatly increased volume of communication and thus stimulates and facilitates trade. It may be remarked that a very large portion of the business done between Canada and America is done by telegraph, and if the rates between England and Canada were sufficiently low, there can no doubt that the volume of trade between these countries would be materially increased.
5. The Marconi system of wireless telegraphy, owing to the small capital outlay necessary, can afford to give low rates, and we do not ask for any subsidy for any portion of the system.
6. A cheap means of communication will enable many settlers to keep in touch with their relations in the mother country, which they are unable to do at

present, owing to the high cable charges, and this will encourage British emigration to British possessions rather than to foreign countries.

7. The result of cheaper press rates will be to increase the knowledge of the different members of the Empire of each other and prevent misunderstandings owing to incomplete reports of facts.
8. A considerable reduction in the Government's annual telegraph bill.
9. Every ship in the British Navy can be in direct communication with the Admiralty, no matter where at sea.
10. All ships at sea, properly equipped, can receive daily news and messages throughout their voyages, originating from any part of the world.
11. The fact that this Company is prepared to agree to a purchase clause at the end of 20 years prevents the establishment of another monopoly and ensures that the system may eventually become State-owned.

These proposals [24] were submitted to the Colonial Office on 10th March 1910 and then to their Cables (Landing Rights) Committee. This committee 'at once recognised the strategic and commercial importance of the proposal' and 'recommended that, in view of the objection to a monopoly in private hands, the stations should be owned by the State'. The issue was also referred to the Imperial Defence Committee, and on 1st June 1911 a sub-committee notified the Post Office that the question was a matter of urgency. Given these affirmations of the merits of the proposals the Post Office agreed to consider the issue of constructing the first six stations. However, it now found itself faced with a dilemma. The only companies that had any experience of long-distance radio working were MIMCC and Telefunken; but the latter company could not be asked to tender because:

1. though Siemens Brothers in England held the licence for the Telefunken system, the system was German and Siemens Brothers was 100 per cent German owned;
2. the Telefunken system infringed the Lodge patent which was now owned by the Marconi company.

Thus, the Post Office had no alternative but to negotiate with the English company. Given this constraint, the Postmaster General appointed a special committee (The Imperial Wireless Telegraphy Committee) of 20 persons to consider the Imperial chain. They recommended at a meeting held on 9th August 1911 that the Post Office should draw up a scheme of terms to be offered to the Marconi Company. Subsequent negotiations led to an agreement that was embodied in a formal contract on 19th July 1912 but this was not ratified by the House of Commons. Another committee, a Select Committee, was constituted in October 1912 to inquire into the history of the agreement. Three months later, on 14th January 1913 the committee simply reiterated the previous view that the Imperial chain was urgent and that the first six stations should be constructed in accord with the 1912 agreement. They recommended further that the government should immediately acquire the sites for them; that the government should be free to use any system from time to time; and that a scientific committee be appointed to inquire into the technology that might be applicable for the implementation of the chain.

The newly formed Scientific Committee reported on 30th April 1913. There were five radio systems which were, possibly, contenders for the Imperial chain, namely:

1. the Marconi system (of MWT) – a quenched spark system with synchronised undamped wave trains which approximated to a continuous wave;
2. the Telefunken system (controlled in England by Siemens Brothers) – another quenched spark system;
3. the Poulsen arc system (controlled in England by the Universal Radio Syndicate), which produced continuous waves and was cheap and easy to maintain (Fig. 14.6);
4. the Goldschmidt high frequency alternator (controlled in England by the Anglo-French Wireless Company, but invented in Germany); and
5. the Galletti system (controlled in England by the Galletti Company) – a synchronised spark system which generated an approximate continuous wave.

Consideration of all these possibilities and the concomitant factors associated with the Imperial chain led the Scientific Committee to one clear conclusion: the stations should be state owned, and there was really no alternative to negotiations being continued with the Marconi company.

A new non-tendered contract, for three stations, was signed with the company on 30th July 1913, and in November 1913 the Postmaster General invited tenders from any radio company for the construction of three further stations. Responses were received from the Universal Radio Syndicate, the Societé Française Radioelectrique, and the Galletti Company but their replies were ill-defined and no demonstrations could be shown.

The first stations of the Imperial chain, those at Leafield and Cairo, were begun in 1914 but when hostilities commenced the government decided to cease all work on the chain. And so the Marconi company's bold plan of 1910 that was acknowledged as being of prime importance to the country came to nought in 1914.

The saga of the Imperial chain is much more complicated than the brief sketch given above: the position was summed up by the *Electrical Review* in 1918 [25]:

Almost from the inception of this great scheme it was made the sport of party politicians; . . . and the erection of the Imperial chain, declared a matter of national urgency by the Committee of Imperial Defence . . . was delayed . . . by the machinations of the political wire pullers. . . . The war broke out before a single station was ready, and it is not surprising that the Government decided not to go on with the scheme; what is a matter for astonishment is that a great department of the Government [the Post Office] attempted to shuffle out of the agreement, and that it continued to shilly-shally and to refuse to admit its liability

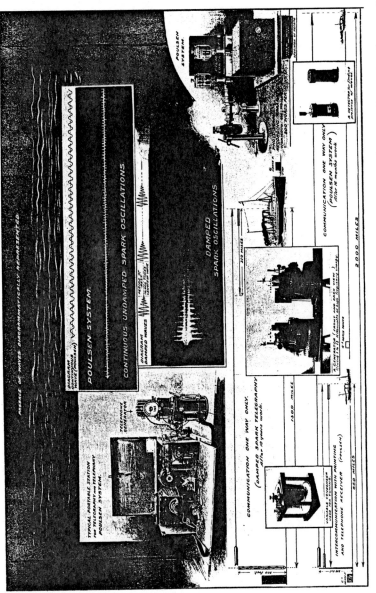

Figure 14.6 'Mr Valdemar Poulsen, who discovered that Duddell's singing arc could be used for telephoning without wires, has now brought his apparatus to very great perfection. The instruments have been used successfully over a distance of 1500 km. By means of a sound-intensifier, normal signals can be read at a distance of a dozen feet from the apparatus, and Mr Poulsen has added a recording instrument which prints the message.'

Source: The Illustrated London News Picture Library.

After the Great War, the Leafield and Cairo stations were completed in 1920 and 1922 respectively. They were each supplied with a 250 kW Elwell-Poulsen arc capable of producing an antenna current of 300 A. The Leafield station operated at a wavelength of 12,200 m and enabled a regular radio service to be established with the USA, India and Egypt.

At this stage in the saga of 'message' communications a digression must be made so that the progress towards a viable system of television may be described.

References

1 STURMEY, S.G.: 'The economic development of radio' (Duckworth, London, 1958), p. 75
2 Ibid, p. 76
3 FLEMING, J.A.: 'The principles of electric wave telegraphy' (Longmans, Green, London, 1908), p. 449
4 Ibid, pp. 451–452
5 Ibid, p. 454
6 ANON.: a report, *The Electrical Review*, 1901, p. 1031
7 RATCLIFFE, J.A.: 'Scientists reactions to Marconi's transatlantic radio experiment', *Proc. IEE*, 1974, **121** 9, pp. 1033–1038
8 MACKEAND, J.C.B., and CROSS, M.A: 'Wide-band high frequency signals from Poldhu', The IEE Conference Publication on '100 years of radio', No. 411, 1995, pp. 26–31
9 BAKER, W.J.: 'A history of the Marconi Company' (Methuen, London, 1970), p. 70
10 Ibid, p. 72
11 Ref. 1, p. 77
12 Ibid, p. p. 83
13 Ref. 3, pp. 455–456
14 Ibid, p. 457
15 Ref. 9, p. 80
16 JOLLY, W.P.: 'Marconi' (Constable, London, 1972), p. 131
17 Ref. 9, p. 98
18 Ibid, pp. 116–117
19 Ref. 1, p. 78
20 Ibid, p. 79
21 Report of the Postmaster General, Cmd. 6495, 1912, Post Office archives
22 Ref. 1, p. 84
23 VYVYAN, R.N.: 'Wireless over thirty years' (George Routledge, London, 1933), p. 31
24 Ref. 1, pp. 87–101
25 ANON.: a report, *Electrical Review*, 1918, p. 265

Chapter 15

Television development, pre-1914

Three years after his first thoughts on television, Campbell Swinton in November 1911, in his Presidential Address [1] to the Röntgen Society, elaborated on his 1908 letter [2] to *Nature* and described an idea for an electronic camera; an idea that was to be implemented in the 1930s and was founded on the utilisation of a cathode-ray tube in the transmitter, as well as in the receiver.

Modern high definition television would not be possible without some version of the cathode-ray tube, a device that owed its origins as a practical laboratory instrument to F. Braun [3] in 1897. Braun's tube was not the outcome of a stroke of genius; rather it represented the consequence of an enquiry into the conduction of electricity in gases that commenced with the work of William Watson [4], an eminent English scientist of the eighteenth century.

Watson's experiments were made possible by a discovery that according to Professor Tyndall put all former ones in the shade and that Dr Priestly [5] called the most surprising yet made in the whole business of electricity. This was the storage of electric charge in a capacitor, called a Leyden jar after the name of the place where the discovery was originated. It was first reported by von Kleist, the discoverer, and Dean of the cathedral in Camin, in a letter dated 4th November 1745, to a Dr Lieberkuhn at Berlin [6].

Watson was particularly interested in the conduction of electricity through rarefied gases and vacuo and communicated his results in 1752 in a paper [7] to the Philosophical Transactions. In his experiments he used a glass tube, one metre in length and 3 in (75 mm) in diameter, fitted with a fixed brass plate at one end. A movable plate could be inserted into the tube and caused to approach the fixed plate. When the glass cylinder was evacuated and the electrodes electrified, there was 'a most delightful spectacle' – the display 'not as in the open air [of] small brushes or pencils of rays, an inch or two in length, but coruscations of the whole length of the tube, and of a bright silver hue'.

Nearly a century later the great Faraday [8] investigated the discharge of electricity through rarefied gases with characteristic care and attention to detail. Fortunately for the development of electrical science, vacuum pumps were available to Faraday for von Guericke had produced the first air pump in 1654 and this had been followed by similar pumps of Boyle, Hawksbee and Smeaton having increased efficiencies. Von Guericke's apparatus had been made of glass but later oil-air pumps were fabricated from metal. The first mercury pump was constructed by Swedenborg in 1722 and other versions were manufactured by Baader in 1784, Hindenburg in 1787, Edelkrantz in 1804 and Patten in 1824. Geissler's mercury pump of 1855 was capable of creating a partial vacuum of 0.002 in (0.05 mm) of mercury but this pump was not available to Faraday, who carried out his experiments at the reduced pressure of 6 in (165 mm) of mercury for some and 5 in (118 mm) for others.

Geissler's vacuum pump and Ruhkorff's induction coil of 1851 made it possible for much larger potential differences to be applied to the terminals of the discharge tube and also allowed much diminished pressures to be obtained. The improved vacuum tubes (often known as Geissler tubes) were utilised in the investigations of Plucker, which he reported in 1858. He noted visible stratifications in the discharge and found they could be modified by a magnet placed outside the tube. In addition the diffused light seen near the negative electrode was found to be concentrated by the action of the magnet, and Plucker [9] concluded that the glow consisted of 'lines of light which, proceeding from the separate points of the positive electrode, coincide with magnetic curves'.

Several weeks afterwards, Hittorf, [10] who was one of Plucker's students, stated that the glow was due to some kind of rays which were given off in straight lines from the negative electrode or cathode. Goldstein, who also observed that the cathode rays, as they were called, were given off in a direction normal to the surface of the cathode, noted that a shadow of an obstacle, placed in the path of the rays, was cast on the wall of the evacuated vessel [11].

Much attention was given to the study of gaseous discharges by various investigators, but it was Sir William Crookes [12] who carried out a systematic experimental investigation to elucidate their properties. He discovered what is now called the Crookes dark space and speculated that it was a region in which the cathode rays had a free path before colliding with the gas molecules – the blue glow being caused by the collisions. His findings were published in several papers in 1878 and 1879.

A few years earlier, in 1871, Cromwell Varley [13] had advanced the idea that the cathode rays were negatively charged particles, but with only partial proof. However, Jean Perrin [14], in 1875, confirmed Varley's hypothesis. At about the same time Clerk Maxwell [15] published his 'Treatise on electricity and magnetism' (1873) and introduced the concept of 'one molecule of electricity'. G. Johnstone Stoney [16], a year later, made an estimate of this elementary charge, and in 1891 suggested the name electron for the natural unit of electricity. By then Schuster [17] in a series of investigations carried out in Manchester from 1884 to 1890 had determined the ratio of the charge-to-mass of the particles which comprised the cathode rays. He achieved this by deflecting the charged particles, with the aid of a magnetic field, into a circular path and obtained a value of e/m of about 1.1×10^6 C/kg.

The most comprehensive enquiry into the properties of the cathode rays was undertaken by Professor J.J. Thomson [18] in the Cavendish laboratory of the University of Cambridge. In 1894 he began a series of brilliant experiments which culminated in the determination of the charge, mass and velocity of the particles in the cathode rays. For the accuracy of measurement he had in mind Thomson constructed discharge tubes in which means were provided to deflect the cathode rays by magnetic and electric fields (Fig. 15.1a). His tube design satisfied the two criteria necessary for a use to be made of cathode rays for measurement purposes; namely, the separation of the electron beam from the other phenomena of the discharge, and a means for measuring the deflection of the rays with certainty and accuracy.

The credit for the invention of the cathode-ray tube (c. r. t.) as a commercial instrument is usually ascribed to Professor Ferdinand Braun, of the University of Strasbourg. In 1897 he published the first account of his tube. It was much larger than Thomson's, although simpler in design and differed from it in a number of details: the use of only one diaphragm B (see Fig. 15.1b) having a circular hole (0.08 ins – 2 mm diameter) instead of a slit; the introduction of a fluorescent screen, S, consisting of a mica plate coated on one side with a fluorescent substance; and the use of magnetic deflection only.

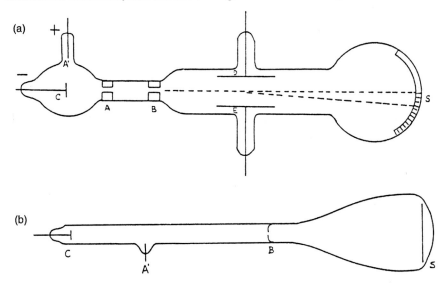

Figure 15.1 *(a) Diagram of the cathode ray tube which J.J. Thomson used in the 1890s to determine the ratio elm for the electron. The cathode, C, anode, A', collimating apertures, A and B, deflecting electrodes, D and E, and the screen, S, are shown.*

Source: Phil. Mag., *1897,* **44,**

> *(b) F. Braun is usually credited with the invention of the cathode-ray tube as a commercial laboratory instrument. The cathode, C, anode, A', collimating apertures, B, and screen, S, are indicated.*

Source: Author's collection.

With this tube the screen showed a bright round spot of light where the electrons bombarded the screen, instead of a line, and thus it was possible to measure the deflection of the spot in any direction across the screen.

Braun employed the tube for investigating the phase relations associated with polarised electrolytic cells. Two years later, in 1899, Zennek [19], in his work on radio circuits and the propagation of radio waves, utilised an improved form of Braun tube having a finer electron beam. Zennek, who was one of Braun's assistants (and later President of the Deutsche Museum in Munich) has described the impact of Braun's invention on his own work: 'When Braun brought out his tube, I was very enthusiastic about it. It was exactly what I had wanted for a long time, a device with which one sees what is going on in the current circuit.'

Much work was carried out by experimenters on the development and application of cathode-ray tubes following Braun's publication in 1897, and so it was perhaps inevitable that someone would suggest the employment of such a tube to reproduce a transmitted image. In a German patent application dated 12th September 1906 M Dieckmann and G Glage [20] described a 'Method for the transmission of written material and line drawings by means of cathode-ray tubes' (Fig. 15.2).

The tracing point, for example a pencil, was attached to a slider, the guide for which was carried by two further sliders capable of moving on two supports mounted at right angles to the first guide. Consequently, the movement of the tracing point was mechanically resolved into two mutually perpendicular components. These were then converted into resistance changes – using the wires f and l – which caused corresponding changes in the currents passing through the wires.

At the receiver the two transmitted currents excited two orthogonal sets of deflection coils, associated with the Braun tube, and thereby established two magnetic fields. The resultant field deflected the electron beam, and the associated luminous spot on the fluorescent screen, in accordance with the motion of the tracing point at the transmitter. Dieckmann and Glage found no difficulty in transmitting drawings and written words in a few seconds with their method.

In an adaptation of the c. r. t. receiver, the inventors designed a small dynamo that provided currents for the deflection coils so as to allow the spot of light to trace a 1.25 in (31.75 mm) square raster, of parallel and equidistant lines, in one tenth of a second.

At the transmitter a fine metal brush moved synchronously with the luminous spot and passed over a sheet metal pattern. When the brush contacted the conductor pattern, a current was sent to the receiver and was used to excite the coils. The magnetic field created deflected the electrons to such an extent that none passed through the hole in the diaphragm. Consequently, those parts of the raster that corresponded with the metal pattern were obliterated, or, as Dieckmann and Glage wrote, 'the production of an exact copy of the pattern in black on a luminous field. As the entire copy was produced [in one tenth of a second] it followed every movement impressed on the pattern' [20].

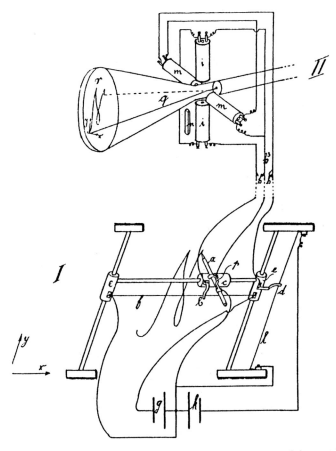

Figure 15.2   The illustration shows a diagrammatic representation of the apparatus of Dieckmann and Glage (1906).
Source: German patent no. 190,102, 12th September 1906.

This was not television, of course, but a form of autographic telegraphy. Nevertheless, as Dr Dieckmann noted: '[The] experiment shows that the cathode ray is well worth the attention of inventors in search of apparatus destitute of inertia' [20].

Boris Rosing [21], a member of the teaching staff of the Technological Institute, St Petersburg, was the first person to put forward a 'seeing by electricity' scheme incorporating a cathode-ray tube. His 1907 patent [22] indicates that he was knowledgeable about some previous work on this subject because he stated that the raison d'etre for his patent was to obviate the defects of receiving equipment which were 'insufficiently mobile and sensitive' for the purpose. Fig. 15.3 illustrates his apparatus. An image of the object or picture, 3, was cast by the lens, 4, and the polyhedral mirrors, 1 and 2, upon a photoelectric cell, 5. The mirrors rotated about mutually perpendicular axes with differing angular

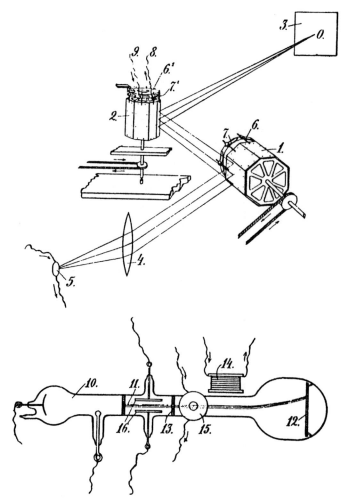

Figure 15.3   Diagrams of Rosing's apparatus of 1907. Mirror-drum scanning was used at the transmitter and cathode-ray scanning was employed in the receiver.

Source: British patent no. 27,570, 13th December 1907.

velocities so that all the points in the plane of the picture or field of view were successively scanned.

At the receiver Rosing reconstituted the image of the object with a Braun tube. Magnetic deflection of the electron beam was employed and the coils for this purpose were excited by currents, controlled by resistors, which varied depending on the movements of the rubbing contacts, 7, applied to each face of the polyhedral mirrors. Variations in the electric field established between the plates, 16, by the photoelectric signals caused variations in the number of electrons that passed through the diaphragm, 13, and hence changes in the brightness of the screen.

Some of Rosing's laboratory notebooks exist and in number three the following entry occurs: 'On 9th May 1911, a distinct image was seen for the first time, consisting of four luminous bands' [23].

After 1911 Rosing did not make any significant contributions to the television problem, though he maintained an interest in the subject. His book on *The electrical telescope* was published in 1923. He certainly appreciated the benefits that television would bring and, in an article in the French journal *Excelsior*, he wrote [24]:

The range of application of the telephone does not extend beyond human conversation. Electrical telescopy will permit man not only to commune with other human beings, but also with nature itself. With the 'electric eye' we will be able to penetrate where no human being penetrated before. We shall see what no human being has seen. The 'electric eye' fitted with a powerful lamp and submerged in the depths of the sea, will permit us to read the secrets of the submarine domain. If we recall that water covers three quarters of the earth's surface, we readily realise the infinite extent of man's future conquests in this portion of his domain, till now inaccessible to him. From now on and in all future times we can imagine thousands of electric eyes travelling over the floor of the sea seeking out scientific and material treasures; others will carry out their explorations below the earth's surface, in the depths of craters, in mountain crevices and in mine shafts. The electric eye will be man's friend, his watchful companion, which will suffer from neither heat nor cold, which will have its place on lighthouses and at guard posts, which will beam high above the rigging of ships, close to the sky. The electric eye, a help to man in peace, will accompany the soldier and facilitate communication between all members of human society.

Rosing's place in the history of television stems from his advocacy of using a cathode-ray tube as the receiving element in a system of distant vision, of 'chopping' the light beam falling onto the photoelectric cell, of resolving the focusing problem by means of velocity modulation and of incorporating photo-emissive cells in such systems.

Campbell Swinton, to whom reference has been made, and who was the first person to suggest an all-electronic television system, could not have been influenced by Rosing's early work in 1907 for Rosing's patent (No. 27 570) was not accepted and published by the Patent Office until 25th June 1908, whereas Campbell Swinton's letter [2] to *Nature* was printed on 18th June 1908.

Swinton's 1911 Presidential Address [1] to the Röntgen Society gave a complete description of his ideas together with a diagrammatic illustration (Fig. 15.4), of the apparatus. In this both the transmitting and receiving cathode ray tubes were of the Braun type and were fitted with cold cathodes and anodes. Though Wehnelt had proposed in 1905 coating a platinum filament with lime so that the necessary electron emission could be obtained with a relatively low anode-cathode voltage, Campbell Swinton in his original diagram showed a tube with a cold cathode. This required a potential difference of about 100,000 V between the anode and cathode structures to give an appropriate beam current.

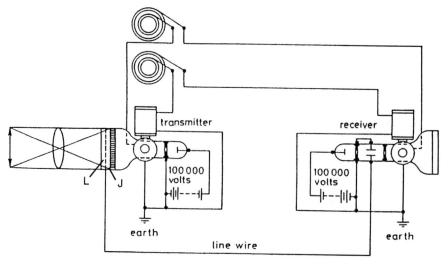

*Figure 15.4   In his presidential address to the Röntgen Society in 1911, Campbell Swinton
described his scheme for television using non-mechanical scanning means at
both the transmitter and receiver.*
Source: *Journal of the Röntgen Society*, 1912, **8** (30).

In both tubes the electron beams were deflected by magnetic fields, using electromagnets excited from a.c. generators. The line scan was at the rate of 1,000 Hz and the frame scan at 10 Hz. Because of the sinusoidal nature of the outputs of the generators the electron beams would describe Lissajous figures on the screen of the transmitting and receiving cathode-ray tubes – a form of scan first put forward in 1880 by Le Blanc.

In the receiving tube the electrons were to produce a luminous image by striking the fluorescent screen and the intensity of the image was to be controlled by applying a voltage, derived from the transmitting tube, to a pair of deflecting plates. With regard to the receiving apparatus, there was nothing of a particularly novel nature. The novel aspect of Swinton's conception lay in the original approach to the generation of the image signal at the transmitter.

He conceived the idea of a mosaic of photoelectric elements onto which the image of the object or scene would be projected by means of a lens and which would be scanned on the side away from the image side by a beam of electrons controlled by line and frame a.c. voltages. In the transmitter c.r.t. the gas tight screen, J, was to be formed from a number of small insulated metallic cubes of a metal such as rubidium, which was strongly active photoelectrically, so that a clean metallic surface would be presented to the electron beam on one side and to a suitable gas or vapour, say, sodium vapour, on the other. A metallic gauze screen, L, parallel to J was to be placed in front of the screen, J.

Campbell Swinton's account of the supposed operation of the system follows, because some doubt has been expressed on whether the integrating feature (i.e. charge storage) of the iconoscope, due to V.K. Zworykin in the 1920s, was implicit in the account.

As the cathode rays oscillate and search out the surface of J they will impart a negative charge in turn to all the metallic cubes of which J is composed. In the case of cubes on which no light is projected, nothing further will happen, the charge dissipating itself in the tube; but in the case of such of those cubes as are brightly illuminated by the projected image, the negative charge imparted to them by the cathode rays will pass away through the ionised gas along the line of the illuminating beam of light until it reaches the screen L whence the charge will travel by means of the line wire to the plate O of the receiver. This plate will thereby be charged – will slightly repel the cathode rays in the receiver; will enable these rays to pass through the diaphragm P, and impinging on the fluorescent screen H will make a spot of light. This will occur in the case of each metallic cube of the screen J which is illuminated, while each bright spot on the screen H will have relatively exactly the same position as that of the illuminated cube of J. Consequently, as the cathode-ray beam in the transmitter passes over in turn each of the metallic cubes of the screen J, it will indicate by a corresponding bright spot on H whether the cube in J is or is not illuminated, with the result that H, within one-tenth of a second, will be covered with a number of luminous spots exactly corresponding to the luminous image thrown on J by the lens M, to the extent that this image can be reconstructed in mosaic fashion. . . .

It is further to be noted that, as each of the metallic cubes in the screen J acts as an independent cell, and is only called upon to act once in a tenth of a second, the arrangement has the obvious advantages over other arrangements that have been suggested, in which a single photoelectric cell is called upon to produce the many thousands of separate impulses that are to be transmitted through the line wire per second, a condition which no known form of photoelectric cell will admit of.

Again, it may be pointed out that the sluggishness on the part of the metallic cubes in J or of the vapour in K in acting photoelectrically, in no way interferes with the correct transmission and reproduction of the image, provided all portions of the image are at rest; and it is only to the extent that portions of the image may be in motion that such sluggishness can have any prejudicial effect. In fact, sluggishness will only cause changes in the image to appear gradually instead of simultaneously [1]. (The question whether these two paragraphs implicitly embrace the charge storage principle has been discussed by the author in *Television, an international history of the formative years.*)

This, then, was the scheme of Campbell Swinton: a most remarkable one when it is borne in mind that at the time it was enunciated radio communication was in its infancy, radio valves were practically unknown, vacuum technology was very primitive, and photoelectric cells were very inefficient.

Campbell Swinton never tried to construct a working model of his 1911 transmitter, and he fully appreciated the difficulties that would have to be surmounted before it could be made to work. With typical honesty he remarked: 'It is an idea only. . . Furthermore, I do not for a moment suppose that it could be got to work without a great deal of experiment and probably much modification' [1]. He was not sanguine about the financial return to an individual who attempted to pursue the solution, but 'if we could only get one of the big research laboratories, like that of GEC or of the Western Electric Company, one of these people who have large skilled staffs and any amount of money to engage on the business, I believe they would solve a thing like this in six months and make a reasonable job of it' [1].

Subsequently, two large industrial research laboratories – those of the Radio Corporation of America (RCA), and Electric and Musical Industries (EMI) – did evolve, independently, electron cameras, known as the iconoscope and the Emitron respectively (see Chapter 20). Dr J.D. McGee [25], the leader of the team at EMI that produced the Emitron camera tube, has written: 'Modern television owes much to the researches and achievements of many distinguished workers, but in essence it has been developed upon the fundamental lines first put forward by Campbell Swinton'.

Since the advancement of television was dependent on work that had been progressing in the field of photoelectricity, some of the results that had been obtained, and which were about to come to fruition during the first and second decades of the new century, are given in the following note. Then, in the next chapter, the development of communications to the war effort is discussed.

Note 1
The early history of photoelectricity

During the period of the late 19th and early 20th centuries when inventors were attempting to seek solutions, using known techniques and devices, to the difficulties posed by the implementation of facsimile systems, wireless telegraphy, and television, other scientists were engaging in fundamental physical researches.

The work of Heinrich Hertz in 1887 provides a convenient and suitable base from which some advances in photoelectricity may be considered.

It was in that year that Hertz performed his classic experiments (described in Chapter 10) on the effect of electrical discharges in one oscillatory circuit upon another similar, but separate circuit. A Ruhmkorff coil, excited by a battery, caused primary sparks to traverse a gap. A second smaller coil produced secondary sparks, about 0.04 in (1 mm) in length, which bridged the gap in a Reiss spark micrometer. In his work Hertz investigated the effect of the parameters of the first circuit on the length of the spark in the micrometer gap. He observed: 'In all of the experiments described, the apparatus was arranged so that the spark of the inductor was visible from the position of the spark at the micrometer gap. If this condition was altered, the same qualitative results were obtained, but the lengths of the secondary sparks appeared to decrease' [26].

Subsequently, Hertz concluded [26] that ultraviolet light from one spark gap could enhance the passage of sparks across a second gap. He did not pursue this subject further and wrote: 'For the present I limit myself to the presentation of these established facts without attempting to advance a theory of how such observed phenomena could occur'.

During the following year Wiedemann and Ebert confirmed Hertz's results – particularly those which showed that the effect was confined to the negative terminal of the irradiated gap.

In the same year, 1888, Hallwachs [27] demonstrated that under the influence of ultraviolet radiation 'negative electricity' left a negatively charged body and followed the electrostatic lines of force. Also, in 1888, Righi [28] found that a polished metal plate and metal grid configuration was capable of producing a current under the action of light. He termed the configuration a photoelectric cell. Righi employed a quadrant electrometer in his experiments but in 1890 Stoletow [29] utilised a high resistance galvanometer and an external source of electromotive force and was able to show that a small continuous current flowed from the grid to the plate (in the positive, conventional sense) when the polished plate was irradiated with light.

The next great advance in photoelectricity was made by the two famous co-workers, Elster and Geitel [30]. They had observed that, of all the metals that had been studied for photoelectric sensitivity, aluminium, magnesium and zinc appeared to give the best results. Hence, they considered that, because these metals were all electropositive, it seemed reasonable to expect similar or better results from metals that were more electropositive than those mentioned above. Elster and Geitel therefore proceeded to investigate the alkali metals, especially sodium and potassium. Unfortunately, these chemically active elements were found to react almost instantaneously with air and water vapour to form oxides and hydroxides which were relatively insensitive. Nevertheless, they persevered with their study of these two metals and having noticed previously that an amalgam of zinc could be used with much greater satisfaction than zinc alone they decided to investigate the photo-activity of amalgams of the alkali elements. Success followed. After much preliminary work they discovered that a fresh dilute amalgam of either sodium or potassium was many times more sensitive than zinc amalgam.

By 1890 Elster and Geitel had published [31] a detailed account of the manufacture of a sodium amalgam photoelectric cell in which the metal electrodes were enclosed in an evacuated, glass vessel. They wrote: 'As may be judged from the description, no provision has been made to admit light of short wavelength. A window of quartz or similar material is not necessary, as light transmitted by the glass proves to be sufficient'.

At about this time (1890), the nature of the negative electricity that left the negative electrode of the cell was not known. Necessarily, much work was undertaken by physicists on the elucidation of its characteristics. Elster and Geitel, Lenard, and Merritt and Stewart and others showed beyond doubt, by 1900, that the photoelectric current was due to the emission of electrons from the negative electrode of the cell.

Other investigations enabled the fundamental laws of photoelectricity to be derived and work on the photoelectric sensitivity of metal surfaces allowed much improved cells to be fabricated. The early cells tended to deteriorate with time unless their cathode surfaces were occasionally renewed, but in 1904

Hallwachs constructed a cell for photometric purposes which was found to be constant in operation over a period of several months. This consisted of an evacuated vessel having a copper plate coated with black oxide as a cathode, but it was not sensitive to visible radiations.

Elster and Geitel continued to make strides and observed that the hydride crystals of sodium and potassium were more responsive than the metals themselves. They also evolved a sensitising procedure which marked a new era in the development of photoelectric cells. If a glow discharge was excited in an alkali cell filled with hydrogen, then the cathode surface was transformed into a colloidal state and became as much as one hundred times more sensitive than the pure element.

From this very brief account it is apparent that when Campbell Swinton put forward his 1911 proposals [1] for an all-electronic television system much fundamental work on the science of photoelectricity had been carried out.

References

1 CAMPBELL SWINTON, A.A.: 'Presidential Address', *J. Röntgen Society*, 1912, **8**(30), pp. 1–5

2 CAMPBELL SWINTON, A.A.: 'Distant vision', a letter, *Nature*, 1908, p. 151

3 BRAUN, F.: 'Ueber ein Verfahren zur Demonstration und zum Studium des Zeitlichen Verlaufes Variabler Strome', *Ann. Phys.*, 1897, **60**, p. 552

4 MOTTELAY, J.P.: 'Bibliographical history of electricity and magnetism' (Charles Griffin, London, 1922)

5 PRIESTLY, J.: 'History and present state of electricity' (London, 1767)

6 Ref. 4, p. 173

7 WATSON, W.: 'An account of the phenomena of electricity in vacuo with some observations thereupon', *Phil. Trans.*, 1752, **147**, p. 362

8 FARADAY, M.: 'Experimental researches in electricity', 1, paras, 1529, (p. 487) and 1554, (p. 494). Also see para. 1523 (p. 485), (London, 1839–55)

9 PLUCKER, J.: 'On the action of the magnet upon the electrical discharge in rarefied gases', *Phil. Mag.*, 1858, **16**, pp. 119–135, 408–418

10 HITTORF, J.W.: 'Ueber die Elektricitatsleitung der Gase', *Ann Phys.*, 1858, **136**, pp. 1–30, 197–234

11 MEYER, H.W.: 'A history of electricity and magnetism' (MIT Press, 1971), p. 227

12 CROOKES, W.: 'On the illumination of lines of molecular pressure and the trajectory of molecules', *Phil. Trans.*, 1879, **170**, pp. 135–164

13 VARLEY, C.: 'Some experiments on the discharge of electricity through rarefied media and the atmosphere', *Proc. Roy. Soc.*, 1871, **19**, pp. 236–242

14 PERRIN, J.: *Compte Rendu*, 1895, **121**, pp. 1130–1136

15 MAXWELL, J.C.: 'Treatise on electricity and magnetism' (Clarendon Press, Oxford, 1892)

16 JOHNSTONE STONEY, G.: 'On the physical units of nature', *Phil. Mag.*, 1837, **11**, p. 384

17 SCHUSTER, A.: 'Experiments on the discharge of electricity through gases. Sketch of a theory', *Proc. Roy. Soc.*, 1884, **37**, p. 317. Also 1887, **42**, p. 371

18 THOMSON, J.J.: 'Cathode rays', *Phil. Mag.*, 1897, **44**, pp. 293–316

19 ZENNECK, J.: Eine Methode zur Demonstration und Photographie von Stromcurven', *Ann. Phys.*, 1899, **69**, p. 838

20 DIECKMANN, M., and GLAGE, G.: 'Verfahren zur Uebertragung von Schriftzeichen und Strichzeichnungen unter benutzung der Kathodenstrahlrohre', DRP 190 102, September 1906

21 GOROKHOV, P.K.: 'History of modern television', *Radio Engineering*, 1961, **16**, pp. 71–80

22 ROSING, B.L.: 'New or improved methods of electrically transmitting to a distance real optical images and apparatus therefor', British patent 27 570, 13th December 1907

23 ROSING, B.L.: Notebook 3, 1911, Archives of the A.S. Popov Central Museum of Communication. Quoted in Ref. 21

24 ROSING, B.L.: Article in the French journal *Excelsior*, c. 1910. Quoted by A. Korn and B. Glatzel in 'Handbuch der Phototelegraphie und Telautograhie', (Nemnich, Leipzig, 1911)

25 MCGEE, J.D.: 'Campbell Swinton and television', *Nature*, 1936, pp. 674–676

26 HERTZ, H.: 'Ueber einen Einglass des ultravioleten Lichtes auf die elektrische Entladung', *Ann. Phys.*, 1897, **31**, pp. 983–1000

27 HALLWACHS, W.: 'Ueber den Einfluss des Lichtes auf electrostatisch geladene Korper', *Ann. Phys.*, 1888, **33**, pp. 301–312

28 RIGHI, A.: 'On some electrical phenomena provoked by radiation', *Phil. Mag.*, 1888, **25**, pp. 314–316

29 STOLETOW, M.A.: 'Sur les courants actino-electrique dans l'air rarefie', *Journal de Physique*, 1890, 2nd series, **9**, pp. 468–473

30 ELSTER, J., and GEITEL, H.: 'Ueber die Entladung negative elektrische Korper durch das Sonnen und Tageslicht', *Annalen der Physik und Chemie*, 1889, **38**(12), pp. 497–514

31 ELSTER, J., and GEITEL, H.: 'Ueber die verwendung des Natrium amalgames zu licht-elektrischen versuchen', *Annalen der Physik und Chemie*, 1890, **41**(10), pp. 161–176

Chapter 16

The Great War years, 1914–1918

At the beginning of 1914 the British Admiralty had 30 shore stations and 435 ships fitted with radio apparatus [1]. It could communicate with, and directly control, its fleet over an immense stretch of sea. Extensive sea trials and fleet manoeuvres in Home waters, and the battle of Tsu Shima between the Russian and Japanese navies when wireless had been used by both sides in the engagement, had established wireless as a tactical and strategic factor of the greatest importance [2]. The Admiralty's well-tested and tried signalling means together with the tireless efforts of Admiral Lord Fisher, when he was First Sea Lord (1904–1910), to modernise and improve the efficiency of the British navy, ensured its ships seemed well-prepared to engage enemy ships.

The position with the British Army was less satisfactory. When hostilities commenced in August 1914 'wireless telegraphy started [the war] as "possibly a useful adjunct to visual and line signalling"; [but], in spite of the valuable work of the Experimental Section at Aldershot and of certain Territorial wireless units, notably the London, Leeds and Scottish sections, it was still an adjunct' [3].

Historically, the Army's interest in wireless telegraphy dates from the period of the Boer war. At the outset of the conflict, the Boers had a total strength of 48,000 skilled horsemen and excellent riflemen disposed against the 27,000 soldiers of the British Army.

In October 1900 the South African towns of Kimberley, Mafeking and Ladysmith, with their British troops, were surrounded by the Boers [4]. Reinforcements, comprising three divisions plus cavalry, were sent from England for the South African ports of Cape Town and Durban. Accompanying the soldiers were six engineers (Messrs Bullocke, Dowsett, Elliot, Franklin, Lockyer and Taylor) of Marconi's Wireless Telegraphy Company, some Royal Engineer sappers, under the command of Captain J.N.C. Kennedy, and five portable wireless stations [5]. Their task initially was to establish ship-to-shore communications to assist in the disembarkation of the troops and then to support them in the field during their expeditions to the besieged towns. On arrival in Cape Town in December 1899 a successful demonstration of the signalling apparatus was given to the General and Staff Officers of the army at Cape Town Castle, but the transmitter-receiver range was just a few hundred yards. Subsequently, three of the sets were allocated to Lord Methuen's column which intended to relieve Kimberly, and two sets were allotted to General Buller's column which was directed to raise the siege of Ladysmith.

Rather ominously, on 17th December, one of the Marconi engineers found he could not communicate between De Aar and the Orange River, some 70 miles (112.6 km) distant, even though an antenna 60 ft (18.3 m) in height and a 'good earth' had been used [6]. Further attempts to prove the functionality of the Marconi equipment proceeded for six weeks but for 'at least half that period most were unserviceable due to cyclonic dust storms which splintered the bamboo [antenna] masts, lightning-induced discharges which overwhelmed the coherers or wind which either was insufficient for the flying of [antenna] kites or ferocious enough to tear away the [antenna] balloons.' The wireless telegraphy system could not give continuous support to the troops in the field and so, on 12th February 1900, the Director of Army Telegraphs ordered the three sets along the Kimberley line to be dismantled, an end which also befell the two sets that accompanied General Buller's forces [6].

The failure of the Marconi equipment has been ascribed by Dr B. Austin to the poor electrical conductivity of the ground of the 'dry sandy plains of the northern Karroo'. He has opined 'that no two wireless installations were ever likely to have been operating on exactly the same frequency ... particularly because the quality of the earth connection was seriously impaired by the nature of the ground itself' [7]. The sets used the method of 'plain aerial working' in which the spark gap was directly in series with the radiating structure. Consequently the radiated frequency was determined by the length of the antenna and the impedance of the ground connection. If these factors were different at the sending and receiving stations signal reception would be difficult or impossible.

As noted previously, the five sets were transferred to ships of the Royal Navy's Delagoa Bay Squadron, which were blockading ports in the region of Lourenco Marques, and gave good service [6].

The first expenditure of money on wireless telegraphy in the British army was in 1901–1902, and the first proper record of experimental work dates from 1902. In the army manoeuvres of 1903, the 1st Army Corps and the 2nd Army Corps each had two wireless telegraphy stations supplied and staffed by the Marconi and Lodge-Muirhead companies respectively (Fig.16.1). Primary training was given to the officers and men. Trials and experiments, with apparatus provided by these firms and Telefunken, led to the evolution of wagon sets and pack sets by 1906.

Wireless work in the army was first entrusted to a section of the 1st Division Telegraph Battalion, Royal Engineers. In April 1907, two Wireless Telegraphy [army] companies were formed. These companies carried out such experimental work as they could in

Figure 16.1   Army vehicle equipped with a 1.5 kW Marconi spark set. Note the ubiquitous poles.
Source: Author's collection.

addition to their duties as field units. This system was not very satisfactory, and on 15th July 1911, the experimental work was separated from the field work by the organization of one Wireless Telegraphy Company and a separate Experimental Section [8].

It soon became evident that the Army, unlike the Royal Navy, was not keeping itself acquainted with recent technology and developments. In August 1912, a special committee was constituted, under the chairmanship of Sir Henry Norman, MP, an enthusiast for new advances in automobiles and wireless, 'To consider the utilization for military services of Wireless Telegraphy and Telephony, with special reference to recent developments in the science, and to report' [9].

The committee held 24 meetings, examined 27 witnesses, studied various catalogues and other documents, inventions and systems, witnessed four practical demonstrations and sought evidence from various persons, including the US Military Attache in London.

A depressing picture emerged from the committee's work: the state of wireless telegraphy in the army was 'so inefficient as to be unreliable, and therefore practically valueless, in time of war' [10]. Practical tests on two wagon sets and two pack sets 'showed failure in every respect'. 'We gathered from witnesses who appeared before us that grave doubts [were] generally felt by senior officers as to the reliability of army wireless as at present organised, and that they would not feel safe in time of war unless an alternative means of communication were provided' [10].

The principal limitations of the equipment stemmed from the unsatisfactory petrol engines used to drive the generators, the necessity to stop an engine when receiving a signal, the need to readjust the detector after every transmission and the 'inefficiency' of the receiving circuits.

Many recommendations were put forward by the Norman Committee, relating to: the general organisation of wireless equipment and service in the army; wireless instruction; technical equipment; future policy with regard to equipment; wireless telephony; army wireless procedure in signalling; and wavelengths. Of the several systems of wireless telegraphy available for purchase, the Committee commended the purchase of two complete Poulsen 4 kW sets of the latest design, for £950 each. The Poulsen system had a number of advantages [11] including: '[the] considerable possibilities of immunity from interference especially in combination with the heterodyne receiver; [the] immediate change to any one of a great number of wavelengths, including very long ones, far longer than could be produced with a spark set with a given aerial, which facilitate[d] working over mountainous country, and render[ed] tapping by the enemy more difficult'.

Summarising its eminently sensible proposals, the Committee recommended the following course of action, inter alia, should be implemented [12]:

1. [The] substitut[ion of] a simpler receiving gear for the present receiving gear in both the wagon and the pack sets (Fig. 16.2).
2. [The] work[ing of] the wagon sets with the operator's limber removed from the limber containing the power unit.
3. The purchase of two Poulsen sets, to be mounted on petrol-electric chassis.

Figure 16.2   Marconi 1.5 kW set mounted in a wagon. The receiver is on the right and the antenna tuning inductance is just inside on the left.
Source: Author's collection.

4. [The procurement of] a hand-operated pack set (Fig. 16.3), as used in the United States Army.
5. [The advice] that no more wagon sets be built.
6. The purchase of a transmitting set of Dubilier pattern for experimental purposes at £50.
7. The purchase of the transmitting portion of an Anglo-French Wireless Company pack set.
8. [The suggestion] that experiments be carried out in the superposition of wireless telegraphy or telephony on the ordinary army field lines.

Alas, the excellent work of the committee does not appear to have been implemented with enthusiasm. When the Great War commenced in 1914 wireless telegraphy had 'no separate organization and but a half-recognised position' [13].

The apparatus available to the army, for use in the field, shortly after the commencement of hostilities consisted of one wagon and nine lorry mounted, 1.5 kW spark sets, and an unknown number of pack sets [14]. Of these, the

*Figure 16.3   Marconi portable wireless telegraph station carried on special pack saddles.
Four horses were required for a complete station.*
Source: Author's collection.

former were used at General Headquarters and Cavalry Headquarters, and the latter with the Cavalry Brigades. Apart from two-way communications, wireless telegraphy sets could be employed in a listening mode to gather intelligence from the enemy. However, there was a problem. Only

one wireless telegraph lorry, after great difficulty, was sent abroad in charge of one officer . . . and two NCOs [who] had to beg personnel from the Service Wireless Section at Boulogne. And yet . . . this set, on 19th August 1914, working in the garden of the Town Hall at Le Cateau, engaged in intercepting enemy messages and giving valuable information to the Intelligence Department concerning his disposition and movements [13].

This success was followed by the dispatch to France, in December 1914, of the first two Marconi direction finding stations. These comprised Bellini-Tosi antennas and Marconi Type 16 crystal receivers; they were erected for tests at Blendecques, near St Omer, on 16th December 1914. Two weeks later, on New Year's day, one station was moved to Abbeville. From each station bearings on German wireless stations could be obtained and their locations pin-pointed. Subsequently from 1st January 1915 weekly maps of these positions were forwarded to the headquarters of British intelligence. The records provided by the stations included the movements of Zeppelins, the changes of German troop deployments, and later the location of enemy aircraft.

In 1914 the combatant flying service had been organised, from 13th March 1912, as the Royal Flying Corps divided into naval and military wings, but shortly before war was declared on 1st July 1914 the wings were allocated to the two older services and became the Royal Naval Air Service for the navy and the Royal Flying Corps for the army.

After this date the whole of the effective strength of the RFC (63 aeroplanes, 105 officers and 95 motor vehicles) was sent to the Western Front as part of the British Expeditionary Force (BEF). The RNAS, which had 41 aircraft and 52 seaplanes, carried out patrol work on the east coast of Britain and established at Dunkirk an offensive force to hinder the activities of the German airship service which comprised five battalions. Initially, because the army had no aircraft to defend the United Kingdom, the Admiralty, at the request of Lord Kitchener, accepted responsibility from September 1914 for the air defence of the country [15].

Of the RFC machines that went to France with the BEF only one was equipped with a wireless set [16], but by early 1915 the first of Marconi's Wireless Telegraph Company's airborne spark transmitters had begun to arrive in France for service with the RFC. At this time a rather cumbersome signalling lamp was the usual means by which communications could be effected between air and ground.

The Royal Navy had been interested in airborne communications from 1912 and had demonstrated its potential in naval manoeuvres of that year. Using a 30 W spark transmitter and a crystal receiver mounted in a Short S41 hydroplane, good signalling had been achieved over a range of c. 6.2 miles (10 km). By 1913, 26 seaplanes had been fitted with French Rouzet transmitters operating in the waveband 91 m–400 m [17]. This particular apparatus was used because it was lighter in weight, for its antenna power output, than that of other manufacturers. However its exposed spark gap was felt by the RFC to be hazardous when housed in petrol engined aircraft and they preferred the British Sterling spark transmitter with its spark gap accommodated in a compact, gas-tight steel box of size 200 mm by 200 mm by 130 mm.

The advantages of airborne sets when co-operating with land forces soon became apparent during the early stages of the war. These advantages included [18]:

1. correcting artillery fire;
2. putting the guns on to fleeting targets;
3. sending tactical information when an attack was in progress, and indicating changes of infantry positions and enemy movements or concentrations;
4. transmitting information about the general activity in the 'back areas';
5. communicating between aircraft in fighting and bomber formations;
6. informing air patrols of the disposition of enemy aerial activities; and
7. warning anti-aircraft defences of the approach of hostile aircraft.

For some of these duties new equipment based on valves would be required.

The first extensive use of airborne wireless occurred during the battle of Festubert in May 1915 [19]. Pilots observed the battle and reported progress by wireless. In addition they spotted for the artillery while they registered their guns on the enemy's trenches and important strong points. Though the attack did not develop nevertheless the wireless equipment worked successfully and the machines maintained contact with their ground stations throughout the operations. The transmitters employed were those produced by Leslie Miller: they had an output of 100 W, and a mass of 66 lb (c. 30 kg).

By September 1915 the training of personnel, operational tactics, and standardisation of equipment had so improved that Lord French in his despatch on the Loos offensive wrote: 'The RFC is becoming more and more an indispensable factor in combined operations. In co-operation with the artillery in particular, there has been continuous improvement both in the methods and in the technical material employed. The ingenuity and technical skill displayed by the officers of the RFC in effecting this improvement have been most marked' [20]. During the week preceding the offensive, over 600 targets were registered by aircraft and though there were c. 20 machines always operating in the salient to c. 60 ground stations with the guns, no serious failure occurred.

Prior to the Loos offensive, the battles of Neuve Chapelle (March 1915) and Festubert (May 1915) had demonstrated the devastating effect of heavy artillery barrages on line communications. In the opening 35 minute barrage of the battle of Neuve Chapelle more shells were fired than had been fired in the whole of the Boer War (1899–1902) [21]. The consequence was an almost complete disruption of the local line telegraph and telephone networks, and hence an inability of the battlefield commanders to exploit and co-ordinate the initial successes achieved because of a lack of communication with the forward advancing units. One historian has written: 'it was glaringly obvious that the breakdown in communications, the inevitable lack of speedy reaction to the situation at the front, the shattering of the telephone lines between observers and guns, had been almost wholly responsible for the frustrations and delays' [22].

This vulnerability of lines to gun-fire persisted throughout the conflict. Assaults could be launched with a bombardment during which a million shells would be fired; and on one occasion on the Western Front an artillery onslaught caused 350 breaks in a one kilometre length of line [23]. It seemed that the intense effect of gun-fire would negate the practical use of wired communications. Moreover, the smoke and dust produced by a bombardment made signalling by signal lamps and semaphoring by flags an impracticability. Carrier pigeons, whistles, message-carrying rockets, cyclists and even messenger dogs were tried with varying successes (Fig. 16.4).

Apart from the problem of breaks the utilisation of land lines for telegraphy and telephony led to problems of intelligence security. The lines could be tapped by the enemy and even if this was not possible the induced ground currents caused by the closeness of the lines to the earth permitted the signals to be read. In 1915 there were frequent reports that the Germans, whose interception abilities were superior to those of the British, were well aware of what was happening on the British side of the Western Front. Battalions relieving others in the line in secrecy and under cover of darkness would be surprised to be greeted the next morning with shouts of welcome from the enemy's trenches. On one occasion the welcome was given by a German cornettist playing the battalion's regimental march [24]. According to one report the Germans were ahead of the British in the use of earth telegraphy (their Pendelunterbrecher was the earliest of the various apparatuses to be used), but from 1916 the British too were employing the technique.

Initially, the British did not have much success with their experiments, which were carried out at Houghton Regis in Bedfordshire [25]. This place was chosen because the chalk soil was similar to that around Arras and Amiens on the Western Front. Tests at the National Physical Laboratory of the French 'Parleur TM2' system, (developed under the direction of General G.A. Ferrie), were also somewhat unsuccessful because of the nature of the soil at Teddington. Eventually, though, the War Office conceded that it was the better system and ordered 500 buzzers. Half of these had been supplied to the British Expeditionary Force by October 1916.

The BEF 'power buzzer', essentially comprised an induction coil and interrupter powered from a small hand-driven alternator. In operation the output from the coil was applied, via insulated cables, to two electrodes buried in the ground and spaced up to 109 yards (100 m) apart, and the signal received, sometimes as far away as 3,280 yards (3,000 m) away depending on the soil conditions, by a similar electrode arrangement, connected to a three-valve amplifier. Owing to its extreme simplicity, portability and ease of installation, the power buzzer was 'very successful in trench warfare' and, like the BF set, was employed in every attack in 1917. Fig. 16.5 shows the distribution of these sets on the Western Front near Lens in August 1918 [26].

(a)

(b)

Figure 16.4   *(a) Intrepid signals motor cycle despatch riders, during the First World War,*
carried not only messages and official mail through dangerous areas, but
homing pigeons as well. The pigeons were later released, carrying messages, to
return to their home lofts. (b) Method of releasing a carrier pigeon from a
porthole in a tank.
Source: Imperial War Museum (a) MH 4048, (b) Q 9247.

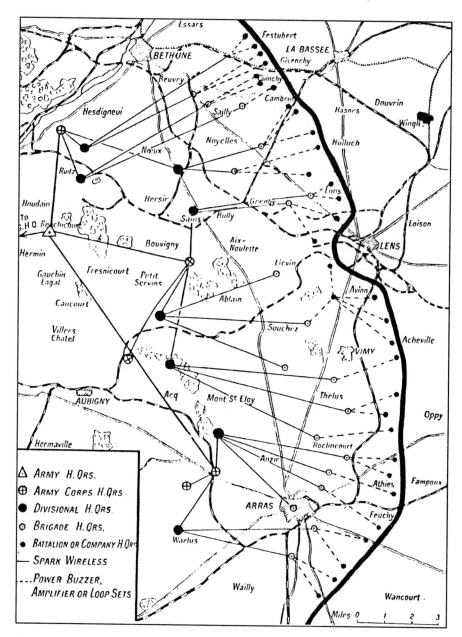

Figure 16.5   *The First Army front just prior to the great offensive of August 1918, showing the wireless telegraph positions.*
Source: *Wireless World*, 1919, **VII**.

From the summer of 1915 the interception of telephone signals had become so serious that the Director General of Signals, General Sir John Fowler, sought the assistance of the War Office. Fortunately, A.C. Fuller, who had been interested in wireless from c. 1910, and who on the outbreak of war had been posted to the 1st Wireless Telephone Company in Aldershot, was able to offer a solution, *vide* British patent no. 2339 of 1916. Fuller's system had the following characteristics [27]:

1. the use of a very weak plain Morse signalling current (interrupted direct current) in the line;
2. the conversion of this current at the receiving station into a 'buzzer' current which could be heard in a telephone; and
3. the utilisation of line screening to prevent the 'buzz' produced at the receiving station from being sent back down the line. (This buzz could be overheard and read by the enemy, but the plain Morse signal could not be read on the telephone, and if it were weak was 'practically untappable'.)

After the war Major Fuller prepared a claim for an award to the Royal Commission on Awards to Inventors. His claim was upheld and he received £4,250.

At a total cost of £255,000, 23,400 Fullerphones (Fig. 16.6), were produced and used from April 1916 by the British Army in the field: some sets were sent to the USA. The advantages of the system were (1) its suitability for signalling in

Figure 16.6   A listening post with a Fullerphone.
Source: Imperial War Museum Q 27,039.

forward areas; (2) the prevention of tapping by enemy listening-in services; (3) the safe use of one wire and earth return; (4) the possibility of simultaneous telephony and telegraphy with the same instrument and line wire; and (5) economy in batteries and consequently in transport.

From early 1916 the principal spark wireless sets of the British army were [28]:

1. the 30 W spark trench set – designed in France during the latter part of 1915 and improved in the Inspection Division, Woolwich, and sent to the Front in large numbers from c. March 1916;
2. the loop set, or Forward Spark 20 W 'B' set operated at wavelengths of either 71 yd (65 m) or 87 yd (80 m) – designed in 1916 especially for unskilled operators and in large scale use from about the autumn of 1917;
3. the early 130 W Wilson set – which was mainly used by Corps Headquarters and also by the RFC for ground station work; and
4. the 1.5 kW set – for use at Army Headquarters.

These sets gave sterling service during the many major engagements of the war. Fig. 16.7 shows the distribution of the wireless telegraph positions for the First Battle of the Somme, September 1916, and the general scheme of spark WT control, in an Army Corps, on which the dispositions given in Fig. 16.7 were based [26]. The company or battalion headquarters signalled the brigade by power buzzer; the brigade communicated with the divisional headquarters by the 50 W spark set; the latter used the 120 W spark set to transmit/receive messages to and from the corps directing station; and the corps directing station utilised the 1.5 kW set for its traffic with the corps headquarters.

The allocation of wavelengths for these various links were [29]:

Spark sets:
71 yd (65 m) and 87 yd (80 m), loop sets; 109 yd (100 m) to 328 yd (300 m), the RAF; 383 yd (350 m), 492 yd (450 m), 601 yd (550 m), trench sets; 656 yd (600 m) to 1094 yd (1,000 m), tanks and cavalry; 1203 yd (1,100 m), GHQ and Army HQ.
Continuous wave sets:
656 yd (600 m) to 2187 yd (2,000 m), artillery, anti-aircraft, scouts and observing parties, tanks, and RAF ground stations.

The difficulties inherent in using line communications highlighted the need to develop further wireless telegraphy and wireless telephony and to evolve a method that would ensure speech security over a telephone line. These systems required reliable valves suitable for operations under arduous military conditions.

Among the valves that gave splendid service, the 'soft' vacuum valves – the Type C receiver valve and the Type T transmitter valve – of H.J. Round, of Marconi's Wireless Telegraph Company, and the 'hard' vacuum valve of the French Telegraphie Militaire must be mentioned. The Round valves were difficult to manufacture as Round has confirmed [30]:

It was probably fortunate in the first year of our work that we used the soft valves because no hard valve had been constructed which can compare with these 'C' type tubes as high

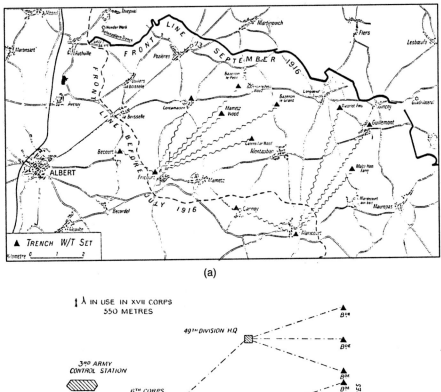

(a)

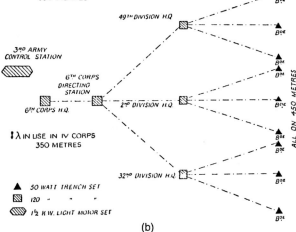

(b)

Figure 16.7 *(a) The wireless telegraph positions during the first battle of the Somme, September 1916. (b) Diagram showing the disposition of 50 W, 120 W, and 1.5 kW sets in an Army Corps.*

Source: *Wireless World*, 1919, **VII**.

frequency magnifiers. These necessitated, however, trained men in their manufacture, and trained operators for their efficient use. . . . Again and again we lost the knack of making good tubes, owing to some slight change in the materials used in their manufacture. A thorough investigation was impossible, as all hands were out on the stations. On several occasions we were down to our last dozen tubes.

Of the French valves employed during the war, the TM (Telegraphie Militaire) tube was immensely successful. Interestingly, the valve came into the hands of the French military in rather an unusual way [31]. A deserter from the French Army named Paul Pichon had settled in Germany, trained as an electrical engineer, and then obtained employment at the Telefunken company. He subsequently became a commercial traveller for the firm and this required him to seek information about the work of Telefunken's foreign competitors, including that of General G. Ferrie, the Technical Director of the French Military Radiotelegraphic Service. On 3rd August 1914 Pichon was in London returning to Germany following a visit to the USA when he experienced feelings of unease about where his loyalties lay. He gave himself up to the French authorities, was promptly arrested and, crucially, was interviewed by Ferrie. Pichon handed over to the General the latest audion valve designed by de Forest and was soon given a position in the Radio-Telegraphique Militaire. After a short time a version of the United States' valve, known as the TM tube, was developed.

War stimulates technological progress. Whereas the total number of audions sold before 1913 was c. 750, in 1913 sales exceeded 500, and in 1914 approximately 6,000 of these valves were sold. Mass production of the TM tube began in October 1915, and by the cessation of hostilities in November 1918 more than one million had been fabricated, mostly by two incandescent lamp manufacturers, Grammont, and Compagnie Generale des Lampes [32].

By 1916, the French TM valve was being fabricated by several British manufacturers, including BTH, Cossor, Edison Swan, GEC-Osram and Metropolitan Vickers, and became known as the R-valve. It was the forerunner of all the early British hard vacuum, low power triodes and continued to be used in new equipment until 1925.

In Germany, too, valves were produced in large numbers by several companies during the conflict. The daily production rate of the Telefunken valve RE16 for example, was c. 1,000.

The principal tube manufacturers in the USA were Western Electric and General Electric, the two companies manufacturing, by agreement, receiving tubes and transmitting tubes respectively. By the end of the war GE had supplied the armed forces with c. 200,000 tubes and Western Electric had provided c. 500,000. Of the latter the 203A (designated the VT-1 by the Signal Corps and the CW-933 by the Navy) was a general purpose tube which could be used as a detector, amplifier, and oscillator [32].

The ready availability to the public, particularly radio amateurs, of all these valves after the war would have a profound effect on the growth of domestic sound and television broadcasting. Table 16.1 lists some valve companies and the dates when they began production: the influence of the 1914–1918 war is readily apparent.

Mention has been made that at the outbreak of the Great War the Royal Flying Corps was a branch of the Army and all provision of wireless apparatus and experiments were the responsibility of the Royal Engineers. However, after

Table 16.1 Early valve companies, 1904–1918

Great Britain	
Edison Swan Electric Company	1904
British Thomson Houston	1916
A.C. Cossor	1916
General Electric Company	1916
Metropolitan Vickers Company	1917
Stearn Electric Lamp Company	c. 1918
Z Electric Lamp Manufacturing Company	c. 1918
France	
Grammont (Fotos)	1915
Compagnie Generale des Lampes (Metal)	1915
Etablissements H. Pilon	c. 1915
Germany	
AEG Telefunken	c. 1912
Siemens and Halske	c. 1912
United States of America	
McCandless	1907
General Electric	c. 1913
Western Electric	c. 1913
De Forest Radio Telephone and Telegraph	1914
Cunningham (Audio Tron)	1915
Moorhead (Electron Relay)	1915
The Netherlands	
NV Philips	1917
Japan	
Tokyo Denki	1917

mobilisation the very small section of the RE engaged in W/T work for the RFC were sent overseas with the result that 'RFC wireless became a dead letter in England except as regards the supply of material' [33]. This condition persisted until early 1915. Fortunately for the RFC, the Marconi company had set up an experimental establishment at Brooklands, Surrey, not far from the Royal Aircraft Factory at Farnborough. When war was declared, the RFC took over the establishment and R.D. Bangay – who before the war had designed a successful airborne transmitter – and the rest of the Marconi staff were seconded to the Corps. Several members, including J.M. Furnival, R. Orme and C.E. Prince (who had been Round's assistant and had been responsible for developing valves for use by the Services and Post Office), would make significant contributions to the RFC. A few months later, in April 1915 Orme and Prince were posted to No. 9 Squadron at Brooklands then under the command of Major H.C.T. Dowding – later Air Chief Marshal Dowding, Commander-in-Chief, Fighter Command. 'Under Major Dowding's energetic guidance, a period of great wireless activity was inaugurated, which was the genesis of all subsequent aircraft wireless work in the RFC at home' [33].

The work at Brooklands was divided into three categories: (1) training wireless officers – 'The Wireless School RFC' was formed in November 1915 and was followed by 'The Wireless and Observers' School in the Spring of 1916; (2) testing wireless materials and methods already available; and (3) experimenting, with the objective of improving the wireless apparatus and the method of use. This work was not without its difficulties. 'most of the apparatus supplied by [the RE] at this time was so crude in design' it attracted criticism; the RE 'appeared to regard Brooklands as purely a testing ground for their ideas, and from the first showed an evident opposition to any attempt of the RFC to originate wireless apparatus for their own use'; and 'no money was available for the purchase of apparatus for purely experimental purposes, and most of the work was carried on with a box of privately owned apparatus brought by Lieut. Prince' [33]. (This state of affairs may be compared to that of the Royal Naval Air Service 'whose purchasing powers were unlimited and experimental work uncurbed'.)

Prince immediately recognised that the most promising line of development 'lay in exploiting the potentialities of the valve both for reception and transmission, rather than in an endeavour to improve the method of spark transmission and crystal reception' [33]. Late in 1915, a satisfactory design of an airborne transmitting set, capable of being employed for both telephony and Morse was realised. The task had not been an easy one. In a claim, to the Royal Commission on Awards to Inventors, that he was the first person to introduce wireless telephony for aircraft, Prince noted: 'Wireless telephony has proved an affair of the greatest importance and magnitude in the world . . . moreover its initiation was not official but was the result of intense personal struggle for its existence and recognition' [33]. He was awarded £650, later increased to £1,000.

A few sets (coded TWA Mk I) were manufactured by a commercial firm and were demonstrated before serving officers, including, in February 1916, Lord Kitchener who was visiting St Omer at that time. Conditions were not conducive to good reception since the pilot had to fly through a snow storm but the Field Marshal was able to hear every word from the aircraft at a distance of 40 miles. He was much impressed.

Nonetheless an 'undefined opposition to experimental work at Brooklands was becoming evident at the War Office' [33] and early in the Summer of 1916 all experimental work at the site, and all wireless work at the Royal Aircraft Establishment, had to cease. During the ensuing reorganisation the War Office in the summer of 1916 transferred its wireless activities to the newly created Signals Experimental Establishment (SEE), Woolwich and in the autumn set up the Wireless Testing Park, Joyce Green; (later from February 1917 at Biggin Hill). Following the formation of the Royal Air Force, from the amalgamation of the RFC and RNAS, on 1st April 1918, the RFC and RNAS staffs from SEE and Cranwell respectively were combined at the Biggin Hill site and the establishment re-named 'The Wireless Establishment, RAF'. A Wireless

Telephony School was formed there (later at Chattis Hill) under Captain J.M. Furnival, one of the Marconi engineers who had moved to Brooklands after the declaration of war.

With the growth of airborne wireless so more and more aircraft could assist the Royal Artillery. At the battle of Loos (September 1915) the number of ground stations with the guns was 60; but during the battle of the Somme (July to September 1916) the number was 600. At the end of 1916 the total wireless personnel comprised 200 officers and 2,000 operators – more than the total personnel of the RFC in August 1914 [34].

As an indication of the effectiveness of airborne wireless, 30 aircraft operating simultaneously with 38 heavy artillery groups, 55 divisional artillery units and 187 heavy and siege battery stations along the 7 mile (11.2 km) front during the battle of Vimy Ridge, achieved the following successes: '256 hostile batteries destroyed; 86 gun pits destroyed and 240 hit; 103 explosions caused; 229 destructive trench shoots; 117 successful registration shoots; 2,843 fleeting target calls received at battery station[s]; [and] 406 artillery observation flights' [16].

By the end of 1917 the equipment developed specifically for the RFC and in use in the field comprised: 'short and medium range air-to-ground spark transmitter[s]; aircraft receivers for Morse and telephony; aircraft and ground telephony and c.w. transmitters; ground spark tonic train and c.w. transmitters and c.w. and spark receivers'. (See Fig. 16.8.) The total strength of the wireless personnel was now 300 officers and 3,760 operators and mechanics.

The corresponding figures at the time of the signing of the Armistice were 520 and 6,200. Between 400 and 500 aircraft had been fitted with wireless and the artillery had been provided with c. 2,000 stations. For the month of September 1918, when the last great offensive was launched, there were '2,100 shoots to destruction; 2,575 flash patrols; 74,201 fleeting target calls from the patrol machines; and 11 wireless failures' [16].

The worth of wireless in aircraft, in time of war, may be assessed from a document, dated 1st June 1918, captured from the enemy [34].

XI Army Corps Staff
South of the Somme
Ic Wing Commander No. 12358/1466   Corps H.Q
Relating to Salvage of Aeroplanes   1–6–18

1. The salvage of aeroplanes must be pursued with still greater vigour. Valuable technical improvements, which have apparently been extensively applied by the enemy, have been lost to us hitherto. Labour and money spent at home on trials might have been saved by earlier information.
2. The enemy has secured a distinct advantage in his successful use in aeroplanes of continuous wave wireless apparatus, which possess great superiority over the spark apparatus.

It is of the greatest importance to us to salvage further enemy wireless apparatus of this description. In this way millions of money will be saved, as we have not so far been

(a)

(b)

Figure 16.8   (a) Combination set (French), 1918. (b) Type E 10-bis set (French), 1918.
Source: Imperial War Museum (a) Q 69126, (b) Q 69128.

successful in constructing a continuous-wave wireless apparatus for aeroplanes which can work without certain disadvantages.

It is therefore the duty of all authorities to ensure the most indefatigable and careful salvage of wireless apparatus of enemy aeroplanes. Every particle is to be salved and collected if it is in any way possible.

Mention has been made of the British army's use of direction finding. Hertz in 1886 had demonstrated that a vertical plane wire loop, and a vertical linear antenna mounted along the focal line of a parabolic cylindrical antenna, had non-circular polar diagrams in a horizontal plane, that is the antennas exhibited directional properties. Later, S.G. Brown (1899), A. Blondel (1903), F. Braun (1904–1906) and Marconi (1905) investigated various antenna arrangements for achieving directionality [35]. Marconi's discovery, in 1905, that a long horizontal antenna could radiate electromagnetic waves in a given direction was followed a year later by a direction-finding system which comprised a considerable number of conductors disposed along the radii of a circle centred on the transmitter. The receiving end of each conductor was connected via a rotary switch to a magnetic detector and earth to enable the direction of maximum signal strength to be determined. This antenna layout was later employed by the Royal Navy.

Further work on loop, or frame, antennas was undertaken by H.J. Round in 1905–1906. He found that the ambiguity in the direction of a source, (θ or [θ + 180°]), where θ is an observed bearing, could be eliminated by the use of an omni-directional antenna working in collaboration with the loop antenna.

Of the several methods of determining the bearing of a signal, that of E. Bellini and A. Tosi was of considerable importance. Their patents were purchased by the Marconi company in 1912, and Dr Bellini joined the technical staff as a consultant. Much rapid development work followed. In their system Bellini and Tosi utilised two static, orthogonal, vertical, triangular shaped antennas, the outputs of which were connected separately to two fixed, orthogonal, field coils. Reception of a signal by the antennas caused currents to flow through the coils to establish a magnetic field. This field was detected by a calibrated, rotating search coil, the output of which was connected to a receiver [36].

The success of the Marconi company's direction finder in December 1914 led to its adoption by the Royal Navy from early 1915. Five direction finding stations from the Shetlands to Kent were set-up to cover the North Sea to enable, by cross bearings, the position of any ship which used its radio to be obtained. Subsequently, by 1917 the Allied d/f network also covered the Mediterranean Sea. There were 21 Italian, 11 French, and 11 British stations in total [37] (Figure 16.9).

An account of the strategic importance of direction finding has been written by Admiral Sir Henry Jackson, who in 1916 was First Sea Lord at the Admiralty, and therefore was responsible for the disposition of the Royal Navy's Grand Fleet.

(a)

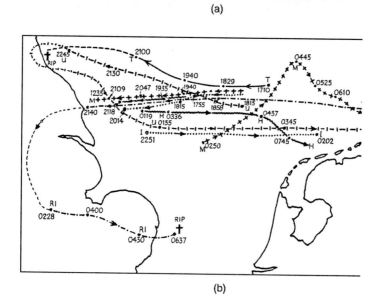

(b)

Figure 16.9   (a) A portable Marconi direction finder station, 1919. (b) German Zeppelin raids plotted by Marconi D.F. stations, 27–28 November, 1916.
Source: The Marconi Company.

Our wireless direction finding stations, under Captain Round, kept careful and very intelligent watch on the positions of German ships using wireless, and on the 30th May 1916 heard an unusual amount of wireless signals from one of the enemy ships which they located at Wilhelmshaven. This was reported to me; the time was a critical and anxious one in the war, and I had also some reasons for expecting that the German fleet might put out to sea during the week. Our fleet was ready at short notice, and had arranged, unless otherwise prevented, to put to sea on the following day for a sweep of the North Sea. But, if the German fleet got to sea first, the chance of a meeting in waters not unfavourable to us was remote; our object was to try to get to sea before or shortly after the Germans, and hitherto we had not succeeded in doing so. Later on in the afternoon it was reported to me that the German ship conducting the wireless had changed her position a few miles to the northward. Evidently she and her consorts had left the basins at Wilhelmshaven and had taken up a position in the Jade River, ready to put to sea. This movement decided me to send our Grand Fleet to sea and move towards the German Bight at once and try to meet the German fleet and bring it to action. This they did with their usual promptitude, and the result was the famous battle of Jutland [38].

The practice of wireless communications – particularly the utilisation of valves – was, as has been seen, greatly advanced by the 1914–1918 war. At the end of the conflict components and valves could be purchased easily by radio amateurs. This encouraged a major electrical engineering manufacturing company to cater for their needs. Sound broadcasting was born.

References

1 DOWSETT, H.M.: 'Wireless telephony and broadcasting' (Gresham Publishing Co., London, 1924), p. 5

2 HARTCUP, G.: 'The war of invention' (Brassey's Defence Publishers, London, 1988), p. 16

3 CUSINS, A.G.T.: 'Development of army wireless during the war', *J. IEE*, **59**, A, p. 763

4 AUSTIN, B.A.: 'Wireless in the Boer war', International Conference on 100 years of radio, Conference Publication No. 411, pp. 44–50

5 Ref. 1, p. 9

6 Ref. 4, p. 46

7 Ref. 4, p. 47

8 'Final report of the Norman Committee', WO 32/8879, PRO

9 Ibid, p. 8

10 Ibid, p. 14

11 Ibid, p. 22

12 Ibid, p. 23

13 Ref. 3, p. 763

14 Ibid, p. 764

15 BURNS, R.W.: 'Air defence, some problems', in BURNS, Russell W. (Ed.): 'Radar development to 1945' (Peter Peregrinus Ltd, London, 1988), pp. 106–131

16 SMITH, T.V.: 'Wireless in the RAF during the war', *Aeronautics*, 1919, pp. 348–350

17 BEAUCHAMP, K.G.: 'History of telegraphy' (Peter Peregrinus Ltd, London, 2001), p. 351

18 'Memorandum on the use of wireless in the RAF', AIR 1/2217/209/33/6, PRO, Kew

19 Ref. 16, p. 348

20 Ibid, p. 349

21 GILBERT, M.: 'The First World War: a complete history' (Henry Holt and Co., New York, 1994), p. 132

22 MACDONALD, L.: '1915, the death of innocence' (Henry Holt and Co., New York, 1995), p. 140

23 KENNELLY, A.E.: 'Advances in signalling contributed during the war' in YERKES, Robert M. (Ed.): 'The new world of science: its development during the war' (Century Company, New York, 1920), pp. 221–246

24 Ref. 2, p. 77

25 Ibid, p. 43

26 SCHONLAND, B.F.J.: 'W/T, R.E.', *The Wireless World,* 1919, **VII**, pp. 174–178, 261–267, 394–397, 452–455

27 Claim of Major A. C. Fuller, Royal Commission on Awards to Inventors, T173/224, PRO, Kew

28 Ref. 3, pp. 765–766

29 Ref. 26, p. 397

30 THROWER, K.R.: 'History of the British radio valve to 1940' (MMA International Ltd, England, 1992), p. 33

31 TYNE, G.F.J.: 'Saga of the vacuum tube' (H. W. Sams, Indianapolis, 1977), chapter 10

32 NEBEKER, F.: 'The role of the First World War in the rise of the electronics industry', ESEJ, 2001 **10**(5), pp. 189–196

33 ORME, R., and PRINCE, C.E.: 'History of R.F.C. wireless from the outbreak of war', AIR 1/733/183/1, PRO, Kew

34 Ref. 16, p. 350

35 KEEN, R.: 'Wireless direction finding' (Iliffe, London, 1922), pp. 759–787

36 Ref. 35, chapter 5, 'The Bellini-Tosi system', pp. 138–183

37 HEZLET, A.: 'The electron and sea power' (Peter Davies, London, 1975), pp, 98–107

38 JACKSON, H.B.: Discussion on a paper presented by Captain H.J. Round, *J. IEE*, 1920, **58**(289), pp. 247–248

Chapter 17

The birth of sound broadcasting

By the middle of 1913, following the purchase by Western Electric of the patent rights of the de Forest audion, engineers of the company had succeeded in fabricating a filament for the valve which had a laboratory life of c. 1,000 hours – a very considerable increase on the average life of c. 50 hours of the de Forest tube [1].

Further developments [2] by Western Electric's engineers – including the utilisation of Wehnelt's oxide-coated cathode, and the redesign of the electrode structure – led to the application of AT&T's tubes as repeaters in the transcontinental line between New York and San Francisco which was opened in January 1915.

The position in England, as noted previously, regarding de Forest's audion patent was rather different to that which prevailed in the US because, although the inventor had taken out a British patent in 1908, he had allowed it to lapse on account of nonpayment of the renewal fees due in January 1911. Hence, the audion was freely available in England and the Fleming patent of 16th November 1904 became a master patent [3].

The Fleming patent was held by the Marconi company and so further advancement in valves took place largely in that company's laboratories. The two principal workers were H.J. Round and C.S. Franklin. They had access to the German Lieben-Reisz tube [4] and, like Arnold and his colleagues at AT&T, concentrated their efforts on the manufacture and application of hard valves, with the aid of Langmuir's vacuum techniques. Because of their company's interests, Round and Franklin directed their attention towards the solution of practical problems.

There was another property of triode valve circuits which undoubtedly hastened the birth of domestic sound and television broadcasting. In 1912 de Forest discovered that the triode valve could be employed in an oscillator to generate electromagnetic waves in addition to acting as a detector and as an amplifier. This was to be a finding of immense significance in the history of broadcasting. The Marconi spark apparatus, the Poulsen arc, and the

Alexanderson alternator, were all expensive and cumbersome, but with de Forest's valve generator and the stimulus provided by the Great War, the progress of continuous wave radio communications rapidly moved forward to the stage where commercial broadcasting could be seriously contemplated shortly after the cessation of hostilities in 1918.

The development of large power valve transmitters was first shown to be within the realms of practical accomplishment when AT&T on 27th August 1915 used a transmitter (employing a bank of up to 500 valves connected in parallel, operating at 50 kHz, and producing a power into the antenna of c. 2 to 3 kW) to send speech signals and phonograph music from the US Navy's radio station at Arlington, Virginia, to Darien in the Panama Canal zone, 2,100 miles (3380 km) distant. About two months later the same transmitting station communicated with the Eiffel tower station in Paris [5].

Later, valve transmitters quickly ousted the spark, the arc and the high frequency alternator types of transmitters and by 1921 the Marconi company had a valve transmitter with a rated output of 100 kW installed at Caernarvon, North Wales [6].

The First World War undoubtedly gave an enormous impetus to the utilisation of valves in signalling systems and so stimulated advances in technique that by 1918 triodes could be manufactured to cover a wide power range and were suitable for both transmitting and receiving purposes; their theory and operation, over the frequency bands used at that time, were both thoroughly understood.

The first broadcasters were the radio amateurs or 'hams', whose hobby it was to communicate with other radio hams. Their activities were curtailed in the United Kingdom during the 1914–18 period so that no interference with essential communications would occur, but in the United States their hobby was given some encouragement by the work of Dr F. Conrad, a well respected engineer of the Westinghouse Electric and Manufacturing Company (WEM), Pittsburg.

During the First World War, WEM was requested by the British Government to engage in work on certain problems associated with wireless telegraph and wireless telephone transmissions. WEM agreed, and in 1916 experimental transmitting and receiving stations were established near the Pittsburg plant and at the home of Dr Conrad, a distance of c. 4.5 miles (7.2 km). Transmission tests were carried out and development work undertaken.

Whether these transmissions were familiar to David Sarnoff, who at that time was the assistant traffic manager of the American Marconi company is not known, but in 1916 he considered the desirability of a public broadcasting service being set up by the Marconi company. In a memorandum to E.J. Nally, the Vice-President and General Manager, Sarnoff outlined his suggestion [7]:

I have in mind a plan of development which would make radio a household utility in the same sense as a piano or phonograph. The idea is to bring music into the home by wireless. ... Should the plan materialise, it would seem reasonable to expect sales of

1,000,000 'radio music boxes' within a period of three years. Roughly estimating the selling price at $75 per set, $75,000,000 can be expected.

The American Marconi company did not heed this prophetic forecast.

After the war ended Conrad re-commissioned his amateur wireless telegraph station and converted it to transmit and receive wireless telephony signals. He found that his broadcasts were being received by a number of radio amateurs. Soon, a community of enthusiasts were in communication with the Westinghouse engineer. Tiring of the endless conversations with members of the group Conrad decided to broadcast phonograph records as a means of entertaining them. For this purpose a constant source of new recordings was required, but as the cost of these was beyond Conrad's means he negotiated with a local phonograph dealer to borrow a steady supply. However, the wily dealer stipulated, not unreasonably, that his store's name should be mentioned when the records were played. To his delight he found that customers were buying many more of the recordings played by Conrad than the others on sale [8].

Interest in the broadcasts grew rapidly, particularly when Conrad's sons – Francis and Crawford – began to assist their father. They introduced local talent and soon there was a demand for radio sets in the community served by Conrad's transmitter. A local department store, the Joseph Horne Company, placed an advertisement in the *Pittsburgh Sun*, on 29th September 1920, and announced their ability to sell crystal sets at '$10.00 up' to radio amateurs. The sets sold quickly and more had to be ordered. The broadcasts flourished.

All of this did not escape the notice of Conrad's employers. As H.P. Davis, the Vice-President of the Westinghouse Electric and Manufacturing Company has recalled [9]:

the programs sent out by Dr Conrad caused the thought to come to me that the efforts that were then being made to develop radio telephony as a confidential means of communications were wrong, and that instead its field was really one of a wide publicity, in fact, the only means of instantaneous collective communication ever devised. Right in our grasp, therefore, we had that service which we had been thinking about and endeavouring to formulate.

Here was an idea of limitless opportunity if it could be 'put across'. . . .

Resulting from this was my decision to install a broadcasting station at East Pittsburg and to initiate this service. This decision, made in 1920, created the present huge radio industry. Not until fall, however, was the equipment ready for operation.

On the 30th September Davis called together his 'radio cabinet', consisting of Dr Conrad, S.M. Kintner, L.W. Chubb and O.S. Schairer, and enquired whether Westinghouse could assemble the station by 2nd November to report the Harding–Cox presidential election. The 'cabinet' agreed that this was possible: station KDKA was licensed by the Department of Commerce on 27th October 1920. It was a very modest affair and comprised a 'tiny penthouse' erected on the roof of the tallest building in the Westinghouse plant. 'It boasted at least two windows, and was flanked by table-like desks, set against the walls on two sides. The engineer sat before the transmitter and the announcer was stationed close by. . . . The hastily constructed station had two 50 W oscillators and four 50 W modulators' [10].

Fortunately, all went well on the occasion of the election and between 500 to 1,000 listeners, equipped with headphones, were enabled to hear the election returns.

For Westinghouse, the enthusiasm for radio broadcasts provided an outlet for the activities of the company's radio section, which had been developed during the war, and allowed the firm to make use of its manufacturing capacity in radio apparatus. Additionally, a source of revenue for the service came from local advertisers who were prepared to pay for transmission time.

The company's initiative touched off an immediate boom in radio broadcasting and in the formation of transmitting stations. Eleven stations were transmitting in September 1921, 23 in October and 32 in December. After the turn of the year the numbers increased considerably: 58, 72, 99 and 187 at the ends of the first four months respectively of 1922 [11]. By 1st May there were 219 registered radio stations in the USA broadcasting a mixture of news, commentaries, weather and stock market reports, lectures and music.

Such was the growth in demand for receivers that manufacturers had difficulty satisfying the public need. One commentator, writing in May 1922 noted:

The rate of increase in the number of people who spend at least a part of their evening in listening in is almost incomprehensible. To those who have recently tried to purchase receiving equipment, some idea of this increase has undoubtedly occurred as they stood perhaps in the fourth or fifth row at the radio counter waiting their turn, only to be told, when they finally reached the counter, that they might place an order and it would be filled when possible. . . . It seems quite likely before the movement has reached its height . . . there will be at least five million receiving sets in this country.

A further 99 stations went on the air in that month (May), and by the end of 1924 there were 530 US stations.

Initially there were no restrictions, no licences and the outcome was chaos. The Wireless Act of 1912 was the only Act on the statute book, but, obviously, it did not make provision for the regulation of a domestic broadcasting industry which did not at that time exist.

The multiplicity of transmitters and the few wavelengths available led to interference, 'a jumble of signals', and 'a blasting and blanketing of rival programmes'. The experience gained during the formative period in sound broadcasting clearly showed the need for a framework of regulations. This view was voiced at the first National Radio Conference, held in February and March 1922: 'the general opinion was that radio communication [was] a public utility and as such should be regulated and controlled by the Federal Government in the public interest' [12]. A bill was debated in Congress in 1922 but was not enacted. In a move to resolve the interference problem Mr H. Hoover, the Secretary of Commerce, decided to act and, exceeding his statutory powers, reallocated the transmitting frequencies of most radio stations in the country. However, in 1926 his authority was challenged in the courts and his action was overturned. Chaos returned. The radio industry, desperate for a solution appealed to the government and as a consequence an Act of Congress was passed which established the Federal Radio Commission.

During this period the Radio Corporation of America (RCA) was formed. Since it played a major role in the progress of both sound and television broadcasting a few words on its formation are appropriate. It was inaugurated on 17th October 1919 to combat the growing influence of the Marconi Wireless Telegraph Company in international communications. After the First World War, the United States had become increasingly aware of the importance of communications for military, commercial and public uses, but at that time the only non-military source of such services available to the country was an organisation owned and controlled by foreign (British) interests, namely, the MWT Company of America. The possible complete dependence upon such a business was a matter of much concern, particularly to the US Navy Department [13], as it was probably the main user and advocate of long distance wireless communications. This apprehension was compounded when the General Electric Company (US), which held some fundamental and extremely vital patents covering the Alexanderson alternator, made it known that it was negotiating exclusive licensing rights to the alternator with the British company [14]. The company wished to buy 24 Alexanderson alternators, 14 for the American Marconi Company and the remainder for the British firm.

When the US Government learned in March 1919 of these negotiations and the consequential prospect of an even tighter control on international wireless communication by the Marconi Company, Admiral Bullard, the Director of Naval Communications, informed General Electric's O.D. Young of the Government's desire for an American dominated organisation. The Government proposed that such a body be set up to provide the essential international communication facilities needed by the United States, and that it should be given the necessary GE licences to use the Alexanderson alternator. GE readily agreed and on 17th October 1919 the Radio Corporation of America was incorporated to operate the appropriate stations and to market equipment, while GE concentrated on manufacturing. A few days later, on 20th November 1919, RCA acquired a controlling interest in the American Marconi Company and the Government's objective was fulfilled [15].

The immediate task facing the new corporation was the establishment of a wireless point-to-point communications service. This service was to be based on the stations that the Government had taken over from the Marconi company during the war – from April 1917 – and which, in February 1920, had been handed over to RCA for commercial use. But, additionally, the expansion of the service demanded the construction of new stations. There were difficulties, however. Many of the most important patents in this field were held by GE, Westinghouse and the AT&T Company. Both GE and AT&T were devoting large resources to vacuum tube and related technology and this led to no less than 20 important patent interference actions from 1912 to 1926. These interferences included the following [16]:

1. Modulating h.f. currents by means of tubes; Nichols and de Forest (AT&T), and White and Alexanderson (GE);
2. Use of a feedback circuit for producing oscillations; Langmuir (GE), and de Forest (AT&T);
3. Structural features of tubes; Langmuir (GE), and Nicholson (AT&T);
4. Suppression of carrier wave; Alexanderson and White (GE), and Arnold and Carson (AT&T);
5. Use of tubes for modulating high frequencies; Alexanderson (GE), and Colpitts (AT&T);
6. Use of tubes for current limiting purposes; Langmuir (GE), and Espenschied (AT&T);
7. Plate modulation of an output tube oscillator; White (GE), and Hartley (AT&T);
8. Suppression of carrier frequency and transmission of one or both sidebands; Alexanderson (GE), and Englund (AT&T);
9. High vacuum; Langmuir (GE), and Arnold (AT&T);
10. Use of alternating current for lighting tube filaments; White (GE), and Arnold (AT&T).

The acquisition of patents by RCA was not easy for no person or firm controlled even a substantial percentage of them. As an illustration of the patent situation at that time, mention can be made of the conclusion of a patent investigation carried out in 1919 by the US Navy. It 'found that there was not a single company among those making radio sets for the Navy which possessed basic patents sufficient to enable them to supply, without infringement, a complete transmitter or receiver' [17].

Thus, the Corporation had perforce to acquire the ownership, or institute extensive cross licensing agreements with all the above mentioned companies, in order to make its commercial operations a success. RCA and GE arrived at congenial arrangements in 1919, and over the next two years similar contracts were drafted between RCA and Westinghouse, AT&T, United Fruit and the Wireless Speciality Company. Internationally, RCA entered into agreements in 1919 with the British Marconi's Wireless Telegraph Company, the French Compagnie Generale de Télégraphie sans Fil, the German Telefunken Company, the Drahtloser Ubersee-verkeht AG, and the Reichspost-ministerium, the Government of Norway, the Kingdom of Sweden, the Imperial Japanese Government, the Republic of Poland and a South American Consortium of British, French and German companies [18].

These agreements, which were to last until 1st January 1945, made provision for each company to have the exclusive right to the use of the other company's patents within the company's prescribed territories, as well as for 'mutual traffic arrangements wherever possible throughout the world'.

By June 1921 RCA had the rights to more than 2,000 patents in the radio field. Its position (in 1923) was described by the US Federal Trade Commission: 'the Radio Corporation of America has acquired all the high-power stations in this country with the exception of those owned by the government, and it has practically no competition in the radio communication field' [1].

As a result of these settlements RCA could operate, though not exclusively, point-to-point radio communications, and market receivers, and GE and Westinghouse had the exclusive right to manufacture these receivers – 60 per cent for GE and 40 per cent for Westinghouse: AT&T retained the exclusive right to manufacture, lease and sell transmitters. The ownership of RCA at this time was as follows: GE, 30 per cent; Westinghouse, 20 per cent; AT&T, 10 per cent; United Fruit Company, 4 per cent; others, 36 per cent [19].

The formation of RCA occurred at a most propitious time – the birth of domestic radio broadcasting. With sales booming the Corporation could afford to establish a research and development section. Indeed, such a section was vital for the well-being of the Corporation. Its products had to be tested and evaluated and new ones had to be developed. Much work had to be carried out as radio broadcasting was in its infancy and television was just becoming a reality. A Technical and Test Department was therefore set up in 1924 with Dr A.N. Goldsmith as its chief engineer. It was located at Van Cortland Park and was based on a staff of about 70 engineers, technicians, administrative and service personnel.

Towards the end of the decade problems arose between RCA and its manufacturing associates, GE and Westinghouse, concerning product standardisation, production scheduling and control, and competitive pressures on profitability. RCA had no effective management control over the costs of the products of its two autonomous manufacturers and the additional mark-up that it imposed for the distribution process created 'significantly increasing profit problems on merchandising only' [20].

Consequently, in February 1929, RCA acquired for $154 million the assets – including a large manufacturing facility in Camden, NJ – of the Victor Talking Machine Company, and by the autumn of the same year agreement [21] in principle had been reached among GE, Westinghouse and RCA that it would become a 'unified, highly self-sufficient organisation'. As part of this undertaking GE and Westinghouse transferred some of their radio facilities and staff to RCA. The official transfer date was 1st April 1930.

Of the research groups that were established at the Camden site, one group under the direction of a former GE engineer, Dr E.W. Engstrom, was called the general research group, and the other, directed by A.F. Murray, was named the research division. Murray's division comprised sections that were concerned with radio receivers, acoustics and television, under the supervision of Dr G. Beers, Dr I. Wolff and Dr V.K. Zworykin respectively. The two groups totalled initially 45 technical specialists.

Returning now to the general issue of broadcasting in the United States of America, there were, in the first half of 1925, 563 broadcasting stations in operation or under construction. Most of them were owned and run by manufacturers of, or dealers in, radio apparatus, department stores, newspapers and radio societies. Except in the case of the societies these organisations provided their services free of charge in order to advertise their businesses. The stations were licensed under a statute passed by the Federal Government in 1912, and administered through the Bureau of Navigation of the Department of Commerce. No fees were payable to the State for licenses and they could be obtained without difficulty.

At this time, 1925, the broadcasting companies in the US were divided into two classes, A and B. Class A stations, of which there were 455, were empowered to broadcast with a maximum power output of 500 W on wavelengths equal to or less than 304 yd (278 m). The 108 Class B stations were authorised to use power outputs of more than 500 W and wavelengths between 306 yd (280 m) and 596 yd (545 m) but had to conform to certain technical and service conditions. Under some circumstances a power rating of up to 5 kW was permitted. Various breaches of the conditions under which a license was issued such as: transmitting a false or fraudulent distress signal; causing wilful or malicious interference; violating any appropriate regulations; employing unlicensed operators; and, of course, utilising transmitting equipment without a license could lead to a fine of up to $500 and the confiscation of the equipment.

Although no licence fee was necessary for radio reception the providers of the programmes required some financial assistance to support their work. Various possibilities existed: endowment, municipal financing, public subscription, sponsoring and advertising [22].

Sponsored programmes were first broadcast in 1922 from station WEAF. They developed rapidly but soon the insidious nature of advertising on radio programmes began to be commented upon.

Anyone who doubts the reality, the imminence, of the problem has only to listen about him for plenty of evidence. Driblets of advertising, most of it indirect so far, to be sure, but still unmistakable, are floating through the ether every day. Concerts are seasoned here and there with a dash of advertising paprika. You can't miss it; every little classic number has a slogan all its own, if it is only the mere mention of the name – and the shrill address, and the phone number – of the music house which arranged the program. More of this sort of thing may be expected. And once the avalanche gets a good start, nothing short of an Act of Congress or a repetition of Noah's excitement will suffice to stop it [23].

And in the same year Hoover opined: 'It is inconceivable that we should allow so great a possibility for service . . . to be drowned in advertising matter'. Nevertheless, advertising persisted.

In the United Kingdom radio broadcasting progress was more cautious and orderly than that in the USA. The early radiotelephony accomplishments of America had given it a lead in the world's markets for this type of equipment, but during the First World War the Marconi's Wireless Telegraph Company had undertaken much work – particularly by H.J. Round – on the development of power valves. This effort had led to the design and construction of 0.25 kW, 1.5 kW, and 2.5 kW transmitter stations. One of the latter, at Ballybunion, Ireland, in March 1919, under the direction of Round, became the first European radio telephony station to be heard across the Atlantic Ocean. Interestingly, whereas the US Arlington transmitter had utilised in excess of 300 power valves, the Ballybunion station employed just two MT1 valves in the oscillator and one MT1 in the modulator [24].

A few months later, in January 1920, the company secured a licence for, and constructed, a 6 kW station in Chelmsford (Fig. 17.1). The station was intended to broadcast on a world-wide basis and to show potential customers that in technical developments the British company was not lagging behind its American counterparts and that only Government restrictions prevented a more widespread usage of broadcasting in Britain. In January 1920 strong signals of good quality speech were received in Madrid.

The station was followed by another of 15 kW on the same site and, from 23rd February to 6th March 1920, two daily half-hour programmes of news and music – using Company employees with musical talent – were broadcast; the wavelength (of 3062 yd (2,800 m)) was that allotted to Marconi's Poldhu station for telegraphically transmitting news to ships. These regular broadcasts preceded those of the KDKA Westinghouse station and provided the first wireless broadcasting service in the world. Many reports of reception aboard ships showed that the range of the service was over 1,000 miles (1609 km), the greatest range being 1,450 miles [25] (2333 km).

Curiously, the reports did not stimulate the imagination of the company's board. Its official policy was that the future of wireless telephony lay in

Figure 17.1   The 6.5 kW experimental transmitter at Chelmsford, 1919.
Source: The Marconi Company.

commercial speech transmission rather than in domestic sound broadcasting, and to further this policy the company inaugurated, on 23rd February 1920 from the Chelmsford station, a wireless telephony news service. It was the first of its kind in the world.

But competition inspires thought and when the Dutch experimental station PCGG, using Marconi equipment, commenced, on 29th April 1920, to broadcast concerts [26] from The Hague a change of outlook was provoked. The well-received broadcasts stimulated much favourable comment from amateurs and newspaper reporters and resulted in a change of attitude of the sceptical board.

This change was hastened by the initiative of the *Daily Mail*, which had been enthusiastic about the Dutch transmissions. The newspaper agreed to sponsor a special broadcast concert by Dame Nellie Melba, the famous Australian prima donna. On 15th June 1920, she travelled to Chelmsford for what she described as 'the most wonderful experience of my career' and sang into the microphone of the 15 kW transmitter. The broadcast was heard at places as far away as St John's, Newfoundland, a distance of 2,673 miles (4301 km). Among her songs were 'Home, Sweet Home', 'Nymphes et Sylvains', 'Addio' from *La Boheme* and 'Chant Venitien'. Her recital had been preceded by a tour of the transmitting station and antenna structure, during which the engineer-in-charge had explained that her voice would be radiated far and wide by the wires at the top of the mast. 'Young man', said Dame Nellie, 'if you think I am going to climb up there you are greatly mistaken' [25].

This broadcast, more than any other, showed what could be achieved by the new medium. The extensive coverage over which good reception could be obtained by relatively simple means, and the widespread interest of the populace, heralded a new form of low cost entertainment for the masses.

Further demonstrations were given when the SS *Victorian* sailed from England to Canada with the United Kingdom delegates to the Imperial Press Conference (which was held in Ottawa on 5th August 1920). The *Victorian's* 3 kW radio transmitter enabled conversations and the exchange of news to be carried on between the delegates and some press representatives at Poldhu, where a 6 kW transmitter had been provided. And during the voyage, the SS *Olympic*, which carried a 1.5 kW set, succeeded in communicating speech directly with the *Victorian* over a distance of 570 miles (917 km). Reports indicated that 'good' speech could be heard up to 1,200 miles (1930 km) distant, and less satisfactory speech at 1,600 miles [27] (2574 km)

Soon, the Chelmsford broadcasts were stopped. By the Wireless Telegraphy Act of 1904 all transmitting stations and receivers of wireless signals were required to possess a licence, the terms and conditions of which were determined by the General Post Office. The licence held by Marconi's Wireless Telegraph Company allowed it 'to conduct experimental telephony transmission', provided that before each occasion a major experiment was carried out special permission was obtained. As part of its brief, the GPO had to ensure that chaos did not result from indiscriminate broadcasts; and it had to take note of the views of the Armed Services as represented by the Wireless Telegraph Board (the most important Services committee concerned with wireless) and the Wireless Sub-committee of the Imperial Communications Committee, which included both civilian and Service members.

To some observers MWTC appeared to be exceeding its licence conditions. One report denounced the Melba broadcast because it represented a 'frivolous' use of a 'national service'. *The Financier*, in August 1920, reported: 'opinion among airmen is practically united against a continuance of the "concerts" given to the world at large by the Chelmsford wireless station. A few days ago the pilot of a Vickers Vimy machine . . . was crossing the Channel in a thick fog and was trying to obtain weather and landing reports from Lympne [airport]. All he could hear was a musical evening'. With such reports there could be only one course of action for the GPO to take: the Marconi company's licence was withdrawn on the grounds that it 'was found that the experiments caused considerable interference with other stations' (Postmaster General in the House of Commons, 23rd November 1920).

However, the time was now ripe for sound broadcasting – the techniques were available and public demand was growing – and if conditions in the UK did not favour a rapid development programme the same state of affairs did not exist elsewhere in several European countries. By the end of 1920 sound broadcasts were being transmitted not only by the Nederlandsche Radio-Industrie, The Hague but also from Paris, and other western European stations, and these could be received in the United Kingdom. The British Thomson-Houston Company, for example, had exhibited a receiver in 1920 that could receive both The Hague and the Paris broadcasts – though the price of c. £30 (about six weeks wages for a skilled male) was probably a deterrent to good sales [28].

Radio amateurs, in particular, were vociferous in their demands for a regular service so that they could proceed with their experimental work. Manufacturers of valves and telephony equipment were also keen for broadcasts to occur: many of them had established costly development and production facilities for such devices and apparatuses during the war and wished to have a continuing market for their goods. It is significant that, of the six guaranteeing firms of the future British Broadcasting Company, three were valve makers (GEC, BTH and Metropolitan Vickers).

Arising from the unrest, negotiations were held in 1921 between the General Post Office and the Wireless Society. The Post Office stated its willingness to licence the Society, but not the Marconi company, for transmissions. It seems the Postmaster General was concerned that if he permitted Marconi's Wireless Telegraph Company to transmit programmes other manufacturers would request similar facilities.

Much unease was expressed and eventually the representatives of 63 wireless societies, on 29th December 1921, petitioned the GPO. They voiced a national resentment that public services such as wireless time and telephony should be left to our neighbours to provide, and that permission to transmit weather reports, news and music by wireless telephony should not be refused to companies competent and willing to do so without interference with the defensive services of the country. The GPO now relented and on 13th January 1922 agreed to MWT being approved to include within the weekly period of half-an-hour already authorised a programme of 15 minutes' telephony (speech and music) in the transmission from [the] Chelmsford station for the benefit of the Wireless Societies [29].

The conditions imposed by the General Post Office were far from generous. Apart from the very limited transmission time, and power (250 W), the station had to close down for three minutes in every ten to allow the engineer-in-charge to listen-in for instructions to close the station if it was causing interference to other services.

On 14th February 1922 the first official radio programme was broadcast, on a wavelength of 700 m (later changed to 400 m), from the station at Writtle, near Chelmsford. Its call sign was 2MT, or in telegraphese, 'Two-Emma-Tock'. The station, erected under the direction of Captain P.P. Eckersley, used a 250 ft (76.2 m) long antenna, of the 4-wire type, suspended from 110 ft (177 m) high masts. The station would be operational until 17th January 1923 [30].

As anticipated, the inauguration of the Writtle station led to other manufacturers seeking similar advantages, and, by the middle of May 1922, more than 20 companies had applied to the GPO for broadcasting licences. The response of the Postmaster General (PMG), Mr Kellaway, was to invite their representatives to a conference at which their views could be expressed. Twenty four firms attended the meeting held at the General Post Office on 18th May 1922. The PMG explained that if licences were granted to each company chaos could occur, and so it was essential that if a satisfactory broadcasting service were to be implemented it should be undertaken by a single authority. He invited the manufacturers to discuss this view among themselves and to establish a single broadcasting authority with which the GPO could negotiate broadcasting matters.

Later, on 28th June 1922, the PMG obtained authority from the Cabinet to offer to the manufacturers: (1) a form of protection to the new broadcasting industry for a period of two years by issuing wireless receiving licences only for sets manufactured by members of the proposed broadcasting Company; (2) a payment to the Company of a half share of the fees for receiving licences [31].

Subsequently the manufacturers formed a committee of seven with the President of the Institution of Electrical Engineers, Sir Frank Gill, as Chairman. By October 1922 a scheme was agreed: a Broadcasting Company would be set-up supported by the various competing companies, and all British manufacturers of wireless apparatus. Inevitably, a committee was appointed to inaugurate this company and invitations to become members were sent out to 400 appropriate firms: 300 accepted. The scheme itself was described by Sir William Noble, the elected chairman of the committee, at a meeting attended by 200 representatives of firms held at the Institution of Electrical Engineers on 18th October. Briefly, the British Broadcasting Company Ltd (BBCo) would be instituted by six of the leading manufacturers of communications equipment, namely: the

Metropolitan Vickers Company, the Western Electric Company, the British Thomson-Houston Company, the Radio Communication Company, the General Electric Company and Marconi's Wireless Telegraph Company. The proposed capital of the BBCo would be £100,000 in £1 shares, of which £60,000 would be underwritten by the 'big six', and any bona fide manufacturer would be eligible to join by depositing the sum of £50 and purchasing one or more shares. Additionally, if more than £40,000 of shares were required by firms other than the 'big six' they would reduce their holdings so as to permit as many shareholders as possible in the new company. Revenue would be obtained from:

1. half the 10s. (50p) licence fee to be subscribed by listeners;
2. a tariff of c. 10 per cent to be paid by the member manufacturers on all receiving sets and certain accessories sold by them.

On 15th December 1922 the BBCo was registered as a public company, and on 18th January 1923 a formal licence was issued to the British Broadcasting Company. The licence gave authority for the establishment of eight broadcasting stations for a period ending on the 1st January 1925; and permitted transmissions to take place between 5.00 p.m. and 11.00 p.m., using the 350–425 m waveband and a maximum power of 3 kW [32].

Broadcasting commenced in London on 14th November 1922, at 2LO, Marconi House, London, and on the following two days at 5IT, the GEC Witton Works, Birmingham, and at 2ZY, the Metropolitan Vickers Works, in Manchester. Other stations followed and before the end of 1923 the United Kingdom was served by nine principal transmitting stations. Of these, seven were of MWT manufacture. By the end of 1924 the BBCo had set up a further ten relay stations, (see Note 1 at the end of this chapter) [33].

In 1922, in the UK, the main patents needed to construct a valve receiver were held by the British Thomson-Houston Company and Marconi's Wireless Telegraph Company. Initially Marconis handled the BTH patents under licence, but later they bought the entertainment rights in them. Then in 1928 the Chelmsford firm sold their patent interest to the Gramophone Company and a patent pool was formed [34].

Before the formation of the British Broadcasting Company, manufacturers who wished to make use of the Marconi Company's patents negotiated licences individually with the company. The company had amassed a wealth of skill in the field of radio communications, and, in the course of its development, had accumulated a large number of patented inventions. In addition Marconis had fully appreciated the importance of a strong holding of master patents, and so, when it became clear that a broadcasting service would be started, they took all possible steps to obtain control of the significant patents. Negotiations were entered into, and agreements were made (or had been made earlier), with Telefunken in Germany and with the Radio Corporation of America, which gave the British concern control of all the essential patents in the above field within the British market. Their policy was to buy any radio patents offered to them.

When the formation of the BBCo was being discussed, the manner by which patents were to be licensed was of some importance. Marconis proposed charging a royalty of 10 per cent on the wholesale selling price of all receivers which were made by manufacturers licensed to employ the Marconi company's patents. This was not acceptable to the other members of the 'big six' but eventually towards the end of 1922 agreement was reached and general licences were available – the royalty under the licence being assessed at 12s 6d (62.5p) per valve holder [35].

The trade disliked the terms of the licence (which became the A2 licence in 1923), for it was felt the royalty charged was excessive. After broadcasting commenced, the manufactured set carried a tariff to the British Broadcasting Company as well as the royalty payable to the Marconi company. Subsequently the scale of tariffs was reduced, and then abolished, but in the meantime the existence of the two payments created a strong inducement to manufacturers to attempt to evade royalties by not taking out a licence. Further difficulties arose with home constructors, and in 1924 Marconis stated that they would not enforce their rights against them because of the complication of doing so [36].

Naturally the patent licences were highly profitable to Marconis, even though considerable evasion of royalties took place. As the price of sets came down, the relative burden of the royalties increased, and, as a consequence of the discontent of the trade with the pricing system, the RMA, in 1927, set up a Royalty Committee [37]. It sent a deputation to MWT Company to persuade the company to reduce its royalties, but without success. Later in the year the RMA again complained to the firm and alleged that manufacturers in Holland, Portugal, Spain and Italy paid no royalties, while in Germany the maximum royalty was 2s 6d (12.5p) per valve holder. When no satisfaction was received, the RMA decided to question the Marconi licence, via the Comptroller-General and the courts, by means of a test case (involving the Brownie firm), and for this they relied on Section 27(e) of the Patents and Designs Act of 1919.

The Comptroller-General considered, in this and another case brought by Loewe, that Marconis were using their monopoly position to prevent the further development of radio in the one case, and to hinder the establishment of a new art in the other (August 1928). Marconis appealed against these decisions of the Comptroller-General, and, in November 1929, the appeal judge found that the company had not abused their monopoly position and were at liberty to carry on with its original royalty terms [38].

Following the success of the Marconi company's appeal, the trade was in a state of considerable frustration and the Royalty Committee of the RMA again approached the company hoping to achieve by persuasion and discussion what it had not obtained by litigation.

The manufacturers were offered a new licence (known as A3), which covered all the relevant patents of the MWT, BTH and Gramophone companies. This licence was issued by a patent pool, although Marconis undertook its administration and it continued to be called the Marconi licence. Under the terms of the licence, the royalty rate was reduced from 62.5p to 25p per valve holder, but as a quid pro quo the licensee had to sign an agreement, running for five years from the 28th August 1929, that he would pay this royalty on all sets produced, whether they used patented devices or not. This was because certain important patents were due to expire in October 1929, and in 1931, but with the above clause these patents would effectively continue to earn royalties until 1934.

The first challenge to the supremacy of the pool came from STC which held a number of patents on loudspeakers, push-pull amplifiers and super heterodyne receivers. Licences to use these patents were granted by STC to manufacturers on a royalty basis, but later in 1929 the company offered its patents to the pool in return for its share of the revenue of the pool. At first agreement could not be reached, thereby putting set makers in a dilemma regarding the licence by which they should manufacture, but by 1st April 1930 an enlarged patent pool was in operation arising from the merger of the Marconi Group with STC [39].

Further trials of strength between the patent pool and organisations outside its control came in 1933, when the Hazeltine Corporation, Philco and the Majestic Electric Company formed a new company in Britain called Hazelpat to licence manufacturers under their patents, and the Philips and Mullard companies announced that they intended to grant licences under all the patents they owned or controlled.

Hazelpat joined the patent pool in September 1983, but Phillips-Mullard did not become one of the grantors until 1938 – after an extended legal battle, involving an infringement suit that reached the Court of Appeal in 1937, and finally, in 1939, the House of Lords.

These competitive ventures by powerful companies against the monopoly of the patent pool highlight one of the weaknesses of the pool system. Patents are not assets that hold for all time, but are of value for a definite period only: they represent wasting assets. Thus, when an important patent expires, the position of the patent pool is considerably weakened, and, unless patents can be continually introduced to reinforce the pool, it must eventually reached a state in which licensees can disregard it – otherwise alternative pools are formed. A patent pool can only maintain its position of power by expanding and embracing new patent holders, but with this expansion in the number of grantors in the pool, there must be a diminution of the share of the proceeds which go to the original grantors. As a consequence, the original grantors may tend to resist any widening of the pool, accepting it only when the need is essential. Apart from the constant requirement to maintain the strength of the patent pool by the acquisition of new patents, to annul the effects of expired patents, there was also a considerable pressure on the pool, from manufacturers, to further this aim so that they – the set makers – could deal with one body only [40].

From the patent owners' position there were two reasons that induced them to offer their inventions to the pool: first, the desire to obtain revenue from the exploitation of the patents, and second, the need to gain recognition as successful inventors. Most patent owners did not wish to risk an infringement of their patented ideas by, say, a large and powerful industrial organisation because such action would demand retaliatory measures via legal proceedings, and such proceedings involving court actions could be 'treacherous', to quote the RMA. The RMA's view was that there was a tendency for all persons with patents of any validity and real application to broadcasting to negotiate and endeavour to enter the pool that way. Indeed one of the pool's difficulties was that so many of those people were not worth having in.

Certainly the large manufacturing organisations welcomed the establishment of the patent pool: GEC considered it was the best thing that happened at the beginning of broadcasting and advocated its use when television broadcasting was started.

On 31st October 1922 the GPO announced it had issued 18,061 radio receiver licences but was unable to cope with the flood of applications. By March 1923 some 32,285 experimental licences were still awaiting processing; but by June 1927 the rate of issuance had so improved that 2,998,220 licences had been approved, see the table below.

Date	Licences issued
December 1923	595,311
June 1924	823,894
December 1924	1,129,578
June 1925	1,387,993
December 1925	1,645,207
June 1926	2,076,230
December 1926	2,178,259
June 1927	2,998,220

The simplistic scheme described above soon highlighted a difficulty. Broadcast receiving licences were given on the assumption that a set bearing the BBCo mark would be used, but this condition did not apply to experimental licences. As a consequence the GPO began to receive large numbers of applications for experimental licences from persons who wished to use home-made sets. Initially, the term 'experimenter' was interpreted liberally, and the view was taken that if a person was sufficiently skilled to construct his own set he was qualified to be given an experimental licence. However, when certain companies began to place kits of parts, for home assembly by the inexperienced, on the market the question arose as to whether this liberality was being abused. Furthermore, many persons owning these home-made sets did not apply for a licence on the grounds that their sets did not comply with the conditions of the broadcasting licence, and that they were not qualified for experimental licences. Various suggestions were put forward to resolve this position, but when no agreed solution could be realised the Postmaster General appointed a committee, on 24th April 1923, under the chairmanship of Major General Sir Frederick Sykes. The committee's terms of reference [41] were to consider;

1. broadcasting in all its aspects;
2. the contracts and licences which have been or may be granted;
3. the action which should be taken upon the termination of the existing licence of the Broadcasting Company;
4. uses to which broadcasting may be put;
5. the restriction which may need to be placed upon its user or development.

The Report of the Committee was presented to Parliament in August 1923 as Command Paper no. 1951. Note 2 at the end of the chapter enumerates its main recommendations. The details need not be considered here, suffice to mention that in 1925 a second Broadcasting Committee was constituted under the chairmanship of the Earl of Crawford and Balcarres [42].

This committee recommended that a public corporation, to be known as the British Broadcasting Commission, acting as Trustee for the national interest, with a monopoly of broadcasting, be set up to take over the broadcasting service from the BBCo when its licence expired at the end of 1926. Although Parliament would retain the right of ultimate control, and the Postmaster General would be answerable to Parliament on matters of general policy, it was envisaged that the Commission would be given the greatest freedom for informing, entertaining and educating the nation's populace by radio. Second, the Crawford Committee recommended that the Commission should consist of from five to seven part-time Governors, nominated by the Crown, who would be persons of integrity and impartiality able to inspire public confidence. On the issue of funding for the new service, the Committee felt that the Commission should receive an adequate proportion of a receiving licence fee to permit the broadcasting service to develop and to operate efficiently – any surplus being retained by the State. The Charter of the Commission should be valid for ten years in the first instance.

All these recommendations were accepted by the Government. Indeed, successive governments have affirmed the Crawford model, appointed Governors of judgement and independence, respected the freedom of the BBC, and funded the Corporation to ensure the provision of a first class broadcasting service.

In other countries the broadcasting arrangements were as briefly described in Table 17.1 [43].

Table 17.2 lists the number of transmitting stations in various countries.

Essentially, the birth of the new form of entertainment took place in 1920 because the conditions necessary for its success were opportune at that time. The discovery of the amplifying and oscillating characteristics of valve circuits in 1912 and their subsequent rapid development for military purposes provided the foundation for the design of the receivers and transmitters which were needed to create a broadcasting service. Economic considerations acted as a catalyst but, most important, there was a strong demand from a section of the public for sound broadcasting transmissions. The early service proved satisfactory in reception, sets could be bought for a few or many pounds, the signals were

Table 17.1 Broadcasting arrangements in various foreign countries

Country and control (legal position)	General nature of scheme	Licence fees trans. and rec.	Other information
AUSTRIA Regulations in regard to broadcasting are laid down in the Enactment of the Austrian Federal Ministry of Commerce and Communications No. 346 of 23rd September 1924.	Broadcasting is conducted under a concession from the Government by a private firm, the 'Austrian Radio Traffic A.G.' in which the State has an interest by the possession of shares and by representation on the administrative and executive Councils of the Company. Only one transmitting station (at Vienna). Relay stations are in course of construction.	*Receiving:* 15,000 crowns or 30,000 crowns according to class of transmitting station. In addition every licensee has to pay an annual 'user' charge to the broadcasting company whether he intends to listen to the broadcast programmes or not.	Licence (obtainable at local PO) covers possession, installation and use of a single set, with essential component parts or to make such parts if none for sale. Dealers have to be licensed and keep records of all sales.
DENMARK Government has assumed control through [the] Minister of Public Works. Licences (issued at Telegraph and Post Offices) are required for all stations.	Government has nominated a Council representing all the interested parties – actors, authors, elocutionists, wireless amateurs, manufacturers, pressmen – to control the amusement and instructive sides of the programmes. All other transmissions are controlled by the State Telegraph Department. Transmitting stations are rented by the Council from the Telegraph Department or the Army.	*Receiving:* crystal set 50p; valve set 75p. Apparatus with loudspeaker in public place £10. Full fees must be paid for any portion of financial year.	On payment of fee for receiving licence mark is issued for attachment to set
FRANCE State monopoly under [the] Minister of Posts and Telegraphs. Decree 24th Nov. 1923 provides that State Authority [is] required to establish wireless sending or receiving stations.	Tour Eiffel – operating experimentally and another in course of erection. Two private stations allowed to broadcast temporarily without Royalty	*Transmitting:* no royalty at present. *Receiving:* If a receiving set is used for public performance (with charge for admission) an annual royalty not exceeding 200 Frs. is charged.	
GERMANY State monopoly administered by Postal Authorities. Licence for sending or receiving sets required by Decree of 8th March 1924.	P.O. erects and maintains stations and leases to broadcasting Cos. 9 main stations, 3 relay stations.	*Transmitting:* co. pays a fixed monthly charge and a further charge according to hours of working. Co. takes 60% of receiving licence fees. *Receiving:* 24 Reichmarks (c. 120p) collected by Postmen in monthly instalments of 2 Reichmarks	

ITALY Broadcasting in Italy is governed by Regulations issued under Royal Decree No. 1234/1226 of 10th July 1924, and Royal decree No. 70/2191 of 14th December 1924. The Ministry of Communications exercises control over stations.	An exclusive concession for 6 years for broadcasting services has been granted to the Societa Anonima Unione Radiofonica (URI). Co. must erect at least 3 stations, 55% of capital must be Italian, and 2/3rds of Directors Italian Nationals. Must use Italian material whenever possible. Hours of service fixed.	*Transmitting:* URI pays annual tax of 15000 lira (c. £113) for each station. *Receiving:* To have and use set – 75 lira (57.5p) of which 18.75p goes to the State and 38.75p to URI.	A tax is imposed on apparatus – which tax must be paid before the apparatus is exhibited for sale. For each set the tax is 15 lira to the State and a minimum of 20 lira for a crystal or 1 valve set (rising to 180 lira for a 5 or more valve set) to the company. Receiving apparatus must be approved by the Minister of Comms. and vendors must keep a register of persons to whom the apparatus is sold.
NORWAY Government has right to all wireless communication including Broadcasting.	Norwegian Broadcasting Co. just formed under concession for 5 years from Govt. Co. entitled to work Broadcasting Stations inside area radius 150 km from Oslo. Technical work of station is performed by Telegraph Dept. Studio work etc done by Co. Co's income consists of 80% of a licence fee of 20 Kr. (about 80p) which is paid by all listeners inside area of Co., plus 10% of value of receiving apparatus and parts sold within area, plus payment for broadcast advertisements. The remaining 20% of licence fee goes to Telegraph Department, also fee on apparatus and parts sold outside area of Co. In addition, the Telegraph Department collect 5Kr per annum from listeners outside area of Co. Dividend of Co. not to exceed 7%.	*Receiving:* 80p	Question of formation of Broadcasting Co. at Bergen, independent of Oslo Co. is being considered, also erection by existing Co. of two relay stations inside area in towns where reception from Oslo is not satisfactory.
ROUMANIA Law dated 25/6/25 governs transmission and reception of wireless signals. Permits are required for all stations.	Broadcasting Co. to which State contributes 60% of capital, and has majority of votes at General meeting and on Council of Administration. 50% of net profits accruing to State form a fund for propaganda and national culture by wireless telephony.	*Transmitting:* £2.91 to £8.74 according to power. *Receiving:* 15p for a crystal set to 24p for a multi-valve set. Licence 30p (crystal set) to 48p (multi-valve set).	Applicants for receiving licences must furnish proof citizenship, age, and good social reputation. Special for installations in schools, factories, etc.

Table 17.1 – continued

Country and control (legal position)	General nature of scheme	Licence fees trans. and rec.	Other information
RUSSIA The Praesidium of Gosplan has issued Ordinances sanctioning erection and exploitation of receiving stations by Private persons. Receiving stations must be registered in the Commissariat for Posts and telegraphs. Transmitting Stations may be installed only by Commisariat's special permission.		*Receiving:* fees are payable according to class – soldiers war invalids & State pensioners 1 rouble, workers and State employees 3 roubles, all other persons 10 roubles.	Broadcasting stations are being erected widely and the installation of receiving sets and loudspeakers in village soviets, workmen's clubs, etc, is being assiduously fostered.
SWEDEN State monopoly.	Broadcasting services and stations of which there are 5 are State monopoly, but the programmes are let out to and provided by Private interests. 9 Relay Stations provided by Radio Clubs.	*Receiving:* 12 Kr (67p). Additional fee (50 to 200 Kr) for loudspeakers.	The Government is considering the erection of a large central broadcasting station similar to Daventry and reducing the licence fee to 10 Kr.
JAPAN Regulations issued by Department of Communications.	3 classes of receiving licence (a) for apparatus passed by electrical laboratory in Tokyo and working on 375 metres, (b) for home made sets, wave up to 400 metres issued to persons with special knowledge of wireless. 1st Broadcasting Station now operating in Tokyo – owned by local broadcasting association. Other applications for broadcasting licences have been received.	*Receiving:* 9p p.a. (or 18 p.a.?) for the Government. This amount is collected by distributors of apparatus in addition to charges levied by the broadcasting associations for service.	
UNITED STATES OF AMERICA Licences for sending stations issued by Department of Commerce in accordance with a Statute passed by Federal Government in 1912. No licence	No difficulty in securing transmitting licence – no fee. 563 broadcasting stations in USA – mostly owned and operated by manufacturers, dealers, stores, newspapers, and wireless societies. Services provided free to advertise business or	*Transmitting:* no fee payable. *Receiving:* no fee or restrictions.	

required for receiving sets. No specific power to regulate use of receiving apparatus except possibly in the case of malicious interference.	advertisements. Two classes of station – 'A': limited to 500 watts power and wavelength 278 metres or below; 455 stations. 'B': Over 500 watts power and wavelengths from 280 to 545 metres – 108 stations. Authority is granted under certain conditions for these stations to use power up to 5 kW.		The fees for 'Dealers listening licences' range from £2 to £5.
AUSTRALIA Wireless telegraphy Act, 1905, administered by PMG. Licences (issued by PO) are required for all stations.	Two categories of broadcasting stations – A and B. 'A' stations receive portion of licence fees. 'B' stations receive no financial aid from State. Class A stations are limited to 1 or 2 in each State. No limit to the number of Class B stations.	*Receiving:* Ordinary – i.e. private house from 87.5p to 137.5p p.a. according to zone. Special – (hotels, etc) £7.50 p.a. to £10 according to zone).	
CANADA Radiotelegraph Act of 6th June 1913, administered by Radiotelegraph [Branch] of Dept. of Marine and Fisheries.	Two types of broadcasting stations. (1) Private commercial (2) Amateur. Stations under (1) may not accept any consideration for service Performed without the consent of the responsible minister. Stations under (2) are owned by radio associations and tolls or fees may not be accepted on account of any service performed. Practically all the stations are conducted for publicity purposes, or propaganda purposes, or to foster the sale of apparatus.	*Transmitting:* Private commercial – £10.42. Amateur – £2.75. *Receiving:* private – 21p. Public commercial – £10.42. No part of these fees is paid to the broadcasting organisations.	associations
NEW ZEALAND Broadcasting service to be carried on by company (in process of formation in April 1925) under agreement (not exceeding 5 years) with Minister of Telegraphs – provision for such agreement made in Post and Telegraph Amendment Act, 1924.	Under the scheme the licensee is required to erect 500 watt broadcasting station in each of the 4 radio districts – Auckland, Christchurch, Dunedin and Wellington. Wireless receiving licences issued at all Money Order Offices. Broadcasting Co. receives £1.5.0 [125p] out of each £1.10.0 [150p] paid for receiving licence and 90% of each dealer's licence.	*Receiving:* £1.50p	All dealers must be licensed, the fees varying from £2 to £10 according to population of district in which dealer resides.
SOUTH AFRICA Regulations (1.1.25) administered by PMG. Licences (issued by PO) are required for all stations.	Transmitting stations to be erected by the broadcaster at his own expense must be able to cover a certain area. No further broadcasting stations are licensed within that area. Period of licence 5 years.	*Transmitting:* £5 p.a. *Receiving:* from £2.25 p.a. (for private residence) to £6.25 p.a. for premises licensed for the sale of liquor).	Broadcaster may hire out receiving sets. Receiving sets must be approved by the PMG.

Table 17.2 Transmitting stations for broadcasting, c. 1925

Country	>50 kW	10–50 kW	5–10 kW	1–5 kW	Small	Total
United States	1	2	15	54	484	556
Australia		1	5	1	16	23
Britain		1		10	10	21
Germany		1		6	14	21
Poland	1					1
France				5	16	21
Canada			1	4	54	59
Austria		1		1	1	3
Spain				2	25	27
Sweden				2	20	22
Argentine				6	2	8
Switzerland				3	4	7

capable of being received by a majority of the population and at a cost it could afford; and so broadcasting went from strength to strength.

In the United Kingdom the major development of the second half of the 1920–1930 decade was the implementation of the Regional Scheme. This was conceived by P.P. Eckersley, the BBCo's Chief Engineer [44], in 1924 and had as its objective the provision of a dual-programme service throughout the UK. At that time the Corporation's nine main stations and several relay stations, together with the high power long-wave transmitter at Daventry, enabled 80 per cent of the population to receive their transmissions, even if simple cheap receivers were used. Only one programme was available to listeners in any particular area except that those who possessed a suitable receiver could also receive the long-wave station as well as the local medium-wave station and thus had a choice of programme at those times when the local station was originating its own broadcasting material.

Eckersley's idea of the Regional Scheme was, specifically, to cater for national and regional programmes: the former being of interest to the whole country and the latter being of local appeal.

The Regional Scheme had to be designed around an allocation of only ten medium wavelengths and one long wavelength under the Prague Plan of 1929 and all the relay stations had to operate on a UK Common Wave of 315 yd (288.5 m), with the exception of Leeds which used an International Common Wave.

In the scheme the transmitters were to be of a much higher output power (up to 100 kW) than had previously been used so as to allow the stations to be more economically run than a larger number of low power stations giving the same coverage, especially as each station had to be staffed. Capital costs would also be lower and the use of increased power would enable the best use to be made of the limited number of frequency channels available.

The development plan of the regional Scheme made provision for five twin-wave stations to be built to cover the following areas:

1. London and the Home Counties
2. Manchester and the industrial north of England
3. Glasgow, Edinburgh and the Scottish Lowlands
4. South Wales and the West of England
5. Birmingham and the Midlands

Each of these stations was provisionally estimated to cost £115,000. Because of the large sums of money that would be involved in implementing the scheme, Reith, the BBC's Director General, thought it wise to have an independent technical committee to examine and report upon the BBC's Chief Engineer's proposals.

The Eccles Committee endorsed Eckersley's plan without suggesting any modifications, but, notwithstanding this, the Post Office withheld its approval as it felt that the important allocation of wavelengths for broadcasting purposes was the responsibility of the Telecommunications Administrations and as there was to be a World Wireless Conference in 1927 the decision regarding the Regional Scheme had to be delayed until the Conference reported. However, on 20th April 1928 the GPO authorised the BBC to go ahead and the construction of the first permanent twin-wave transmitting station was commenced [45].

It had been decided that London and the South-Eastern Counties should have the first of the regional transmitters, and much thought was given to the location of the site. There were a number of conditions to be satisfied. The site had to be:

1. within about 15 miles (24.1 km) of Oxford Street, so that the signal would be strong enough in Central London to permit the continued use of existing sets;
2. far enough from populated areas to avoid blanketing a significant number of receivers;
3. far enough from the coast to avoid wasting a large part of the radiated energy over the sea;
4. in an open position to avoid absorption of energy by neighbouring buildings; and
5. accessible to a Post Office cable route.

Brookmans Park, 36 acres (14.6 ha) in extent, 16 miles (25.7 km) north from Charing Cross and 414 ft (126 m) above sea level, was chosen for the first site and work started there in July 1928. It was anticipated that the construction work would take about one year, but owing to a severe frost in the early part of 1929 the first transmitter (356 m, 45 kW), was not put into service until 21st October 1929, and the second transmitter (261 m, 67.5 kW), was not completed and brought into service until four and a half months later.

It is worth noting that the 1924 concept of a national and a local radio service was applied by the BBC to the post-war development of its television service.

The 1920–1930 decade was a period of much change in communications. Apart from the introduction of sound broadcasting, the decade was noteworthy for (1) the great expansion of long distance telephony in Europe, (2) the engineering of the short-wave beam system, and (3) the advancement of facsimile transmission. These topics form Chapter 18.

Note 1
The British Broadcasting Company's stations, August 1925 [37]

City/town	Date of opening	Wavelength
High power station (maximum power 25 kW)		
Coventry	27th July 1925	1,750 yd (1,600 m)
Main broadcasting stations (maximum power 3 kW)		
Aberdeen	10th October 1923	541 yd (495 m)
Belfast	15th September 1924	480 yd (439 m)
Birmingham	16th November 1922	522 yd (477 m)
Bournemouth	17th October 1923	420 yd (384.5 m)
Cardiff	13th February 1923	386 yd (353 m)
Glasgow	6th March 1923	461 yd (422 m)
London	14th November 1922	395 yd (361 m)
Manchester	15th November 1922	409 yd (374 m)
Newcastle upon Tyne	24th December 1922	442 yd (404 m)
Relay broadcasting stations (maximum power 200 W)		
Bradford	8th July 1924	266 yd (243 m)
Dundee	12th November 1924	363 yd (332 m)
Edinburgh	1st May 1924	359 yd (328 m)
Hull	15th August 1924	367 yd (336 m)
Leeds	8th July 1924	379 yd (347 m)
Liverpool	11th June 1924	341 yd (312 m)
Nottingham	16th September 1924	320 yd (292.5 m)
Plymouth	28th March 1924	365 yd (333.5 m)
Sheffield	16th November 1923	257 yd (235.2 m)
Stoke on Trent	21st October 1924	261 yd (239 m)
Swansea	12th December 1924	526 yd (481 m)

Note 2
The recommended scheme of the Broadcasting Committee (1923) [41]

Controlling authority

That a Broadcasting Board should be established by statute to assist the Postmaster General in the administration of broadcasting and to advise him on important questions concerning the service.

Operating authorities

That the broadcasting service should not be operated by a Government Department, but that those entrusted with the service should work under Government licence.

That it is desirable that the operation of the existing service by the British Broadcasting Company should be continued for a definite period, subject to agreed modifications to the Company's licence, but that, subject to existing rights, the Government should keep its hands free to grant additional licences, and should consider various alternatives for the operation in the future, either by the Company or by other authorities, of local or relay stations in addition to large stations.

Financial provisions

That no part of the cost of broadcasting should fall on the taxpayer, but that the Government should not endeavour to make a profit on the administration of the service.

That the bulk of the revenue should be obtained from the receiving licence fee, which should be retained at 10s [50p] a year, subject to consideration of a reduction in the event of more revenue being received than is sufficient to carry on an adequate service.

That instead of 5s as much as 7s 6d [37.5p] of the 10s fee might be allocated under any new scheme to meet the cost of broadcasting, subject to a sliding scale under which the payment per licence would decrease as the number of licences increases.

That certain supplementary sources of revenue should be the subject of early consideration.

Conditions of receiving licences

That in place of the present broadcast and experimental receiving licences a uniform and simple type of licence be issued and placed on sale at Post Offices without any formalities, containing a clause forbidding improper use of back-coupling on pain of withdrawal of the licence, but no other limitation on the apparatus allowed to be used.

That effective measures be taken to enforce such a licence, and that certain additional statutory powers be obtained to strengthen the Postmaster General's hands.

References

1 MACLAURIN, W.R.: 'Invention and innovation in the radio industry' (Macmillan, New York, 1949), p. 90

2   FAGEN, M.D. (Ed.).: 'A history of engineering and science in the Bell System' (Bell Telephone Laboratories, 1975), p. 364

3   STURMEY, S.G.: 'The economic development of radio' (Duckworth, London, 1958), p. 33

4   THROWER, K.R.: 'History of the British radio valve to 1940' (MMA International, England, 1992), p. 32

5   ESPENSCHIED, L.: 'The origin and development of radio telephony' *Proc. IRE*, 1937, **25**(9), p. 1102

6   BAKER, W.J.: 'A history of the Marconi Company' (Methuen, London, 1970), pp. 204, 207

7   Quoted in BITTING, R.C.: 'Creating an industry', *J. SMPTE*, 1965, **74**, pp. 1015–1023

8   ARCHER, G.L.: 'History of radio to 1926' (Arno Press, New York, 1971), p. 199

9   Ref. 8, pp. 200–201

10 Ref. 8, p. 203

11 Ref. 8, pp. 241, 393–397

12 ANON.: 'Revised memorandum concerning broadcasting by wireless telephony in the United States of America', Minute 1695/1926, Post Office Records and Archives

13 Ref. 1, p. 101

14 Ref. 6, pp. 180–181

15 Ref. 8, pp. 160–174

16 Ref. 1, pp. 97–98

17 Quoted in Ref. 1, p. 105

18 Ref. 1, pp. 108–109

19 ARCHER, G.L.: 'Big business and radio' (American Historical, New York, 1939), p. 8

20 BURNS, R.W.: 'Television, an international history of the formative years' (IEE, London, 1998), pp. 405–407

21 RCA Annual report for 1929, RCA Archives, Princeton

22 SIEPMANN, C.A.: 'Radio, television and society' (Oxford University press, 1950), pp. 8–10

23 ANON.: a report, *Radio broadcasting Magazine*, November 1922

24 Ref. 6, p. 184

25 Ibid, p. 185

26 ANON.: a report, *The Times*, 30th January 1922, p. 8

27 Ref. 6, pp. 186–187

28 Ref. 3, p. 138

29 Ref. 3, p. 139

30 ANON.: 'Historical summary of the broadcasting services in the United Kingdom and resume of events prior to appointment of Committee', Broadcasting Committee papers, Post 33/1697–8, Minute 1695/1926, Post Office Records and Archives

31 Ref. 3, pp. 140, 143

32 DOWSETT, H.M.: 'Wireless telephony and broadcasting' (Gresham, London, 1924), p. 62

33 Command Paper 1822 of 1923, HMSO

34 Ref. 6, p. 200

35 Ref. 3, p. 215
36 ANON.: a report, *Wireless Trader*, February 1924, p. 391
37 Ref. 3, pp. 216–217
38 Ref. 3, pp. 218–219
39 Ref. 3, p. 224
40 Notes of a meeting of the Television Committee held on 12th July 1934. Evidence of Messrs Paterson and Heather on behalf of the General Electric Company Ltd, Minute 4003/1935, Post Office Records and Archives
41 Command Paper 1951, 1923, HMSO
42 Command paper 2599, 1925, HMSO
43 'Broadcasting arrangements in the Dominions and certain Foreign Countries' Post 33/1697–8, Minute 1695/1926, Post Office Records and Archives
44 ECCLES, W.H.: Presidential Address, *J. IEE*, 1927, **65**, p. 10
45 PAWLEY, E.: 'BBC engineering 1922–1972' (BBC Publications, 1972), pp. 83–101

Chapter 18

Some important developments in the 1920s

1. Long distance telephony

Prior to 1924 the expansion of the long distance telephone system on the Continent of Europe had been on a slower scale compared to that of the United States of America. The prime reason for this state of affairs was the radical difference in the political conditions of the two geographical regions. Long distance communications across the national boundaries were dependent upon the provision of a proper organisation and the adoption of certain practices. More particularly, it was necessary to standardise the methods of construction, upkeep and operation of the international circuits and to ensure their security.

In the July 1921 issue of *Annals des Postes Télégraphes et Téléphones* a paper [1] written by Monsieur M.G. Martin was published entitled 'Long distance telephony in Europe'. Martin made the suggestion that an association should be established by the various European telecommunications administrations for the purpose of constructing, operating and maintaining long international telephone circuits. A similar proposal was put forward, independently, by Mr F. Gill of the International Western Electric Company in his 1922 Presidential Address [2] to the Institution of Electrical Engineers (UK) when he spoke on 'Electrical communication – telephony over considerable distances'.

As a result of these pronouncements Monsieur P. Laffont, the Under-Secretary of State for Posts and Telegraphs in the French government, convened a meeting, early in 1923, in Paris, of the representatives of the telephone administrations of Belgium, France, Great Britain, Italy, Spain and Switzerland. Their discussions ranged over the whole ground of the problem and a number of recommendations were adopted unanimously and unreservedly. These recommendations [3] were later approved by the administrations of the six countries and the stage was set for practical progress. The recommendations were grouped under six headings, namely: (1) administration; (2) transmission – engineering construction; (3) engineering maintenance, including removal of faults; (4) traffic; (5) programme of immediate additional construction work; and (6) preliminary programme of further construction work.

On administration, the delegates agreed to the formation of a permanent 'International Consulting Committee for International Telephone Connections' and further recommended:

1. that the greatest attention should be given to the selection, to the installation, and to the maintenance of apparatus and installations used for the equipment of trunk circuits intended for long distance international telephony; and
2. that telephone administrations should equip themselves with the necessary measuring apparatus required for the proper supervision and for the proper maintenance of the installation.

For the European telephone network it was intended that telephone speech signals should be transmitted, without frequency translation, by quadded cables and terminal equipment arranged so that independent signals could be sent over the side and phantom circuits of the telephone multiplex system.

The first proposal for a telephone multiplexing system was made by F. Jacob [4] in 1882. He put forward the idea, illustrated in Fig. 18.1a, whereby using a Wheatstone bridge arrangement an additional telephone signal, C, could be superposed on a pair of circuits A' and B' known as the side circuits, without the additional signal interfering with the signals, A and B, carried by the side circuits. Effectively the two conductors of circuit A' and the two conductors of circuit B' act as the two conductors of another circuit, called the phantom circuit, which can be used for the transmission of an extra signal.

Referring to Fig. 18.1a the current from C divides into two equal parts by means of the resistance bridge and each part is transmitted over one wire of the pair used for A. A similar arrangement is used to combine the two components at the far end and to provide a return path over the pair of wires employed for B. Since the phantom currents flow in the same direction over the two wires of the pair, there is no mutual interference between the phantom circuit and the side circuit. Thus, the two pairs of wires of the quadded cable can actually carry three independent signals.

Jacob's use of resistors for the derivation of the phantom circuit allowed the phantom signal current to be switched into two equal parts but introduced considerable attenuation into the transmission system. In 1886, J.J. Carty [4] suggested a solution based on the use of transformers (sometimes known as repeating coils) in place of the resistors (Fig. 18.1b). His scheme overcame the major loss of the Jacob configuration but difficulties were experienced in attempts to make repeating coils with satisfactory balances. Thus the concept of the phantom circuit, though simple was not immediately implemented commercially and for many years it remained a scientific curiosity.

During the period 1873–1901 O. Heaviside had published a series of papers [5] in the *Electrician* and in the *Philosophical Magazine* on the theory of electrical transmission and had shown in 1893 that the addition of a series inductance to a

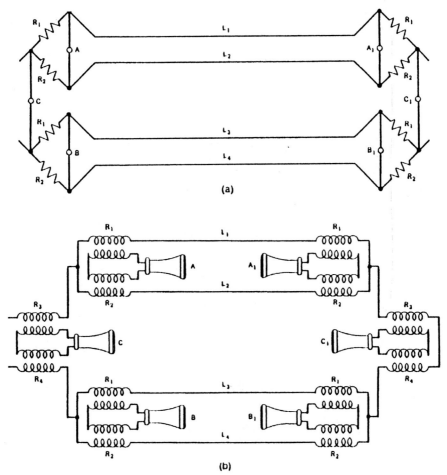

Figure 18.1   Development of the phantom circuit: (a) phantom circuit based on concepts of F. Jacob, intended for the simultaneous operation of three telephone circuits over two metallic circuits; (b) phantom scheme of J.J. Carty using induction or repeating coils instead of resistors. A, B and C are telephone sets.

Source: *A history of Engineering and Science in the Bell System*, M.D. Fagen (Ed.), (Bell Telephone Laboratories, Inc., 1975).

transmission line could lead to a decrease in its attenuation constant. Stone, as early as 1894, had proposed the utilisation of additional continuously distributed inductance for the same purpose but it was Campbell and Pupin who, independently, in 1899 advanced the notion of lumped loading, with series connected loading coils, as being more practical [6].

The coils used in the early experiments were of the air-core, solenoid type. They were large and unsatisfactory for commercial telephone practice since as Heaviside remarked, 'inductance coils have resistance as well, and if this be too great the remedy is worse than the disease' [5]. Furthermore, because of their size the coils established magnetic fields which could interfere with other telephone circuits. By April 1901 the toroidal, or doughnut-shaped, coil had been developed. This has the property that the magnetic field produced is mostly confined within the core material. Initially the coils were wound on cores made up of many miles of very fine, lacquer-insulated iron wire, but by c. 1916 this type of core had been replaced by a powered iron core. This was manufactured from insulated iron granules that were compressed into toroidal rings. Later cores consisted of special heat-treated iron alloys, 'the permalloys', having larger permeabilities.

The development work on loading coils led to improvements in the construction of repeating coils and the application of phantoming. However its widespread utilisation did not come about until coils could be made for loading the phantom circuits as well as the side circuits. This was accomplished in 1910 with a set of three coils, one for each side circuit and one for the phantom (Fig. 18.2). By 1911 the basic problems of loading and of phantoming both open-wire lines and underground cables had been solved, though refinements were needed to achieve a high degree of balance.

It was soon realised that the application of loading could yield large economies on long distance cable circuits, and by the end of 1907 c. 60,000 loading coils had been installed on c. 138,000 km of cable in the US telephone network. Further expansion led to the use of about 1,250,000 coils on c. 2,570,000 km of cable circuits and c. 402,000 km of open wire lines in the Bell System. The number of loading coils utilised by ISEC is not known but in the United States of America in 1974 the number of loading coils manufactured was in the order of 14.5 million.

With the growth of the long distance lines in Europe, following the establishment of the International Consulting Committee for International Telephone

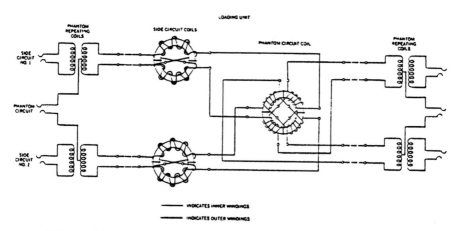

Figure 18.2   Bell System standard method of loading phantom circuits and their side circuits.

Source: Fagen, op. cit.

Connections, it was essential that good telephone engineering practices should be implemented. Loading and phantoming were adopted, and the star-quad cable was extensively used. This cable had the plane of the two wires of one side circuit at right angles to the plane of the wires of the other side circuit. Provided this configuration was rigorously maintained there was theoretically no coupling between the pairs. In practice the cables had to be balanced.

With any telephone circuit it is, of course, essential that the speech signals transmitted by a given circuit should not be heard in adjacent circuits; that is, the 'cross talk' between telephone circuits should ideally be zero. Cross talk between telephone circuits in close proximity is due mainly to differences in the resistances of the two wires of a pair and to electric and magnetic inductive effects [7]. The former effect is important only in connection with cross talk between side circuits and the associated phantom and, while for non-loaded open-wire circuits the two inductive effects are approximately of equal importance, in adjacent cable circuits the electric inductive effect predominates. An accurate indication of the magnitude of this effect may be readily obtained from capacity unbalance measurements [8].

The distributed capacitances of a quad (excluding the direct capacitances between the two wires of a pair), may be represented as shown in Fig.18.3a. This circuit arrangement can be transformed to give the configuration of Fig. 18.3b from which, using the Wheatstone bridge principle, it is apparent that cross talk will not occur from one pair of coils, say AB, to the other pair, CD, when C_{AD}/C_{AC} is equal to C_{BD}/C_{BC}, and mutatis mutandis, for the elimination of cross talk between a side circuit and the phantom circuit. Hence the measurement in the field of the capacitances of a quad, and its subsequent balancing, was of great impact in the commissioning of the long telephone lines of the European network. For this purpose a capacitance bridge was necessary. Though several bridges existed and were used in the field they were not ideal. Fortunately, A.D. Blumlein, the greatest British electronics engineer of the 20th century, who was employed by ISEC, designed the definitive bridge [9] for the purpose. It became known as the transformer ratio arm bridge and was widely used for measuring the electrical properties of components [10].

Telephony developed rapidly in Europe following the setting-up of the CCITC. Fig. 18.4 shows, for the year 1925, the existing and projected long-distance cables and associated central offices [11]. In North America too long-distance telephony developed quickly during the 50 year period following Bell's patent, but this was not the position elsewhere in the world (see Table 18.1).

Further statistics are presented in Table 18.2.

Some indication of the enormous growth of communications, generally, in a developed country is provided by the following statistics (for the year ending 31st December 1925) for the United Kingdom: inland telegrams, 52,110,000; cablegrams and wireless messages, 11,314,000; letters, 3,500,000,000; printed papers, 1,710,000,000; newspapers, 165,000,000; postcards, 465,000,000; telephone calls, 929,000,000; trunk calls, 86,000,000; and Continental calls, 480,000.

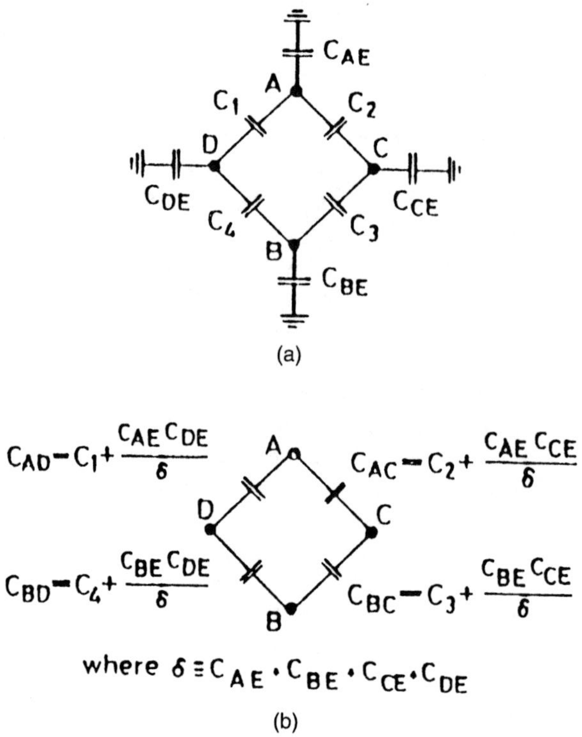

$$C_{AE}$$

C_1 A C_2

D C

C_{DE} C_4 C_{CE}

C_3

B

C_{BE}

(a)

$$C_{AD} = C_1 + \frac{C_{AE} C_{DE}}{\delta} \qquad A \qquad C_{AC} = C_2 + \frac{C_{AE} C_{CE}}{\delta}$$

D C

$$C_{BD} = C_4 + \frac{C_{BE} C_{DE}}{\delta} \qquad B \qquad C_{BC} = C_3 + \frac{C_{BE} C_{CE}}{\delta}$$

$$\text{where } \delta \equiv C_{AE} \cdot C_{BE} \cdot C_{CE} \cdot C_{DE}$$

(b)

Figure 18.3 (a) The distributed capacitances associated with a quad. (b) The equivalent circuit.
Source: The author.

Table 18.1 The number of telephones in use throughout the world

Continent	31st Dec. 1925	31st Dec. 1926
Africa	165,500	168,500
Asia	912,500	963,000
Australasia	530,000	578,500
Europe	7,475,000	8,020,000
North America	18,240,000	19,120,000
South America	402,000	427,000
United Kingdom	1,357,000	1,511,000

2. The short-wave beam system

A vertical conductor when used as an antenna has, by symmetry, an omni directional polar diagram in the horizontal plane. In point-to-point wireless transmission where radiation must be propagated from a transmitter A to a

Figure 18.4 Map showing the existing and projected cables of the long distance European telephone network.

Source: Author's collection.

receiver B it is desirable that all the radiated energy should be directed in the given direction, i.e. along the line AB, since energy not so directed is wasted. Hertz showed that electromagnetic waves obey the laws of optics and can be directed: for several of his experiments he utilised a pair of cylindrical parabolic reflectors, along the focal lines of which he mounted linear oscillators (Fig. 10.1).

Table 18.2 Miscellaneous telephone statistics, c. 1925 [12]

Country	Telephones per 100 persons	Telephone conversations/ capita/annum	Percentage of the world's total no. of telephones	Conversations/ instrument/ annum
United States	14.2	190.8	62	1340
British Empire			11.2	
Germany	3.9	30.1	9.2	764
United Kingdom	2.8	22.5	4.9	813
Canada	11.6		4.1	
France		20.3	2.5	1220
Japan		29.9	2.1	3250
Sweden	6.9	105.6	1.6	1620
Australia	5.5	47.7	1.3	868
Denmark	9.0	130.7	1.2	1440
Argentina			0.66	
Italy		9.0	0.66	2090
Russia			0.58	
Belgium		21.1	0.53	1190
Poland			0.46	
Spain			0.40	
Switzerland	4.8	37.7		777

A similar arrangement was employed by Marconi during some of his work in the late 1890s. The advantage of the method was given by him in his 1899 The IEE paper:

By means of reflectors it is possible to project the waves in one almost parallel beam which will not affect any receiver placed out of its line of propagation, whether the said receiver is or is not in tune or syntony with the oscillation transmitted. This would enable several forts, or hill tops, or islands to communicate with each other without any fear of the enemy tapping or interfering with the signals. . . . There exists a most important case to which the reflector system is applicable, namely to enable ships to be warned by lighthouses, light-vessels, or other ships, not only of their proximity to danger, but also of the direction from which the warning comes.

During the c. 1899 beam transmission tests Marconi succeeded in signalling over distances of approximately 1.8 miles (3 km).

In place of cylindrical parabolic reflectors fabricated from sheet metal, the use of vertical wires or rods arranged along a curve the horizontal shape of which is a parabola was suggested by de Forest in a 1904 patent. This configuration was later used by C.S. Franklin.

In June 1915 Marconi was commissioned in the Italian Army with the modest rank of Lieutenant [13]. Though Italy before the commencement of the Great War was bound to Germany by the Triple Alliance [14], the country maintained its neutrality for the first few months of the war. Then on 26th April 1915 Italy, with Britain and France, signed the Treaty of London [15] and agreed to enter the war on the Allied side in return for a promise that the lands of the Tyrol, Trieste and North Dalmatia would be ceded to Italy in the event of an Allied victory. One month later, on 24th May, Italy declared war on Austro-Hungary. War on Germany was declared on 28th August 1916.

These events must have given Marconi great relief from any doubts he may have had about where his loyalties lay. His initial army task was to inspect the mobile wireless stations used at the Front and to suggest means for improving their effectiveness. The work included investigating the possible military applications of wireless in aircraft and led in September 1915 to the first Marconi apparatus being fitted into an Italian two-seater military biplane. For this purpose long wave transmission and reception (which require very long antennas) are quite inappropriate. Consequently there was a need to consider the utilisation of much shorter wavelengths. It is known that some small antennas and metal reflectors about a metre square were fabricated in the Marconi works in Genoa and that experiments were conducted, by Marconi and his friend Solari, in a long corridor of the hospital in Genoa where Marconi was being treated for tonsillitis [16].

Shortly afterwards, in 1916, the investigations were extended, with the co-operation of the Regia Marine, to include short-wave communication between ships at sea. In this work Marconi was 'most valuably assisted' [17] by C.S. Franklin, an extremely able engineer with MWT. Their efforts led to a transmission range of six miles (9.65 km) between two battleships. The apparatus used has been described by the great inventor [17]:

During my tests in 1916, I used a coupled spark transmitter, the primary having an air condenser and spark in compressed air. By these means the amount of energy was increased and the small spark gap in compressed air appeared to have a very low resistance.

The receiver at first used a crystal receiver, whilst the reflectors employed were made of a number of strips or wires tuned to the wave used, arranged on a cylindrical parabolic curve with the aerial [antenna] in the [f]ocal line.

The transmitting reflector was arranged so that it could be revolved and the effects studied at a distance on the receiver.

Mr Franklin has calculated the polar curve of radiation into space, in the horizontal plane, which should be obtained from reflectors of various apertures. . . . The calculated curves agree very well with the observed results.

Among the advantages of low power short wave transmission and reception, mention must be made of the possibility of ensuring secrecy when communicating sensitive tactical information from one ship to another, since a narrow beam width and short range effectively provide security of operation. (An antenna's beam width to the half power points is proportional to the wavelength of the radiation and inversely proportional to the aperture of the radiating structure.)

Further tests, by Franklin, were continued at Caernavon during 1917 when ranges of more than 19.8 miles (32 km) were readily obtained with a 3 m (100 MHz) improved compressed air spark gap transmitter, an antenna and a reflector having an aperture of 2λ by 1.5λ, and a receiver without a reflector. In 1919 Franklin extended his investigations by employing valves working on a wavelength of 15 m (20 MHz); the objective being to evolve a directional radiotelephonic system. Soon clear speech was being received in Holyhead, 19.8 miles (32 km) away from the Caernavon transmitter. Afterwards, the mail boats operating across the Irish sea enabled the reception distance to be extended. It was found that good signals could be received all the way over to the Irish coast and into Kingstown Harbour at a distance of 78 miles (125 km) from Caernavon.

By the end of the war Franklin's investigations at Caernavon, and later at Inchkeith and Portsmouth, were sufficiently advanced to persuade him to suggest to Marconi that the apparatus should be tested between London and Birmingham, a distance of 97 miles (156 km), using a wavelength of 15 m (20 MHz). It seems that, Marconi being sceptical of success being attained, Franklin backed his judgement with a £5 wager. Success was achieved with a transmitter power of just 700 W, whereupon Marconi graciously parted with £5 [18].

From the commencement of his work in the United Kingdom Marconi had conducted experiments aboard several ships – the *San Martino* (1897), the *Ibis* (1899), the *Philadelphia* (1902), the *Carlo Alberto* (1902), the *Lucania* (1903), and the *Principessa Mafalda* (1910). Now, with the need to extend the short wave investigations to greater distances another ship fitted with a well-equipped laboratory had to be either commissioned or purchased. The latter possibility was preferred, to give Marconi the freedom of action that was necessary in making observations at extensive distances in different directions. 'I want to live at sea', he wrote to a friend. 'I shall sell my house in Rome and get in its place a yacht, which will in future be my home. There I shall be able to study and experiment without the fear of unwelcome interruption' [19].

And so, with the approval of the Italian Minister of Marine Affairs, Marconi bought from the British Admiralty a former enemy vessel that previously had been commandeered for war service. The vessel was the *Rovenska*, a handsome steam yacht built at Leith, in 1904, for the Archduchess Maria Therese of Austria. After the purchase the yacht was re-registered as the *Elettra* and was repaired and refitted at Birkenhead during the months of January and February 1920. The *Elettra* displaced more than 700 tons (711,000 kg) was 220 ft (72.2 m) in length, 27.5 ft (9.2 m) in beam, and carried sufficient coal for 12 days cruising at a speed of 10 knots [20] (5.1ms).

On 14th August 1920 the UK's Defence of the Realm Act regulations concerning wireless apparatus and transmission were rescinded, and three days later the UK's Radio Research Board was appointed, together with four sub-committees, to consider wireless matters. Representatives from the Post Office, the Admiralty, the War Office, and the Air Ministry served on the RRB over which Admiral Jackson presided.

All was now ready for an intensive and extensive investigation of short wave transmission and reception by Marconi and his staff.

To aid their endeavours Franklin at Marconi's behest established an experimental short wave transmitting site at Poldhu. It comprised basically a 12 kW (input) transmitter and parabolic reflector. Various antenna arrangements could be tested with the reflector in or out of service, and means were provided to permit the wavelength of the radiated wave to be altered from 97 m downwards. Another very able MWT engineer, G.A. Mathieu, designed the necessary receiving equipment for use on the *Elettra* [21].

Tests began on 11th April 1923 with the main antenna beam pointing in a south westerly direction. The *Elettra* sailed from Falmouth and measurements of received signal strength were taken as the ship steamed past the west coasts of France, Spain and Portugal, Gibraltar, Tangier, Casablanca, Funchal, Madeira, and finally the Cape Verde Islands, situated 2,318 miles (3,730 km) from Poldhu. Observations of the signal strength were recorded during the day and night, with the Poldhu reflector in and out of use, and the times of fading and the presence of static were noted. A curious phenomenon was soon discovered. As the *Elettra* sailed south the signals initially attenuated, but after several hundred kilometres they began to increase in amplitude. Also it was found that reception at night was much better than during the day [22].

Fortuitously for Marconi's experimental study of short-wave propagation at this time much research was being conducted on the mechanisms of absorption, refraction (including reflection as a limiting case) and polarisation of electromagnetic waves in the upper atmosphere. It is now known that conducting layers, known collectively as the ionosphere, exist in the upper atmosphere and that these layers arise because the ultra-violet radiation from the sun ionises the gas molecules, which constitute the higher regions of the atmosphere, to produce free electrons and ions. Theory shows that an ionised medium has a refractive index (n) which is dependent on the frequency of the electromagnetic wave which propagates through it, and on the number of electrons per unit volume (N). The latter quantity is a function of the intensity of the incident ultra-violet radiation and so N varies throughout the day and from day to day. Experimental measurements have shown that the ionosphere is a non-homogeneous medium and comprises several layers which have been designated by the symbols D, E and F, in order of height. At times the F layer splits into two layers called F_1 and F_2. They vary in intensity and in height with diurnal and seasonal variations in the ionosphere, and extend from 31 miles (c. 50 km) to 248 miles (c. 400 km) above the earth's surface. (See Note 1.)

The presence of these layers explains the phenomena that Marconi observed. As an electromagnetic wave from a transmitting antenna travels in a direction inclined to the earth's surface so it will enter the ionosphere. Here the wave will be refracted by an amount determined by n which is equal to $\sqrt{(1 - 81N/f^2)}$. For a range of values of n the refraction, i.e. the bending, of the wave's direction is sufficient to return the wave to earth. The distance from the transmitter to the region where the waves return to earth is the skip distance. Thus as the *Elettra*

sailed away from Poldhu reception was initially obtained by means of the surface wave but as the distance increased this wave was attenuated until it could no
longer be detected; then at a greater distance the downward wave reflected from
the ionosphere was received.

The ionosphere is a very complex region with constantly changing properties.
Fading of the received signals can occur necessitating changes of transmitter
frequency and or diversity reception. Nevertheless, as Marconi observed after his
1923 cruise: 'The strength of the signals we received . . . at night, left no doubt in
my mind that their practical range was very greatly in excess of that distance
[Poldhu to the Cape Verde Islands]' [23].

There seemed to be a prospect that a new era of wireless communication
was about to be opened-up, an era that would enable trans-oceanic and trans-
continental communications to be implemented with modest transmitter
powers. Until the Cape Verde cruise the general belief was that 'long wavelengths plus high powers equals long ranges'. (The Caernavon transmitter
operated at 14,000 m with a transmitter power of 200 kW; and the Coltano
station in Italy had an output of 500 kW.)

Following the successful 1923 cruise, a new series of tests was undertaken,
in 1924, between Poldhu and *Elettra* based on the propagation of electromagnetic waves having wavelengths of 92 m, 60 m, 47 m, and 32 m [24]. The
results of these tests, and information from wireless experimenters in Canada,
Australia and elsewhere, showed clearly that the ranges achieved in the
1923 tests were not freak results, and that short-wave beam wireless was
realisable. Indeed, the experimental data was so encouraging that Marconi,
on 30th May 1924, decided to attempt to telephone to Sydney, Australia. This
feat, using the 92 m Poldhu transmission but without a reflector, was a complete
success.

All of this posed a dilemma for Marconi. Mention has been made in Chapter
15 to the difficulties associated with his company's pre-First World War Imperial
Wireless Scheme to link wirelessly the Dominions and the Colonies of the
British Empire to the United Kingdom. After the war, in 1919, Marconi's Wireless Telegraph Company, with a wealth of experience of long distance communications behind it, advanced another proposal to the British Government to
establish direct wireless connections between India, Australia, South Africa and
the United Kingdom based on high power, long-wave stations and low power
feeder stations [25]. The Company's initiative was not well received.

Several Government committees were appointed in 1920, 1923 and 1924
to consider the matter. The Norman Committee of 1920 was ill-disposed to any
suggestion that a private company should be granted any semblance of a
monopoly in the operation of a wireless network which covered the Empire. In
preference the committee argued the need for a chain of stations – at intervals
of 5,182 miles (3,220 km) – to be managed by the Post Office. Four years later the
Donald Committee's Report, which was presented to Parliament on 28th
February 1924, put forward an essentially similar view. Its principal recommendations were [26]:

1. the Post Office should own and operate all wireless stations in Great Britain which communicated with the Empire, with a partial exception in the case of Canada, for which a service run by private enterprise already existed; and
2. private enterprise should be free to develop wireless communications with countries which were not in the British Empire.

However, prior to 1924, two Dominions, Australia and South Africa, had awarded contracts to MWT for the provision and erection of super-power long-wave stations and now (April/May 1924), while the vexed questions of ownership and operation of the British stations were still being debated, MWT informed the Empire countries of the prospect that these costly stations could be replaced by much cheaper short-wave beam stations. The capital savings would be substantial. Shaughnessy has given the cost of one 273 yd (250 m) mast as from £12,000 to £15,000. Since 12 such masts with antennas and earth systems would be needed for a high-power station the total cost of the radiating structure would be between £150,000 to £180,000 (1925 prices). This cost together with the price of the transmitter and auxiliary equipment and services would lead to an overall cost of between £400,000 and £500,000 per station; and to implement a service two stations would be required. In addition the interest and depreciation charges on these sums would be very appreciable and could amount to from £70,000 to £96,000 [27]. (The purchasing power of £1 in 1925 was comparable to £30.54 in 1996.)

MWT's disclosures caused much surprise. Nonetheless, the company's reputation was such that Australia, Canada and South Africa quickly came to decisions in favour of the short-wave beam system. The British Government prevaricated for a while but on 2nd July 1924 it too found favour with the new system although, to safeguard the integrity of the Empire and the overseas possessions in the event of conflict, it confirmed the need to advance the super-power long-wave Post Office station at Rugby. Subsequently, the Post Office entered into a contract with MWT for the company to provide a beam station to transmit to Canada – the station, if successful, to be capable of extensions to operate to South Africa, India and Australia simultaneously. The terms of the contract were stringent [28].

1. The price to be paid for the first station was to be the capital cost plus 5 per cent establishment charges and 10 per cent contractors' profits, the maximum figure not to exceed £35,120. [Shaughnessy in his 1925 paper mentions a price of £50,420 for the main station and £31,406 for each extension.]
2. The Post Office was to pay MWT a royalty of 6.25 per cent of gross traffic receipts as long as any Marconi patents essential for the working of the station were in use.
3. The station had to be completed within 26 weeks of the site becoming available.
4. The Government would only be liable for an initial payment if certain minimum guarantees were fulfilled during a seven day demonstration. The payment was then to be 50 per cent of the cost. If the station succeeded in

meeting the guarantees for a further six months under Post Office operation, a further 25 per cent was to be paid, with the final balance of 25 per cent payable after a further six months' satisfactory working.
5. To qualify for any payment at all, messages had to be exchanged with the Canadian station seven days per week for at least 18 hours per day (duplex) at a speed of not less than 100 words per minute in both directions.
6. If the station failed to win Post Office approval during the stipulated period, it had to be cleared from the site entirely at [MWT's] expense and any payments already made were to be refunded to the Government.

On 18th October 1926 the sending and receiving beam stations, erected at Bodmin and at Bridgewater respectively for the UK–Canada link had been submitted to seven days consecutive working and had satisfactorily fulfilled the conditions mentioned above.

The stations of the Imperial beam system became operational on the following dates [29]: Bodmin to Canada, 24th October 1926; Drummondville, (Canada) to England, 25th October 1926; Grimsby to Australia, 8th April 1927; Ballan (Australia) to England 8th April 1927; Bodmin to South Africa, 5th July 1927; Klipheuvel (South Africa) to England, 5th July 1927; Grimsby to India, 6th September 1927; and Kirkee (India) to England, 6th September 1927.

The implementation of the Empire beam system was an astounding engineering achievement. Its completion was a great credit to the engineering prowess of MWT, and to the genius of Franklin and the staff, especially Mathieu, who had undertaken the necessary design. The stern conditions of the Post Office's contract were not only met, but as Marconi reported: 'It has since the agreement been found that in regard to stations communicating with Australia, South Africa, and India, a daily average of over 20 hours of high speed communication is attainable, and that the spread of the beam is much narrower than had been specified in the Government requirements' [29].

But not everyone was happy with the new communications medium. The cable companies began to lose money as a consequence of the reductions in the full-rate charges per word to and from Great Britain. For 1928 the rates were: [Command Paper 3163]

	Australia	*South Africa*	*India*
Cable rates, January 1927	2s 6d (12.5p)	2s 0d (10p)	1s 8d (8.33pp)
Present cable rates	2s 0d (10p)	1s 8d (8.33p)	1s 5d (7p)
Beam wireless rates	1s 8d (8.33p)	1s 4d (6.66p)	1s 1d (5.42p)

Thus the Beam full-rate was cheaper by 7d (2.92p) to 10d (4.17p) per word than the old rates. Following a price reduction by the cable companies the difference was 4d (1.66p) per word, on each of the three services, between the cable and wireless charges for full-rate traffic; a uniform difference of 2d (0.83p) for deferred traffic, and differences ranging from 3d (1.25p) to 0.5d (0.21p) for letter telegrams and Press telegrams. Naturally, the cheaper Beam rates led to traffic moving away from the cable services. For two periods in 1927 and 1928 the traffic transmitted by the Beam services has been estimated at the following rates per annum:

	September–November 1927	*March–May 1928*
Canadian service	5 million words	6 million words
Australian service	8 million words	9 million words
South African service	9 million words	9 million words
Indian service	9 million words	10.5 million words

This transfer of traffic to the Beam services away from the cable services, together with the fall in the cable rates, led to a loss of revenue for the cable companies and caused some of them to be placed in a parlous financial state. There was a suggestion that 'those responsible for the Cable Companies might be pressed, unless a satisfactory means of obviating the effect of acute competition could be provided, to liquidate their undertakings at once and distribute their large reserves among their shareholders, rather than remain in operation and dissipate their resources' [Command Paper 3163]. The Pacific Cable Board, by way of illustration, suffered a revenue loss of c. £80,000 in 1927–28 and as this was of concern to the Dominion Governments who had subsidised the service they quite reasonably became alarmed at the future prospect and requested the British Government to enquire into the state of affairs that had arisen.

As a consequence 'The Imperial Wireless and Cable Conference' was convened in 1928 under the chairmanship of The Rt. Hon. Sir John Gilmour, Bt., DSO, MP, Secretary of State for Scotland, with representatives from Canada, the Commonwealth of Australia, New Zealand, the Union of South Africa, the Irish Free State, India and the Colonies and Protectorates.

Their terms of reference were: 'To examine the situation which has arisen as a result of the competition of the Beam Wireless with the Cable Services, to report thereon and to make recommendations with a view to a common policy being adopted by the various Governments concerned'.

Thirty four meetings were held from 16th January 1928 and on 6th July 1928 the Conference issued its Report. In it some stress was placed on the need to retain the cable systems for strategic purposes and as a safeguard against the fading and occasional prolonged interruptions to which the wireless services were subjected. Moreover, the Conference was concerned about the threat from foreign enterprises to secure an increased share in the control and operation of world communications. Thus both the cable and wireless services had to co-exist.

There were five ways by which the Governments concerned could deal with this problem, namely: by non-intervention; by a subsidy; by a minimum revenue guarantee; by a polling scheme; and by a fusion of interests – 'To amalgamate so far as possible in one undertaking all the cable and wireless interests conducting communications between the various parts of the Empire so as to secure unity of control and unity of direction'. The latter course was chosen. Actually, negotiations for a merger between the Eastern Telegraph Companies and

Marconi's Wireless Telegraph Company had commenced before the Conference held its first meeting, and on 14th March 1928 the Chairmen of the two Companies informed the Conference of their agreement for a fusion of the interests of the Companies through the medium of a proposed Holding Company.

Subsequently, the Conference recommended a scheme that would have the following objects: '(a) to secure, as far as [was] possible, all the advantages to be derived from unification of direction and operation; (b) at the same time to preserve for the Governments concerned control over any unified undertaking which [might] be created, so as to safeguard the interests of the public in general and of the cable and wireless users in particular; and (c) to secure these desiderata at the minimum of cost to the Governments concerned'.

The scheme would be based on the formation of (1) a 'Merger Company' (which became known as Cables and Wireless Ltd) that would acquire all the ordinary shares of the Eastern, Eastern Extension and Western Telegraph Companies, and all the ordinary and preference shares and debentures of the Marconi Wireless Telegraph Company; and (2) a 'Communications Company' (which became known as Imperial and International Communications Ltd) to which the Cable and Marconi Companies would sell all their communications assets in exchange for shares. The Board of Directors of the two companies would be identical, with two of the Directors – one of whom had to be Chairman of the Communications Company – approved by His Majesty's Government at the suggestion of the Cable Companies.

And so, with the merger, from 8th April 1929, there came into operation the finest system of world communications ever to exist under the control of a single organisation. From 1934 the names of the Merger and Communications Companies became Cable and Wireless (Holding) Ltd and Cable and Wireless Ltd respectively.

3. Picture transmission

The development of electronics and radio communications during the First World War had a most profound effect on the realisation of practical picture transmission services. Prior to 1914, Korn, Belin, Dieckmann, Thorne-Baker and others (see Chapter 8) had had to endeavour to evolve facsimile systems without the aid of electronic amplifiers, oscillators, detectors, modulators and all the other circuit variants that diode and triode valves make possible. At the end of the war, these circuits had been fully demonstrated and used, and the components had become available to professionals and amateurs alike. Not surprisingly, Korn, Belin, and Dieckmann, each with c. 20 years experience of picture transmission experimentation, together with other 'lone' inventors, including C.F. Jenkins, and D. von Mihaly, embraced the new technology and incorporated it into their post-war work. Experimental transmissions of picture material across the Atlantic ocean were made by Belin [30] and Korn [31] in 1922 (Fig. 18.5); Jenkins [32] demonstrated his radio system in Washington, D.C., in

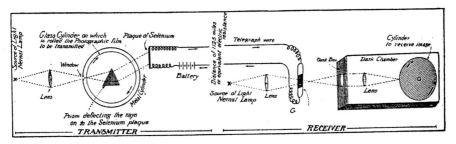

Figure 18.5   *Diagram showing the principles of operation of Korn's 1906 transmitting and receiving instruments for reproducing photographs at a distance. G is a Geissler tube which modulates the light flux, from the Nernst lamp, onto the photographic film in the receiver.*
Source: The Illustrated London News Picture Library.

1923; the American Telephone and Telegraph Company mounted a trial of well-engineered picture transmission equipment operating over telephone lines in 1924 [33]; and in the same year photoradiograms were sent from London to New York by Marconi's Wireless Telegraph Company in collaboration with the recently formed Radio Corporation of America [34].

The tests carried out by AT&T, and by MWT and RCA, with their vast laboratory and workshop resources, their extensive experience and knowledge, and their considerable financial backing effectively brought an end to the efforts of the 'lone' inventors in the speciality of facsimile transmission. A similar situation would manifest itself about a decade later in the field of high definition television. (After c. 1936 only Electric and Musical Industries, the Radio Corporation of America, and the (Farnsworth) Television Laboratories were able to offer all-electronic high definition television systems for sale.) The days of an individual inventor working on 'seeing by electricity' in a private laboratory/ workshop with inadequate funding, and with just some meagre literature scattered in various magazines and journals to assist them, were now coming to an end. From the early 1920s the large industrial organisations of the American Telephone & Telegraph Company, the General Electric Company, the Radio Corporation of America, the Westinghouse Electric and Manufacturing Company, Marconi's Wireless Telegraph Company, and Telefunken, and, from the late 1920s, the Baird and the Farnsworth companies, Electric and Musical Industries Ltd and Fernsehen, would dominate the subject.

The interest of the Bell System in telephotography dates from the beginning of the 20th century. In 1903 staff of AT&T produced a comprehensive report on 'Transference of images by electricity' [35]. It was recognised at the time that the state of scientific and engineering knowledge and technology was inadequate to meet the formidable challenge posed. Further consideration of the problem was given by AT&T's engineers during the First World War following requests from press associations for private-line picture services. By 1920 AT&T's exploratory work was sufficiently promising to justify a large scale investigation, and for the next three years scientists and engineers of the company's Department of Development and Research and at Western Electric's headquarters collaborated closely on the implementation of an overall system design (see Fig. 18.6). This work led to a private demonstration in May 1923 when images were sent over a telephone circuit from New York to Yama Farms, a distance of c. 60 miles (96.5 km).

A public demonstration of the preliminary commercial design was given on 8th May 1924 when highlights of the Republican National Convention were sent between Cleveland and New York [33]. Another system was set up, also in 1924, between Chicago and New York to cover the Democratic National Convention in Chicago. The 5 in × 7 in (12.7 cm × 17.8 cm) images were published in newspapers in these cities on the same day as they were received. Their speedy publication emphasised the time advantage of the method over the slow process of sending photographs by mail. However, the *New York Times* noted that the picture quality was 'restricted by interference and fading' [36].

Further development work led to an increase in the resolution of the images from 25 to 40 lines per cm and a change from the variable line width method to the variable density method. Fig. 18.7 shows details of aspects of the new design. During the latter half of 1926 modified machines were installed in New York, Chicago, Cleveland, Boston, Atlanta, St Louis and Los Angeles to test the market demand for a transcontinental commercial telephotograph service [37]. The scanning time was about 7 minutes but, of course, the total time required from the taking of a photograph at the sending end to the printing of the photograph at the receiving end was in the order of 45 minutes.

Subsequent analysis of the system (known as type A), from the beginning of 1927, by AT&T's Long Lines Department, led to the decision by Bell System management that a R&D programme should be initiated by the recently formed Bell Telephone Laboratories which would lead to an improved system (later designated Type B).

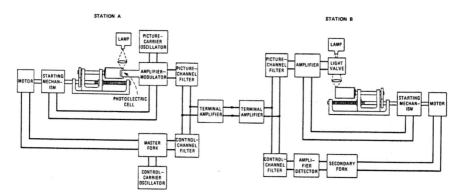

Figure 18.6   Block diagram of AT&T's picture transmitting and receiving system.
Source: Fagen, op. cit.

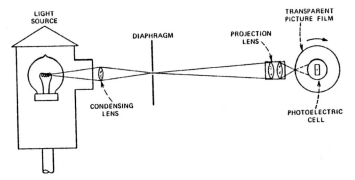

Cross section of sending-end optical system.

Photoelectric cell of type used in picture transmission.

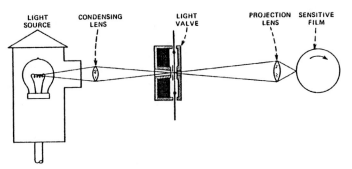

Cross section of receiving-end optical system.

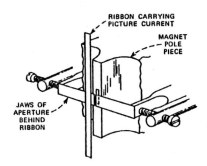

Details of light valve.

Figure 18.7 Details of some of the devices used in AT&T's system.
Source: Fagen, op.cit.

Of the other facsimile schemes tried during the 1920s some mention must be made of the RCA system which was developed under the leadership of R.H. Ranger [38]. His objective was to base his system on a dot-dash method. He wrote (in 1926): 'the reason that the Morse code still exists is because it is the most economical means of getting a given amount of words from one point to another, in the shortest time, with the least power, over the greatest distance and through the greatest amount of interference'.

Ranger appreciated that newspaper illustrations used a half-tone process, in which 25.6 dots per cm and 5 tonal values per dot (i.e. 128 photo-pulses per cm) could adequately reproduce an image of a scene or object, and sought to evolve a means whereby just 65 'photo-pulses per inch' (25.6 photo pulses per centimetre) could be used to analyse and synthesise an image. In his pulse width modulation method, the analogue signal from the photo-electric cell of the sending picture unit was used to modulate the pulse width of the individual pulses of a constant amplitude pulse train. At the receiver the variations of the width of the received pulses changed the tonal gradations produced from almost white (with very narrow pulses) to almost black (using very long pulses). During the transmission 'the radio transmitter [was] on even for the shortest dot, [and so] we have taken full advantage of this telegraphic supremacy to carry a picture to the greatest distance'.

On 6th July 1924 the first long distance trial of the system took place when a photograph of Secretary Hughes was sent from New York to England and then back to New York. The first public trans-Atlantic demonstration occurred in November 1924. Picture signals were sent from London to the Caernarvon radio station, transmitted to Riverhead, Long Island, USA, and then sent by line to RCA's New York office [34]. The image was reproduced in Ranger's 1926 *Proc IRE* paper. However, the very 'grainy' quality of the picture stressed the need for much more R&D work.

Some improvements were made and from 1926 the system was operational between Radio House, London (Marconi's Wireless Telegraph Company's main telegraph station) and New York. The material copied included cheques, wills, contracts, blueprints and specimen signatures. Two limitations of the service were soon manifested, namely, the length of time taken to send an image, and the poor resolution of the reproduced picture. The first constraint made the cost of transmission prohibitively expensive and the second led to disappointments; either a reappraisal of the scheme or an entirely new approach was required [39].

Meanwhile MWT's super high-speed Morse telegraph was achieving much success; by 1927 demonstrations had been given at which speeds of 3,000 words per minute had been recorded. Since certain aspects of this development, including the use of the Kerr cell as a high speed shutter for polarised light, seemed to have an application to facsimile work, a decision was made to engineer new copying and recording instruments. On 3rd November 1929, before a large gathering of press representatives, the Marconi-Wright facsimile apparatus was demonstrated. An image was transmitted from a beam station at Rocky Point, USA and was received at the beam station at Somerton, Somerset, UK. Three minutes was the average time for sending a photograph, and therefore, as the equipment could be run continuously, the electrical transmission of entire newspapers from one country to another could be realistically contemplated.

By 1934 the state of the art had progressed to the stage where Cable and Wireless were able to inaugurate a facsimile service between England and Australia. The Marconi-Wright system was highly adaptable: not only documents and photographs but also motion picture film could be electrically transmitted. On 23rd October 1934 when the winners of the England–Australia air race landed in Melbourne, film of the event was shown in British cinemas just a few hours later.

Unfortunately, as with the long distance sending of Morse and speech, the facsimile service was not immune to the effects of changes in the ionosphere and from time to time the service suffered degradation. Diversity transmission and reception and error detection and correction methods had to be developed to minimise the deleterious effects but consideration of these topics is outside the scope of this book.

4. Domestic wireless pictures

In May 1926 W. Watson and Sons Ltd, a highly reputable firm of manufacturing opticians of High Barnet, (near London), wrote [40] to the British Broadcasting Company and suggested that the Company should transmit to the general public still pictures by radio using a system, invented by T. Thorne-Baker, known as the Izon wireless picture system.

Watson offered to supply the appropriate apparatus for the BBC's Daventry transmitting station and invited the Company to set aside a time, two or three times a week when broadcasting was not normally in progress (say, between 11.00 p.m. and midnight) for the above purpose. Watsons further mentioned that, in their opinion, attachments which would allow domestic radio receivers to reproduce the transmitted pictures could be manufactured and sold for approximately £30. They considered the BBC should engage speakers, to talk about particular broadcast pictures, and enquired whether the Company would be willing to pay half the engagement fees.

Thorne-Baker, from the time of his introduction to photo telegraphy (see Chapter 8) had always been interested in developing a simple portable picture reproducer which could be manufactured cheaply and be used by many different sections of the community including the armed forces, the police, the newspaper industry and the general public. Consequently, he had abandoned the commercially adopted methods of driving and synchronising two mechanically independent rotating cylinders by means of synchronous or phonic motors, tuning fork oscillators and the like and had concentrated his attention on the task of achieving the same purpose with gramophone-type clockwork motors.

The above constraint introduced two problems:

(1) the reduction of the frictional losses associated with the driving mechanisms to enable the small output torque machines to drive the transmitting and receiving cylinders for approximately 3.5 minutes; and

(2) the synchronising of two non-electrically driven motors.

Thorne-Baker's solution [41] to the first of these problems was to replace the 'half-nut' lead screw carriage arrangement, which had been employed in many apparatuses, by a stylus carrier that incorporated a knife-edged wheel travelling in a screw thread. Second, he developed a synchronising device [42] based on a freely swinging pendulum having a period of about one second. This was fitted with contacts, which formed part of an electric circuit, and permitted an electromagnetic catch to be energised at the end of each complete swing of the pendulum. In the unenergised state of the catch, the motion of the cylinder (which had a speed slightly greater than one revolution per second), was stopped at the end of every rotation but when the catch was actuated, the cylinder was able to recommence rotating. Thus the speeds of both the transmitting and receiving cylinders were controlled by two *independent* pendulums, unlike Bain's system which used two *interactive* pendulums, (see Chapter 8). Ideally what was required was a system in which the position of the recording cylinder was checked by means of a transmitted synchronising pulse at the end of every revolution. Korn had investigated this problem in some detail during the first decade of the 20th century.

The solution to the replacement of Thorne-Baker's pendulum synchronisers (which militated against a small portable apparatus) and the automatic control of the receiving cylinder, was suggested by a Captain Otho Fulton, who had been influenced by the success that Korn had achieved with his method. In the version [43] that was eventually sold to the public under the trade name *Fultograph*, the cylinder was 4 inches (101.6 mm) in length and 1 15/16 inches (49.2 mm) in diameter, and, as the total traverse of the screw was 3 13/16 inches (96.8 mm), the maximum size of picture that could be reproduced was 3 3/4 inches by 5 1/2 inches (95.25 mm × 13.97 mm). At the normal running speed of 50 rev/min, the time taken to receive the 19.7 lines/cm picture was approximately 3 3/4 minutes.

To promote the general exploitation of Thorne-Baker's Izon scheme, a company called Wireless Pictures Ltd was established in January 1927 (with a share capital of £100) [44]. The three directors were Captain Otho Fulton, F.W. Watson Baker and V. Sheridan. The prospectus for the new company appeared in the national press on 17th September 1928 and the public was invited to subscribe to the issue of 775,000 ordinary shares of 4 shillings each at par. Because of the 'very heavy over subscription' the shares had to be allotted on a basis that allowed applicants only 10 per cent of their requested number of shares when this number exceeded 7,500. For small applications, the allotment varied from zero to 50 per cent [45].

Meanwhile, much consideration had been given by the British Broadcasting Corporation and the General Post Office to the May 1926 proposal of W. Watson and Sons Ltd. This consideration (which is dealt with at length in one of the author's papers [46]) led to a trial of the system held from 1st August 1928 to 14th August 1928. During the eight days of tests, approximately 100 pictures were transmitted from the BBC's transmitting stations, 5XX and 2LO. Of these, 86 were 'good transmissions, many of them being extremely good' [47]. In view of the simplicity of the apparatus and its ease of operation the Post Office felt the received images were very satisfactory and accordingly assented to the commencement of an experimental service [47].

The new service began at 02.00p.m. on Tuesday, 30th October 1928. Two pictures were broadcast, one of the King and another of a topical cartoon. *The Times* noted [48]: 'Both were very distinct, and that of the King was an amazingly good presentation. The line drawing of the cartoon made a sufficiently exacting test of the synchronising gear, out of which the Fultograph emerged successfully'. Subsequently picture transmissions were effected daily, with the exception of Sundays and Mondays, and normally four subjects were dealt with during the 25 minute programme period. Apart from the inaugural service, newspaper and picture agencies were associated with the provision of the subjects. The *Daily Express* provided these from 31st October to 3rd November inclusive.

Notwithstanding the initial good press which the experimental service received at this time, there were some ominous clouds on the horizon. First, there was some difficulty in receiving the Daventry (5XX) signals in many areas [48]; second, sales of the Fultograph machine were proving to be disappointing [49]; and third, the Company's problems were being compounded by production delays. In order to sell their receiver at a price that would suit the average person's pocket, Wireless Pictures (1928) Ltd had entered into 'fairly large contracts' but because of manufacturing setbacks these receivers were not available to the shops during the Christmas spending period [50].

The BBC appreciated the company's difficulties and agreed in December to guarantee the picture transmissions for a year from 31st October 1928. However, by July 1929 the Company had sold only 700 Fultograph receivers: rather less than half were in the hands of the general public – the balance belonged to traders who used them for demonstration purposes [51].

An interesting application of the Fultograph took place in 1929 during a preliminary flight of the R101 airship when weather maps were received aboard the craft. On this occasion the maps were transmitted from the Rugby radio station, which was operated by remote control from the Cardington airship base where the maps had been prepared.

Several reasons, in addition to those indicated previously, can be advanced to account for the public's lack of interest in purchasing picture receiving apparatus:

1. the prices of £22.75 and £24.75 were equivalent to approximately eight

weeks' earnings (before tax), for the working man, (the average yearly income for employees in 1928 was £144.90), and so the cost of a Fultograph machine was a significant part of a person's annual salary;

2. a daily newspaper could be bought for a penny (0.416p) and possibly as many as twelve photographs obtained thereby;

3. some of the broadcast picture programmes were not 'too satisfactory', as the Company admitted in July 1929 [50]; and

4. in March 1929, the Postmaster General had informed the Baird Development Television Company, (and the general public for his letter to that company had been published in *The Times*), that he was anxious that facilities for further television developments should be granted and would agree to a station of the BBC being used for this purpose outside broadcasting hours [52].

The last point had a most inhibiting effect on the commercial fortunes of Wireless Pictures (1928) Ltd and the Company's shares suffered 'a severe slump in price' with the consequence that the uncalled capital of the Company (amounting to £190,000), which was 'very badly needed', could not be obtained [50]. In the meantime, their 'rather meagre' working capital had been almost expended (by July 1929) owing to:

1. the necessary carrying on of the Company's work;
2. the purchase of 3,000 receivers valued at about £35,000.

These impediments to progress led the BBC to conclude, with regret, that it would not be justified in continuing the picture transmissions beyond the end of October 1929 and informed Wireless Pictures (1928) Ltd accordingly. Naturally, the BBC's decision distressed the Company, which was 'extremely anxious' to secure an extension of the transmission period for the following arguments:

1. the Company was negotiating for the sale of its overseas rights in several countries and hoped, in particular, to effect a sale in the United States of America and Canada (a picture transmission service had commenced on 21st November 1928 from Konigwusterhausen, Germany);

2. the Company was making some slow progress in its endeavours to interest the Air Ministry, the Admiralty, the War Office, Scotland Yard and newspaper offices in the Fultograph;

3. further broadcast facilities would enable the Company to liquidate its stock holding;

4. negotiations for an amalgamation with the Baird company were taking place; and

5. the Company was responsible for £235,000 share capital.

These representations were heard sympathetically by the BBC, consistent with a logical view of the situation and a proper consideration of the Corporation's duty to its listeners [51]. But the Corporation was unable to grant an extension of time.

Wireless Pictures (1928) Ltd was voluntarily liquidated on 31st July 1930 and dissolved on 12th July 1935 [44]. Essentially the service failed to achieve success because contemporaneous developments in the field of television were actively being pursued in several countries and the anticipated entertainment potential of television greatly exceeded that of still-picture broadcasts. These developments are considered in the next chapter.

Note 1
The early history of ionospheric physics

When Marconi, on 12th December 1901, transmitted signals from Poldhu, Cornwall and received them at St Johns, Newfoundland there was some speculation concerning the path that the electromagnetic (e.m.) waves had followed. Because of the earth's curvature the elevation of the Atlantic Ocean at the mid-point of the great circle route joining the two places is 124 miles (c. 200 km) higher than the mid-point of a straight line joining Poldhu and St Johns. Rectilinear propagation, atmospheric refraction, and diffraction could not explain the observed result.

A. Kennelly [53], of the USA, and O. Heaviside [54], of the UK, independently in 1902 suggested a propagation mechanism. Kennelly wrote: 'On waves that are transmitted . . . to distances that are large by comparison with 80 km, it seems likely that the waves may . . . find an upper reflecting surface in the conducting rarefied strata of the air'. The reflecting surface became known as the Kennelly–Heaviside layer.

This hypothesis was not immediately accepted, but, from 1912, experimental work began to corroborate its validity. W.H. Eccles in 1912 showed that the effective velocity of some e.m. waves increases at certain heights of the atmosphere and that the waves experience 'ionic refraction' and bend around the earth as they travel. In the same year L. de Forest and L.F. Fuller [55] of the Federal Telegraph Company, using the company's newly installed Poulsen arc transmitters, observed phenomena on the San Francisco to Honolulu radio link which could only be explained by assuming an interference of two coherent waves. De Forest estimated the height of the reflecting layer and noted: 'If the reflecting layer is half way between the stations its height is 62 miles [c. 100 km]'. Fuller interpreted his results as being due to interference fringe patterns formed by the interference of a sky wave and a ground wave.

Further confirmation of a reflecting layer came from T.T. Eckersley's work, in 1921, on radio direction finding. His carefully conducted experiments showed that 'night errors' in radio direction finding could only be interpreted by postulating the existence of such a layer.

The first direct evidence of the ionosphere was obtained by E.V. Appleton and M.A.F. Barnett [56] on 11th December 1924 and 17th February 1925. They persuaded the British Broadcasting Corporation to vary (at a uniform rate in a given time, e.g. 10 s to 30 s) the frequency of the unmodulated carrier of its Bournemouth transmitter, during which time the antenna current remained substantially constant. In Oxford, Appleton and Barnett employed a 4-stage h.f. amplifier coupled to an Eindhoven galvanometer to observe the received signal. The anticipated effects were noted and ascribed to coherent interference between the direct and indirect waves propagated from Bournemouth. For a wavelength change from 421 yd to 429 yd (385 m to 392 m), the average number of interference fringes was 4.5; a change from 421 yd to 432 yd (385 m to 395 m) produced 7.0 interference fringes. From these findings Appleton and Barnett deduced that the virtual height of the Kennelly–Heaviside layer was about 50 miles (80 km).

Further experiments were conducted to exclude the possibility of an indirect horizontal path, and to determine the height of the layer from the downward angle of arrival at the receiver. For reception Appleton and Barnett used a T antenna (the horizontal portion of which was at right angles to the plane of polarisation of the e.m. waves) and a single turn loop antenna having its plane in the plane of polarisation of the e.m. waves. The two antennas had horizontal polar diagrams which were omni-directional and figure-of-eight respectively. By comparing the signals from these antennas it was possible to deduce the vertical arrival angle of the reflected wave. Other observations at increasing ranges gave arrival angles that were consistent with the Kennelly–Heaviside layer (or E-layer) being at a height of 50 miles to 62 miles (80 km to 100 km). A further layer, now called the Appleton or F-layer, was detected at 149 miles (c. 240 km). It reflects radio waves of a frequency higher than those which are reflected by the E-layer.

From 1927 to 1932 Appleton, aided by Hartree, established a magneto-ionic theory to explain the physics of the ionosphere. For all his work in this field, which enabled forecasts to be made of the optimum frequencies for long-distance short-wave communications as a function of transmitter and receiver positions, Appleton was awarded the Nobel Prize for Physics in 1947.

References

1   MARTIN, M.G.: 'Long distance telephony in Europe', *Annales des Postes Télégraphe et Téléphones*, June 1921
2   Gill, F.: 'Electrical communication – telephony over considerable distances', Presidential Address, *J.IEE*, 1922, **61**, No. 313, pp. 1–5
3   Report on 'European long distance telephony', *Electrical Communications*, 1923, pp. 64–80
4   FAGEN, M.D.: 'A history of Engineering and Science in the Bell System' (Bell Telephone Laboratories, 1975), pp. 235–240
5   Heaviside, O.: papers in *Electrician* (London), 1873–1901
6   Ref. 4, pp. 241–252
7   Report on 'Elementary theory of capacity unbalance in quadded cables and of capacity unbalance sets', Crosstalk practices bulletin, No.407, section 6-C, AT&T, 24th April 1929, pp. 1–11

8 Morris, A.: 'Some aspects of the electric capacity of telephone cables', *Post Office Electrical Engineers Journal*, 1927–28, **20**, pp. 43–51

9 Melling, H.: 'Capacity unbalance sets', Report R3111/2513/ H M, International Standard Electric Corporation, 22nd February 1929, pp. 1–5, personal collection

10 BURNS, R.W.: 'The life and times of A.D. Blumlein' (Institution of Electrical Engineers, London, 2000), pp. 68–72

11 CRUICKSHANK, W.: 'Telegraphy and telephony', *J. IEE*, 1926, **64**, p. 158

12 ECCLES, W.H.: 'Inaugural address', *J. IEE*, 1927, **65**, pp. 7–9

13 JACOT, B.L., and COLLIER, D.M.B.: 'Marconi – master of space' (Hutchinson, London, undated), p. 150

14 LLOYD GEORGE, D.: 'War memoirs of David Lloyd George' (Odhams Press, London, undated), vol. 1, pp. 9, 14

15 Ref. 14, vol. 2, p. 1544

16 JOLLY, W.P.: 'Marconi' (Constable, London, 1972), p. 227

17 MARCONI, G.: 'Radio telegraphy', *Proc. IRE*, 1922, **10**, pp. 215–238

18 BAKER, W.J.: 'A history of the Marconi Company' (Methuen, London, 1970), p. 217

19 Ref. 13, p. 159

20 Ibid, p. 160

21 Ref. 18, p. 218

22 Ref. 18, p. 219

23 Ref. 13, p. 186

24 Ref. 18, p. 220

25 Ref. 18, p. 206

26 Ref. 18, p. 211

27 SHAUGHNESSY, E.H.: 'Chairman's Address', *J. IEE*, 1925, **63**, p. 64

28 Ref. 18, pp. 214–215

29 Ref. 13, p. 193

30 BELIN, E.: 'Transatlantic radio-telephotography of written or printed characters and of drawings', *Onde Electrique.*, 1922, **1**, pp. 271–283

31 BENINGTON, A.: 'Transmission of photographs by radio', *Radio News*, 1922, **4**, pp. 230, 369–372

32 DAVIS, W.: 'Seeing by radio', *Popular radio*, 1923, **3**, pp. 266–275

33 ANON.: 'Pictures by wire sent with success for the first time', *New York Times*, 20th May 1924

34 ANON.: 'Pictures sent from London here in 20 minutes', *New York Times*, 1st Dec. 1924

35 Ref. 4, pp. 784–790

36 ANON.: 'Pictures by radio restricted by interference and fading', *New York Times*, 25th May 1924

37 Ref. 4, p. 786

38 RANGER, R.H.: 'Transmission and reception of photoradiograms', *Proc. IRE*, 1926, **14**, pp. 161–180

39 Ref. 18, p. 253

40 REITH, J.F.W.: letter to DALZELL, R.A., 4th June 1926, Post Office bundle Post 33/2371, 10060/1928

41 THORNE-BAKER, T.: British patent no. 248,836, 11th September 1925

42   THORNE-BAKER, T.: British patent no. 249,020, 11th September 1925
43   HAYNES, F.H.: 'The Fultograph', *Wireless World*, 24th October 1928, pp. 557–560
44   Wireless Pictures Ltd., company file, BT31/32987. Public Record Office
45   ANON.: a report, *The Financial News*, 22nd September 1928
46   BURNS, R.W.: 'Wireless pictures and the Fultograph', *IEE Proc*, 1981, **128**, Pt. A, No. 1, pp. 78–88
47   GILL, A.J.: a report, 14th August 1928, Post Office bundle Post 33/2371, 10060/1928
48   Minutes, Control Board, 20th November 1928, BBC file T12
49   Minutes, Control Board, 3rd December 1928, BBC file T12
50   Wireless Pictures (1928) Ltd.: letter to the BBC, 12th July 1929, BBC file T12
51   REITH, J.F.W.: letter to Secretary, GPO, 15th July 1929, Post Office bundle Post 33/2371, 10060/1928
52   BURNS, R.W.: 'British television, the formative years' (Institution of Electrical Engineers, London, 1986), chapter 5
53   KENNELLY, A.: 'On the elevation of the electrically-conducting strata of the earth's atmosphere', *Electrical World and Engineers*, 1902, **39**, p.473
54   HEAVISIDE, O.: 'Telegraph theory', *Encyclopaedia Britannica*, **33**, 10th edition, p. 215
55   TUVE, M..A.: 'Early days of pulse radio at the Carnegie Institution', *Jour. of Atmos. and Terr. Physics*, 1974, **36**, pp. 2079–2083
56   APPLETON, E.V., and M.A.F. BARNETT.: 'On some direct evidence for downward atmospheric reflection of electric ray', *Proceedings of the Royal Society*, Series A, 1926, **109**, pp. 621–641

Chapter 19

The rise and fall of low definition television, c. 1920–c. 1930

As described in Chapter 16, the First World War gave an enormous impetus to the utilisation of valves in signalling systems and so stimulated advances in technique that by 1918 triodes could be manufactured to cover a wide power range and were suitable for both transmitting and receiving purposes; their theory and operation, over the frequency bands used at that time, were both well understood.

Accordingly, when interest in television was revived a few years later by scientists and inventors in the UK, the US, France, Germany and elsewhere, the basic components of a distant vision system were available. The principles of scanning an object or image by means of apertured discs, lensed discs and mirror wheels had been expounded by Nipkow (1884), Weiller (1889), and Brillouin (1891) and others; methods and apparatuses for synchronising two non-mechanically coupled scanners had been suggested by many workers and demonstrated in facsimile transmission systems; much development work had taken place on photoelectric cells; the means for amplifying the weak currents obtained from these cells seemed to be available; and now, at the end of the 1910–1920 decade, the subject of radio communications had evolved to the state where commercial broadcasting could be seriously contemplated. And if sound signals could be propagated by radio, then surely vision signals too could be transmitted.

Undoubtedly the growth of commercial radiotelephony and domestic broadcasting influenced the progress of television. The time was opportune for further attempts at the practical implementation of a television system. Whereas from 1911–1920 only a few isolated attempts (Table 19.1) had been made to investigate, on an experimental basis, the subject of 'distant vision', from early in the 1920–30 period determined efforts to advance television were made. Initially these endeavours were mainly those of individuals working in isolation from others. J.L. Baird of the UK, C.F. Jenkins of the USA, E. Belin of France and D. von Mihaly, a Hungarian working in Germany, were four of the principal early investigators in this period. For a short time in 1923, V.K. Zworykin

Table 19.1 *Summary of the principal proposals made during the 1910–1920 decade for television*

Date	Name	Opticoelectrical transducer	Transmission scanner	Transmission link	Electrooptical transducer	Receiver scanner	Synchronisation means	Remarks
2.03.1911 (30.11.1911) British patent	B.Rosing Russia	photoelectric receiver – not specified	two rotating mirror drums scanning in orthogonal directions	two signal wires plus four synchronising wires	not specified	two mirror oscillographs, the axes of which correspond to the axes of the mirror drums	the mirror drums generate synchronising signals which drive the receiver's oscillographs	the patent is wholly concerned with synchronising means and states: 'It has been found impracticable to obtain necessary synchronism . . . in view of the enormously high speeds of the mechanisms'
7.11.1911 (lecture) January 1911 (paper)	A.A. Campbell Swinton, UK	mosaic of photosensitive cubes (e.g. of rubidium) in contact with sodium vapour	electron beam, of cold cathode-ray tube, magnetically deflected	signal wire and two synchronising wire	the brightness of the luminous spot on the fluorescent screen of a c.r.t. is modulated by the electron beam	the electron beam of a cold c.r.t.: the beam is magnetically deflected	a common source of line and frame scanning signals for both the transmit and receive	
25.06.1914 (25.06.1915) British patent	S.L. Hart UK	any form of photosensitive cell: a selenium bridge is mentioned	a rotating multilens drum, the axis of which oscillates through a small angle: the axes of the lenses are tilted with respect to each other	one wire for both the vision and the synchronising signal	a discharge tube and an external electromagnet, excited by the vision signal, which deflects the line of the discharge	the same as the transmitter scanner	line and frame pulses are generated ano combined with the vision signal	a form of interfacing is described. This seems to be the first patent to state a method for combining the vision, line and frame signals; no evidence of experimentation

Date	Name / Country	Detector	Scanning	Signal link	Light source / modulator	Receiver scanner	Synchronisation	Remarks
13.07.1914 (presentation) 27.07.1914 (paper)	G. Rignoux France	a mosaic of 64 selenium cells (each of which has a relay in series with it)	a rotating commutator samples the relay field (the relays operate when the selenium cells are illuminated)	two signal wires plus four synchronising wires	an arc lamp and a Faraday effect modulator which uses carbon tetrachloride	a single mirror wheel	not described	images of the letters H, T, L and U were reproduced on a screen; no half-tones could be reproduced and the images were faint: The commutator ran at 450 r.p.m.
01.04.1915 (08.12.1915) French patent	A. Voulgre France	a glass ampoule containing an amalgam of sodium and rubidium	the scanning system uses three moving belts – each of which has transverse slots – and a rotating slotted disc (the slots being radial)	two signal wires	a mercury vapour lamp fed from the output of a transformer the input to which is the vision signal	the same as the transmitter scanner	not described: the motors driving the scanners at the transmitter and receiver must have constant speeds	a rather impractical scheme: an attempt to ease the problem of synchronisation; no evidence of experimentation
07.12.1917 (16.10.1923) US	A.M. Nicolson UK	a photocell – not specified	a small oscillating mirror is supported by two orthogonal wires and scans spirally by means of two electromagnets	a radio link is described	the brightness of the luminous spot on the fluorescent screen of a c.r.t. is modulated by the electron beam	the electron beam of a c.r.t. plus two pairs of deflecting plates	the synchronisation and vision signals modulate a carrier wave: they are separated at the receiver	no evidence of experimentation
18.02.1919 French patent	D. von Minaly Hungary	a mosaic of selenium cells (each of which has a coil in series with it) and a common battery	a rotating coil has a voltage induced into it when it scans the bank of coils associated with the selenium cells	two signal wires	a bank of lamps, each of which is connected to a coil	a rotating coil connected to the signal wires causes the relays of the bank of lamps to operate	a pendulum escapement driven through a worm gear	much experimental work was undertaken by Minaly

Table 19.1 – continued

Date	Name	Opticoelectrical transducer	Transmission scanner	Transmission link	Electrooptical transducer	Receiver scanner	Synchronisation means	Remarks
10.09.1919 (25.07.1922) US patent	H.K. Sandell US	a linear array of n selenium cells	a single mirror drum scans the image over the linear array of cells (the width of the scanner is the same as the width of the linear array of cells)	n transmitters and n receivers; transmitter-receiver combination is tuned to a different carrier frequency	a linear light source and a linear array of n small mirrors each of which is deflected by a small electromagnet	a single mirror wheel causes the light reflected from the mirrors to be reflected onto the screen	not described	the width of the scanners is the same as the width of the linear arrays; no evidence of experimentation
18.08.1920 Russian patent	C.H. Kakourine Russia	a photoelectric cell	a Nipkow disc	a radio link	an electromagnetic-ally operated shutter	a Nipkow disc	not clear from patent	an impractical, naive scheme as described in the patent; no evidence of experimentation
24.08.1920	H.C. Egerton US	a photoelectric cell	a single oscillating mirror driven by two orthogonal vibrating motors	a wire or radio link for the combined (vision and synchronising) signal	a lamp, plus a screen of varying transparency, and a movable mirror controlled by the vision signals	same as transmitter scanner	synchronising signals generated at the transmitter and sent over the link to the receiver	the picture area could be scanned in groups of scannings, each group consisting of two scannings in succession, the different scanning of a group traversing different paths; no evidence of experimentation

pursued some personal work on an all-electronic television camera while at the Westinghouse Electric and Manufacturing Company, USA, but the only determined effort by an industrial/government organisation was that initiated at the British Admiralty Research Laboratory, Teddington, UK, in 1923 [1].

From 1925 this situation changed. Bell Telephone Laboratories, of the American Telephone and Telegraph Company began an ambitious programme of work that led to an impressive demonstration in April 1927 of well-engineered apparatus for the transmission and reception of television images by land-line and radio links [2]. Later in the USA, General Electric, Westinghouse Electric and Manufacturing Company, and the Radio Corporation of America, in addition to a number of smaller companies, also began to be associated with television projects; while in Germany both Fernseh A.G. and Telefunken were active in this field by the end of the decade [3]. Leading companies in the UK adopted a rather reserved position on television matters until 1930, and before then only the Baird companies (Television Ltd., Baird Television Development Company and Baird International Television Ltd.) vigorously engaged in the pursuit of 'distant vision' research and development [4]. The Marconi Wireless Telegraph Company, and the Gramophone Company started their television activities in 1930. (See Reference [3] for more details on these developments.)

Of the work of Baird, Jenkins, Belin and von Mihaly, that of Baird (Fig. 19.1) from 1923 to 26th January 1926, when, for the first time anywhere a crude form of television was demonstrated, was remarkable. Baird's life and work during this period is unique in the annals of British 20th century electrical engineering. Here was a 35-year-old man with no extended experience of research and development work, no workshop or laboratory facilities, no

Figure 19.1   John Logie Baird (1888–1946).
Source: Radio Rentals Ltd.

scientific apparatus of any sort, no employment and no external source of funding, no access to acknowledged expertise or experience, and only one friend in Hastings to give encouragement, seeking in a small room in a suburban house the solution to a problem that had defied the efforts of inventors and scientists around the world for approximately fifty years, and which Dr C.V. Drysdale, the Superintendent of the Admiralty Research Laboratory, described in 1926 as 'extremely difficult' [5]. That Baird succeeded is to his everlasting credit. It was a most astonishing and outstanding accomplishment which has not been paralleled since.

Unlike his fellow countrymen Lord Kelvin and James Clerk Maxwell, Baird was not an intellectual giant. But, as with Marconi, he possessed qualities of perseverance, enterprise and inventiveness which enabled him to succeed where others had failed [6].

John Logie Baird was born on 13th August 1888, 14 years after the birth of Marconi, in the town of Helensburgh, a small seaside resort situated approximately 22 miles (35.4 km) north west of Glasgow [7].

Like the Italian pioneer, Baird was brought up in a comfortable middle-class professional household. His father was the minister of the local church and an intellectual of some merit. He was 47 years older than John Logie, in an era when any man over 50 had become an elderly, pompous figure. Here, again, a similarity exists between the two inventors, for Guglielmo's father was 48 years old when the radio experimenter was born [8].

Baird's acquired interest in science, while a schoolboy, seems to have been self-generated for no form of science was taught at the school, Larchfield, in Helensburgh that he attended [9]. Nevertheless, he engaged in various experiments and projects in his spare time. He installed electric lighting in his home [10], The Lodge, at a time when such an event could make news in the local press; he constructed a small telephone system so that he could easily contact his friends [11]; he tried to make a flying machine [12]; and he fabricated selenium cells [13].

This curiosity in science appears to have greatly influenced Baird's post-Larchfield education, for he rejected his father's request to enter the ministry and enrolled at the Royal Technical College, Glasgow, in 1906, to follow a course in electrical engineering. Eight years later Baird was awarded an Associateship of the College. An examination of the course curriculum shows that the subject timetables for the second and third years of the mechanical engineering and electrical engineering programmes (following a common first year) were almost identical [14].

This fact had an important bearing on Baird's work on television, for he had a penchant for designing and inventing devices which had a mechanical basis rather than an electrical foundation. Baird displayed considerable ingenuity and innovativeness in the fields of optics and mechanics and produced many patents on aperture disc, lens disc, and mirror drum scanning mechanisms, but only a few on electronic devices or systems. Electronics was not Baird's forte. Neither electric telegraphy nor wireless telegraphy formed part of the course programme.

Baird was 26 when he left the Royal Technical College (RTC). He tried to enlist in August 1914 and when he was declared unfit for service entered Glasgow University as a BSc student. He stayed for six months but did not sit the examinations.

Subsequently, he obtained work as an assistant mains engineer with the Clyde Valley Electrical Power Company (CVEP) [15]. This job entailed the supervision of the repair of any electrical failure in the Rutherglen area of Glasgow, whatever the weather, day or night.

Throughout his life Baird was subjected to colds, chills and influenza which necessitated lengthy periods of convalescence. His studies at the RTC were constantly interrupted by long illnesses and his numerous absences from his employment as an assistant mains engineer because of illness militated against any promotion in the company. Because of this he disliked the job and eventually resigned.

Actually, Baird's departure from the CVEP Company was hastened by his entrepreneurial exploits during the period 1917–19. In 1917 boot polish was difficult to obtain. Baird seized the opportunity to enliven his existence by registering a company and employing girls to fill cardboard boxes with his own boot polish [16]. This venture possibly escaped the notice of his employees, but the next did not.

Baird had always suffered, and always did suffer, from cold feet, and on the principle of capitalising on one's deficiencies, he devised an undersock – consisting of an ordinary sock sprinkled with borax [17]. He arranged its commercial exploitation with such a degree of business acumen and skill that, when he sold the enterprise twelve months later, he had made roughly £1,600 – a sum of money that would have taken him 12 years to earn as an engineer with the CVEP.

By 1919 the future television pioneer appeared to be on the threshold of a lucrative commercial life. Unfortunately, continuous good health was not a blessing bestowed on Baird and, during the winter of 1919–20, he suffered a cold that entailed an absence of six weeks from his venture. He decided to sell out and, following glowing accounts from a friend of the possibilities that seemed to exist in the Caribbean, travelled to the West Indies. His stay there was short and unprofitable and he returned to London in September 1920 [18].

Again he set about establishing a trading business and for the next two years dealt in honey fertilisers, and coir-fibre. Again the business was successful, but again another illness caused him to remain in bed for several weeks, 'the business meanwhile going to bits', and when his cold did not improve he sold his undertaking [19].

Later in 1922, on returning to good health, another trading enterprise, based on soap, was started. The concern flourished so much that Baird imported large quantities of soap from France and Belgium and formed a limited liability company. But another illness compelled him to sell out and convalesce. He went to Hastings where a friend from childhood lived [20].

Television development was not initially in Baird's mind when he settled there, for in some autobiographical notes he related how, when his health improved, he attempted to invent a pair of boots having pneumatic soles [21] and also a glass safety razor. The author, in his biography of Baird [22], has postulated that Baird's interest in television was stirred by reading an article; 'A development in the problem of television' by N. Langer in the *Wireless World and Radio Review* issue of 11th November 1922. Langer's paper was optimistic in tone and endeavoured to indicate the lines along which progress could be made. (See Fig. 19.2.)

It is interesting to recall, as noted previously, that Marconi was led to pursue his life's work when he read, while holidaying at Biellese in the Italian Alps, an obituary describing Hertz's experiments [23].

Whatever the source of Baird's inspiration, the solution to the television problem seemed to him to be comparatively simple. Two optical exploring devices rotating in synchronism, a light sensitive cell and a controlled varying

Figure 19.2   Cartoon published in the Hastings and St Leonards Observer, *26th January 1924. It refers to Baird's early television experiments which he carried out in Hastings from 1923.*
Source: The Hastings and St Leonards *Observer.*

light source capable of rapid variations in light flux were all that were required, 'and these appeared to be already, to use a Patent-Office term, known to the art' [24]. Baird, however, appreciated the difficult nature of the problem: 'The only ominous cloud on the horizon,' he wrote, 'was that, in spite of the apparent simplicity of the task, no one had produced television' [24].

Both Baird and Marconi commenced their investigations at advantageous times for, in addition to the ideas put forward by others, the technology existed for narrow-band television broadcasting in the one case and narrow-band wireless communications in the other. And both inventors commenced their experiments in private residences. Marconi had two large rooms at the top of the Villa Grifone set aside for him by his mother, and Baird made use of various rooms that he rented when staying in Hastings and elsewhere.

The approaches of Baird, Jenkins [25] and Mihaly [26], in 1923, to their work were individualistic. Jenkins was a well known inventor and a person of considerable means. He had produced important inventions in the field of cinematography and was able to design and manufacture apparatus of some complexity. His early rotary scanners consisted either of specially ground prismatic discs or costly lensed discs. Mihaly, an experienced patent expert and engineer, used a fragile, oscillating mirror scanner, together with tuning forks and phonic motors for synchronising purposes. Neither Jenkins nor Mihaly achieved great success.

Baird's approach to the television problem necessarily, because of his impecunious state, had to be entirely different from those of his contemporaries. Still, undaunted by the formidable difficulties that faced him, he commenced his experiments in 1923 by collecting bits and pieces of scrap material and assembling them into a scanning system. The constraints imposed by his finances severely limited the type of investigations he could carry out, but nonetheless he pursued his objective with dogged determination, ingenuity and resourcefulness. A Nipkow disc could be made from a cardboard hat box, the apertures could be formed using a knitting needle or pair of scissors, electric motors could be obtained cheaply from scrap metal merchants and bull's-eye lenses could be bought at low cost from a cycle shop [27] (see Fig. 19.3).

'The first televisor', wrote Moseley [28] (Baird's staunchest supporter), 'had the ingenuity of Heath Robinson and a touch of Robinson Crusoe. Baird described it as having the saving grace of simplicity'. Surprisingly, by the beginning of 1924, Baird was able to transmit, electrically, silhouettes of objects over short distances, and from this time Baird's efforts began to receive increasing publicity [29].

An appreciation of Baird's equipment can be obtained from the impression gained by the editor of a semi-popular science journal who paid a visit to Baird's Frith Street, London, address in April 1925.

I attended a demonstration of Mr Baird's apparatus and was very favourably impressed with the results. His machinery is, however, astonishingly crude and the apparatus in general is built out of derelict odds and ends. The optical system is composed of lenses

Figure 19.3   Baird in his Frith Street laboratory, c. 1925.
Source: The Royal Television Society.

out of bicycle lamps. The framework is an unimpressive erection of old sugar boxes and the electrical wiring a nightmare cobweb of improvisations. The outstanding miracle is that he has been able to produce any result at all with the indifferent material at his disposal [30].

But results were achieved. Images of simple objects such as letters of the alphabet were transmitted with a certain degree of clarity. 'The hand appeared only as a blurred outline, the human face only as a white oval with dark patches for the eyes and mouth. The mouth, however, can be clearly discerned opening and closing and it is possible to detect a wink' (Baird, April 1925). Two months later, Baird's contemporary, Jenkins, transmitted the silhouette images of a moving model windmill vane.

Baird at this time was experiencing difficulties with the poor transient response of his selenium cell and was not able to reproduce half tones. A number of experimenters and inventors had previously encountered or appreciated this property of the cell and had advanced various solutions to overcome its defect, but these were unsuitable for television purposes. Baird's solution had a simplicity characteristic of his early work. He added to the cell's output current a current proportional to the first derivative of the output current and obtained a much improved response. Success followed. In October

1925 the dummy's head, which Baird employed as an object, showed up on the screen not as a mere smudge of black and white, but as a real image with details and gradations of light and shade.

Baird has left a most interesting account [31] of the immediate events that followed his awareness that he had televised an object for the first time ever.

I was vastly excited and ran downstairs to obtain a living object. The first person to appear was the office boy from the floor below, a youth named William Taynton and he rather reluctantly consented to subject himself to the experiment. I placed him before the transmitter and went into the next room to see what the screen would show. The screen was entirely blank and no effect of tuning would produce any result. Puzzled and very disappointed I went back to the transmitter and there the cause of the failure became at once evident. The boy, scared by the intense white light, had backed away from the transmitter. In the excitement of the moment I gave him half a crown [12.5p] and this time he kept his head in the right position. Going again into the next room I saw his head on the screen quite clearly.

A public demonstration to 40 members of the Royal Institution was given on 26th January 1926 at 22 Frith Street [32].

Looking back at Baird's achievement it may seem that his choice of transmitting and receiving Nipkow disc scanners, neon lamp and electrically compensated selenium cell, and his method of obtaining synchronism between the two scanners based on the use of a synchronous motor and a.c. generator was obvious and that it really was surprising that no one had experimented with this particular combination of system components before c. 1923. The possibilities for component selection and invention, however, were very great.

Almost every conceivable type of scanner was suggested in the period 1877–1926 and beyond. There were vibrating mirrors, rocking mirrors, rotating mirrors, mirror polyhedra, mirror drums, mirror screws and mirror discs; there were lens discs, lens drums, circles of lenses, lenticular slices, reciprocating lenses, lens cascades and eccentrically rotating lenses; there were rocking prisms, sliding prisms, reciprocating prisms, prism discs, prism rings, electric prisms, lens prisms and rotating prism pairs; there were apertured discs, apertured bands, apertured drums, vibrating apertures, intersecting slots, multispiral apertures and ancillary slotted discs; there were cell banks, lamp banks, rotary cell discs, neon discs, corona discs, convolute neon tubes and tubes with bubbles in them; there were cathode ray tubes, Leonard tubes, X ray tubes, tubes with fluorescent screens, gas screens, photoelectric matrices, secondary emitting surfaces, electrostatic screens and Schlieren screens.

One of the most extraordinary features of the evolution of practical systems of television is that for approximately 50 years, from the first 1878 suggestions for seeing by electricity to 1930, no-one anywhere had sought to compare the efficiencies of the various mechanical scanners. Indeed the technical literature for this period is notable for the almost complete absence of mathematical analyses of the merits or otherwise of the different methods. This was a period of empiricism in television matters and was unlike the situation that prevailed when Marconi commenced his work in 1895. Then, the subject of electro-magnetic wave propagation had been well founded by the theoretical work of Maxwell, and the experimental investigations of Hertz, Lodge and a few others. This situation did not prevail in television until 1930 when Moller [33] and Kirschtein [34] showed, in separate papers, that an analysis of the optical per-formance of a scanner was a relatively straightforward matter based on the elementary principles of illumination and physics, and simple mathematics.

Of all the many different mechanical scanners listed above the Nipkow disc was the most robust, the most versatile in operation, the simplest and the cheapest. It could be made from cardboard or metal, in a workshop or home, by the unskilled or the skilled, for a few pence or many pounds.

Nipkow's invention was greatly developed: the scanner was used in 30, 50, 60, 120, 180 and 240 line television systems; it was run in air and in vacuum; it was constructed with single spirals, multi-turn spirals and spiral segments; it was designed to produce orthodox scanning, dis-contiguous scanning, graduated scanning, overlapping scanning and, when pairs of discs were employed, inter-laced scanning, and bilateral scanning; it was produced in apertured-hole and lensed form; it was used by amateurs and professionals alike. It was utilised for many years in telecine equipment and enabled films to be shown with a degree of clarity and brilliance that rivalled that of iconoscope cameras.

The Nipkow disc was used by the American Telephone & Telegraph Company for their ambitious and spectacular television tests in April 1927. During the succeeding years, the Nipkow disc formed an essential part of the early tele-vision schemes of Mihaly, Karolus, Fernseh AG and Telefunken in Germany; of AT&T, Sanabria, Jenkins, GEC and others in the USA; of the Baird com-panies, the Marconi Wireless Telegraph Company and HMV in Great Britain; and of several others. Probably, Baird's success in using the Nipkow disc, and the wide publicity that he received from January 1924, influenced others to adopt it.

Baird's October 1925/January 1926 achievement may be weighed by consider-ing the work of the British Admiralty Research Laboratory (ARL) on television [35]. The objective of the Admiralty in conducting experiments in this field was 'for spotting [the fall of shot] at sea with the use of aeroplanes'. A university-trained research scientist and others commenced the investigation in 1923. In January 1925, Dr C.V. Drysdale described the problem as difficult but felt that it could be solved with 'money and staff'. About 17 months later he had modified his opinion and referred to the 'extreme difficulty' of finding a practical solution. An inspection of ARL's television equipment, which included a photoelectric cell fabricated by the National Physical Laboratory (NPL), was undertaken by two representatives of the Air Ministry, on 27th May 1926. During their visit an image of an object consisting of a grid of three bars of cardboard, each about 0.25 inch (6.35 mm) in width and 0.25 inch (6.3 mm) apart, was transmitted. Although transmitted light was used the object 'could just be recognised at the receiving end, but the reproduction was very crude' [36,37].

Following his successful but crude demonstration of television Baird (as with

Marconi) wished to capitalise on his invention and develop and extend it to the stage where it would form the basis of a public television service. This objective required a substantial sum of money. Previously, in 1923, Baird had tried to obtain some support from a major manufacturer of communication equipment, but without success. An advertisement in *The Times* for assistance led to a poor response; and though some financial sponsorship was given by a Mr W.E.L. Day [38] this was insufficient for Baird's needs. In 1926 the only source of money for his purpose seemed to be the general public. But the general public had to be given an indication that investment in the new method of broadcasting was worthwhile.

The prospect facing Baird in 1926 was much more daunting than that which faced Marconi in 1895–96. Marconi had the advantage that his invention had an immediate application in military and naval operations and, when his demonstrations before Service officers proved successful, his future seemed assured. Additionally, Marconi had obtained valuable support for a short but critical period from W.H. Preece, the Engineer-in-Chief of the General Post Office, who had himself been experimenting on signalling through space without wires.

On the other hand, Baird's invention had no immediate application to warfare, or to safety, and he received no patronage or encouragement from the one body that could assist him, the British Broadcasting Corporation. Baird's financial resources from 1923 to 1927 were minimal, and so, necessarily companies had to be formed that would attract investment from the general public.

Three adventitious factors aided Baird's work in the period following his October 1925 demonstration and prior to the introduction of the 30-line experimental television service in 1929. First, the appointment of O.G. Hutchinson (Fig. 19.4) as Baird's business manager towards the end of 1925; second, the state of the stock market in 1927; and third, the association of S.A. Moseley with Baird and the companies in which he had an interest from 1928 to 1933.

The appointment of Hutchinson enabled Baird to concentrate on laboratory work. As a consequence, he was able to devote all his energy and inventive skill to the achievement of 'firsts' without being encumbered by the need to attend to business matters: he was able to demonstrate a number of applications of his basic scheme before any other person or industrial organisation succeeded in doing so. These demonstrations attracted favourable comments from several notable scientists, many newspapers (including the *New York Times*) and two Postmaster Generals. As an illustration of these comments the *New York Times*, on 11th February 1928, noted (apropos Baird's transatlantic television experiment):

His success deserves to rank with Marconi's sending of the letter 'S' across the Atlantic, the first intelligible signal ever transmitted from shore to shore in the development of transoceanic radio telegraphy. As a communication, Marconi's 'S' was negligible; as a milestone in the onward sweep of radio, of epochal importance. And so it is with Baird's first successful effort in transatlantic television ... All the more remarkable is Baird's

Figure 19.4   O.G. Hutchinson, Stookie Bill and Baird in the Frith Street workroom,
c. December 1925.
Source: Mr R.M. Herbert.

achievement because he matches his inventive wits against the pooled ability and vast resources of the great corporation physicists and engineers, thus far with dramatic success [39].

Unfortunately Hutchinson's business and publicity methods were justifiably regarded with some suspicion and concern in some quarters, notably the BBC, and these caused Baird to suffer some criticism.

There is little doubt that Hutchinson engaged in gross exaggeration to advance Baird television. Nevertheless it has to be said that he brought the accomplishments of Baird to the attention of the populace. Hence, when the shares of the Baird Television Development Company and Baird International Television Ltd became available for purchase in 1927 and 1928, respectively, there was no shortage of buyers.

The formation of these companies was aided, probably, by the stock market conditions that prevailed at that time. According to Moseley: 'Television was born at a time when company promoting was running rather wild. Speculators, greedy for fat profits and quick returns, were ready to gamble on very slim chances' [40].

Moseley's association with the Baird companies was of great importance to the furtherance of the low definition system in the UK. He had commenced his career as a journalist in 1910, had been the Cairo correspondent of the *New*

York Times, and had been the special correspondent of the *Daily Express*, *Evening News*, *Sunday Express*, *Daily Herald*, and *People*; he had written 14 books by 1929, including *Money making in stocks and shares*, and *The small investor's guide*; he had (unsuccessfully) contested a Parliamentary seat as an ILP candidate in 1924; he had produced criticisms of radio plays and considered himself to be Britain's first radio critic; and he was an experienced broadcaster. He knew everybody in Fleet Street and was on cordial terms with the various BBC chiefs and had numerous contacts in Parliament and elsewhere. He played a central and crucial role in the development of television in Britain prior to his departure from Baird Television Ltd in 1933 [41].

Baird was quite content to allow Moseley to look after his business interests and a bond of friendship was established between them shortly after Moseley's introduction to Baird in 1928, which lasted until the latter's death in 1946. Unfortunately for Baird, (and possibly Baird Television Ltd), Moseley resigned his directorship in 1933. The part played by Moseley in advancing Baird's interest was stressed in the BBC's letter [42] of regret to him (on hearing of his resignation): 'Although there has not always been agreement either in policy or in method, it should be recognised that your consistently active advocacy has been an important, perhaps the decisive, factor in the progress that Baird Television has made to date.'

On the other hand, Marconi's principal rivals in the United States, de Forest and Fessenden, were not so well adjusted personally as either Baird or Marconi and did not team up with such skilful operators as Moseley or Isaacs. Marconi, de Forest and Fessenden were associated with the three most important American communications firms, namely Marconi Wireless Telegraph Company of America, de Forest Wireless, and National Electric Signalling. Only the Marconi companies that were well managed survived, although each of the three communication concerns was organised around one outstanding inventor.

The case histories of each of these inventors, and the importance that all of them placed on the acquisition of a patent holding, were most probably known to the widely read Baird. As with Marconi, de Forest and Fessenden, Baird had to devote a great deal of time and labour to amass a collection of patents [43] which would place his companies in a favourable position commercially, and until c. 1930 he engaged in this task almost single-handedly.

Not surprisingly, Baird had little time for writing scientific papers and engaging in extensive field trials. He tried to anticipate, and be the first to implement, every likely development and application of the new art. Daylight television, noctovision, colour television, news by television, stereoscopic television, long distance television, phonovision, two-way television, zone television and large screen television were all demonstrated, in a rudimentary way, by Baird during a hectic four-year period of activity from 1926–1930. (See Reference [4].) Additionally, these demonstrations of 'firsts' were required to stimulate and maintain the interest of members of the public. Their support was vital to the well-being of the Baird companies, and to the initiation of a television broadcasting service.

Baird's plans for television were as ambitious and extensive as those of Marconi for marine wireless communications. By the late 1920s Baird had accumulated a considerable number of patents and hoped to establish his system in many countries including, of course, the United Kingdom. He wished to establish a dominant position in this country, but the rather pushing methods employed by Hutchinson and Moseley caused antagonism with the broadcasting monopoly. The lack of enthusiasm shown by the BBC towards Baird's low definition system was a source of much concern and frustration to Baird and his supporters, and resulted in delays in the execution of their desires [44].

The monopolistic position of the Corporation during this period was obviously a considerable obstacle to Baird's aspirations. Essentially the BBC was not interested in participating in the advancement of television on the basis of a system that could not reproduce images of, say, a Test match at Lord's or tennis at Wimbledon: the BBC considered that low definition television was inappropriate to its services. As a consequence, the BBC's policy towards Baird's work was necessarily negative in outlook and did not conduce to the rapid advancement of Baird's ambitions.

In America and elsewhere facilities for television broadcasting were given by broadcasting stations in the late 1920s, but in Britain the Chief Engineer of the BBC opposed the use of the BBC's stations for this purpose [45]. This opposition led a former Postmaster General to write:

It is understood that the Chief Engineer of the BBC holds the view that progress cannot be made along the lines so far pursued. He may be right. On the other hand he may be wrong. In any case the road to further experiment ought not to be closed. If the apparatus proves to be valueless, it will find no patrons and the question will solve itself [46].

The opposition of the BBC to the use of its stations by Baird has its parallel in the history of the Marconi organisation. Although Preece had given Marconi much needed support in the early stages of his work, Austen Chamberlain, as Postmaster General, took quite a different attitude. He saw the Marconi company as a potential competitor of the government-controlled telegraph industry, and, at first, stubbornly refused to allow the Marconi overseas service to utilise the Post Office's telegraph lines. Later, an agreement between the Post Office and the Marconi company was signed on 11th August 1904 and facilities for wireless telegraphic traffic were granted to it [47].

In many respects Baird's early commercial activities and difficulties mirrored those experienced by Marconi. Both inventors were keen to exploit their initial successes, and, as their work could be furthered by obtaining funds from the public, companies were established. The Wireless Telegraph and Signal Company Ltd was formed in 1897 the year following Marconi's visit to England, and Television Ltd was registered in 1925, the year of Baird's successful demonstration of television. Later, further companies were floated. These incorporated the name of the inventor in each case. None of the public companies was initially rewarding for their shareholders; Marconi's Wireless Telegraph Co. Ltd. did not pay any dividends from 1897 to 1910 [48]. Both inventors felt that the creation of overseas interests was necessary for the advancement of their plans: in 1899 the Marconi Wireless Telegraph Company of America was set up and was followed by the Marconi International Marine Communication Co. Ltd. in 1900. In 1927 Baird Television Development Company was established, and in 1928 Baird International Television Ltd was launched.

The original capital of the Wireless Telegraph and Signal Company was £100,000 [49], that of the Baird Television Development Company was £125,000 and was subscribed largely by wealthy individuals. Presumably, in each case the subscribers wanted a speculative investment in the new communication systems. Marconi was handsomely rewarded for his enterprise and received £15,000 in cash and 60 per cent of the original stock in exchange for nearly all his patent rights. Similarly Baird Television Development Company acquired the sole rights to exploit the inventions of Baird in the United Kingdom by paying the purchase price of £20,000 to Television Limited, which had Baird as one of its founder members.

Moreover, both Marconi and Baird encountered difficulties in advancing their commercial interests. Aggressive tactics were adopted by the two companies to further their objectives, and much good will towards Marconi and Baird was dissipated later by this approach. Sir William Preece, said in 1907: 'I have formed the opinion that the Marconi Company is the worst managed company I have ever had anything to do with. ... Its organisation is chiefly indicated by the fact that they quarrel with everybody' [50].

Eventually, the efforts of Hutchinson and Moseley to obtain transmission facilities were successful and on 30th September 1929 the Baird experimental television service, transmitted by the BBC, was inaugurated [51]. The broadcasts lasted until 22nd August 1932 when they were taken over by the BBC and formed the BBC's low definition television service (Fig. 19.5).

Of all the demonstrations of television given in the 1920s, none surpassed in technical excellence those mounted by the Bell Telephone Laboratories (BTL), of the American Telephone and Telegraph Company (AT&T) [2]. The Laboratories were formed in 1925, when the engineering department of Western Electric was reorganised and became Bell Telephone Laboratories, with a total staff of approximately 3,600.

The results that were obtained by BTL in 1927 are particularly important historically because they were the best that could be expected with the technology as it existed at that time. With its vast resources in finance and equipment, and in staff expertise and experience, BTL was uniquely able to demonstrate what could be engineered in the field of television. Subsequently, colour television and two-way television systems were implemented in 1929 and 1930 respectively, all at great cost. From 1925 to 1930 (inclusive) AT&T approved the expenditure of $308,100 on low definition television – a sum far in excess of anything available to any of the 'lone' television workers of whom Baird, Jenkins, Mihaly, Belin and Karolus were in the vanguard.

After 1930 the Laboratories continued its work on television, but without

Figure 19.5   Control room of the television studio at 16 Portland Place. The mirror drum camera (far right) is pointing into the studio. The drum in the foreground was used for presenting and terminating programmes.
Source: The BBC.

achieving successes of the type being manifested contemporaneously by RCA of the US and EMI of the UK, despite the allocation of $592,400 to the work from 1931 to 1935 (inclusive) [52]. Thereafter, television research and development declined and ceased, sometime in 1940, to be part of BTL's interests.

The American Telephone & Telegraph Company (AT&T) commenced its experimental study of the television programme when 'it began to be evident that scientific knowledge was advancing to the point where television was shortly to be within the realm of the possible' [53]. The company was of the opinion that television would have a real place in world-wide communications and that it would be closely associated with telephony. It was certainly well placed to advance television, not only because of the extensive facilities of the newly formed BTL but also because of the experience acquired in the R & D work which had made transcontinental and transoceanic telephony and telephotography possible.

In January 1925 development work under the direction of Dr H.E. Ives [54] had been completed on a system for sending images over telephone lines, and R&D resources and expertise existed for a new scientific venture. Dr Ives and Dr Arnold, the Director of Research, agreed that the next problem to be undertaken was television. 'At Arnold's request,' wrote Ives [55], 'I prepared and submitted to him on 23 January 1925, a memorandum surveying the problem and proposing a programme of research.'

Ives's memorandum discussed the characteristic difficulties of securing the requisite sensitiveness of the pick-up apparatus, the wide bandwidths that from his experience of picture transmission were indicated as necessary for television, the problem of producing enough modulated light in the received image to make it satisfactorily visible, and the problem of synchronizing apparatus at the sending and at the receiving ends of the transmission link. The memorandum concluded with a proposal for 'a very modest attack' on the problem, capable, however, of 'material expansion as new developments and inventions materialised'.

Ives felt that these difficulties could be examined by utilizing a mechanically linked transmitter and receiver, each incorporating a Nipkow disc scanner operating on a 50 lines per picture, 15 pictures per second standard. A photographic transparency, later to be superseded by a motion picture film, would be used at the sending end, together with a photo-electric cell and a carbon arc lamp. At the receiving end Ives proposed the use of a crater-type gaseous glow lamp. His plan was thus based on the transmission of light through the 'object' rather than on the reflection of light from an opaque body. A sum of $15,000 was approved for the project.

By 7th April 1927 the system (Fig. 19.6) was ready to be demonstrated. The demonstration using a wire link, consisted of the transmission of images from Washington, DC, to the auditorium of the Bell Telephone Laboratories in New York, a distance of over 250 miles (402 km). During the radio demonstration, images were sent from the Laboratories' experimental station 3XN at Whippany, New Jersey, to New York City, a distance of 22 miles (35.4 km). Reception was by means of two forms of receiver. One receiver produced a small image of approximately 2.0 inches × 2.5 inches (50.8 mm × 63.5 mm), which was suitable for viewing by one person. The other receiver gave a large image of nearly 24 inches × 30 inches (609.6 mm × 762.0 mm) for viewing by an audience of considerable size [56].

Ives and his colleagues used a Nipkow disc with 50 apertures for scanning purposes. They arrived at this figure by taking as a criterion of acceptable image quality the standard of reproduction of the half-tone engraving process in which it was known that the human face can be satisfactorily reproduced by a 50-line screen. Thus, assuming equal definition in both scanning directions, 2,500 elements per picture had to be transmitted at a rate of 16 pictures per second. The frequency range needed to transmit this number of elements per second was calculated to be 20 kHz.

A spotlight scanning method [57] was adopted to illuminate the subject, the beam of light being obtained from a 40A Sperry arc (Fig. 19.7). Three photoelectric cells of the potassium hydride, gas-filled type were specially constructed and utilised to receive the reflected light from the subject. At that time they were probably the largest cells that had ever been made and presented an aperture of 120 square inches (77420 mm^2).

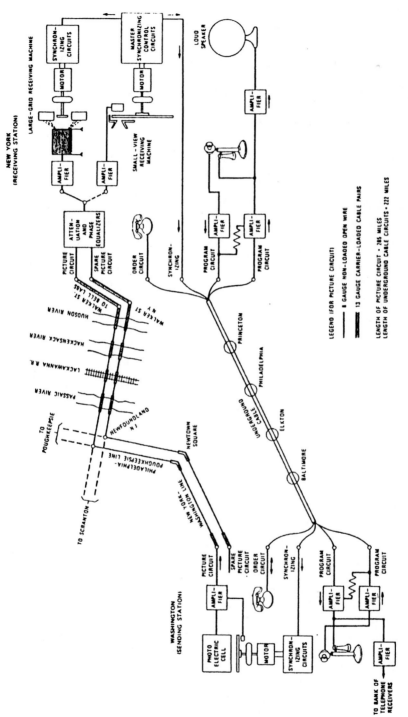

Figure 19.6   Schematic diagram of the line and radio circuits used in the 1927 Bell Telephone Laboratories' television demonstration.
Source: AT&T Bell Laboratories.

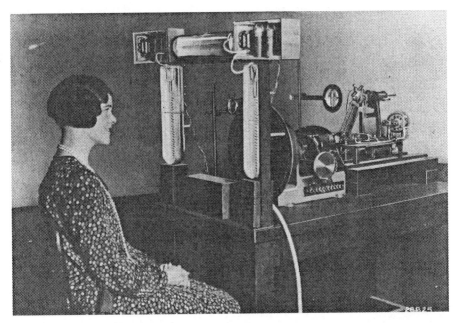

*Figure 19.7   View of Bell Telephone Laboratories' 'spotlight' scanner. The light source
(on the right), the focusing lens, the Nipkow disc type scanner, and large
photo-electric cells are clearly shown.*
Source: AT&T Bell Laboratories.

For reception a disc similar to that at the sending end was used together with a
neon glow lamp. The disc had a diameter of 36 in (91.4 mm) and synthesised the
2.0 × 2.5 inch (50.8 mm × 63.5 mm) image. Another form of receiving apparatus
comprised a single, long, neon-filled tube bent back and forth to give a series
of 50 parallel sections of tubing. The tube had one interior electrode and 2,500
exterior electrodes cemented along its rear wall. A high frequency voltage
applied to the interior electrode and one of the exterior electrodes caused the
tube to glow in the region of that particular electrode. The high frequency
modulated voltage was switched to the electrodes in sequence from 2,500 bars
on a distributor with a brush rotating synchronously with the disc at the transmitting end. Consequently, a spot of light moved rapidly and repeatedly across
the grid in a series of parallel lines, one after the other, and in synchronism
with the scanning beam. With a constant exciting voltage the grid appeared
uniformly illuminated but when the high frequency voltage was modulated by
the vision signals an image of the distant subject was created. To transmit the
vision, sound and synchronizing signals three carrier waves were employed:
1,575 kHz for the image signals, 1,450 kHz for the sound signals and 185 kHz for
the synchronization control signal [58].

   The first demonstration consisted of the transmission of an image of, and an
address by, Herbert Hoover, Secretary of Commerce, from Washington to New

York, over telephone lines. The second demonstration by radio comprised three events: first an address by E.L. Nelson, a BTL engineer; second, a 'vaudeville act' featuring 'a stage Irishman, with side whiskers and a broken pipe, [who] did a monologue in brogue' [2], and then, after a quick change, returned with a blackened face and made a few quips in dialect; and, finally, a short humorous dialect talk.

The received images were subject to some fading and ghosting and occasionally appeared in the negative but in general they impressed the audience. 'It was if a photograph had suddenly come to life and began to talk, smile, nod its head and look this way and that,' said one observer [2].

The April 1927 demonstrations were undoubtedly the finest that had been given anywhere even though no especially novel features had been incorporated into the various systems. They established standards from which further progress could be measured. Moreover, the publication in October 1927, in the *Bell System Technical Journal* (volume 6, pages 551–653), of five detailed papers on the factors that led to Ives's group success enabled other workers to ponder on whether their own ideas and practices were likely to lead to similar favourable outcomes. Certainly, the Bell Laboratories equipment could be further developed. The large screen grid display was 'very much inferior' to that of the gaseous discharge lamp; the person being televised had to sit in a semi-darkened room; there was a need to dispense with the separate synchronizing channel; and there was a requirement for more detailed images. In addition to these considerations the policy of AT&T towards television advancement in the Bell System had to be defined.

F.B. Jewett, the President of BTL, believed that Ives's group should proceed as vigorously as possible with the preliminary work concerned with the development of public address television apparatus, without there being any definite commitment [59]. In addition, Jewett felt the Laboratories should carry on, as adequately as possible, whatever fundamental work would be necessary to safeguard the company's position and advance the art along lines that were likely to be of interest to the company. This mandate gave Ives ample scope to investigate a quite wide range of television problems. He seized the opportunities made available to him and during the next three years daylight television, large screen television, television recording, colour television and two-way television were all subjected to the group's scrutiny and engineering prowess [60].

The televising of objects illuminated by natural daylight by the method of direct scanning was demonstrated on 10th May 1928, following detailed laboratory studies [61]. A press show was given on 12th July 1928. This contained scenes of a sparring match, a golf exhibit and other movements [62].

The evolution of the equipment for public address television proceeded along several paths and by February 1928 Ives could 'guarantee' that a face could be reproduced so as to be 'very satisfactorily recognizable' [63].

In 1927 Hartley and Ives patented methods for projecting televised images by photographing the received image with a cine camera and developing the film images with the minimum of delay [64]. They also patented the generation of television signals from rapidly processed cine film.

Ives advocated using transmitted light (through film), rather than reflected light (from a scene), to ease the solution of the television problem [65]. He noted: 'It may be pointed out that the use of a moving picture film as the original moving object is equivalent to a very great amplification of the original illumination brought about by the photo-chemical amplification process involved in the production and subsequent development of the photographic latent image.' This intermediate film system of television was employed in the 1930s by Fernseh AG and Telefunken in Germany, and by Baird Television Ltd at the London television station, Alexandra Palace.

On 27th June 1929 colour television [66] was shown by Bell Telephone Laboratories to an invited gathering of scientists and journalists. Ives employed three signal channels so that the three colour signals could be sent simultaneously from a transmitter to the receiver. An advantage of this arrangement was that the same scanning discs and motors, synchronising equipment and light sources, and the same type of circuit and method of amplification were used as in the monochrome scheme. The only new features were the form and disposition of the specially devised photocells at the sending end and the type and grouping of the neon and argon lamps at the receiving end.

An account of the demonstration, in which the transmission was over lines, was published in *Telephony* on 6th July 1929. The display [67]:

opened with the American flag fluttering on a screen about the size of a postage stamp. The observer saw it through a peep hole in a darkened room. The colours reproduced perfectly. Then the Union Jack was flashed on the screen and was easily recognised by its coloured bars.

The man at the transmitter picked up a piece of watermelon, and there could be no mistake in identifying what he was eating. The red of the melon, the black seeds and the green rind were true to nature as were the red of his lips, the natural colour of his skin and his black hair.

Of the projects approved by Jewett one pertained to two-way television (Fig. 19.8). This was established between the main offices of the AT&T Company at 195 Broadway, New York, and the Bell Telephone Laboratories at 463 West Street, New York, and was demonstrated on 19th April 1930 [68]. It consisted essentially of two complete television transmitting and receiving sets of the kind employed in the 1927 one-way television scheme. Spotlight scanning, Nipkow discs and neon lamps were still incorporated but with several improvements. Two discs, each containing 72 holes to give double the image detail as compared with the 50-hole discs of the 1927 apparatus, were utilised at each end of the line links, one for image analysis and the other for image synthesis. In addition to the photoelectric cells and neon lamps each 'ikonophone' booth had a concealed microphone and loudspeaker.

When the two-way system was withdrawn from service it had been seen by more than 17,000 people. A novel application was observed when two deaf

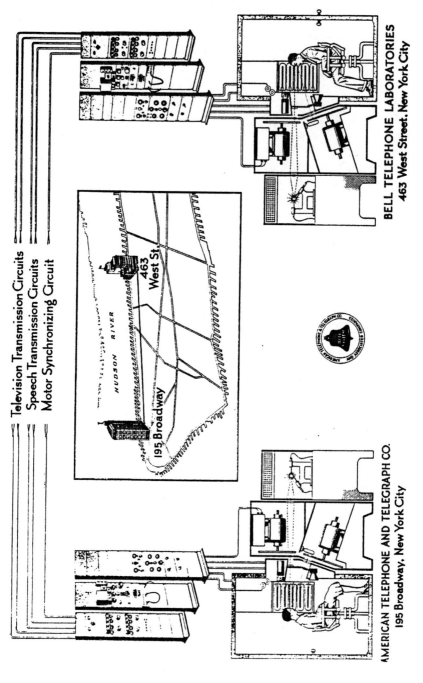

Figure 19.8 Diagrammatic layout of Bell Telephone Laboratories' two-way television.
Source: AT&T Bell Laboratories.

persons carried on a telephone conversation by reading each other's lips. The cost of providing the service, by the New York Telephone Company, was estimated to be $15,350 per year, excluding the cost of the technical operation and maintenance, which was borne by Bell Telephone Laboratories [69].

Following the completion of the two-way link, Ives undertook an important appraisal of the progress that had been made by his group and attempted to define the course of action to be implemented for the future advancement of television. His prognosis was gloomy in outlook [70]. For Ives the statement of the problem that had to be solved was simple.

An electrically transmitted photograph 5 in × 7 in (127 mm × 178 mm) in size, having 100 scanning strips per inch, has a field of view and a degree of definition of detail which, experience shows, are adequate (although with little margin) for the majority of news events pictures. It is undoubtedly a picture of this sort that the television enthusiast has in the back of his mind when he predicts carrying the stage and the motion picture screen into the house over electrical communications channels.

The difficulty of achieving this desirable result was readily apparent. In the photograph the number of picture elements is 350,000, and at a repetition speed of 20 per second (24 per second had now become standard with sound films) this meant the transmission of 7,000,000 picture elements per second and a bandwidth of 3.5 MHz for the system on a single sideband basis. Ives compared the criteria for high definition television and the results that had been obtained in America, and observed: 'All parts of the television system are already having serious difficulty in handling the 4,000-element image [70].' (This was the number of image elements used in the 72-line picture of the two-way television link.)

The obstacles that had to be overcome before a high definition system could be implemented were found in the use of the scanning discs at the transmitter and receiver, the photoelectric cells, the amplifying systems, the transmission channels and the receiving lamps. Ives noted that the disc, while quite the simplest means for scanning images of a few elements, was entirely impractical when really large numbers of image elements were in question, and wrote: 'As yet however, no practical substitute for the disc of essentially different character has appeared' [70].

Turning next to the photocells, there were, in 1930, two types of cell that could be utilised for television: the gas-filled cell, which had a good sensitivity but poor frequency response; and the vacuum cell, which was much less sensitive than the gas-filled cell, although it was free from its failings. The self-capacitance of the cells and the associated wiring and amplifier caused the high frequencies to be attenuated relative to the lower frequencies and consequently equalising circuits with their attendant problems of phase adjustment, together with more amplification, were needed. But Ives observed that amplifiers capable of handling frequency bands extending from low frequencies up to 100,000 Hz or more gave serious problems.

The communication channels, either radio or wire, also posed grave difficulties for high definition television and its related bandwidth specification.

In radio, fading, different at different frequencies, and various forms of interference stand in the way of securing a wide frequency channel of uniform efficiency. In wire, progressive attenuation at higher frequencies, shift of phase, and cross-induction between circuits offer serious obstacles. Transformers and intermediate amplifiers or repeaters capable of handling the wide frequency bands here in question also present serious problems.

Finally, at the receiving end of the system the neon glow could not follow satisfactorily television signals well below 40,000 Hz and, in the case of the 4,000-element image the neon had to be assisted by a frequently renewed admixture of hydrogen, which again could not be expected to increase the frequency range indefinitely. With the receiver disc, as at the sending end, increasing the number of image elements rapidly reduced the amount of light in the image and, with a plate glow lamp of given brightness, the apparent brightness of the image is inversely as the number of image elements [70].

These considerations led Ives to one clear conclusion: 'The existing situation is that if a many-element television image is called for today, it is not available, and one of the chief obstacles is the difficulty of generating, transmitting, and receiving signals over wide frequency bands' [70]. A partial solution was to employ multiple scanning and multiple channel (zone) transmission.

Ives's multi-channel experimental set-up [70] is illustrated in Fig. 19.9. This used scanning discs with prisms over their apertures, so that, at the sending end, the beams of light from the successive holes were diverted to different photo-electric cells. At the receiving end, the prisms enabled beams of light from the three lamps to be deflected in a common direction.

Ives found that his three-channel apparatus yielded results strictly in agreement with the theory underlying its conception and observed that the 13,000-element image was a marked advance over the single-channel 4,000 image.

Even so, the experience of running a collection of motion picture films of all types is disappointing, in that the number of subjects rendered adequately by even this number of image elements is small. 'Close-ups' and scenes showing a great deal of action, are reproduced with considerable satisfaction, but scenes containing a number of full length figures, where the nature of the story is such that the facial expression should be watched, are very far from satisfactory. On the whole the general opinion . . . is that an enormously greater number of elements is required for a television image for general news or entertainment purposes [70].

This point had been appreciated by a few workers for several years. Swinton had outlined an all-electronic system of television in 1911 [71], and V.K. Zworykin had patented a version of such a system in 1923. When, in 1926, Farnsworth started his work on television it was on the basis of an all-electronic scheme. Again, in Great Britain, a similar approach was to be adopted in 1931 by Electric and Musical Industries Ltd. (EMI). Subsequently, Zworykin,

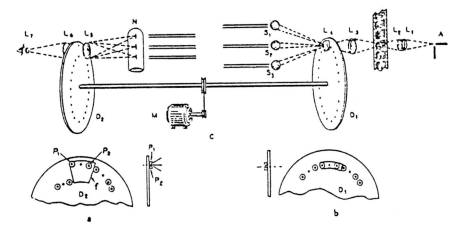

Figure 19.9   *Schematic diagram of Bell Telephone Laboratories' three-zone television system: (a) receiving end disc with spiral holes provided with prisms, (b) sending end disc with circle of holes with prisms, (c) general arrangement of apparatus.*
Source: *Journal of the Optical Society of America,* 1931, **21**, pp. 8–19.

Farnsworth, and EMI satisfactorily demonstrated all-electronic television before the end of 1935.

Following this account of the thoughts and work of Ives and his group, it is now possible to give an assessment of the early thoughts and work of Baird and his group. A comparison of their efforts is shown in Table 19.2. The table does not serve to determine whether Baird or a staff member of Bell Telephone Laboratories was the first to patent or demonstrate a particular aspect or method of television. Rather, it is a basis to indicate Baird's sound appreciation of the low definition television problem and to show that, in the 1920s, his ideas and his implementation of those ideas were consonant with the thoughts and activities of a well endowed and well staffed research and development organisation. Thus, both groups demonstrated low definition 'spotlight' television, large screen television, long distance television, colour television, daylight television and zone television. Neither Baird nor Ives felt that cathode-ray tubes were appropriate for television reception in the 1920s. Both Baird and Ives used glow discharge tubes and Nipkow discs in their receivers, and both television pioneers considered employing very large discs of up to 8 ft (2.44 m) to 10 ft (3.05 m) in diameter. Commutated lamp displays, the use of arcs, the intercalation of images, trans-Atlantic television transmission and reception, and the means to improve the performance of photoelectric cells, were all examined by Baird and Ives. The agreement between the views and strategies of the two groups is quite remarkable, and especially so when the great variety of possible devices and methods that could have been used is appreciated.

Table 19.2 Some contributions by Bell Telephone Laboratories and by J.L. Baird to the development of television in the 1920s

Use of/demonstration of	Baird	Bell Laboratories
1 Nipkow disc	from 1923	from 1925
2 Glow discharge lamps	various experiments 1924	various experiments 1925
3 Means to reduce time lag of photocells	from 1925; used derivative of photocell current; patent 270 222, 21 October 1925	from 1925–6; used C–R coupling circuit to enhance high-frequency gain, internal memorandum 27 February 1926
4 Coloured filters on lamps to reduce discomfiture of persons televised	various experiments in 1925; demonstration of infrared radiation 23 November 1926	various experiments in 1925; mentioned in internal memorandum, 26 August 1925
5 Large Nipkow discs	utilised discs up to 8 feet (2.44 m) in diameter at some time during the period 1923–5	advantage of using discs up to 10 feet (3.05 m) in diameter mentioned in an internal memorandum, 27 July 1925
6 Spotlight scanning	employed from 1926 to 1936; patent 269 658, 20 January 1926	employed from 1925–6; US patent applied for on 6 April 1927; UK patent 288 238, 18 January 1928
7 Two-way television	patent 309 965, 19 October 1927	UK patent 297 152, 17 June 1927; demonstrated from 9 April 1930 to 31 December 1932
8 Transatlantic television	demonstrated 9 February 1928	suggested as a publicity event in an internal memorandum 4 May 1927
9 Intercalated images to improve resolution	patent 253 957 1 January 1926; various experiments 1924	mentioned in internal memorandum 9 September 1927
10 Colour television	demonstrated 3 July 1928	demonstrated July 1928
11 Large-screen television	demonstrated 28 July 1930	demonstrated 7 April 1927
12 Daylight television	demonstrated June 1928	need to work on natural-light scanning mentioned in an internal memorandum 4 May 1927; demonstrated 16 July 1928
13 Commutated lamp bank/display	patent 222 604, 26 July 1923, demonstrated 28 July 1930	demonstrated 7 April 1927

14 Use of arcs	demonstrated January 1931	mentioned in internal memorandum 11 February 1929; experiments 1929
15 Zone television	demonstrated 2 January 1931; patent 360 942, 6 August 1930	described in a paper by H.E. Ives, 'A multichannel television apparatus', *J. Opt. Soc. Am.*, 1931, 21; demonstrated 1930–1

The challenge to evolve television systems having 'enormously greater number[s] of elements' to produce high definition images was taken up by several organisations, but of these only the schemes of the Radio Corporation of America (RCA) and Electric and Musical Industries (EMI) were of sufficient merit to enable world-wide high definition television to prosper. Effectively, by 1955, the growth of the world's television networks was founded on just two systems, namely, the USA's 525-line and the UK's 405-line systems. They were engineered by RCA and EMI respectively. In Chapters 20 and 21, the steps that led to the achievements of these companies in the television field are considered.

References

1 BEATTY, R.T.: 'Report on television', parts 1 and 2, 34pp, December 1925, AIR 2/S24132, PRO

2 BURNS, R.W.: 'The contributions of the Bell Telephone Laboratories to the early developments of television' (Mansell, London, 1991), pp. 181–213

3 BURNS, R.W.: 'Television, an international history of the formative years' (Institution of Electrical Engineers, London, 1998), pp. 242–282

4 BURNS, R.W.: 'John Logie Baird, television pioneer' (Institution of Electrical Engineers, London, 2001)

5 LEFROY, H.P.: minute, 9th June 1926, AIR 2/S24132, Public Record Office

6 Ref. 3, see Tables 1.1, 6.1, 7.1, and 15.1

7 BURNS, R.W.: entry on 'J.L. Baird', in 'New dictionary of national biography' (Oxford University Press, in preparation)

8 MARCONI, D.: 'My father, Marconi' (Frederick Muller, London, 1962), p. 8

9 MOSELEY, S.A.: 'John Baird' (Odhams Press, London, 1952), pp. 27–8

10 BAIRD, J.L.: 'Sermons, soap and television. Autobiographical notes' (Royal Television Society, London, 1988), p. 21

11 Ibid

12 Ref. 9, pp. 33–5

13 Ref. 10, p. 21

14 Ref. 4, note 2, p. 28

15 Ref. 10, pp. 28–30

16 WATKINS, H.: 'How television began – Part 13', *London Home Magazine*, c. 1962

17 Ref. 10, pp. 30–2

18 Ref. 4, pp. 20–3
19 Ref. 4, pp. 24–5
20 Ref. 4, pp. 26–7
21 Ref. 10, p. 42
22 Ref. 4, pp. 31–3
23 Jolly, W.P.: 'Marconi' (Constable, London, 1922)
24 BAIRD, J.L.: 'Television', *Journal of Scientific Instruments*, 1927, **4**, pp. 138–43
25 Ref. 3, pp. 195–205
26 Ref. 3, pp. 242–254
27 Ref. 4, pp. 37–40
28 Ref. 9, p. 63
29 ANON.: 'Television. Amateur scientist's invention. Secret plans. Outline of objects transmitted', *Daily News*, 15th January 1924
30 EDITOR.: Note by the Editor, *Discovery*, April 1925, p.143
31 BAIRD, J.L.: Broadcast lecture on WMCA and WPCH Radio, New York, 18th October 1931
32 Ref. 4, pp. 90–1
33 MOLLER, R.: 'Das Weillersche Spiegelrad', *Fernsehen*, 1931, **2**, pp. 80–97
34 KIRSCHSTEIN, F.: 'Nipkowscheibe oder Spiegelrad', *Fernsehen*, 1931, **2**, pp. 98–104
35 Ref. 4, pp. 93–103
36 LEFROY, H.P.: minute, 9th June 1926, AIR 2/S24132, Public Record Office
37 HORSLEY, E.C.: 'Report on inspection of certain television apparatus made at the ARL', A.M.L. report no. T.V.1, June 1926, 4pp, AIR 2/S24132, Public Record Office
38 Ref. 4, chapters 2 and 3
39 ANON.: report, *New York Herald Tribune*, 11th February 1928
40 MOSELEY, S.A.: 'The private diaries of Sydney Moseley' (Max Parish, London, 1960), p. 292
41 Ref. 4, pp. 156–59
42 Ref. 40, pp. 320–1
43 Ref. 4, p. 166, and pp. 397–403
44 BURNS, R.W.: 'British television, the formative years' (Institution of Electrical Engineers, London, 1986), chapter 4, 'The BBC view', pp. 94–108
45 Ref. 4, chapter 7, 'Television and the BBC', pp.155–176
46 SAMUEL, Rt. Hon. Sir H.: letter to the Postmaster General, 14th January 1929, Post Office minute 4004/33
47 BAKER, W.J.: 'A history of the Marconi Company' (Methuen, London, 1970)
48 Report of a meeting of Marconi's Wireless Telegraph Co. Ltd., *The Times*, 29th June 1910, p. 20
49 Ref. 47, p. 35
50 Report of the Select committee on Radio Telegraphic Convention (HMSO, London, 1907), pp. 232–234
51 Ref. 44, chapter 6, 'The start of the experimental service, 1929', pp. 132–147; and chapter 7, 'The low definition experimental service, 1929–1931', pp. 148–175
52 Ref. 2, p. 209

53 'Remarks by Frank B. Jewett at the television demonstration', *Bell Laboratory Record*, May 1927, p. 298

54 FINDLEY, P.B.: 'Biography of Herbert E. Ives', internal report, date unknown, AT&T Archives

55 IVES, H.E.: 'Television: 20th anniversary', *Bell Laboratory Record*, 1947, **25**, pp. 190–3

56 IVES, H.E.: 'Television', *Bell System Technical Journal*, October 1927, pp. 551–9

57 GRAY, F., HORTON, J.W., and MATHES, R.C.: 'The production and utilisation of television signals', *Bell System Technical Journal*, October 1927, pp. 560–81

58 NELSON, E.L.: 'Radio transmission for television', *Bell System Technical Journal*, October 1927, pp. 633–53

59 H.S.R.: 'Television', memorandum for file, 18th March 1966, Case Book No. 1538, Case File 20348 and Case File 33089, Vol. A, pp. 1–9, AT&T Archives

60 Ref. 4, pp. 230–235

61 GRAY, F., and IVES, H.E.: 'Optical conditions for direct scanning in television', *Journal of the Optical Society of America*, 1928, **17**, pp. 423–34

62 ANON.: 'Television shows panchromatic scene carried by sunlight', *New York Times*, 13th July 1928

63 IVES, H.E.: memorandum to H.D. Arnold, 18th February 1928, case File 33089, pp. 1–5, AT&T Archives

64 HARTLEY, R.V.L., and IVES, H.E.: 'Improvements in or relating to television', British patent no. 297078, application date 19th March 1928

65 IVES, H.E.: 'Television', memorandum for file, 10th July 1925, Case File, 10th July 1925, Case File 33089, Vol. A, pp. 1–2, AT&T Archives

66 IVES, H.E.: 'Television in colour', *Bell Laboratory Record*, July 1929, pp. 439–44

67 ANON.: 'Television in colour successfully shown', *Telephony*, 1929, **97**, pp. 23–5

68 ANON.: '2-way television in phoning tested', *New York Times*, 4th April 1930

69 FARNELL, W.C.F.: memorandum to J. Mills, 23rd June 1930, WCFF-EO, AT&T Archives

70 IVES, H.E.: 'A multi-channel television apparatus', *Journal of the Optical Society of America*, 1931, **21**, pp. 8–19

71 CAMPBELL SWINTON.: A.A.: 'Presidential Address', *Journal of the Röntgen Society*, 1912, **8**, pp. 1–5

Chapter 20

The birth of high definition television

During Bell Telephone Laboratories' programme of work on television, Ives's group had undertaken some investigations (in 1926) on the appropriateness of utilising a cathode-ray tube in a television receiver. Images of simple objects, such as a letter A, a bent wire and so on had been received on a modified cathode ray tube. The received picture signals had been impressed on an extra grid in the tube and controlled the intensity of the electron beam incident on the fluorescent screen. In one type of tube, the grid was close to the hot filament and the picture signals acted on the beam before it reached the accelerating field. In a second type of tube, the cathode beam passed through two parallel wire gauzes just before striking the screen and the image signals were applied across these two gauzes. Both types of tube reproduced images of the simple objects just mentioned, but they did not reproduce more complex objects, nor did they show half-tones in a satisfactory way.

Baird never utilised cathode ray tubes in the 1920s. His view on their application was given in an article published in *The Draughtsman* January 1928. 'The use of the cathode ray', he wrote, 'is beset with the greatest difficulties, and so far, no practical success has been met with in its application.'

Dr R.T. Beatty, of the Admiralty Research Laboratory, held a similar opinion. 'Although this method sounds attractive the practical difficulties are so great that little progress has been made by experimenters on these lines.' These difficulties were delineated in a report which he wrote in December 1925 [1].

1. Since the angular displacement of the spot is not proportional to the voltage applied, and owing to the curvature of the fluorescent screen, considerable distortion of the picture takes place. This is increased by the fact that the deflection voltages are sinusoidal instead of being linear functions of the time.

2. The modulations produced in the anode voltage for the purpose of increasing the luminosity cause radial displacements of the spot, whereby serious confusion of the picture is produced.

3. The successive pictures will not register unless the amplitudes and frequencies of the simple harmonic motion deflection voltages are regulated with great accuracy.

Until these three sources of error have been removed the oscillograph method seems to rank as inferior to the [mechanical] devices already described.

These problems, associated with the then gas-filled cathode-ray tubes, were to become known to other experimenters, including Farnsworth and Zworykin, and so efforts were made to evolve vacuum cathode ray tubes. Some later workers, for example, Bedford and Puckle [2], and von Ardenne [3], sought to ameliorate the effects of gas-filled tubes by using velocity modulation rather than intensity modulation of the electron beam.

On the 18th November 1929, V.K. Zworykin gave a lecture on his new 'kinescope' or picture tube, to a meeting of the Institute of Radio Engineers (IRE) [4]. His efforts to develop a vacuum picture receiving tube had been successful and represented an important step forward in the slow march towards an all-electronic television system. Ives was not impressed though. A demonstration of the kinescope had not been given at the IRE meeting and the account of Dr Zworykin's work was 'chiefly talk'. 'This method of reception', wrote Ives, 'is old in the art and of very little promise. The images are quite small and faint and all the talk about this development promising display of television to large audiences is quite wild' [5].

With the leader of BTL's television group holding such an opinion it was perhaps inevitable that R&D effort on the evolution of an all-electronic television scheme would not be a major part of the group's activities. Instead, in his May 1931 memorandum [6] on 'The future programme for television research and development', Ives confirmed his support for mechanical scanning by suggesting three special projects all based on this mode of reconstituting an image. These were: (1) the demonstration of reception from an aeroplane; (2) the demonstration of direct scanning of some major outdoor event; (3) the demonstration of reception on film and thence projection in the theatre. All these recommendations if approved would have been based on existing principles and technology. Ives did not propose the investigation of electronic cameras or electronic receiving tubes, and yet when his group eventually embarked on such a programme an important advance was made.

Of all the contributions made by individuals towards the realisation of an all-electronic television system none were of greater importance than those of Dr V.K. Zworykin. His invention and development of the iconoscope, and his development of the cathode ray tube as a television display tube (which he called a kinescope) were outstanding in conception and execution. The tubes were essential components of the Radio Corporation of America's high definition television system of the 1930s, and the iconoscope was the forerunner of a family of electronic camera tubes manufactured by RCA.

Vladimir Kosma Zworykin [7] was born on 30th July 1889 at Murom, Russia, which is 220 miles (354 km) east of Moscow. He was the youngest of seven children and became interested in electrical devices when he was only nine years of age. His father operated boats on the Oka River and from this age Zworykin began to spend his summer vacations aboard the craft and eagerly helped his father with the various electrical repairs which had to be undertaken from time to time.

After graduating from high school, Zworykin enrolled at the University of St Petersburg determined to become a physicist. This did not accord with his father's wishes because he felt that Russia's rising new industries offered a richer future in engineering than in physics, and so Zworykin was persuaded to transfer to the Imperial Institute of Technology. Here he remained for six years, from 1906–1912.

Zworykin loved student life, even during the restlessness and repression that characterised the last years of the Czarist government. He was a keen and diligent student, unlike some of his fellows, who 'tried to evade their laboratory work' to the annoyance of Professor Rosing, and as a consequence Rosing became friendly with the young engineering student. Zworykin assisted Rosing with his work on distant vision conducted in Rosing's private laboratory, a 'little cubby hole' in the basement of the Artillery School, which was situated across the street from the Institute [8]. Their relationship developed into a close friendship and Zworykin found Rosing to be not only 'an exceptional scientist but [also] a highly educated and versatile person'.

During the 'glorious three years' (1910–1912) when he aided Rosing [9], much of the apparatus had to be manufactured by them, including the photo-cells and all the glass vessels needed for their work. Of these times, Zworykin has written [10]:

At that time the photocells . . . were in their infancy and although potassium photocells were described in the literature, the only way to have them was to make them ourselves. Vacuum technique was very primitive and it required a tremendous amount of time to obtain the vacuum needed. The vacuum pumps which we had were manually operated and quite often we had to raise heavy bottles of mercury up and down for hours at a time in order to produce a vacuum. Electronic amplifying tubes had just been discovered by de Forest and our reconstruction was very inefficient. We were struggling to improve it ourselves. Even the glass for the bulbs was not suitable – it was very brittle and therefore difficult to work with; we had to learn to be glass blowers ourselves. Still, at the end of my association with Professor Rosing, he had a workable system consisting of rotating mirrors and a photocell on the pick-up end, and a cathode ray tube with partial vacuum which reproduced very crude images over the wire across the bench.

Zworykin's love for physics took him to Paris, following his graduation in 1912 from the Institute, and there he worked on X rays under the guidance of Paul Langevin.

When the Great War broke out Zworykin returned to Russia and was drafted immediately. He was sent, a few months later, to the Grodno fortress near the Polish frontier, but after a year and a half was transferred to the Officers Radio School [7]. There, he was commissioned and began teaching soldiers how to operate and prepare electrical equipment. A further period of time was spent in the Russian Marconi factory, which was constructing radio equipment for the

Russian Army and was situated on the outskirts of St Petersburg. He was attached to the factory as an inspector of radio equipment [10].

In 1917 the Russian Revolution started and Zworykin, fearing that it would disrupt his scientific career, decided to leave his native land. At first he could not obtain the necessary permission and the United States refused him a visa. For months he wandered around Russia to avoid arrest during the chaos of the civil war between the Reds and the Whites. Then, when an Allied expedition landed in Archangel, in September 1918, to aid Russia's northern defences against the Germans, Zworykin decided to make his way to that town. Pleading his case with an American official, Zworykin told him something about the work he could do in advancing television [7]. He was given a visa and arrived in the USA on 31st December 1918.

His first requirement, of course, was to seek employment. After several unsuccessful applications, a locomotive company called Baldwin recommended him to the head of research at Westinghouse's factory in East Pittsburgh. Zworykin was hired at a salary of $200 per month and a promise that his salary would be increased the following year. When this year came there was an announcement that due to the hard times everyone's salary would be reduced by 10 per cent.

His first task was to assemble vacuum tubes on the production line. 'The assembly was cumbersome and took a lot of time. There was a tremendous number of rejects – about 70%. I did this for about three months and almost went crazy', Zworykin has written [9]. He then devised an apparatus to make and test simultaneously 100 tube filaments for the WD 11 tube which had an oxide-coated cathode. During this work an explosion occurred that burnt his right hand and resulted in him spending some time in hospital – and filing a claim for damages. The episode had a fortunate outcome for Zworykin because he was given an opportunity to work on television. It seems Westinghouse felt responsible for the accident and allowed him to engage in his beloved field of endeavour in order to 'humour him' [8].

Zworykin subsequently spent a year developing a high vacuum cathode-ray tube, but in 1920 he left Westinghouse following a dispute over some patents relating to the WD 11 valve [8]. He obtained employment with the C&C Development Company in Kansas City but soon received a most attractive proposal from Westinghouse. Zworykin had previously greatly impressed O.S. Schairer, the manager of the Westinghouse patent department, and he persuaded the new manager of the research laboratory, S.M. Kintner, to extend an invitation to Zworykin to return to the company. According to Zworykin, he was offered a three-year contract at about three times his former salary [10]. Under the terms of the contract Zworykin would retain the rights to his prior inventions with Westinghouse although the firm would hold an exclusive option to purchase his patents at a later date. Zworykin accepted the offer and recommenced his employment with Westinghouse in February/March 1923.

On arrival he was asked by Kintner to suggest a suitable research project. Naturally Zworykin mentioned television. He was permitted to engage in the field that had held his attention about a decade previously. Zworykin worked rapidly and on the 8th October 1923 submitted a plan to the patent department for an all-electronic television system (Fig. 20.1). His proposed camera tube was not too dissimilar to that of Campbell Swinton. However, there was one major difference which indicates that Zworykin was not then acquainted with the very important charge storage principle. Campbell Swinton's photoelectric target comprised a mosaic of rubidium cubes whereas Zworykin's target was based on a uniformly deposited layer of photoelectric material.

Zworykin applied for a patent [11] on his television system on 29th December 1923, but it was not granted until 23rd December 1938. The patent that was based on Zworykin's electronic camera tube, a later version of which became known as the iconoscope, was the subject of a number of interference actions. (A brief history of these has been given in the author's book; see Reference [12].)

Figure 20.1 illustrates Zworykin's system. Its most salient component was a very thin aluminium oxide film supported by a thin aluminium film on one side and a photosensitive coating with a large transverse resistance on the other. The image of the object was projected through a fine wire collector grid, in front of the aluminium oxide film, onto the photosensitised side of the film, while a

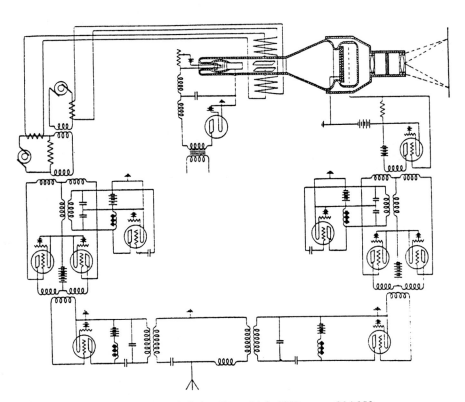

Figure 20.1 Circuit diagram included in Zworykin's 1923 patent 214 059.
Source: British Library.

high velocity electron beam scanned the opposite side. Illuminated portions of the photoelectric layer, which charged up negatively by photoemission to the collector between successive scans, were momentarily shorted to the aluminium coating or signal plate by the scanning beam penetrating to the insulating substrate. This resulted in a signal pulse, proportional to the illumination at the scanned element, in the signal plate and collector circuits. The process depended on bombardment-induced conductivity, a phenomenon investigated at a much later date by Pensak.

There is no evidence that any effort was made in 1923 to reduce the patent to practice, and until July 1924 it appears that Zworykin worked on other projects.

In April 1924 Campbell Swinton had three articles [13] published in the *Wireless World and Radio Review* on 'The possibilities of television with wire and wireless'. These papers probably stimulated much interest and may have led General Electric in the US, Karolus in Germany and Takayanagi in Japan to commence, in 1924, their investigations of television. Significantly, electronic camera tube patents were applied for in 1924 by L. and A. Seguin of France, G.J. Blake and H.J. Spooner of the UK, A. and W. Zeitline of Germany and by K.C. Randall the USA (Table 20.1).

'Zworykin certainly learned of the Campbell Swinton articles from the Westinghouse patent library, and he was able to use [them] to persuade Westinghouse to reduce his patent application to practice as soon as possible' [14]. He began his practical work in June 1924. There were many difficulties to be overcome. Nevertheless Zworykin, using a modified Western Electric type 224A cathode ray tube, worked with enthusiasm and in 1925 felt confident that his system could be favourably demonstrated to Kintner. He was 'very impressed by [its] performance' [10], but to further the work more effort, space and financial resources were needed. It was decided to show the system to H.P. Davis, the general manager of the company.

In the late summer/early autumn of 1925 Davis, Kintner and Schairer witnessed a demonstration of Zworykin's electronic television scheme. A small cross was held in front of the transmitter cathode-ray tube and its image appeared on a screen in the end of the receiver cathode ray tube. The image of a pencil was also transmitted and received. Although the images were dim, of low contrast and of poor definition, Schairer was 'deeply impressed' but, unfortunately for Zworykin, Davis was not. Zworykin has described the consequence of the test: 'Davis asked me a few questions, mostly how much time I had spent building the installation, and departed saying something to Kintner which I did not hear. Later I found out that he had told him to put this "guy" to work on something useful' [8].

Still, the laboratory experience gained by Zworykin had a positive attribute, it enabled him to prepare another patent – one which was technically superior to his patent of 1923. The new application [15] differed from the 1923 submission in two important respects. First, the target comprised a mosaic of discrete photosensitive elements (globules) and, second, both the camera and display cathode ray tubes contained 3-colour screens. Of these changes the first was of

Table 20.1 Summary of the proposals made during the period 1911–1930 for electronic television cameras using cathode-ray tubes

Date	Name	Type of cathode	Line scanning	Frame scanning	Type of raster	Focusing means	Light sensitive target	Remarks
07.11.1911 (lecture) Jan. 1912 (paper)	A.A. Campbell Swinton UK	cold	magnetic field	magnetic field	Lissajous	use of aperture to give a collimated beam	mosaic of photosensitive cubes (e.g. of rubidium) in contact with sodium vapour; the target is double sided, i.e. the electrons strike one side and the photons the opposite side	'...it must be distinctly understood that my plan is an idea only, and that the apparatus has never been constructed.'
23.08.1921 (28.06.1922) French patent	E.G.Schoultz France	thermionic	not applicable	not applicable	spiral by means of coils 20 Hz	not described	the light sensitive plate has a coating of potassium, or thallium sulphate, or selenium; the tube is double sided, i.e. the photons and electrons have opposite directions	the first patent on electron camera tubes: a parabolic mirror, mounted in the tube, reflects an image of the object onto the p.e. surface; a small hole in the mirror allows the electron beam to scan the surface; no record of experimentation

Table 20.1 – continued

Date	Name	Type of cathode	Line scanning	Frame scanning	Type of raster	Focusing means	Light sensitive target	Remarks
29.12.1923 (20.12.1938) US patent	V.K. Zworykin US	thermionic	electric field 1 kHz	magnetic field 16 kHz	Lissajous	use of aperture to to give a collimated beam	a double sided p.e. plate; a thin sheet of aluminium foil is coated with aluminium oxide; on this potassium hydride is deposited in the form of small globules; the overall thickness 'need not exceed half mil'	a grid collects the photo-emission which may be intensified by the use of argon vapour and ionisation; (cf. Campbell Swinton) claims 1, 13, 14, 15, 16, 17, and 18 refer to storing elements; curiously the patent description makes no mention of these
08.02.1924 (06.09.1924) French patent	L. and A. Seguin France	cold	electric or magnetic field	electric or magnetic field	substantially parallel lines or spiral, means not described	use of aperture to give a collimated beam	a thin selenium plate	a small mirror mounted in the tube reflects an image of the object onto the selenium plate; a single-sided tube; no evidence of experimentation
28.02.1924 (28.05.1925) UK patent	G.J. Blake and H.J. Spooner UK	thermionic	electric or magnetic field	electric or magnetic field	probably Lissajous – not explicitly stated	use of aperture to give a collimated beam	a plate having a very thin coating of selenium	a single sided tube; no evidence of experimentation
18.03.1924 (17.07.1930) German patent	A. and W. Zeitline Germany	cold	electric or magnetic field	electric or magnetic field	probably Lissajous	not stated explicitly	a plate having a coating of potassium or other suitable substance	the patent describes a double sided tube; no evidence of experimentation

Date/source	Author						Plate	Notes
10.04.1924 (03.09.1935) US patent	H.J. McCreary	thermionic	electric field	electric field	Lissajous	use of aperture to give a collimated beam	a double sided photoelectric plate; the insulated plate has a large number of minute pins imbedded in it; embedded on the faces of the plate are conducting grids; the image side is coated with potassium hydride or selenium	the patent mentions that both sides of the plate may be coated with potassium hydride, selenium or other suitable substance; also three camera tubes and appropriate filters may be combined to enable colour television to be produced
09.04.1924 & 16.04.1924 23.04.1924 (papers)	A.A. Cambell Swinton UK	thermionic	electric or magnetic field	electric or magnetic field	Lissajous or parallel lines	use of aperture to give a collimated beam	a mosaic of photosensitive cubes in contact with sodium vapour	basically the same as his 1911 proposal but updated to take account of current (1924) practice; no evidence of experimentation
11.07.1924 (28.02.1928) US patent	K.C. Randall US	thermionic	electric field 1 kHz	magnetic field 16 Hz	Lissajous	use of aperture to give a collimated beam	a double sided photoelectric plate; a thin sheet of aluminium foil is coated with aluminium oxide; on this an alkali metal (e.g. potassium hydroxide) is deposited	the camera tube is identical to that shown in Zworykin's 1923 patent; the object of Randall's patent was to monitor the operation of substations from a central control station

Table 20.1 —continued

Date	Name	Type of cathode	Line scanning	Frame scanning	Type of raster	Focusing means	Light sensitive target	Remarks
05.04.1925 (15.09.1927) German patent	M. Dieckmann and R. Hell	not applicable	magnetic field 500 Hz	magnetic field 10 Hz	probably Lissajous	not applicable	cathode coated with potassium or rubidium	an image dissector tube; Hell claimed a tube was constructed but operation was not possible
27.05.1925 (11.12.1928) US patent	C.A. Sabbah US	thermionic	not applicable	not applicable	spiral, by use of two pairs of deflecting plates 16 Hz	use of aperture to give a collimated beam	a double sided photoelectric plate; the transparent plate is coated on one side with a thin semi-transparent film of photoelectric material (caesium, sodium or potassium)	Lissajous raster also mentioned; no evidence of experimentation
13.06.1925 (13.11.1928) US patent	V.K. Zworykin							an adaptation of Zworykin's 1923 patent for colour television. The system uses Paget colour filters
25.08.1925 (03.02.1931) US patent	T.W. Case US	thermionic	electric or magnetic field 6 kHz	electric or magnetic field 10 Hz	probably Lissajous	use of aperture to give a collimated beam	the inner surface of the end of the cathode-ray tube is coated with a semi-transparent opaque layer of some conducting photoelectric material (e.g. potassium)	both double-sided and single-sided operation mentioned; no evidence of experimentation

04.12.1926 (04.11.1930) US patent	F.W. Reynolds US	thermionic	electric or magnetic field 100 Hz	electric or magnetic field	substantially parallel lines or spiral	use of beam focusing anode	a multiple unit photoelectric element comprising a large number of closely compacted relatively fine and long glass or quartz tubes; a film of rubidium or potassium is deposited on the inner walls of the tubes	the photoelectric elements permit the passage of the cathode-ray beam in proportion to the light activation; the discrete space charges of the photo-electrons successively present a varying impedance to the cathode-ray beam; no evidence of experimentation
07.01.1927 (26.08.1930) US patent	P.T. Farnsworth US	not applicable	electric field 5 kHz	electric field 10 Hz	substantially parallel lines 500 lines per raster	not stated	the cathode is a fine mesh screen covered with a light sensitive material (sodium, potassium or rubidium)	an image dissector tube
10.07.1928 (British patent void)	K. Tihany Hungary	thermionic	electric field 3.2 kHz	electric field 8 Hz	substantially parallel lines	use of apertures and magnetic field to give a uniform collimated beam	the photosensitive layer of the image plate could be of (1) a photoelectric material (e.g. an alkali metal, (2) a photo-conductive substance (e.g. setenium, (3) crystals in which a change of state occurs	the British patent had 127 claims and described many different types of image carrier; however, the complete specification was not accepted and the patent became void, double-sided tubes were described; no evidence of experimentation

Table 20.1—continued

Date	Name	Type of cathode	Line scanning	Frame scanning	Type of raster	Focusing means	Light sensitive target	Remarks
22.06.1928 (05.09.1929) UK patent	C.E.C. Roberts UK	not applicable	electric or magnetic field	electric or magnetic field	probably Lissajous	not stated	the cathode may be formed of one or more different photoelectric materials	a single-sided tube of the image dissector type; several arrangements shown; several tubes were constructed but could not be made to operate satisfactorily
24.08.1928 (31.10.1929) UK patent	D.N. Sharma UK	not stated explicitly	electric or magnetic field	electric or magnetic field	substantially parallel lines or spiral	use of aperture to give a collimated beam	a small opaque region is produced in a transparent or semi-transparent 'composite' plate when the region is struck by the electron beam	an image of the object is formed on the surface of the plate and as the opaque spot traverses the image it is reflected spot by spot onto a photoelectric cell; no evidence of experimentation
26.11.1928 (21.04.1936) US patent	P.T. Farnsworth US	not applicable	magnetic field	magnetic field	substantially parallel lines	uniform magnetic field	the cathode is a photosensitive film deposited on a metallised surface supported on a glass plate	an image dissector tube having two apertures (one large and one small) to produce a signal current which comprises a low frequency component and a component covering the entire desired frequency range; object: to improve sensitivity and detail

01.06.1928 (02.09.1929) US patent	Associated Telephone and Telegraph Co	thermionic	electric field	electric field	Lissajous	use of aperture to give a collimated beam	the double sided image plate comprises a large number of short, parallel, insulated conductors; two metallic grids coated with e.g. potassium hydride or selenium make contact with the two faces of the plate	the wires are arranged to be orthogonal to the faces of the plate; the British patent hints at some attempt at experimental practice but apart from this there is no other evidence of experimentation
26.11.1929 (14.07.1931) US patent	W.J. Hitchcock US							a photoelectric control element is scanned by an electron beam from one of two cathodes
01.05.1930 (17.06.1941) US patent	V.K. Zworykin US	thermionic	electric field		parallel lines	use of aperture to give a collimated beam		

fundamental importance since it would allow the camera tube to function according to the principle of charge storage.

The immediate effect of the 1925 demonstration was the relocation of the television project to Dr F. Conrad, who, as mentioned previously, had participated in the engineering of Westinghouse's KDKA transmitter station, and who was highly regarded by Davis. Conrad's approach to his new task was to return to more conventional methods of scanning an object. He devised a system of television known as radiomovies in which 35 mm motion picture of film was scanned to provide a source of video signals [16].

By August 1928 the system had been developed to the state where it could be demonstrated to senior members of the company. Such was the interest and progress of television in the UK, Germany, France and the US at this time that the display of Westinghouse's television scheme attracted the scrutiny of senior executives from RCA, NBC and GE, as well as from Westinghouse. Among the spectators were D. Sarnoff, the President of RCA, Dr A.N. Goldsmith and E Bucher of RCA; M.H. Aylesworth and E.B. Taylor of NBC; Dr E.F.W. Alexanderson and Dr W.R.G. Baker of GE; and H.P. Davis, L.W. Chubb, O.S. Schairer, S. Kintner and Dr F. Conrad of Westinghouse.

During the laboratory demonstration on 8th August 60-line images, obtained from film, were transmitted, 2 miles (3.2 km) by land line from the film scanning apparatus, to the KDKA transmitter where they were radiated back to the laboratory. Three frequencies were specified, *viz:* 2.00 MHz, 4.762 MHz and 3.33 MHz for the vision, sound, and synchronising signals, respectively. A 60 lines per picture, 16 pictures per second standard was adopted, and the vision signals were generated by means of a modified 35 mm standard cine film projector and a caesium photocell. At the receiver the mercury arc lamp and scanning disc enabled 'the radio pictures to be thrown upon a ground glass or screen, the first time this [had] been done with television apparatus'. One year later, on 25th August 1929, KDKA began broadcasting radiomovies on a daily basis [17].

In 1928 all radio and television research carried out by General Electric, the Westinghouse Electric Manufacturing Company and, later, the Radio Corporation of America was on behalf of RCA. Each firm had engineers working independently on the elucidation and engineering of the principles of television. However, since RCA was primarily a sales organisation, the influence of Sarnoff on developments was especially strong.

Of the senior executives in GE, WEM and RCA, Sarnoff was particularly forward looking. Just as he had foreseen, and expounded on, in 1916, the need for radio music boxes, so now (in 1928) he outlined his views on the prospects of television. His thoughts were expressed in an article, 'Forging an electric eye to scan the world', published on 18th November 1928 in the *New York Times* [18]. He wrote: 'Within three to five years I believe that we shall be well launched in the dawning age of sight transmission by radio'. He predicted that television broadcasting would be classified as radiomovies (following the policy adopted by WEM), and as radio television, in which vision signals would be generated directly – as was being done, for example, by Baird.

Sarnoff's use of the expression electric eye is intriguing. In the summer of 1928 he had visited France where it was known that television experimention with cathode ray tubes had been conducted by E. Belin and Dr F. Holwek, by Dr A. Dauvillier, and by G. Valensi. It appears that while in Paris Sarnoff, who closely followed television developments, saw something relating to television which provoked an interest in electronic television. He knew that Zworykin was a proponent of cathode-ray television and instructed him to visit Europe to determine 'the status of ideas applicable to cathode-ray television there'. The visit was to be, possibly, the most valuable ever to be undertaken by a television pioneer since RCA's television endeavours were to be greatly advanced by the knowledge and hardware that Zworykin acquired and by the appointment of an engineer (G.N. Ogloblinsky) of some considerable ability. (From 1925 Zworykin had been working on the recording and reproduction of sound on cine film. This work had considerably enhanced his standing in WEM.)

He sailed for Europe on 17th November 1928 with a schedule to visit England, France and Germany. Of these visits the one Zworykin made to the Laboratoire des Etablissements Edouard Belin was of prime importance. Here he met E. Belin, Dr F. Holweck, G.N. Ogloblinsky and P.E.L. Chevallier, and was shown one of Belin's and Holweck's latest continuously pumped, all-metal cathode-ray tube display tubes [19].

During the 1920s three methods of focusing the electron beam in a cathode-ray tube (c.r.t.) were investigated: gas focusing, magnetic field focusing and electric field focusing. The latter method resembles the focusing of a light beam by a lens system [20], (see Fig. 20.2).

Among the early investigators of electrostatic focusing were W. Rogowski and W. Grosser who applied, in December 1923 for a patent [21] for a c.r.t. based on this type of focusing; A. Dauvillier [22] who studied the potential of such a method for television purposes and filed a patent in February 1923; Holweck and Chevallier whose experimentation led to a patent [23] application dated 4th March 1927; and R.H. George [24] of Purdue University.

Zworykin immediatley recognised the superiority of Holweck's and Chevallier's c.r.t. (compared to the Western Electric Type 224 that employed gas focusing) and the likelihood that their approach to the design of the c.r.t. would have a great impact on the evolution of an all-electronic television system. Some modifications to the construction of the tube would be necessary before a practical version suitable for the domestic market could be manufactured but, nevertheless, so convinced was Zworykin that the Holweck and Chevallier tube held the promise of providing the essential foundation for his cherished ideal that he sought to purchase the tube from the Laboratoire des Etablissements Edouard Belin. He was successful and on 24th December 1928 arrived back in the US with the latest Holweck and Chevallier metallic, demountable cathode ray tube, and a Holweck rotary vacuum pump. The c.r.t. was one of the most valuable devices in the television field ever brought to the shores of the USA. As

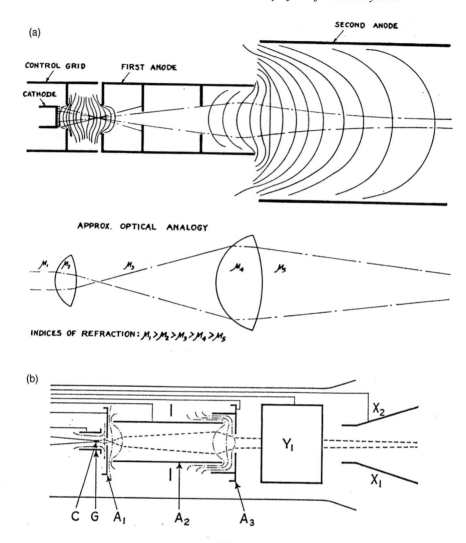

Figure 20.2 (*a*) *Diagram showing the analogy between the focusing of a beam of electrons by the use of electron lenses, and the focusing of a beam of light by optical lenses.*

Source: *Journal of The IEE*, 1933, **73**, p. 443.

(*b*) *The elements of a cathode-ray tube: cathode, C, control grid, G, anodes, A_1, A_2, and A_3, the Y deflection plate, Y_1 and the X deflection plates, X_1 and X_2.*

Source: Chapman and Hall.

Zworykin's biographer has written [25]: 'Zworykin's trip to Paris changed the course of television history'. The consequence of his journey would be the engineering of an electron gun, with electrostatic focusing, which would be at the heart of all camera and display tubes built and sold by the RCA and some of its licensees.

On his return to New York, Zworykin reported to Kintner, now the vice President of Westinghouse. He suggested that since he (Zworykin) had gone to Europe on behalf of RCA rather than the Westinghouse Electric Manufacturing Company he should go to New York and discuss his work with David Sarnoff, Vice-President and General Manager of RCA [8].

The meeting took place in January 1929. 'Sarnoff quickly grasped the potentialities of my proposals,' Zworykin later wrote [26], 'and gave me every encouragement from then on to realise my ideas.'

Recalling Sarnoff, 'a brilliant man without much education' [9], Zworykin has recounted the following anecdote [10]:

Sarnoff saw television as a logical extension of radio broadcasting, which was already commercially prosperous. His parting question, after my presentation, was to estimate how much such a development would cost. I said I hoped, with a few additional engineers and facilities, to be able to complete the development in about two years and estimated that this additional help would cost about a hundred thousand dollars. This of course was too optimistic a guess; as Sarnoff has stated that RCA had to spend many millions before television became a commercial success.

Fortunately, Sarnoff was inveigled by Zworykin's enthusiasm and arranged for Westinghouse to give Zworykin additional finances, staff and equipment. Work on the new television system commenced in February 1929.

Good progress was made and Zworykin was able to present a lecture on his system (Fig 20.3) on 18th November 1929. The production of moving images, obtained from a film scanner-photo cell unit, and their reception and display, using a kinescope, were described [4]. Conrad had been working on film scanning using vibrating mirrors from 1925 and as the principles of mechanical scanning were well understood it was natural that, lacking a suitable camera tube, Zworykin should have chosen this method of generating images for the tests with the kinescope.

By November 1929 Zworykin's team had constructed six television receivers. He was permitted to use the KDKA transmitter three nights a week from 02.00 a.m. to 03.00 a.m. to broadcast carrier waves, modulated by vision signals and synchronising pulses, derived from the cine film projector equipment, and sent to KDKA by land line.

At other times the station radiated, on a daily basis, television signals generated by Conrad's film projector. The received signals could be displayed using Zworykin's kinescope. With this the images had a size of 3 inches by 4 inches (76.2 mm × 101.6 mm) and were 'surprisingly sharp and distinct' notwithstanding the 60 lines per picture standard utilised [27].

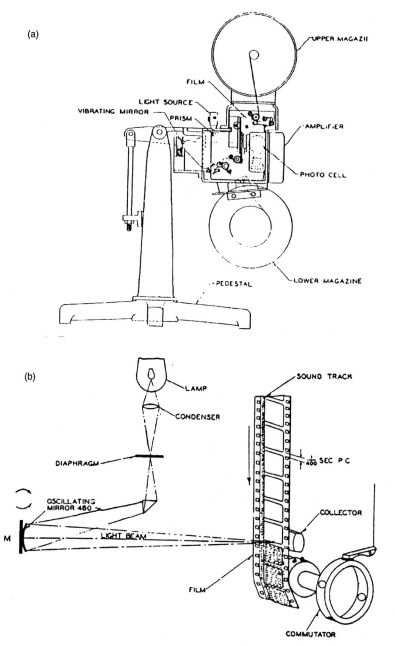

*Figure 20.3   (a) Details of the modified standard cine film projector showing the location
of the photocell, light source and vibrating mirror. (b) Diagram showing the
manner in which the whole surface of the picture is explored by the light
reflected from the vibrating mirror.*

Source: *Radio Engineering*, 1929, **9**.

Zworykin's, and his team's, work in 1929 showed that the non-mechanical method of image synthesis had the following desirable attributes [4]:

1. the image was visible to a large number of people at once, rather than to only one or two as with some mechanical scanners, and no enlarging lenses were required;
2. there were no moving parts and hence no noise;
3. the picture was brilliant enough to be seen in a moderately lit room;
4. the framing of the picture was automatic;
5. the use of a fluorescent screen, aided the persistence of the eye's vision;
6. the motor previously used, together with its power amplifier, was redundant, and the power consumed by the kinescope was no greater than that used in ordinary vacuum tubes;
7. 'the inertialess electron beam was easily deflected and could be synchronised at speeds far greater than those required for television'.

These advantages were also known to Philo Taylor Farnsworth, who was born on 19th August 1906 on a farm near Buckhorn, Utah and who, at 20 years of age, embarked on his life's work, the evolution of an all-electronic television system [28]. He was an inventor of the 'Marconi' type for he did not complete a formal degree course in engineering and commenced his experiments in one of the rooms of his apartment. Unfortunately for Farnsworth his electronic camera tube, which he called an 'image dissector', did not embrace the principle of charge storage and hence did not pose a serious commercial threat to Zworykin's system which was based on the iconoscope (see below).

There can be no doubt that the short period (from February 1929 to November 1929) it took Zworykin [29] and his team to engineer the kinescope resulted from his acquisition of the Chevallier c.r.t., and the work of R.H. George of Purdue University. This work was presented, in March 1929, at a meeting of the American Institution of Electrical Engineers, and was published in July 1929. George's hard vacuum cathode-ray oscillograph featured a special hot cathode electron gun, and made use of a new electrostatic method of focusing the electron beam. His solution to the problem of devising a satisfactory means of producing and focusing a high intensity beam over a wide range of accelerating voltages constituted the chief contribution of his investigation [24].

Chevallier applied, in France, for a patent [30] to protect his c.r.t. invention on 25th October 1929: Zworykin's c.r.t. US patent application [31] was dated 16th November 1929. Both patents were for c.r.ts of the hard vacuum, hot cathode, electrostatically focused type, with grid control of the fluorescent spot intensity and post deflection acceleration.

The prior disclosures by George and Chevallier did not enable Zworykin, and RCA, to establish an undisputed patent claim to the use of grid control, electrostatic focusing and post deflection acceleration. Since these characteristics were essential to the proper functioning of the kinescope it was necessary for RCA to purchase the rights of George's and Chevallier's patents.

Zworykin's hard vacuum c.r.t. (the kinescope) was well suited to adaptation as a camera tube. Basically, the fluorescent screen had to be replaced by a mosaic target plate, and an additional collecting electrode incorporated into the tube. At first, however, when Zworykin and Ogloblinsky (the former Chief Engineer of the Laboratoire des Etablissement Edouard Belin who joined Zworykin in July 1929) began their work on the iconoscope (the name given by Zworykin to the camera tube) they used the demountable, metallic tube that Zworykin had obtained in Paris. This had the advantage, compared to a glass tube, that it could be dismantled, the electrode configuration/mosaic modified and the tube re-assembled for testing without the need to construct new tubes.

In his 1923 iconoscope patent Zworykin specified potassium hydride as the photoelectric substance. Now, in 1929, following the 1928 discovery by Koller [32], the new, highly sensitive caesium-silver oxide photoelectric surface could be deposited onto his target plate and fabricated into a mosaic that exhibited charge storage.

With a single scanning cell, as used in the Nipkow disc-photocell arrangement and also in Farnsworth's image dissector tube, the cell must respond to light changes in $1/(Nn)$ of a second where N is the frame rate and n is the number of scanned elements. For N equal to ten frames per second and n equal to 10,000 elements (corresponding to a square image having a 100-line definition), the cell must react to a change in light flux in less than ten millionths of a second. But if a scanned mosaic of cells is used each cell has 100,000 millionths of a second in which to react, provided that each cell is associated with a charge storage element. Theoretically, the maximum increase in sensitivity with charge storage is n, although in practice the 1930s iconoscopes only had an efficiency of about five per cent of n.

This most important principle is implicit in a patent [33], dated 21st May 1926, of H.J. Round, of the Marconi Wireless Telegraph Company. The principle is not described in the body of Zworykin's 1923 patent application.

Abramson, Zworykin's biographer, has written [34]: 'Zworykin's patent lawyers included "a bank of condenser elements" [charge storage elements] on September 30, 1931, as claim 32, which ultimately became claim 13 (VKZ patent file, 1923 application). Unless one knows the background to this patent it appears that Zworykin did apply for a patent with "charge storage" in December 1923, when in fact this was added in 1932.'

Subsequently, in November 1931, an interference action was initiated between Zworykin and Round. The examiner of interferences awarded priority of invention to Round on 28th June 1935, and on 26th February 1936 the district court affirmed the examiner's verdict. Because the principle of charge storage was of such fundamental importance in the operation of the iconoscope, RCA purchased Round's patent.

In January 1930 Zworykin and his group moved to RCA Victor at Camden, New Jersey. His main task was to engineer a reliable, sensitive, high definition electronic camera tube. At this time the only other activity, anywhere, on such camera tubes was that being conducted by Farnsworth at his Green Street laboratories in San Francisco. Since Zworykin had been admirably rewarded when he had visited the Laboratoire des Etablissement Edouard Belin, Paris, it was reasonable to anticipate that a visit to Farnsworth's laboratory would likewise be beneficial. Actually, Zworykin's visit was brought about by an adventitious factor.

1930 was a difficult year for speculative business ventures generally. The effects of the recession were having an adverse influence on financial returns, and since the Farnsworth system appeared to need much further development before profits could be realised some of the supporters of the Farnsworth Television Laboratories were keen to sell out. The firm of Carroll W. Knowles Company had obtained an option to buy the stock of the laboratories and wished to have a technical assessment of its worth. It sent an invitation – without Farnsworth's knowledge – to RCA to inspect the laboratories and its patents. As a consequence Zworykin spent several days from 16th to 18th April 1930 appraising the apparatus. He was astonished by a technician's success in sealing a disc of optical glass to an image dissector tube. Zworykin had been told by Westinghouse and RCA engineers that such discs could not be fused onto glassware.

After the visit Zworykin wrote a report on what he had seen. It was read by Dr E.F.W. Alexanderson, the chief engineer of RCA. He opined [35]:

Farnsworth has evidently done some very clever work but I do not think that television is going to develop along these lines. However, this is a question that can be settled only by competitive experimentation and I think that Farnsworth could do greater service as a competitor to the Radio Corporation by settling this provided that he has financial backing. If he should be right, the Radio Corporation can afford to pay more for his patent than we can justify now, whereas, if we buy his patents now it involves a moral obligation to bring this situation to a conclusion by experimentation at a high rate of expenditure. I feel that we can use our experimental funds to better advantage and that we should not assume such responsibility.

Following the establishment of the Camden research groups, Zworykin and Ogloblinsky tried to fabricate a two-sided camera tube (Fig. 20.4) but when 'the technical difficulties associated with the making of [the] targets with the high degree of perfection required [became] very great' they decided in May 1931 to experiment with single-sided targets.

Meantime, a trial, held on 15th July 1930, of the television systems of the GE television group, led by Alexanderson, and of the former Westinghouse television group, led by Zworykin, showed the superiority of the electronic approach over that of the mechanical approach to the solution of the television problem. A few days later a decision was agreed to appropriate 90 per cent of the budgeted research funds to Zworykin's television system and just ten per cent to that of GE.

The Radio Corporation of America was one of the first to recognise the necessity for conducting investigations in the (43–80)MHz band and during the 1930 to 1940 decade it carried out a number of field tests, in 1931–32, 1933,

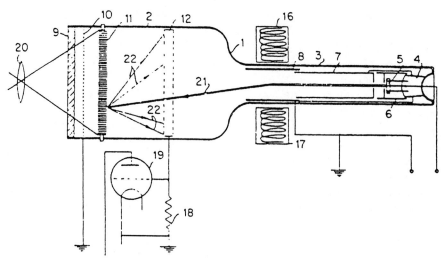

Figure 20.4   *Early (EMI) double-sided mosaic signal generating tube.*
Source: Thorn EMI Central Research Laboratories.

1934 and 1936–1939. These tests were conducted with the same thoroughness and engineering excellence as the Bell Telephone Laboratories' 1927 trials. An important feature of RCA's enquiries was the use of a systems approach, an approach concerned with the formulation and understanding of the individual units that make up the whole system and with the interconnection of the units to form an integrated system.

The 1931–1932 test was based on the cathode-ray tube television system then being developed by the research branch of the company and the results were published in a series of papers in the Proceedings of the Institute of Radio Engineers for December 1933 [36,37,38,39].

There were many points to be investigated:

1. the number of lines per picture for good and interesting television;
2. the number of pictures per second for flicker free reproduction;
3. the propagation characteristics of very high frequency television signals and their susceptibility to interference signals;
4. the use of cathode ray tube reception;
5. the use of studio and cine film scanners for signal generation;
6. the design of circuits having bandwidths appropriate to medium definition television; and
7. the synchronising of the receiving apparatus with the transmitter.

RCA chose as its location for the tests the Metropolitan area of New York and installed its studio and transmitting equipment in the Empire State Building (on the 85th floor, 1,001 ft (305 m) above street level). The short wave transmitting antenna was fixed to the top of the airship mooring mast 1,250 ft ((381 m) above street level). By the end of 1931 the system had been put into position and testing started during the first half of 1932.

Figure 20.5 shows a block diagram of the equipment layout. It was essentially unambitious in concept for the cathode-ray tube receiver had been demonstrated in 1929, the mechanical scanner of the Nipkow type had been in use for several years, as had the spotlight scanning of both studio and film scenes, and some work had been carried out on the transmission and reception of short waves.

However, no other organisation by 1931–32 had engaged in such a large scale test of this type and made its findings known, and hence RCA, with its firm judgment that television was inevitable, was compelled to proceed with its own research and development programme. As Engstrom, the group's leader, observed: 'The equipment used . . . was in keeping with the status of television development at that time' [36].

The basic parameters of the system were selected to be 120 lines per picture, sequentially scanned, and 24 pictures a second.

A number of receivers were placed in and around New York, many of them in the homes of the technical personnel, in order to collect subjective viewer reaction data as well as objective engineering measurements. Separate sound and vision transmitters, widely spaced in frequency, were used to simplify the apparatus, and both line synchronisation and frame synchronisation were employed.

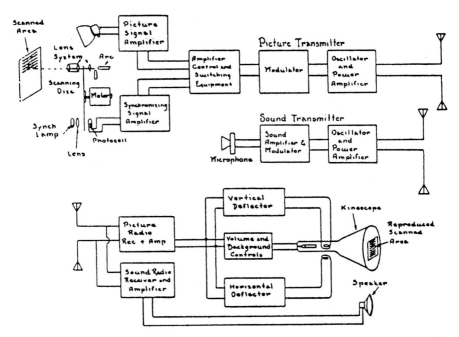

Figure 20.5 Schematic diagram of RCA's experimental television system (1931–1932).
Source: *Proc. of the IRE*, 1933, **21** (12), pp. 1652–1654.

Engstrom summarised some of the major observations and conclusions of the 1931–32 field test as follows [36]:

The frequency range of (40–80) MHz was found [to be] well suited for the transmission of television programmes. The greatest source of interference was from ignition systems of automobiles and airplanes, electrical commutators and contactors, etc. It was sometimes necessary to locate the receiving antenna in a favourable location as regards signal and source of interference. For an image of 120 lines, the motion picture scanner gave satisfactory performance. The studio scanner was adequate for only small areas of coverage. In general, the studio scanner was the item which most seriously limited the programme material. Study indicated that an image of 120 lines was not adequate unless the subject material from film and certainly from studio was carefully prepared and limited in accordance with the image resolution and pick up performance of the system. To be satisfactory, a television system should provide an image of more than 120 lines. . . . The operating tests indicated that the fundamentals of the method of synchronising used were satisfactory. The superiority of the cathode-ray tube for image reproduction was definitely indicated. With the levels of useful illumination possible through the use of the cathode-ray tube, the image flicker was considered objectionable with a repetition pregnancy of 24 per second.

During the period of RCA's field tests Zworykin and his group had been endeavouring to engineer a single-sided target plate camera tube (the iconoscope). In such a tube the plate consisted of a very thin (between one and 75 μm thick) mica sheet, onto one side of which a mosaic of minute silver globules, photosensitised and insulated from one another, was deposited; and onto the other side of which a metal film (known as the signal plate) was deposited. Each globule was capacitatively coupled to the signal plate and hence to the input stage of the video amplifier. When an optical image was projected onto the mosaic each photosensitive element accumulated charge by emitting photoelectrons. The information contained in the optical image was stored on the mosaic in the form of a charge image. As the electron beam uniformly scanned the mosaic in a series of parallel lines the charge associated with each element was brought to its equilibrium state ready to start charging again. The change in charge in each element induced a similar change in charge in the signal plate and, consequently, a current pulse in the signal lead. The train of electrical pulses so generated constituted the picture signal (Fig. 20.6).

Although promising results were obtained just one month after Zworykin's group transferred its efforts to the fabrication of signal-sided plates much experimentation was required before the preparation of the mosaic was perfected. As with many successful undertakings, an adventitious factor aided the group. On one occasion Essig accidentally left one of the silvered mica sheets in an oven too long and found, after examining the sheet, that the silver surface had broken up into a myriad of minute, insulated silver globules. This, of course, was what was needed: by October 1931 good results were being obtained with the new processing method. According to Iams, the first satisfactory tube (no. 16) was fabricated on 9th November 1931. One of the early experimental tubes was employed by the National Broadcasting Company for three years but

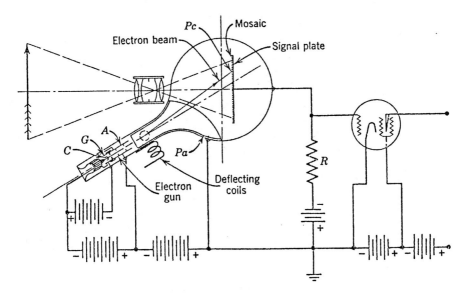

Figure 20.6   Diagram of the iconoscope.
Source: Professor J.D. McGee.

general adoption of the iconoscope had to wait until 1933 when it could be manufactured to have uniform and consistent properties.

With progress being made in the development of electronic scanners, it was to be expected that RCA's second field test would embrace an appreciation of the device. In the New York tests the major limitation to adequate television performance had been the studio scanning apparatus, since the lighting in the studio had been of too low an intensity to give a satisfactory signal-to-noise ratio; only when a motion picture film was being scanned could a reasonable ratio be achieved. Fortunately, the sensitivity of the iconoscope was sufficient to allow a further increase in the number of lines scanned per picture, in addition to permitting outdoor, as well as studio, scenes to be televised.

RCA's second experimental television system (of 1933) employed a picture standard of: 240 lines per picture, sequentially scanned at 24 pictures per second, with an aspect ratio of 4:3 (width to height). The testing programme was comprehensive in concept and execution: line and radio links; film, studio and outside broadcasts; and relay working were all included (Fig. 20.7).

Among the group's findings Engstrom noted, in a paper [40] published in November 1934:

1. The use of the iconoscope permitted transmission of greater detail, outdoor pick-up, and wider areas of coverage in the studio. Experience indicated that it provided a new degree of flexibility in pick-up performance, thereby removing one of the major technical obstacles to television. . . . The iconoscope type pick-up permitted a freedom in subject material and conditions roughly equivalent to motion picture camera requirements.

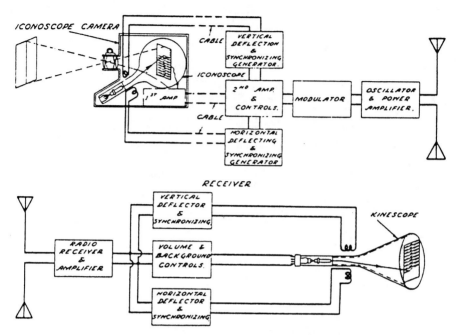

*Figure 20.7   Block diagram showing the television system which incorporated an
iconoscope and a kinescope.*
Source: *Proc. of the IRE*, 1934, **22**(11), pp. 1241–1245.

2. The choice of 240 lines was not considered optimum, but all that could be
    satisfactorily handled in view of the status of development. . . . The increase
    of image detail widened very considerably the scope of the material that
    could be used satisfactorily for programmes. Experience with this system
    indicated that even with 240 lines, for critical observers and for much of the
    programme material, more image detail was desired. The desire was for both
    a greater number of lines and for a better utilisation of the detail capabilities
    of the system and lines chosen for the tests. [Fig. 20.8 shows a cartoonist's
    view of programme material.]
3. As in the New York tests, much valuable experience was obtained in con-
    structing and placing in operation a complete television system having
    standards of performance abreast of research status. Estimates of useful
    field strengths were formulated. The need for a high power television
    transmitter was indicated.

From 1931 RCA's only competitor, apart from Farnsworth, in the all-
electronic television field was Electric and Musical Industries (EMI) of the

Figure 20.8 'The future of television . . . How the cartoonists saw it in the "thirties"'.
Source: *The Daily Mirror*, June 1932.

United Kingdom. The company's R&D effort led to the inauguration of the world's first, high definition, public television service on 2nd November 1936, and is considered in the next chapter.

References

1 BURNS, R.W.: 'Early Admiralty interest in television', *The IEE Conference Publication of the 11th The IEE Weekend Meeting on the History of Electrical Engineering*, 1983, pp. 1–17
2 BEDFORD, L.H., and PUCKLE, O.S.: 'A velocity modulation television system', *J.IEE*, 1934, **75**, pp. 63–82
3 VON ARDENNE, M.: 'Improvements in television', British patent no. 387,087, application date 21st December 1931
4 ZWORYKIN, V.K.: 'Television with cathode-ray tube for receiver', *Radio Engineering*, 1929, **9**, pp. 38–41
5 IVES, H.E.: memorandum to H.P. Charlesworth, 16th December 1929, Case File 33089, p. 1, AT&T Archives
6 IVES, H.E.: 'Future programme for television research and development', memorandum for file, 18th May 1931, Case File 33089, pp. 1–11, AT&T Archives
7 BINNS, J.J.: 'Vladimir Kosma Zworykin', in 'Those inventive Americans' (National Geographic Society, 1971), pp. 88–95
8 'Interview with Dr Zworykin on 3rd May 1965', transcription of tape, pp. 1–25, Science Museum, UK
9 ANON.: 'Vladimir Zworykin. The man who was sure TV would work', *Electronic Design*, 1977, **25**(18), pp. 112–115
10 ZWORYKIN, V.K.: 'Electronic television at Westinghouse and RCA', unpublished paper, 13pp., RCA Archives
11 ZWORYKIN, V.K.: 'Television system', US patent 2 141 059, 29th December 1923
12 BURNS, R.W.: 'Television, an international history of the formative years' (The Institution of Electrical Engineers, London, 1998), pp. 381–382
13 CAMPBELL SWINTON, A.A.: 'The possibilities of television with wire and wireless', *The Wireless World and Radio Review*, 9th April 1924, pp. 51–56; 16th April 1924, pp. 82–84; 23rd April 1924, pp. 114–118
14 ABRAMSON, A.: 'Zworykin, pioneer of television' (University of Illinois Press, Chicago, 1995), p. 49
15 ZWORYKIN, V.K.: 'Improvements in or relating to television systems', British patent 255 057, application date 3rd July 1926 (UK); US patent 1 691 324, filed 13th July 1925
16 Ref. 14, p. 67
17 ANON.: 'Television from film goes on the air', *The New York Times*, 25th August 1929, p. 15:7. Also: 'Cathode-ray television receiver developed', *Sci. Am.*, February 1930, p. 147; and ' "Crystal globe" reception', *Wireless World*, 27th November 1929, **25**, p. 595
18 SARNOFF, D.: 'Forging an electric eye to scan the world', *The New York Times*, 18th November 1928, p. 3:1

19 Ref. 14, p. 71
20 Ref. 12, pp. 386–387
21 ROGOWSKI, W., and GROSSER, W.: DRP patent no. 431,220, 4th February 1927
22 DAUVILLIER, A.: 'Procede et dispositifs permettant de realiser la télévision', French patent 592 162, filed 29th November 1923; and first addition 29 653, filed 14th February 1924
23 Ref. 14, pp. 71–75
24 GEORGE, R.H.: 'A new type of hot cathode oscillograph and its application to the automatic recording of lightning and switching surges', *Trans. AIEE*, July 1929, p. 884
25 Ref. 14, chapter 6
26 ZWORYKIN, V.K.: 'The early days: some recollections', *Television Quarterly*, 1962, 1(4), pp. 69–72
27 Ref. 14, p. 83
28 Ref. 12, pp. 356–376
29 ZWORYKIN, V.K.: 'Television with cathode-ray tube for receiver', 9th September 1929, a report, Westinghouse Electric and Manufacturing Company, archives
30 CHEVALLIER, P.E.L.: 'Kinescope', US patent 2 021 252, filed 20th October 1930
31 ZWORYKIN, V.K.: 'Vacuum tube', US patent 2 109 245, filed 16th November 1929
32 KOLLER, L.R.: 'S. 1 photocathode', *Phys. Rev.*, 1930, **36**, p. 1639
33 ROUND, H.J.: 'Improvements in or relating to picture and like telegraphy', British patent 276 084, 21st May 1926
34 Ref. 14, p. 248
35 ALEXANDERSON, E.F.W.: memorandum to H.E. Dunham, 4th June 1930, GE Archives, p. 175
36 ENGSTROM, E.W.: 'An experimental television system', *Proc. IRE*, 1933, **21**(12), pp. 1652–1654
37 ZWORYKIN, V.K.: 'Description of an experimental television system and the kinescope', *Proc. IRE*, 1933, **21**(12), p. 1655–1673
38 KELL, R.D.: 'Description of experimental television transmitting apparatus', *Proc. IRE*, 1933, **21**(12), pp. 1674–1691
39 BEERS, G.L.: 'Description of experimental television receivers', *Proc. IRE*, 1933, **21**(12), pp. 1692–1706
40 ENGSTROM, E.W.: 'An experimental television system', *Proc. IRE*, 1934, **22**(11), pp. 1241–1245

Chapter 21

EMI and high definition television

On 2nd November 1936 the world's, first, public, regular, high definition television service was inaugurated at Alexandra Palace, London [1]. The decision to establish the service was made by Parliament following the submission to it of the report of the Television Committee. This Committee was constituted in May 1934 'to consider the development of television and to advise the Postmaster General on the relative merits of the several systems and on the conditions – technical, financial, and general – under which any public service of television should be provided'.

The Committee, chaired by Lord Selsdon, a former Postmaster General, comprised Colonel A.S. Angwin and Mr F.W. Phillips of the General Post Office, Vice Admiral Sir Charles Carpendale of the BBC, Mr O.F. Brown of the Department of Scientific and Industrial Research, and Sir John Cadman of the Anglo–Persian Oil Company [2].

Lord Selsdon and his colleagues worked with commendable speed and tendered their recommendations, a total of 17, to the Postmaster General, the Right Honourable Sir Kingsley Wood, on 14th January 1935. During their work the committee had examined 38 witnesses, had received numerous written statements from various sources regarding television and had visited Germany and the USA to investigate television developments in those countries.

The principal conclusion and recommendation of the committee was that 'high definition television had reached such a standard of development as to justify the first steps being taken towards the early establishment of a public television service of this type' [2].

Marconi–EMI Television Company Ltd and Baird Television Company were the two companies invited to submit tenders for studio and transmitting equipment and for a short period from 2nd November 1936 to 13th February 1937 both companies transmitted television programmes, on an alternate basis from the London television station. Subsequently, until the commencement of hostilities in September 1939, only the Marconi–EMI equipment was in operation.

The setting up of the high definition service was a splendid achievement for, until 1931, low definition systems, operating on a 30-line or 60-line standard, predominated for public use, and the first demonstration of rudimentary television had been given by Baird just five years previously on 26th January 1926.

Prior to the realisation of high definition television a partial solution to the problem of medium/high definition television had been to employ multiple scanning and multiple-channel transmission. Interestingly, in 1931, the Gramophone Company (sometimes called HMV, after His Master's Voice records), Baird Television Ltd, and Bell Telephone Laboratories all demonstrated multi-channel television: independent advancements in television engineering were now following convergent paths [3].

HMV's interest in television effectively dates from October 1929 when a 'case opening' report on the subject was prepared [4]. Mechanical scanning was regarded as the more practical method of analysing and synthesising images but the advantages of cathode ray scanning systems in both transmission and reception was noted. A few months later, in January 1930, a Television Section was set up in the Research Department and an allotment of £800 was given for work, which was on mechanical scanning lines, for the period January to June 1930.

In April 1930 A. Whitaker, of the Advanced Development Department, visited RCA and was given demonstrations of 60-line television using cathode ray tube displays. While the definition of these was still not satisfactory for the commercial market, nevertheless the brightness of the cathode ray tubes used was 'so remarkably in advance of anything previously known', that Whitaker returned with the belief that this method of reception was worthy of progression. Indeed, 'the prospects of this line of attack seemed so bright' that in the second half of the year, July to December 1930, a further allotment of £800 was requested. An additional sum of £1200 for work on the transmission system was also sought. This followed an HMV report that recommended the utilization of the 3 m to 5 m short-wave band for the system. Earlier, the General Post Office had been approached and asked to waive some of the restrictions that had been applied to experimental transmissions [4].

During the second half of the year (1930) some investigations were carried out by HMV on cathode ray tube reception. The correctness of this course of action was confirmed when G.E. Condliffe, of HMV, visited RCA's Camden works in November 1930. These visits by Whitaker and Condliffe to RCA entirely changed their views regarding television receiving apparatus. Before April 1930 HMV had adjudged cathode ray tube reception as 'an interesting theoretical scheme but probably impracticable owing to the instability and low luminosity of the tubes available at that time'. After their visits and the demonstrations of the new tubes, that had enhanced brilliance, focus and efficiency, they were converted to the opinion that the RCA way was the right one. As Whitaker noted in a letter to the Vice President of RCA, Dr Goldsmith: 'Our decision to change our line of attack was caused by these RCA demonstrations . . . and as we had no wish to start repeating all the work which [had] been so excellently carried out in America, we decided to investigate the possibility of using a very considerably greater number of pictures elements than [had] been possible' [5].

Such a plan necessitated a wide bandwidth and therefore a very high radio carrier frequency. In his letter of 22nd August, Whitaker informed Goldsmith that HMV were 'at present' working with wavelengths from 2 yd to 5 yd (2 m to 5 m) although the radio section had recently been experimenting as far down as 12 in (30 cm). Goldsmith replied on 4th September 1930: 'I am indeed glad to learn that your research group is working along somewhat different lines from those which are being carried forward in Camden. The field is so new that duplication of effort would be uneconomical. . . . I note with some astonishment that you propose to use 250 [kHz] side bands [6].'

The Gramophone Company's attitude to the television problem was cautious and sensible. After a watching brief from 1926 the company naturally was keen to avoid some of the pitfalls encountered by others. Apart from the visits to RCA, representatives of the company had been to the Baird Television Company's studios in Long Acre, London, and had had discussions with O.G. Hutchinson, J.L. Baird's business manager, about current and future trends. Moreover HMV had subjected each Baird patent specification to a careful examination for claims which might subsequently 'be awkward in operating a television system' [7].

The company also had sought confirmation, from both the General Post Office and the BBC, of statements made by the Baird companies concerning future licences and wavelength allocations [7].

On the practical side HMV wished to advance from a position where some knowledge of the properties of the essential components of the television system had been accumulated by carefully conducted experimental studies. The photo-electric cell and the light-valve were two of these components and hence laboratory work was initiated on investigations of photo-emissivity [8] and the Kerr effect [9]. These were pursued by W.F. Tedham and W.D. Wright respectively and reports were prepared by the researchers in September and November 1930.

Meanwhile, C.O. Browne, a colleague of Tedham, was giving some thought to a system of five-channel television. Browne's report [10], dated 20th September 1930, indicates that the Gramophone Company was anxious to become a competitor in the infant but growing television industry for he wrote: 'In the proposed system of television as far as possible known results and data are utilised with a view to producing a workable system in a short time with a reasonable chance of success. It is for this reason largely, that a number of transmission channels are to be used'.

Browne succeeded in his task and his plan was engineered in the remarkably short time of approximately four months. The equipment was demonstrated, in January 1931, at the Physical and Optical Society's Exhibition, London. Five channels were chosen and each image of the cinematograph film used at the sending end was scanned at the rate of 12.5 per second. The line standard was 150 lines per second [11]. (See Fig. 21.1.)

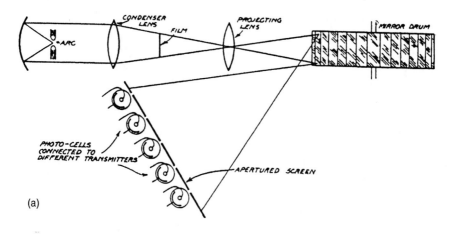

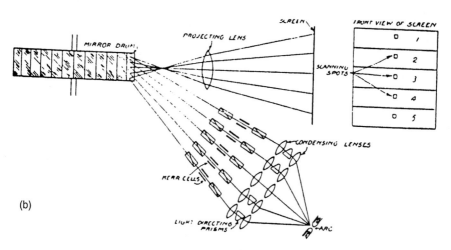

Figure 21.1   *(a) Schematic diagram of the HMV company's five-zone television transmitter. Effectively, five 30-line television systems were combined to give a 150-line image. (b) The diagram illustrates the disposition of the elements of the five-zone television receiver. The five picture signals were applied to five Kerr cells; these modulated the light flux from the arc and enabled the image to be synthesised. Mirror drum scanning was used at both the transmitter and receiver.*

Source: *Journal of The IEE*, 1932, **70**, pp. 340–349.

The demonstrations created great interest and queues of people had to wait outside the small theatre, which the Gramophone Company had set up to view the new system. G.A. Atkinson, the film critic of the *Daily Telegraph*, referred to the notable gain of the system and wrote: 'It marks a considerable technical advance on any system yet demonstrated, especially in the direction of bringing television rapidly into use for entertainment purposes' [12]. The *Evening Standard* noted that television now seemed to be within sight of realization [13].

During the exhibition the Gramophone Company showed televised images projected onto a screen measuring 24 in × 20 in (60.9 cm × 50.8 cm). No longer were 'head and shoulders' shown but instead the audience saw images of buildings, soldiers marching, cricketers walking on and off the field, and so on.

Everything was easily recognisable. [A London] tram car showed up so clearly that its number on the front was decipherable without difficulty. The pictures were steady on the screen. They were in good focus at very short range. They would probably stand enlargement up to four or five times the size of the screen actually used. . . . The general effect was that of looking at a performance of miniature films lacking full illumination [12].

Three months after the Gramophone Company's demonstrations EMI was formed to acquire the ownership of the Gramophone Company Limited and the Columbia Graphophone Company Limited. The research groups of the two companies merged [14].

EMI's directors agreed that HMV's television development effort should continue to be an item of the new company's R&D programme. One of the first questions that had to be tackled was whether this work should proceed on mechanical or electronic lines. Mechanical scanners had the advantage that they had been made successfully, whereas electronic scanners were still in the development phase. On the other hand, electronic scanning had many potential advantages for high definition television, and HMV's exploratory work on television had been based on the strategy that the company should aim to develop an effective cathode ray tube picture receiver, but should leave any endeavour to fabricate a television camera to others.

EMI felt that the most promising attack on the problem of home television should be on the lines initiated by HMV. There were thought to be diverse factors that made it desirable for EMI to proceed independently of RCA: these have been discussed at length in one of the author's books, *Television, an international history of the formative years* and are not repeated here [15].

Actually, the programme was soon modified. While the view had been taken that the company's business was in the field of receiving, and not transmitting, apparatus it was essential to have, for the testing of the receivers, a source of video signals. Brown's 150-line equipment was unnecessarily complicated, with its multi-channel requirements, and was not really suitable for this aspect of the work. Consequently, four months after the merger, work started on the design of a single picture channel system based on a 120 lines per picture, 24 pictures per second, mirror drum film scanner [16]. By the end of August 1931 Wright had constructed a cathode ray tube receiver, using saw tooth scanning, and in a report dated 18th August he had recorded: 'Some good receptions of the Baird television broadcasts have been obtained, but not consistently from day to day, nor during any one broadcast' [16].

In September 1931 A.G.D. West, of EMI's Research and Design Department visited Zworykin's RCA laboratory in Camden, New Jersey. West was shown the current results of RCA's television work and reported [17]: 'television is on the verge of being a commercial proposition. They [RCA] intend to erect a transmitter on top of a New York skyscraper in the autumn of 1932'. West had seen a televised image about 6 in × 6 in (15.2 cm × 15.2 cm) in size and had observed that its quality was comparable to that of an ordinary cinema picture viewed from a position near the extreme back of a large theatre. The planned selling price of the receiver was to be c. $470 (c. £100) and all the receiving apparatus for sound and sight was to be contained in a single cabinet of the size of an ordinary radiogram. The vision and sound wavelengths were to be 6 yd (6 m) and 4 yd (4 m) respectively.

West's report had an immediate effect: EMI reviewed its position concerning television. As Whitaker noted: 'Television is apparently coming rather more quickly than even the most optimistic of us have considered likely during the last couple of years' [17]. Consequently there had to be a reappraisal of EMI's television project. On the one hand, in the UK, the Baird companies had achieved a great deal of favourable publicity for their 30-line system; the BBC had reluctantly given the Baird companies some degree of official recognition by allowing them to use the BBC's transmitters for experimental broadcasts during non-programme hours; there seemed to be a prospect that the Baird companies would be granted permission for their equipment to be housed in the BBC on a permanent basis, and additionally it had been stated that short Baird transmissions would take place during normal BBC broadcasts. On the other hand, the depressed business conditions of 1931 had led to a curtailing of EMI's expenditure on television and because of this EMI could not give the demonstration at the end of the year which it had originally planned; there was a possibility that, if Baird's system became firmly established in the BBC, future television standards – and 'there were at least a dozen points which [needed] to be considered in great detail with a view to securing standardisation' – could be influenced by the standards of the Baird system; and there was a 'most urgent necessity' for EMI to establish 'a real demonstrated competition to the Baird system, which at the same time would give EMI prestige in television' [17].

For Whitaker there appeared to be only one short term solution to the problem of gaining a position of authority in television matters and that effectively was to buy in expertise. RCA was further advanced than EMI in its work on 'seeing by electricity' and so Whitaker suggested that approval should be given for the expenditure of roughly $50,000 (c. £13,000) for the purchase of transmitting equipment (including 4 m and 6 m transmitters for sound and vision, studio equipment and demonstration receiving sets) and for the expenditure of not more than £2,000 for the installation of this equipment in England. Whitaker's figure of $50,000 was based on an approximate quotation he had received from RCA. A later quotation received on 2nd December 1931 gave a total figure of $83,732 [18].

EMI's Executive Committee was persuaded by Whitaker's arguments and decided to recommend to the Board of Directors the proposals outlined by him [19]. However, this was adamantly opposed by Mr I. Shoenberg, the Head of the Patent Department.

By October 1931 it had become clear that technical cooperation with RCA was going to be 'very difficult and unreliable' [20]. Fortunately, the RCA rights on the transmitting side belonged to the Marconi Company and a favourable offer had been received by EMI from the company for the hire of radio transmitting apparatus. Hence the idea of attempting to work in parallel with RCA faded out and Whitaker's proposition was not implemented. Instead a programme of work on television began independently of RCA [21], except for the patents to which EMI was entitled [22].

The challenge facing EMI, and that was being faced by Farnsworth and by Zworykin's group at RCA, was immense [23]. Photo-electricity, vacuum techniques, electron optics, the physics of the solid state and of secondary electronic emission, and wide-band electronics and radio communications, were all in a rudimentary state of development. Many fundamental investigations would have to be undertaken before a high definition television system could be engineered.

To further their work, EMI was granted a licence [24] on the 8th December 1931 allowing them to establish, for experimental purposes, a wireless sending and receiving station at the Hayes premises of the company [25]. In compliance with this licence EMI could operate within the following bands: (1) c.w. and telephony, 62.01–61.99 MHz; (2) television 44.25–43.75 MHz; and at a power of up to 2 kW into the antennas.

The practical association between EMI and Marconi's Wireless Telegraph Company Limited, which was to prove so beneficial to both companies, began late in 1931 when MWT was given a contract for the supply of a 400 W, 44 MHz transmitter, complete with a modulator, for used with EMI's film scanner: the units were delivered to EMI in January 1932 [26,27,28].

Progress was now swift and on 11th November 1932 Shoenberg invited the BBC's chief engineer, Mr N. Ashbridge, to a private demonstration both in the transmission and reception of television. 'In my humble opinion', wrote Shoenberg, 'they would be of quite considerable interest to you' [29].

Ashbridge visited the Hayes factory on the 30th November and was shown apparatus for the transmission of films using four times as many lines per picture and twice as many pictures per second as Baird's equipment. He was impressed and thought the demonstrations represented by far the best wireless television he had ever seen and felt they were probably as good as or better than anything that had been produced anywhere else in the world. He wrote [30]:

there is not the slightest doubt that a great deal of development, thought and expenditure [has] been expended on these developments. Whatever defects there may be they represent a really remarkable achievement. In order to give some idea of the cost of such work, I might mention that the number of people employed is only slightly less than that in the whole of our research department.

The actual demonstration consisted of the transmission of a number of silent films, over a distance of approximately 2 miles (3.2 km), by means of an ultra short wave transmitter using a wavelength of 6 m and a power of about 250 W.

On the quality of the images Ashbridge reported;

The quality of reproduction was good, that is to say one could easily distinguish what was happening in the street scenes and get a very fair impression of such incidents as the changing of the guard, the Prince of Wales laying a foundation stone and so on. A film showing excerpts from a play was in my opinion not so good although it was possible to follow what was going on all over the stage. On the other hand excerpts from a cartoon film were definitely good. I think they could have given a better demonstration had they been in possession of better films. The ones they showed had been in use for several years. The size of the screen is about 5 in × 5 in (12.7 cm × 12.7 cm) but they have a second machine which magnifies this by about four times in area. The quality of the reproduction can be compared with the home cinematograph but the screen is smaller [30].

EMI was very keen that some form of television service should be started on ultra short waves, and, following up Ashbridge's visit to Hayes, Mr Alfred Clark, the Chairman of the firm, paid a visit to the BBC to have discussions [30] with the Director General, J.F.W. Reith. Clark was anxious to know what television standards would be adopted for television. He hoped Reith would say that the number of pictures per second and the number of lines per picture would be 25 to 30, and 120 to 180, respectively. EMI would then have been in a position to have started an experimental service, with equipment in the BBC's Broadcasting House, on ultra short waves early in 1933, and probably before Bairds were in a position to do so.

The emergence of a competitor in the form of EMI caused J.L. Baird and his associates much unease. They could not acknowledge for some considerable time that EMI's television system was being engineered by British workers in a British factory using British resources. For them the Radio Trust of America (i.e. RCA), through its associated companies in London, was the mainspring of EMI's progress. A very noticeable bitterness is evident in the letters emanating from the Baird companies, during the early 1930s, on the progress, and support from the BBC, of the Hayes company. Baird was always ready to point out that his firm could match the steps towards high definition television that were being made by EMI and that therefore the pioneer company should be preferred. The controversy [31], which is described and discussed in copious detail in the author's book *British television, the formative years* [1] is not repeated here.

Following demonstrations [32,33] of the Baird Television Ltd and EMI systems on the 18th and 19th of April 1933, a conference was held on 21st April 1933, at the Post Office, between the BBC and the Post Office representatives. It was agreed that [34]:

1. the EMI results were vastly superior to those achieved by the Baird Company;
2. the results were incomplete because of the different transmission methods (line and wireless) used in the two cases;

3. further tests by wireless in a town area were essential to determine the range of reception and the effect of interference;
4. whatever system of synchronisation was adopted in the first instance for a public service might be liable to standardise the type of receiving equipment;
5. a test of one system could not be a reliable judgement on the results achieved by the other.

Meanwhile, from c. May 1932 W. Tedham and J.D. McGee had been undertaking many investigations on the chemistry of the preparation of electronic camera mosaic signal plates and on the physics of the mechanisms operating at the surfaces of the plates. Both single-sided and double-sided signal plate camera tubes had been constructed, and patents obtained [35] (Fig. 21.2).

In July 1933 Zworykin presented a paper [36] on his electronic camera tube at an IEE meeting, held in London. McGee attended the lecture since he was especially interested to learn about the practical construction of the iconoscope but he noted, when reviewing the situation at this time: 'it was quite clear that [we] could learn nothing from Zworykin's paper'. No practical information was

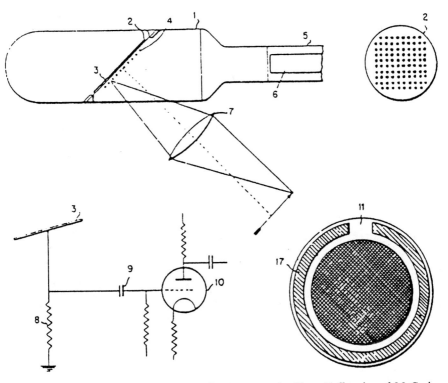

Figure 21.2   Single-sided mosaic target plate camera tube (from Tedham's and McGee's patent of 25th August 1932).
Source: British Library.

given and the preparation of the mosaic by a 'special process' was not described. Apart from the knowledge that successful development was feasible, McGee's group had to develop all the necessary techniques themselves.

One effect of Zworykin's lecture was the further recruitment of staff into Shoenberg's R&D team. By June 1934 the Research Department totalled [37] 114 persons including 32 graduates, nine of whom had PhDs, despite the fact that PhDs were not particularly common in the early 1930s, and ten had been recruited direct from Oxford and Cambridge Universities. The department also included A.D. Blumlein, the greatest British electronics engineer of the 20th century – see the author's book *The life and times of A.D. Blumlein* [59].

Funding for the television project was amply sufficient. Shoenberg succeeded in persuading the EMI Board to invest about £100,000 per year in EMI's R&D work on television. Any equipment considered necessary to advance the research projects was agreed at once. Initiative was encouraged. The expertise and funding of EMI's Research Department represented an ominous situation for Baird Television Ltd which, initially, could not match EMI's staff and financial resources.

The enlargement of the R&D department soon led to favourable results. Of the first 10 experimental camera tubes fabricated by McGee's group, no. 6 gave 'a very presentable picture' on 24th January 1934. Five days later, the all-electronic television camera was demonstrated, for the first time, to the Company Chairman, Mr A. Clark, and the Director of Research, Mr Isaac Shoenberg. 'Reasonable' tubes were being made by February 1934. When tube no. 14 was constructed it was so much better in picture quality and sensitivity that a very experimental camera directed through the window of the laboratory, on 5th April 1934, enabled a daylight outside broadcast picture to be obtained [38].

The iconoscope, or Emitron as EMI called their version of this type of camera tube (Fig. 21.3) had several defects which had be overcome before a practical camera could be marketed. Given a single-sided mosaic onto which both the scanning electron beam and an image of the scene to be televised are incident, it is obvious that either the optical image must be projected normally onto the mosaic target and the scanning beam projected obliquely, thereby giving rise to keystone distortion of the raster, or the converse. In either case a considerable depth or field is required, either of the electron lens or of the camera lens respectively.

In 1934–35 the target area of the mosaic was 5 in × 4 in (12.7 cm × 10.2 cm). This called for a long (6.5 in (16 cm)) focal length lens, to cover it. At full aperture, f/3, the optical depth of field was insufficient for the second of the above alternatives. However, by lengthening the electron gun and by stopping-down, the electron lens could be designed to give a uniform beam focus even when the beam was incident obliquely onto the mosaic. For this reason the arrangement given in Fig. 21.3 was chosen. Necessarily the line scanning signals had to be modulated to give a constant width scanning raster on the mosaic.

The analyses by RCA and EMI of these constraints led naturally to the same

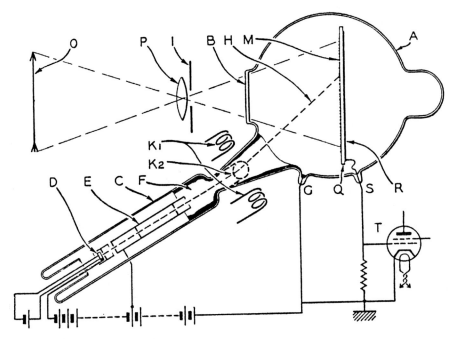

Figure 21.3   The Emitron television camera.
Source: Professor J.D. McGee.

configurations and shapes of their camera tubes, and to the innuendo that EMI had copied Zworykin's iconoscope. This imputation probably gained further credence because of the known RCA association with EMI. But, McGee [38] has stated 'categorically that there was no exchange of know-how between the two companies in this field during the crucial period 1931 to 1936'. Dr G.H. Lubszynski [39], Dr L. Broadway, Mr I. Shoenberg, Mr S.J. Preston, Mr C.O. Browne, N.E. Davis and Mr A.D. Blumlein have also confirmed the independence of McGee's team from any RCA influence. This matter is discussed in the author's book, *Television, an international history of the formative years* [3].

After McGee's group had fabricated their first batch of Emitrons it was found that the signals obtained tended to become submerged in great waves of spurious signals associated with some secondary emission effects. Fig. 21.4A illustrates the spurious signals superimposed on the wanted pictured signals and Fig. 21.4E shows the signals required. To effect the necessary transformation from A to E correcting signals Fig. 21.4B, which became known as 'tilt' and 'bend', had to be electronically generated and added to both the line and frame signals to annul the unwanted signals (Fig. 21.4C). The large spurious pick-up signals that appeared during the line and frame fly-back periods were also suppressed (Fig. 21.4D) and appropriate synchronising signals inserted (Fig. 21.4E). 'Great credit', recalled Shoenberg in 1952, '[was] due to Blumlein,

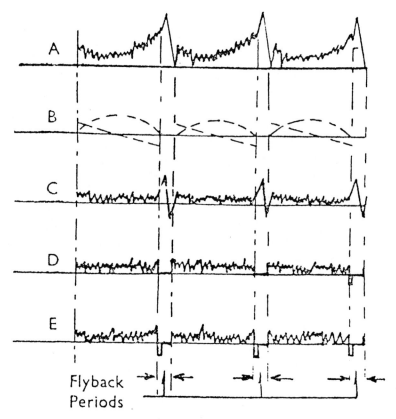

Figure 21.4   Spurious signals from the Emitron
Source: Professor J.D. McGee.

Browne and White for the resourcefulness by which they devised tilt, bend and suppression circuits to combat this evil [22].

Strangely, Zworykin did not refer to these signals, or to keystone distortion in his 1933 IEE paper [36].

At about this time, 1934, Blumlein and McGee invented [40] the ultimate solution – cathode potential stabilisation – to the problem of the spurious signals. The method of, and a proposed apparatus for, cathode potential stabilisation (c.p.s.) of the target were patented by Blumlein and McGee on 3rd July 1934.

Cathode potential stabilisation has several important advantages:

1. the utilisation of the primary photo-emission is almost 100 per cent efficient, thus increasing the sensitivity by an order of magnitude;
2. the signal level generated during the scan return time corresponds to that of zero illumination, or 'picture black', in the picture;

3. the spurious signals are eliminated; and
4. the signal generated is closely proportional to the image brightness at all points; that is, it has a photographic gamma of unity.

These advantages became of great importance in the operation of signal generating tubes such as the c.p.s. Emitron (Fig. 21.5) the image orthicon, the vidicon, the plumbicon and every one of the Japanese photo-conductive tubes up to the advent of the charge-coupled devices.

Following the 1939–45 war c.p.s. operation of camera tubes became the norm. The first, public, outside broadcasts with the then new c.p.s. Emitrons were from the Wembley Stadium and the Empire Pool during the 14th Olympiad held in London in 1948. The cameras were c. 50 times more sensitive than the existing cameras and enabled a 'wealth of detail and remarkable depth of field' to be obtained. Much enthusiasm for British television prevailed. Lord Trefgarne, Chairman of the Television Advisory Committee opined: 'In its television transmissions from the Olympic Games, the BBC has just given the world a striking visual demonstration of the technical excellence of the British system, using the latest British cameras, British transmitting equipment and British receivers. The results have been the admiration of our overseas guests, including the Americans.'

Figure 21.5   CPS cameras in the Empire Pool, Wembley; 1948 Olympic Games.
Source: Thorn EMI Central Research Laboratories.

However, when cathode potential stabilisation was put forward in July 1934 there were very considerable practical difficulties in implementing the principle. At that time the Television Committee, chaired by Lord Selsdon, was taking evidence from Baird Television Ltd, EMI, and others, and it appeared the Committee would recommend the early establishment of a system of medium/ high definition television in the UK. Because of this, and the improvements in image quality being obtained from the experimental Emitron tubes, Shoenberg decided to concentrate McGee's group's efforts on these tubes.

Many visits to EMI were made by important persons during the formative period of EMI's system. These included Sarnoff, of RCA; the Prime Minister; Ashbridge and Kirk of the BBC; Vandeville of MWT; Reith, the Director General of the BBC; representatives of the General Post Office, and others. Of these private demonstrations, that given to Ashbridge in January 1934 is of some importance since his account [41] of his visit is extant. He was very impressed. 'The important point about this demonstration is, however, that it was far and away a greater achievement than anything I have seen in connection with television. There is no getting away from the fact that EMI have made enormous strides.'

The demonstration had consisted of the transmission of films, with 150 lines per picture, from the EMI factory at Hayes to the recording studios in Abbey Road, a distance of approximately 12 miles (19.3 km).

This was by means of an ultra short wave transmitter with a power of 2 kW on a wavelength of approximately 6.5 m. The results were extremely good and there was no question in my mind that programme value was considerable. The receivers used appeared to be in a practicable form and looked very much like large radiograms. On the other hand, it has to be said that the aerial arrangements were very elaborate, being directional in order to cut out interference [41].

Shoenberg told Ashbridge the policy of the EMI was to develop television energetically since they believed there was a great commercial future for the firm that was first in the field with something practicable. Two months later the Marconi–EMI Television Company Limited was formed 'to supply apparatus and transmitting stations' [42,43].

In an attempt to settle the rival claims of EMI and Baird Television Ltd, Reith on the 15th March 1934 wrote [44] to Kingsley Wood, the Postmaster General, and proposed a conference 'between some of your people and some of ours to discuss the future arrangements for the handling of television'. Reith thought there were three aspects to discuss: the political, 'using the term in a policy sense and for want of a better one', the financial and the technical. Kingsley Wood agreed [45].

The informal meeting [46], was held at the General Post Office on 5th April 1934. A number of general questions were examined by the BBC and GPO representatives, including:

1.   the method of financing a public television service;
2.   the use of such a service for news items and plays;

3. the relative merits of some of the systems available including those of the EMI, Baird, Cossor and Scophony companies;
4. the arrangements necessary to prevent one group of manufacturers obtaining a monopoly on the supply of receiving sets;
5. the possible use of film television to serve a chain of cinemas.

With two rival companies campaigning for the creation of a television service – EMI for a new BBC station and Baird Television for a station of its own – it was agreed by the conference that a committee should be appointed to advise the Postmaster General (PMG) on questions concerning television. The BBC representatives were keen that this committee should be established as soon as possible since difficult questions were arising, and would continue to arise, and they thought it would be helpful for the BBC and the GPO to have the weight of the authority of a committee behind them in any decision they might take. The PMG concurred and the Television Committee was constituted. The Chairman was Lord Selsdon.

By September 1934 the committee had so advanced its deliberations [47] that it was able to commence the preparation of its report, and, but for the absence of Sir John Reith (whom the committee wished to interview), towards the end of 1934, the committee's report would have been produced before January 1935 (the actual month of publication).

Some of the committee's time was, of course, spent studying EMI's proposed high definition system. Shoenberg told Lord Selsdon that 'the reasons which [had] guided [EMI] in the choice of the fundamental features of [its] system appear[ed] as convincing as the proof of a theorem in Euclid'. Lord Selsdon riposted by saying: 'I think the answer is that they [EMI] have picked the plums out of the pudding, in so far as they can find any plums'. Shoenberg challenged the Chairman to give him 'a more logical system' [43].

Any system of television is characterised not by the use of a particular type of camera, nor by the use of electrical or mechanical scanning, but by the specification of the system and the transmitted waveform. As Blumlein said [48]: 'The fixation of a waveform may almost certainly imply the use of particular types of apparatus at the transmitter, but such apparatus is largely subservient to the standard chosen'.

Of all the decisions that Shoenberg had to face the most difficult concerned the specification of this standard. McGee has recalled [23]:

There was, for example, a long debate as to whether d.c. or a.c. amplification should be used – and then, if a.c. how could the d.c. level (black level) be re-established? And as the picture definition was increased the bandwidth would be increased, so that the radio transmission frequency necessary also increased, limiting the range to approximately line of sight. Would this be acceptable? Then, should positive or negative modulation be used? And as the brightness of pictures increased, flicker became a serious problem so that the alternatives of sequential or interlaced scanning had to be decided. These problems landed squarely on Shoenberg's desk; to paraphrase a well known saying: 'the buck stopped there'.

As the number of picture lines crept up from 120 to 180 and then to 240, the required picture-signal bandwidth increased; this increased progressively the problems of amplifiers, of transmitters, of tubes and of achieving adequate service area. It was clear that the quality of the picture increased very noticeably as the number of lines increased. Since it was possible to scan an electron beam at these much higher speeds, it was natural, indeed inevitable, that the advantages of the electronic system should be exploited to the maximum. But what was the practical maximum? It would clearly be difficult, if not impossible, to push the number of lines much further than 240. No one knew what disastrous snags we might meet if we attempted to reach still higher definition. Higher definitions were terra incognito, and not just the rather obvious line of development that it may now [1971] appear . . .

To us, then young men, it was a challenge and an adventure, but to Shoenberg it must have been a very worrying problem. On him fell the responsibility to the company and to his staff to make the right decision.

And having made the decision the task of delineating the television waveform that fulfilled the specification belonged to Blumlein. Among the research and development team of vacuum physicists and circuit engineers, Blumlein's role was that of overall system engineer – he was primus inter pares in the team.

He outlined the reasons for the choice of the 405-line waveform (see Fig. 21.6) in an important paper [48] on 'The transmitted waveform' which he read at a meeting of the Institution of Electrical Engineers in April 1938. The paper was one of three on the Marconi–EMI television system: the others were 'The vision input equipment' by C.O. Browne; and 'The radio transmitter' by N.E. Davis and E. Green.

The EMI team thought it was essential to transmit the direct component so that the receiver would accurately reproduce the brightness values of the original picture although the practice in the United States of America was to use the alternating component only.

In 1952 Shoenberg recalled [22]: 'It seems strange now that this matter should ever have been one for controversy, but at the time there were cogent arguments on both sides and in the end I made the decision in favour of restoration.' This decision meant that definite carrier values represented black, white and the synchronising signal. There was no mean carrier value as this depended upon the picture brightness and so the system of transmission was analogous to telegraphy rather than telephony.

On the question of the number of lines per picture necessary for entertainment purposes, several studies had been carried out during the period 1929–33. The most important of these was that undertaken in 1933 by E.W. Engstrom [49], of the RCA Victor Company. For part of his investigation, Engstrom used an ingenious technique to make cine films having a detail structure equivalent to television images. His films included: (1) head and shoulders of girls modelling hats; (2) close-up, medium and distant shots of a baseball game; (3) medium and semi-close up shots of a scene in a zoo; (4) medium and distant shots of a football game; (5) animated cartoons; and (6) titles.

His findings were based on: first, the resolving properties of the eye; and,

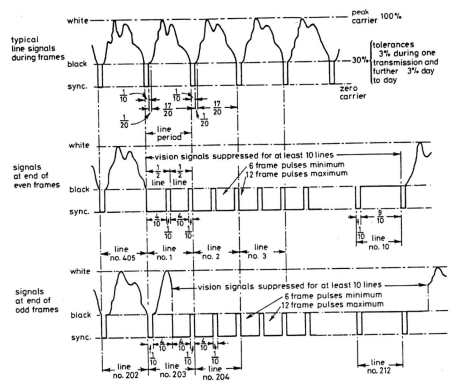

Figure 21.6   The video waveform of the 405-line television system.
Source: *Journal of The IEE*, 1938, **83**, pp. 758–766.

second, practical viewing tests of the cine films which were specially prepared to give pictures having 60-, 120-, 180-, and 240-line structures, and normal projection print quality.

In his work Engstrom set as his standard the ability of the eye to see the elements of detail and picture structure. Another less exacting standard was the ability of images having various degrees of detail to tell the desired story. Taking as a standard the information and entertainment capabilities of the 16 mm home movie film and equipment, Engstrom estimated the television images in comparison as:

60 scanning lines	entirely inadequate
120 scanning lines	hardly passable
180 scanning lines	the minimum acceptable
240 scanning lines	satisfactory
360 scanning lines	excellent
480 scanning lines	equivalent for practical conditions.

His conclusions agreed effectively with those of Wenstrom [50], particularly with regard to the number of lines per picture needed to gave excellent reproduction. Both investigators found that 360 to 400 lines per picture were required for this purpose. This finding was published generally in December 1933 and so was known to EMI.

When EMI was invited to give evidence [43,51] to the Television Committee in June 1934, it submitted a memorandum outlining the characteristics that would be suitable for a commercial broadcast television system. Of the 11 points listed, the only one that was later altered concerned the number of scanning lines per picture, namely 243.

By February 1935 [23]:

Shoenberg had made what was probably the biggest – and I [McGee] consider the most courageous decision – in the whole of his career: to offer the authorities concerned a 50 frames/second, 405 lines/picture television system. Remember that this meant a 65% increase in scanning rate and a corresponding decrease in scanning beam diameter in the c.r.t., a nearly three-fold increase in picture-signal bandwidth; and – worst of all – a five-fold decrease in the signal/noise ratio of the signal amplifiers and this lists only a few of the resulting problems.

The cynic may say that this was a piece of gamesmanship planned to overwhelm our competitors. But no one who knew Shoenberg or who was aware of the real state of technical development at that time would give this idea a moments credence. No – it was the decision of a man who, having taken the best advice he could find, and thinking not merely in terms of immediate success, but rather of lasting, long term service, decides to take a calculated risk to provide a service that would last . . .

To us this decision was a stimulating challenge. To Shoenberg it must have been a heavy and worrying burden. In later years, he often recalled how colleagues in the higher management of the company had seriously questioned his decision, and had warned him that should he fail to fulfil his contract it would be disastrous for the company.

Yet I cannot remember that he ever showed his worries to us at all obviously. The nearest perhaps was one day when things were particularly sticky. That day he finished up a rather depressing review of our progress with the comment 'Well gentlemen, we are afloat on an uncharted ocean and God alone knows if we will ever reach port'.

Mr S.J. Preston, formerly Patent's Manager of EMI, has observed [52] that Shoenberg made his choice to adopt 405 line television 'knowing that receivers available at that time could not be expected to deal with the full bandwidth which 405 line scanning would require but his view was that it was better that the early pictures should be somewhat lacking in definition along the line so that later developments in receiver design which he was sure would take place could be usefully employed without any change of standards'.

EMI was led to consider interlacing when their cathode-ray tube receivers began giving brighter pictures and it was obvious that 25 Hz flicker would be unacceptable in practice [48]. Prior to the beginning of 1934 the company had been working on a 180-line picture, with sequential scanning at 25 frames per second, using cathode-ray tubes which were not at all bright. With a poorly illuminated picture at the receiver, flicker was not very objectionable, but as the brightness of the screen improved the flicker could not be tolerated

In early television, the investigators were limited to the utilisation of a channel bandwidth of the order of 10 kHz, and, as the bandwidth for television reproduction is proportional to the square of the number of lines employed and directly proportional to the picture scanning rate, a compromise had to be decided upon between the need to use a high picture rate for flicker-free viewing and a large number of lines for good definition. Baird adopted a 30 lines per picture, 12.5 pictures per second standard, but naturally this gave rise to some image flicker. The Gramophone Company also used the same picture rate in 1931, but when cathode-ray tubes began to be made and used for television reproduction in the early 1930s, capable of giving bright pictures, the low picture rate was found to be unsatisfactory for prolonged viewing. Television engineers increased the rate to 25 pictures per second (or 30 pictures per second in the USA), and, in 1934 EMI proposed, to the Television Committee, the adoption of interlaced scanning. This suggestion enabled the frame rate to be increased to 50 frames per second, while the picture rate was maintained at 25 pictures per second, and paralleled the practice that had been adopted in the motion film industry [53]. But, there is an important difference between the projection of a cine film and the reproduction of a television image. In the former case the whole of the picture is shown at any instant of film projection, whereas in the latter the image is built up line by line. The employment of the technique used in the film industry to achieve a high frame rate clearly could not be used and so the principle of interlacing was adopted. The interlaced scanning system that EMI adopted was that devised by R.C. Ballard of RCA and patented by him in 1933.

Of the other factors that had to be defined before a waveform specification could be prepared, the direction of line and frame scanning was arbitrarily chosen as left to right and top to bottom, to accord with normal writing practice. The system that was adopted has all the lines scanned in the same direction at a uniform speed, each line being followed by a fly back: the method necessitates the transmission of distinct synchronising and vision signals.

In cinematographic practice, in the 1930s, it was usual to increase the contrast of the recorded scene by a factor, called gamma, having a value between 1.5 and 2.0, presumably to make up for lack of colour. Since this same effect was required in television, the question arose as to whether this increase should be applied at the transmitter or the receiver. Blumlein concluded from his analysis of the problem that it was advantageous to transmit pictures with unity gamma and make any correction at the receiver. His arguments led to the decision that the transmitted signal should, as far as possible, be undistorted in frequency or amplitude, and should be representative of the brightness of successive elements of the picture scanned with a constant scanning velocity.

The ratio of picture width to picture height was defined as 5:4, as a compromise between 1:1 to give maximum picture area on the face of a circular cathode-ray tube and 3:2 popular in photography. From the number of lines per picture, the number of pictures per second, and the aspect ratio, the bandwidth required was calculated to be 2.5 MHz.

The specified picture standards enabled the television waveform to be characterised, but there were several important issues still to be resolved:

1. whether the effect of interfering signals in producing white spots on a television picture (as in 'positive' modulation) would be preferable to the effect of such signals in greatly disturbing the picture scanning (as in 'negative' modulation); and
2. the nature of the line and frame synchronising signals.

The actual number of lines per picture, 405, was chosen because of several additional considerations [54], namely: (1) an odd number of lines per picture was required for interlacing; and (2) the number of lines per picture had to be compounded from integers preferably less than ten for ease of signal generation. Of the odd integers from 343 to 405, only 343 ($= 7 \times 7 \times 7$), 375 ($= 5 \times 5 \times 5 \times 3$), and 405 ($= 5 \times 9 \times 9$) satisfy these points. For reasons given previously Shoenberg chose 405.

This standard was tried in EMI's laboratories and gave sufficiently promising results with the experimental Emitron to be confirmed by Shoenberg as the standard EMI would offer to the BBC. It was good enough to last nearly 50 years.

Some of the important features of the Marconi EMI television waveform are as follows (they are quoted from Blumlein's 1938 IEE paper [48])

1. d.c. modulation

    The picture brightness component (or the d.c. modulation component) is transmitted as an amplitude modulation so that a definite carrier value is associated with a definite brightness. This has been called 'd.c. working' and results in there being no fixed value of average carrier, since the average carrier varies with picture brightness.

2. vision modulation

    The vision modulation is applied in such a direction that an increase in carrier represents an increase in the picture brightness. Vision signals occupy values between 30% and 100%. The amount by which the transmitter carrier exceeds 30% represents the brightness of the point being scanned.

3. synchronising modulation

    Signals below 30% of carrier represent synchronising signals. All synchronising signals are rectangular in shape and extend downwards from 30% peak carrier to effective zero carrier.

Blumlein's role in the development of EMI's television system was much appreciated by Shoenberg. In November 1934 he presented Blumlein with a gold pocket watch.

Following acceptance by Parliament of the Television Committee's report [2], and the recommendations that

1. A start should be made by the establishment of a service in London with two television systems operating alternately from one transmitting station;
2. Baird Television Limited and Marconi–EMI Television Company Limited, should be given an opportunity to supply, subject to conditions, the necessary apparatus for the operation of their respective systems at the London station.

the two contracting companies submitted their proposals to the newly formed Television Advisory Committee [55] (TAC) in February 1935. They were based on standards of 240 lines/picture, 25 pictures/s; and 405 lines/picture, 50 frames/s, interlaced 2:1 to give 25 pictures/s respectively. The press notice that outlined the plans of Baird Television Ltd and Marconi–EMI Television Ltd was not issued until the morning of 7th June 1935.

Following the publication of the press notice on the adoption of the two standard system, invitations to tender were issued – the date by which tenders had to be returned being 4th July 1935. One of the most important of the issues that had be considered by the TAC concerned the system of vision waveform generation to be used for film transmission, studio scenes, and outdoor broadcasts, and the lighting intensities required for each of these different modes of television.

In its tender M–EMI [56] proposed to utilise six cathode ray scanning tube cameras for all three activities – four for studio use and two for scanning film – and considered that adequate illumination for studio working would be produced by 18 kW of roof lighting and 6 kW of directional lighting.

Baird Television Ltd's response [57] to the same section of the tender was much more extensive and highlighted the cumbrous nature of its equipment vis-a-vis the highly mobile Emitron camera. The company said that film transmission would be carried out initially by means of disc scanning and then almost immediately by electron image scanning using an electron image camera (of the Farnsworth type). For studio scenes, the spotlight method with a scanning disc was suggested for the transmission of close-ups, while for studio scenes of all types, the intermediate film apparatus was recommended. These systems needed substantial power inputs for the lighting, namely, 28.5 kW for the supply arc for the two telecine transmitters and the intermediate film transmitter; 31.5 kW for the supply arc of the spotlight transmitter; and 94.4 kW for the studio lighting for the electron camera.

An additional disadvantage of the intermediate film process was the cost of the 35 mm film stock and processing chemicals which amounted to £48 per hour or £12 per hour using split 35 mm film stock. Against this, the cost of servicing the Emitron cameras with tubes was £2.50 per transmission tube-hour. Subsequently, following some modifications, the tenders were accepted by the TAC.

The general layout of EMI's vision input equipment is illustrated in schematic form in Fig. 21.7. In the system the frequencies of both the line scanning and

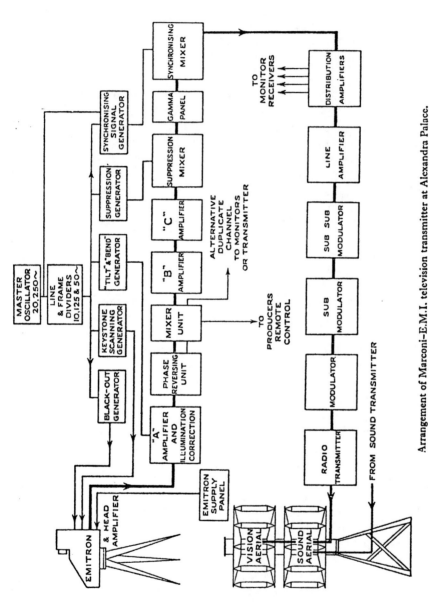

Arrangement of Marconi–E.M.I. television transmitter at Alexandra Palace.

Figure 21.7   Schematic layout for the control equipment of the six-camera and three-picture channel television system.
Source: *Journal of The IEE*, 1938, **83**, pp. 767–782.

frame scanning circuits were synchronised to the mains power supply frequency (normally 50 Hz). This was achieved by means of a master electronic oscillator operating, under mains frequency control, at twice line scan frequency (i.e. 20,250 kHz), and a series of electronic frequency dividers. From these line scan signals (at 20,250/2 = 10,125 kHz) and frame scan signals (at 20,250/5/9/9 = 50 Hz) were obtained.

The outputs of the dividers enabled the frequencies of the various electronic pulse and waveform generators to be locked together. Such generators were required to allow the electron beam of the Emitron, to scan linearly over the signal mosaic plate; to black out the Emitron's output signal during the fly-back period; to provide tilt, bend, and suppression signals to annul the deleterious effects of the secondary emission from the mosaic plate; and to provide the necessary line and frame synchronising signals.

The extent of the studio and camera control apparatus was influenced markedly by the television production director's need to provide a variety of attractive programmes. Six camera channels were available of which any two could be used in conjunction with two film scanners for the transmission of a continuous programme of film. All the channels were identical and therefore interchangeable, an essential advantage when 'live' studio scenes were being televised and reliability of presentation was of the utmost importance. Furthermore all the cameras were synchronised thereby enabling the producer to fade from one picture to another without resorting to a change-over of synchronising signals. An added advantage of this arrangement was the facility that allowed two or more pictures to be superimposed for special effects.

From the picture channel the vision signal was fed to the vision transmitter. This comprised two parts. The radio frequency section consisted of a crystal–controlled oscillator and a number of multiplying stages for generating a signal at the carrier wave frequency, followed by a cascade of six power amplifiers. The final power amplifier was grid modulated and gave a continuous output 17 kW, which was the power radiated during the white parts of the picture. The modulator section, which included four vision-frequency power amplifier stages, provided a 2 kW output vision signal having a bandwidth which extended from zero to 3 MHz, and an amplitude of 2000 V.

More than 500 (sic) valves were utilised in the vision apparatus. Meticulous design ensured that the overall vision system was flexible in studio operation and reliable in transmission use. An important feature was the provision of means to monitor the performance of the many sub-systems. Altogether the design and engineering of the 405-line high definition, all-electronic television system was an outstanding achievement given the rudimentary nature of electronics and camera tubes at the beginning of the 1930–40 decade. In just four years, from January 1931, when the HMV multi-channel apparatus was demonstrated, to the spring of 1935, when a full-scale studio equipment that answered the requirements for the 405-line service was completed, the bandwidths of the electronic circuitry had increased almost 30 times. EMI's R&D staff had had to engineer the system effectively *ab initio*. Amplifiers of various types (head amplifiers, video amplifiers, radio frequency amplifiers, and power amplifiers), modulators, pulse generators, pulse forming circuits, equalisers, filters, attenuators and special power supplies were developed. A great deal of innovative activity was needed to implement the system and much credit must be accorded to Shoenberg and his brilliant team of physicists and engineers.

In this endeavour Blumlein played not only a leading and inspirational role, but in addition he evolved many of the circuits that were used. Browne, in the references to his paper[58] on the 'Vision input equipment', listed 17 relevant British patents; of these Blumlein was associated with nine. Other members whose names appeared on the 17 patent specifications were E.L.C. White (6), C.O. Browne (5), J. Hardwick (2), M. Bowman Manifold (3), F. Blythen (2) and E.C. Cork (1).

Of course, the total number of EMI patent applications, relating to high definition television during its formative period, much exceeded the number quoted by Browne. Blumlein's patents total 52 from the time (early 1933) when he became actively involved in television to the date (2nd November 1936) of the opening of the London television station. This represents on average slightly more than one patent application per month for the whole of the four year period – a most impressive contribution. Blumlein's television patents from 1933 to c. 1939 range over many disparate fields of inquiry. A classification of the patents leads to the following analysis: cathode ray tubes (including camera tubes) 9; cathode ray tube circuits 7; DC restoration 7; AGC circuits 6; power supplies 10; modulation 2; miscellaneous television circuits 8; antennas and cables 15; and miscellaneous electronic circuits 11; a total of 75, or an average of c. 11 per year. He was 'an extraordinarily able and creative engineer' who was 'as creative as half a dozen of the best'. See the author's book: *The life and times of A.D. Blumlein* [59].

Installation of M–EMI's and BTL's television equipment at Alexandra Palace, the site of the London television station [59], was carried out during the period December 1935 to August 1936. Extensive building operations were necessary to accommodate the heavy machinery, and sound and vision transmitters, and various rooms had to be converted for use as studios, dressing rooms, control rooms and offices.

Major structural changes were effected in the south east tower of the Palace to support the steel antenna mast. All the existing floors, and the windows on the south and east sides of the tower, were removed: fire resistant floors and staircases were then constructed to provide five floors of offices, and bay windows were added to increase the lighting of the office spaces. The additional floors and staircases were supported by steel beams and the brickwork of the tower was tied horizontally by steel bars to provide a solid foundation on to which the mast could be fabricated.

The mast, shown in Fig. 21.8, was square in section at its base and tapered up to a height of 105 ft (32.0 m) above the tower, the sides of the square being 30 ft (9.1 m) at the base and 7 ft (2.13 m) at the top of the tapered section. Above

Figure 21.8   The Alexandra Palace antennas.
Source: The BBC.

105 ft the mast had an octagonal section, the distance from one face to the face opposite being 7 ft.

The mast carried both sound and vision antennas. They were of similar design and each antenna array comprised eight wide band dipoles, evenly spaced around the octagonal section of the mast, with eight reflectors mounted between the dipoles and the mast. The antennas were connected to the transmitters via two special 5 in (12.7 cm) diameter concentric feeders.

The opening ceremony of the London television station on the 2nd November 1936 was a most modest affair. Although the station provided the world's first, high definition, regular, public television broadcasting service, the inaugural programme [60], arranged by the BBC and approved by the Television Advisory Committee (TAC), lasted hardly one quarter of an hour. Just a few short speeches lasting four minutes each, by Mr Norman, the Chairman of the BBC,

Major Tryon, the Postmaster General, and Lord Selsdon, the Chairman of the Television Advisory Committee comprised the proceedings. The ceremony was witnessed by many invited guests from the BBC, the GPO, the Alexandra Palace trustees, Baird Television Ltd (BTL), Marconi–EMI Television Ltd and various other bodies.

The experiences gained by the British Broadcasting Corporation of the operation of both television systems under service conditions from 2nd November to 9th December were described in an important report [61] written by Gerald Cock, the BBC's Director of Television. It proved highly damaging to the interests of BTL; indeed it meant the end of the company as a supplier of television studio and transmitting equipment for the Corporation's stations and studios. Cock stated that the Marconi–EMI equipment proved capable of transmitting both direct and film programmes with steadiness, and a high degree of fidelity.

Its apparatus being standardised throughout, reproduces a picture of consistently similar quality and requires only one standard of lighting, make-up, and tone contrast in decor. Its studio control facilities are convenient and comparatively simple. It has proved reliable, and has already established a large measure of confidence in producers, artists and technicians. Outside broadcasts and [those] of multi-camera work have added considerably to the attractions of programmes. With improved lighting, additional staff and studio accommodation, single system working by Marconi–EMI would make a service of general entertainment interest immediately possible. (Fig. 21.9)

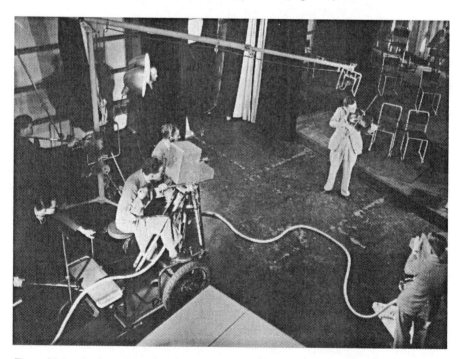

Figure 21.9　Studio A at the London station, showing two Emitron cameras in use.
Source: The Marconi Company.

Unfortunately, the Director of Television found it difficult to say anything complimentary about Baird Television Ltd's system. The programmes were being transmitted under practically experimental conditions and the prospect of anything approaching finality in the studio stages of transmission seemed remote.

Alterations in apparatus were constantly taking place. Breakdowns, with little or no warning, and, even more serious, sudden, unexpected, and abnormal distortions are a frequent experience. In such cases, it is difficult and embarrassing to make a decision to close down, since there is always the possibility that faults may be corrected within a short time. This inevitably leads to criticism of television by those who may only have observed it in adverse conditions.

In studio operations, the BTL system made use of the spotlight scanner, the intermediate film method, the Farnsworth electronic camera, and the telecine scanner for 35 mm commercial film. The first three systems each required a different technique in lighting and make-up to add to the difficulties of the producers and cameramen. Cock's views on these methods were as follows:

1. Spotlight scanner

   This apparatus is limited to double portrait reproduction. Distortion of picture tone and shape still appears to be intrinsic and unavoidable. No reading is possible in a spotlight studio, nor could any artist depending upon looks and personality be expected to televise by this method. The result is a caricature of the image televised.

2. Intermediate film method

   This is extremely intricate, and depends upon so many processes that it causes continual anxiety. It is inflexible and rigid in operation, being confined to 'panning' in two planes. Changes of view can only be effected at the cost of lens changes and 'black outs'; otherwise the picture is static. Its quality is variable; the delay action is extremely inconvenient for timing and other production purposes. The maximum continuous running time is at present limited to approximately 16 minutes, which adds to the difficulty of arranging programmes. The cost for film alone is £12 per hour (rehearsal or performance). Sound, recorded on 17.5 mm film and subject to development at high speed, is invariably of bad quality, whole sentences having occasionally been inaudible. It is consequently an unsuitable method of presenting any programme item in which quality of music or speech is important. Mechanical scanning produces line bending and twisting in a variable degree. Black and white contrast is however generally good.

3. Farnsworth electron camera

   These cameras have quite recently been improved, but future progress is likely to be seriously handicapped by destruction of the Baird research plant and technical research records in the Crystal Palace fire. Bending and twisting of the lines are pronounced. The cameras are in a somewhat primitive stage of development and are still without facilities for remote (outside) or 'dissolved' work. Breakdowns have been frequent. Electronic cameras do not appear likely seriously to compete with Emitrons at any rate for a considerable time. They have advantages over other Baird apparatus in being instantaneous in action; in permitting good sound transmission; in mobility; and in the elimination of the mechanical scanning. At present their operation seems somewhat precarious.

4. Telecine

Originally, this apparatus gave the best picture obtainable by the Baird system and possibly (flicker apart) was better than the Marconi–EMI for reproduction from standard film. Line bending and twisting have since been noticeable due perhaps to mechanical scanning and the difficulty of maintenance in a first-class condition. Its contribution to programmes is limited by the present restricted use of 35 mm film.

With Cock's and other reports [62] in mind, the TAC recommended to the Postmaster General the adoption of Marconi–EMI's transmission standards as the standards for the London television station. The discontinuance of the BTL system would date from 2nd January 1937 [63].

The BBC's decision probably caused no astonishment. Spotlight scanning had been employed in the USA, France, Germany, the UK, and elsewhere from the late 1920s and its limitations were well known. Again the intermediate film process had been much investigated and developed by Fernseh AG and others during the first half of the 1930–1940 decade and certain deficiencies of the process had been highlighted at the Berlin Radio Exhibitions. Furthermore Farnsworth's electron camera was based on the image dissector tube which lacked charge storage [64] and hence could not compete in sensitivity with camera tubes such as the Emitron and the iconoscope that utilised the charge storage principle. These points were known generally at the beginning of 1937.

Effectively those countries that, in 1937 and subsequently, aspired to operating high definition television broadcast stations had to use studio equipment based on the Emitron and the iconoscope cameras. Only EMI and RCA, and their licensees, world-wide, manufactured such equipment: and so, during the formative years of television growth, prospective television administrations, perforce, had to found their nascent television systems on the work of RCA and EMI [65]. Pre-war, the influence of these companies was particularly evident in France, the USSR, Japan and Germany.

The United States of America's Radio Manufacturers Association (RMA) was one of the relevant bodies that was influenced by the Marconi–EMI system, and, more particularly, by the Blumlein waveform. During the summer of 1937 engineers of the American Hazeltine Service Corporation established a temporary laboratory in England for the purpose of making a survey of television.

The survey included observing the transmissions and making measurements of the received signals with special equipment designed for the purpose. Visits were made to Alexandra Palace and to receiver manufacturers, and talks were held with several television engineers. This work led to an important paper [66], by Lewis and Loughren, published in the magazine *Electronics* in October 1937.

The object of the authors was to determine whether certain aspects of the 'present practice of the proposed television standards in the United States [were] wise'. There were three US standards (on the polarity of the transmission, the transmission of the DC or background component, and the shape, amplitude and duration of the synchronising pulses) which were the exact reverse of the corresponding standards employed in the UK. Clearly it was highly desirable that the RMA's standards should be based on sound practical experience.

Lewis and Loughren were most complimentary about British television

First let it be said that the British pictures are remarkably good. They are steady; they are brilliant; they have an exceptional amount of detail. . . . That [the] British standards constitute a major improvement over present American practice is an inescapable conclusion because television is technically successful and an accomplished fact in England . . . We cannot avoid the fact that the situation in the United States is much less favourable. Unless changes are made in the type of signal which is now being used for experimental transmitters, American receivers will be more expensive, more difficult to service and will give performance inferior to British receivers. . . . It will be in steadiness and control that the American picture will suffer. . . . In viewing the Alexandra Palace transmissions on a large variety of different receivers there were practically no cases of faulty synchronisation [66].

As an indication of the stability of synchronisation of UK sets, Lewis and Loughren noted that reception of the Alexandra Palace transmission at 80 miles (129 km) (which was beyond the optical horizon) gave an image that 'was not visible except as a hazy movement of light on a grille of noise', yet 'it was easily demonstrated that the grille [raster] was synchronised'. The authors found that even the most excessive automobile static that resulted in 'a snowstorm' on the screen failed to have any disturbing effect on the synchronisation.

In their paper Lewis and Loughren extolled the virtues of positive modulation and DC transmission and observed: 'It would not be putting it too strongly to say that the level of black is a definite foundation upon which the structure of British television is built. . . . A final confirmation of the practical nature of the British standard is that the other European countries are adopting its principal features.'

Among the major industrial firms, apart from the Hazeltine Service Corporation, which sent representatives to Europe to collect data that could be helpful in the formulation of US national standards were RCA, CBS, and Bell Laboratories [67,68].

Further work by the RMA Standards Committee in July 1938 led to several additional recommendations to its 1936 standards. DC transmission of the brightness of the televised scene was specified – black in the picture was to be represented by a definite carrier level, as in British practice [69]; the radiated electromagnetic wave had to be horizontally polarised; and details of the line and frame synchronising pulses were delineated – the frame synchronising pulse had to be serrated and include equalising pulses, again as in British practice.

A detailed description of the factors that affected the development of television in France, Germany, Japan, the USA, the USSR and other countries has been given elsewhere by the author, suffice to mention that many systems were in operation by the commencement, in September 1939, of the Second World War (see Appendix 3).

After the cessation of hostilities in 1945, television broadcasting progressed in many countries. By 1953 just four systems had reached the standard of excellence required for national television broadcasting services. They were all-electronic, high definition systems (see Table 21.1) based on 405, 525, 625 and 819 lines per picture; (the 625 (USSR) system was a variant of the 625 (CCIR) system). Since the 525-line and 625-line standards are closely similar, and the 405-line and 819-line standards are congruent, it follows that effectively the growth to 1955 of the world's television broadcasting networks was founded on the US's 525-line and the UK's 405-line systems. In the main, these stemmed from the endeavours of RCA and EMI. Thus, of the myriad of suggestions that were advanced from c. 1879 for 'seeing by electricity' only those of two

Table 21.1 Television system parameters, 1953

System	405	525	625 (CCIR)	625 (USSR)	819
Number of lines per picture	405	525	625	625	819
Video bandwidth (MHz)	3	4	5	6	10.4
Channel width (MHz)	5	6	7	8	14
Sound carrier relative to vision carrier (MHz)	−3.5	+4.5	+5.5	+6.5	−11.15
Sound carrier relative to edge of channel (MHz)	+2.5	−0.25	−0.25	−0.25	+0.10
Interlace	2:1	2:1	2:1	2:1	2:1
Line frequency (Hz)	10 125	15 750	15 625	15 625	20 475
Frame frequency (Hz)	50	60	50	50	50
Picture frequency (Hz)	25	30	25	25	25
Sense of vision modulation	positive	negative	negative	negative	positive
Level of black as % of peak carrier	30	75	75	75	25
Sound modulation	AM	FM	FM	FM	AM

Source: International Telecommunications Union. Quoted in 'Television, a world survey' (UNESCO, 1953) p. 6

well-endowed companies, each having highly skilled and talented research scientists and engineers, operating in an environment conducive to the achievement of progress, succeeded.

At the commencement of the Second World War the principal high definition television broadcasting systems operated at radio frequencies in the very high frequency (VHF) band (i.e. 30 MHz to 300 MHz), albeit at the lower end – less than 50 MHz – of this band. Some isolated investigations utilising higher frequencies than this had been undertaken by the early 1930s, but by the end of the decade determined efforts were being carried out to employ not only the upper portion of the VHF band but also the UHF (300 MHz to 3,000 MHz) band. This activity stemmed from the appointment in January 1933 of Adolf Hitler, as Chancellor of the Third Reich, and the associated re-formation of the German Luftwaffe, and the necessity in the United Kingdom to set-up swiftly an early warning radar system able to detect the approach of hostile aircraft. During the war, both VHF and UHF techniques were much developed. They led to a vast increase in the portion of the electromagnetic spectrum that could be employed for communication purposes and enabled new communication systems, such as, for example, satellite communications to be effected. An introduction to the history of the emergence of some of these new technologies for communications is given in the following chapter.

References

1   BURNS, R.W.: 'British television, the formative years' (Institution of Electrical Engineers, London, 1986), Chapters 13, 14, and 15
2   Report of the Television Committee, Cmd. 4793, HMSO, January 1935
3   BURNS, R.W.: Television, an international history of the formative years', (Institution of Electrical Engineers, London, 1998), pp. 309–332
4   WHITAKER, A.: 'Television development of HMV', 17th February 1938, EMI Archives
5   WHITAKER, A.: Letter to Dr A.N. Goldsmith, 22nd August 1930, EMI Archives
6   Goldsmith, A.N.: Letter to A Whitaker, 4th September 1930, EMI Archives
7   History sheet on Baird television, Baird II, BM22, pp. 1–3, EMI Archives
8   TEDHAM, W.F.: 'Report on investigation of the factors affecting emission in photoelectric cells'. Report GD2, 17th September 1930, EMI Archives
9   WRIGHT, W.D.: 'Report on Kerr cells', Report GC5, 26th November 1930, EMI Archives
10 BROWNE, C.O.: 'Proposed system for five channel television'. Report GC1, 20th September 1930, EMI Archives
11 BROWNE, C.O.: 'Multi-channel television', *J.IEE*, 1932, **70**, pp. 340–349
12 ANON.: 'Great advance in television sets. Nearing practical success. Broadcast of films and plays. When all may see the Derby', *Daily Telegraph*, 6th January 1931

13 ANON.: 'Success of new tests today. Film of everyday life projected. Ride on a bus. Everything clearly recognizable'. *Evening Standard*, 6th January 1931

14 Ref. 3, pp. 431–435

15 Ref. 3, pp. chapter 19, pp. 431–478

16 WRIGHT, W.D.: 'Report of television progress to date', 31st August 1931, pp. 1–10, EMI Archives

17 WHITAKER, A.: 'Note on television', 1st October 1931, pp. 1–3, EMI Archives

18 Cable from Victor, Camden, received 2nd December 1931, EMI Archives

19 Executive Committee minute no. 20851, 1st October 1931, EMI Archives

20 MITTEL, B.: 'Television publicity', 16th February 1936, p. 3, EMI Archives

21 WHITAKER, A.: 'Note on a visit to Engineer-in-Chief's department, General Post Office [on] 21st October 1931', 23rd October 1931, pp. 1–3, EMI Archives

22 SHOENBERG, I.: Discussion on 'The history of television', *J.IEE*, 1952, **99**, Part IIIA, pp. 41–42

23 McGEE, J.D.: '1971 Shoenberg Memorial Lecture', *R. Telev. Soc. J*, 1971, **13**, (9)

24 WISSENDEN, J.W.: Letter to EMI Ltd, 8th October 1931, BBC file T16/65

25 'Statement re-television', 10th October 1931, p. 1, EMI Archives

26 CONDLIFFE, G.: 'Re-television – present programme', 14th December 1931, pp. 1–2, EMI Archives

27 CONDLIFFE, G.: 'Television transmitter', 4th January 1932, p. 1, EMI Archives

28 DAVIES, N.E.: 'Marconi-EMI television 1931 to 1937', Appendix 2, 24th April 1950, Marconi Historical Archives

29 SHOENBERG, I.: Letter to N. Ashbridge, 11th November 1932, BBC file T16/65

30 ASHBRIDGE, N.: Report on television demonstration at EMI, 6th December 1932, BBC file T16/65

31 Ref. 1, pp. 240–301

32 PHILLIPS, F.W.: Letter to Sir J.F.W. Reith, 10th April 1933, BBC file T16/42

33 SIMON, L.: memorandum on Baird and EMI demonstrations, 27th April 1933, Minute 4004/33, Post Office Records Office

34 Notes on a meeting held at the GPO, 21st April 1933, BBC file T16/42

35 Ref. 3, pp. 456–459

36 ZWORYKIN, V.K.: 'Television with cathode-ray tubes', *J.IEE*, 1933, **73**, pp. 437–451

37 ANON.: Laboratory staff, research department, 12th June 1934, EMI Archives

38 MCGEE, J.D.: 'The early development of the television camera'. Unpublished manuscript, p. 16, The IEE Library

39 LUBSZYNSKI, G.: 'Some early developments of television camera tubes at EMI Research Laboratories', *IEE Conf. Pub.*, 1986, (271), pp. 60–63

40 BLUMLEIN, A.D., and McGEE, J.D.: 'Improvements in or relating to television transmission systems', British patent 446 661, 3rd August 1934

41 ASHBRIDGE, N.: Report on television, 17th January 1934, BBC file T16/65
42 WHITE, H.A.: Letter to Sir J.F.W. Reith, 23rd March 1934, BBC file T16/65
43 Notes of a meeting of the Television Committee held on 27th June 1934. Evidence of Messrs Clark and Shoenberg on behalf of EMI and the Marconi–EMI Television Co. Ltd., minute 33/4682, Post Office Records Office
44 REITH, Sir J.F.W.: Letter to Sir Kingsley Wood, 15th March 1934, BBC file T16/42
45 KINGSLEY WOOD, Sir H.: Letter to Sir J.F.W. Reith, 20th March 1934, BBC file T16/42
46 Notes on 'Conference at General Post Office, 5th April 1934', Minute Post 33/4682, Post Office Records Office
47 Ref. 1, pp. 302–350
48 BLUMLEIN, A.D.: 'The transmitted waveform', *J.IEE*, 1938, **83**, pp. 758–766
49 ENGSTROM, E.W.: 'A study of television image characteristics', *Proc. IRE*, 1933,21(12), pp. 1631–1651
50 WENSTROM, W.H.: 'Notes on television definition', *Proc. IRE*, 1933, 21(9), pp. 1317–1327
51 Notes of a meeting of the Television Committee held on 8th June 1934. Evidence of the Marconi–EMI television Company represented by Messrs. Shoenberg, Condliffe, Blumlein, Agate, Browne and Davis, minute 33/4682, Post Office Records Office
52 PRESTON, S.J.: 'The birth of a high definition television system', *Telev. Soc. J.*, 1953, **7**
53 Ref. 1, pp. 377–383
54 Ref. 3, pp. 505–506
55 Television Advisory Committee, minutes of the first meeting, 5th February 1935, Post Office bundle 5536
56 Electric and Musical Industries, response to questionnaire of the Technical Sub-committee on 'Proposed vision transmitter', Minute Post 33/5533
57 Baird Television Ltd, response to questionnaire of the Technical Sub-committee on 'Proposed vision transmitter', Minute Post 33/5533
58 BROWNE, C.O.: 'Vision input equipment', *J. IEE*, 1938, **83**, pp. 767–782
59 BURNS, R.W.: 'The life and times of A.D. Blumlein' (Institution of Electrical Engineers, London, 2000), pp. 201–205
60 Television Advisory Committee, minutes of the 32nd meeting, 15th October 1936, Post Office bundle 5536
61 COCK, G.: 'Report on Baird and Marconi–EMI systems at Alexandra Palace', TAC paper no. 33, 9th December 1936
62 A.C. Cossor Ltd.: letter to the Secretary (TAC), 13th November 1936, Post Office bundle 5536
63 Draft of public announcement, Post Office bundle 5536
64 Ref. 3, pp. 618–620
65 Ref. 3, pp. 537–541, 610–615
66 LEWIS, H.M., and LOUGHREN, A.V.: 'Television in Great Britain', *Electronics*, October 1937, pp. 32–35, 60, 62

67 'Report of the Fall Meeting, Rochester 1937', *Electronics*, December 1937, pp. 11–15, 67–71

68 ANON.: 'Reviewing the video art', *Electronics*, January 1938, pp. 9–11

69 FINK, D.G.: 'Perspectives on television: the role played by the two NTSCs in preparing television service for the American public', *Proc. IEEE*, 1976, **64**(9), pp. 1322–1331

Chapter 22

The emergence of new technologies

During his classic experiments on electromagnetic wave propagation at the Technische Hochschule, Karlsruhe, Hertz generated radio waves by means of the oscillatory discharge of a capacitor through a loop of wire and a spark gap. A similar circuit acted as a detector. The spark gaps and circuits used by Hertz radiated at wavelengths of 50 cm to 60 cm. Later, in 1894, Righi succeeded in producing electromagnetic waves having wavelengths in the range between 20 mm and 125 mm. Further work by several investigators, including Lodge, Fleming, and Nicols and Tear, extended the range and by 1924 Glagoweda-Arkadiewa, utilising a novel form of apparatus, was able to generate waves from 50 mm to 82 μm in wavelength – albeit at low power levels [1].

With the invention of the audion, or triode valve, by de Forest in 1906, and the consequential development of valve circuits, alternative methods of generating metric radio waves became available. At the end of the 1910–1920 decade regenerative oscillators of the conventional type and electronic oscillators of the Barkhausen-Kurz and Gill-Morrell types had been described.

In oscillators of the latter group the grid potential is positive with respect to the potentials of the anode and cathode. Four forms of oscillations can occur, namely, those

1. within the cathode-anode space having a frequency independent of the external circuit constants;
2. within the cathode-anode space having a frequency dependent on the external circuit constants;
3. within the grid-anode space having a frequency independent of the external circuit constants; and
4. within the grid-anode space having a frequency dependent on the external circuit constants.

Oscillations of types 1 and 3 are generally associated with the names of Barkhausen and Kurz, while oscillations of types 2 and 4 are named after Gill and Morrell.

Table 22.1 summarises some of the progress in metric and centimetric wave technology to 1930 [1].

By 1930 a theory, based on extensive experimental work, of the four positive grid oscillators had been advanced by Hollman and steps were being taken to utilise Barkhausen-Kurz oscillators in communications systems. Uda, in 1930, described telegraph and telephone transmission experiments using 50 cm radiation, and, in the same year, Beauvais reported the results of his tests on radio telephone transmissions at wavelengths of 15 cm to 18 cm. Transmitting and receiving parabolic reflectors of 120 cm diameter and 20 cm focal length were employed, the transmitting antenna being mounted on the Eiffel tower. Strong telephone signals were received at distances between 14 km and 23 km.

Table 22.1   Ultra short wave progress

Experimenter(s)	Year	Wavelength achieved (cm)
Regenerative oscillators		
White	1916	600
Gutton and Touly	1919	200–400
van der Pol	1919	375
Southworth	1920	110–260
Holborn	1921	300
Gutton and Pierret	1925	50–200
Kruse	1927	41–500
Englund	1927	100–500
Yagi	1928	60–200
Bergmann	1928	80
Ritz	1928	300
Esau and Hahnemann	1930	300
Brown	1930	200
Barkhausen–Kurz oscillators		
Barkhausen and Kurz	1920	43–200
Gill and Morrell	1922	200–500
Scheibe	1924	30–330
Grechowa	1926	18
Hollmann	1929	20–140
Uda	1930	50
Beavais	1930	15–18
Magnetron oscillators		
Breit	1924	60–150
Yagi	1928	15–100
Forro	1929	30–65
Okabe	1929	5–40
Okabe	1930	3–15

A somewhat similar demonstration was given by A.G. Clavier and his associates of the Laboratoire Central de Télécommunications, Paris, on 31st March 1931 when 3 m diameter parabolic reflector antennas, mounted on the cliffs above St Margaret's Bay, near Dover, transmitted and received telephone messages, (using a 17.6 cm carrier), from Calais. Three years later, STC and its associate company in Paris, Le Material Telephonique, established a microwave link between aerodromes at Lympne, Kent and St Inglevert, Pas de Calais, France. It was put into operation on 26th January 1934 and provided both telephone and teleprinter services [2] (Fig. 22.1).

Barkhausen-Kurz oscillators were also employed by Butement and Pollard in their 1931 work on the reflection of radio waves, and later by Gutton in his work on an obstacle detection system. Such a system, engineered by the Societe Francaise Radio-Electrique, was fitted to the French liner SS *Normandie* in 1935. The obstacle detector used a 16 cm wavelength transmitter valve mounted along the axis of a 0.75 m diameter parabolic reflector, and a similar arrangement as a receiver but with the valve configured as a detector. The system was employed to detect icebergs [3].

Electronic oscillators of the magnetron type date from 1921. In that year A.W. Hull of the General Electric Company (USA) published a paper [4] in which he developed the equations of motion for electrons emitted from a cylindrical cathode, and moving towards a coaxial cylindrical anode, under the influence of the anode-cathode electric field and a uniform magnetic field parallel to the common axis of the two electrodes (Fig. 22.2a).

Hull, during World War I, had been associated with the advancement of the dynatron circuit (which has a negative resistance characteristic), and had considered the effect produced on the dynatron valve's anode current by modulating currents (such as those from a microphone) passed through a large number of turns of wire wound around the valve.

Hull regarded the deployment of a magnetic field as just another means of controlling the anode current of a valve, in addition to the use of filament temperature control and grid potential control. In his autobiography he mentioned the motivation for his invention: 'The radio patents were about equally divided between the different companies and there seemed to be no way in which any one of them could proceed with a development. I conceived the idea of breaking this stalemate by developing a magnetically controlled tube which would be free of these patents' [5].

Initially Hull viewed his magnetron as an electronic switch controlled by a magnetic field but later experiments by him led to the valve being used as an amplifier and as an oscillator. By 1925 Hull was able to generate an output of 8 kW at 30 kHz, with an efficiency of 69 per cent.

In 1924 A. Zacek [6], of Prague, demonstrated that oscillations could be established in simple diode types of magnetrons biased at the critical (cut-off) magnetic field, with a frequency close to the cyclotron frequency. He succeeded in operating his valve at 29 cm. Further work on this mode of magnetron operation was reported in 1928 by H. Yagi, of Sendai, Japan [7,8].

Figure 22.1   (a) This parabolic reflector, 3 m in diameter, was set up on the cliffs at St Margaret's Bay, Dover, for the world's first public demonstration of microwave transmission on 3rd April 1931. (b) Marconi and a UHF antenna which he used to conduct experiments in radio telephony between the Vatican and Castel Gandolfo, 15 miles (24.1 km) distant.

Sources: (a) Northern Electric. (b) The Marconi Company.

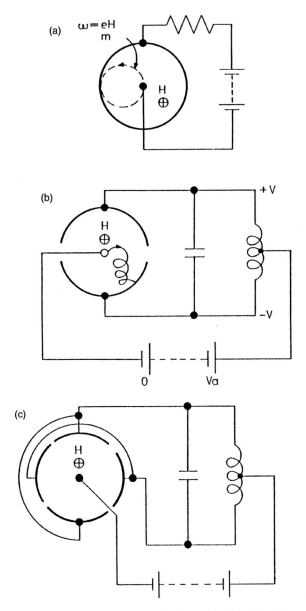

Figure 22.2 *(a) Simple diode type of magnetron biased at the critical (cut-off) magnetic field, with a frequency close to the cyclotron frequency. Investigated by Hull (1921) and Zacek (1928). (b) The negative resistance mode of magnetron action was postulated to exist by Habann (1924), and was confirmed by Okabe (1929). (c) Posthumus in 1934 investigated the travelling wave mode of magnetron oscillation.*

Source: *The GEC Journal of Research.*

A negative resistance mode of magnetron action was postulated to exist by E. Habann [9], of Jena, who in 1924, following a theoretical study, suggested that the cylindrical anode of the diode magnetron should be cut into two hemi-cylindrical segments (Fig. 22.2b) as one of the several possible electrode configurations which would lead to negative resistance characteristics. Habann's conclusion was confirmed in 1929 and 1930 by the investigations of K. Okabe [10], a co-worker of Yagi. Okabe succeeded in generating centimetric waves down to 3 cm wavelength with a split-anode magnetron. His work attracted world-wide interest in the practical possibilities of the magnetron, and from the early 1930s workers in Europe (including E.C.S. Megaw of the GEC Research Laboratories, UK), Russia, Japan and the USA advanced concepts on the utilisation of the magnetron for short wave generation.

A third mode of oscillation was also found to exist in the split-anode magnetron and this was investigated by K. Posthumus [11]. His acknowledged evolution in 1934 of the 4-segment magnetron (Fig. 22.2c) and the rotating field theory of operation by which he explained its advantages over a similar 2-segment system was an outstanding contribution. Of the three possible modes of oscillation in magnetrons – cyclotron, negative resistance, and travelling wave – the last has proved to be the most efficacious for practical microwave magnetrons.

Posthumus's researches and findings appear to have stimulated much work. Megaw [12] has recorded that when he presented the results of the first study of the subject, in the UK, before The IEE in 1933 the literature comprised a dozen papers: by 1939 it had multiplied more than tenfold.

Of all the factors that led to the evolution of practical (commercial) metric wave and centimetric wave electronics, the most important were the appointment, in January 1933, of Adolf Hitler as Chancellor of Germany and the subsequent re-arming of Germany. This heralded an ominous situation for the democracies of Europe. In 1933 no German combat aircraft were constructed but in 1934 840 were built, followed by 1923 in 1935. On 1st March 1935 Berlin informed the world that the German Luftwaffe had been re-created. With its 20,000 officers and men and 1888 aircraft the new service became a potential threat to peace and caused much concern in the United Kingdom [13].

The re-armament led, in February 1935, to the constitution of the Committee for the Scientific Survey of Air Defence (CSSAD) and to the commencement of the development of an early warning radar system known as CH (Chain Home). For various reasons, which need not be discussed here, the system was based on known technology. Indeed all the devices, circuits and systems – wide-band amplifiers (of all types), pulse generators and pulse shaping circuits, the wide-band modulator, transmitter, antenna, and transmission line, the hard vacuum cathode-ray tubes, the power supplies, and test equipment – that had been developed and engineered for the 405-line high definition television system were applicable to the Chain Home radar. There was a technology transfer from television engineering to radar engineering. Approximately ten years later, following the enormous growth of microwave electronics, a technology transfer from radar to television, and communications generally, would take place. This transfer would enable the ultra high frequencies (300 MHz–3,000 MHz) as well as the very high frequencies, (30 MHz–300 MHz) to be used for microwave communications, including the relaying of television programmes, music, voice, and data signals; and for the control and telemetering of satellites and interplanetary probes.

Typical operating conditions of the CH radar transmitter were [14]:

carrier frequency	20 to 30 MHz
peak power	350 kW (later 750 kW)
pulse repetition frequency	25 Hz and 12.5 Hz
pulse length	20 µs

The planning, development, engineering, manufacturing, installation and commissioning of the Chain Home (CH) radar stations, and the training of the staffs to operate the stations and to interpret and act upon the cathode ray tube displays, in the short time of just four years was a remarkable achievement. Although CH was, by modern standards, a rudimentary radar system it served the United Kingdom well. By the beginning of the Second World War the entire east and south east coasts had radar coverage. During the Battle of Britain the CH radar system provided crucial information, about the disposition of the enemy's aircraft, to the RAF's fighters. Without CH it is doubtful whether the United Kingdom would have survived as a sovereign nation.

RAF Fighter Command was formed on 6th July 1936 with Air Marshal Sir Hugh Dowding as the first Air Officer Commander-in-Chief. He worked tirelessly to integrate the growing numbers of men (and later women), airfields, aircraft, radar, balloons, observers, headquarters and communications into a coherent and efficient organisation. This was tested in the summer of 1937 when an 'enemy' force of Bristol Blenheim bombers had to be intercepted by a force of RAF fighters. It was found that none of the RAF's biplane fighters was able to intercept the new, fast, twin-engined Blenheims without being in a position of advantage above them. In his report Dowding observed: 'The exercise again demonstrated the self-evident fact that fighters which cannot overtake enemy bombers are quite useless' [15].

Unfortunately the majority of the RAF's fighter force comprised obsolescent biplane fighters such as Gloster Gladiators and the like. Even in 1938, the year before the Second World War commenced, Fighter Command's force of aircraft comprised 666 slow biplane fighters and just 93 fast monoplane aircraft [16]. Consequently the question arose as to the tactics that could be employed by the biplane fighters in attacking fast monoplane bombers. One tactic suggested was known as 'bombing the bomber'. The idea advanced was that the fighters should drop bombs on the bombers which would explode in their midst. Since neither contact fuses or time delay fuses were considered practicable, work was initiated on proximity fuses.

Two types [17], the acoustic proximity fuse and the photoelectric proximity fuse, were developed but each fuse had inherent limitations. Among these, the acoustic fuse was responsive to shell bursts as far away as 900 m, and the photoelectric proximity fuse was inoperative at night. Fortunately in May 1940 the need for proximity fuses for anti-aircraft (AA) shells and anti-aircraft rockets, as well as bombs, had been identified and various schemes, based on radio principles, were in the course of consideration. For AA shell use special miniature valves and ancillary components able to withstand linear accelerations of up to 20,000 g were required. In 1940 these valves and components did not exist.

The increasing numbers of German aircraft and U-boats being built demanded the development of means to detect them, i.e. air-borne and ship-borne radars. Clearly, these had to utilise much smaller antennas and therefore had to operate at higher frequencies than those used for CH, namely at frequencies in the very high frequency (30–300 MHz) and ultra high frequency (300–3,000 MHz) bands. Again, for these needs special valves had to be designed and fabricated.

In September 1936 an airborne radar group, comprising initially Dr E.G. Bowen, Mr A.P. Hibberd and Dr A.G. Touch, was formed at the Bawdsey Research Station [18]. At first their work was handicapped by the lack of a small very high frequency transmitter valve suitable for working in a 200 MHz air-borne transmitter. However, by illuminating a target plane from a 6 m ground transmitter, the group in June 1937 demonstrated that reflections from the target aircraft could be detected by an antenna-receiver system in another aircraft. The system was known as RDF 1.5.

The history of the UK's Services' interest in valves for very high frequency and ultra high frequency working effectively commences sometime in 1932 when the Air Ministry found difficulty in obtaining from the commercial valve manufacturers the special valves that were needed for the distinctive needs of the Service [19]. In 1932 a meeting was held at the Admiralty's HM Signal School, which was attended by representatives of the three Services, to discuss the possibility of the establishment of a central valve development organisation. The only Service valve laboratory in existence was that at HM Signal School and so the meeting agreed that the Air Ministry should put to HM Signal School problems in valve development which required solution. One year later the valve section at HM Signal School was slightly increased to permit valve development and fabrication work to be undertaken for the Radio Research Board.

In May 1936 C.S. Wright, the Director of Scientific Research (DSR), Admiralty, represented to the Board of Admiralty his concern with the state of R&D on very short wavelengths. Wright's unhappiness was possibly based on four facts [20]. First, in 1921 Brigadier General W. Mitchell, Director of Military Aviation and Assistant Chief of Air Staff (USA), had demonstrated in practical sea trials that the mightiest capital ships were highly vulnerable to bombing attacks from hostile aircraft. (His conclusions were vindicated many times in the Second World War.) Second, it was manifest in 1936 that Germany was rearming. Third, Germany was displaying aggressive military tendencies – on the 7th March 1936 Hitler had ordered his troops (35,000 men) into the demilitarised zone of the Rhineland and had occupied the main towns. And, fourth, the only means of locating hostile aircraft at sea was by the use of 7 × 50 binoculars.

There was thus an urgent need for a ship-borne early warning surveillance system. The CH radar warning system being developed at the Bawdsey research Station worked in the HF (high frequency, 3 MHz–30 MHz, band) and employed 106.7 m transmitting antenna masts and 73.2 m receiving antenna masts and was totally unsuitable for adaptation for use on board ships. There had to be a move towards the employment of very high frequencies (30 MHz–300 MHz) and ultra high frequencies (300 MHz–3,000 MHz) to permit naval radar antennas to be effected.

The Board of Admiralty was persuaded of the validity of its DSR's arguments and gave its approval for the DSR to enter into negotiations with a commercial firm. Discussions were held with GEC, STC, and the Metropolitan Vickers Company and subsequently an Admiralty contract was placed, in 1937, with GEC for the development of valves and components suitable for communications at wavelengths below one metre [21].

Twelve sets of equipment for operation in the range 40 to 60 cm were manufactured by GEC Telephone Works: most of them had been fitted to HM ships by the commencement of hostilities in 1939. The power generated was 20 W. Modulation of the carrier wave was carried out by square wave switching of the anode voltage of the output valve, a split-anode magnetron valve (Fig. 22.3).

The contract led to very close cooperation between the Admiralty and GEC and, on 30th November 1938 J.F. Coales and E.C.S. Megaw of GEC showed by tests on a glass magnetron (E821), designed for 150 W c.w. operation at 300 MHz, that such a valve could be operated under pulsed conditions. The pulse output of 1.5 kW at 37 cm was limited principally by the emission of the tungsten filament which was overrun for the experiment [22].

During 1937 the Samuel triode type 4316, (sometimes known as the 'door knob'), became available; it could work at 200 MHz. With this valve the Bawdsey Research Station airborne radar group constructed the first British Air Surface Vessel (ASV) radar and showed that signals from a 2,000 ton freighter, at a distance of six to eight kilometres from the radar transmitter, could be detected [23].

The following year the Sub-committee on Air Defence Research invited Sir Frank Smith, the Secretary of the Department of Scientific and Industrial Research, to examine the question of the manufacturing resources for the production of the special valves required for RDF (Radio Direction Finding), the precursor name for radar. Sir Frank Smith's committee, in October 1938, decided there was an urgent demand for R&D, on an increased scale, to fabricate valves of greater output at the very high frequencies which had to be utilised for RDF.

This urgency arose from the ominous political situation in Europe in 1938. The Anschluss of Austria, by Germany, had taken place on 15th March 1938, and, following the Munich crisis, the Sudetenland had been annexed on 29th September 1938. These offensive moves by the Third Reich heralded a

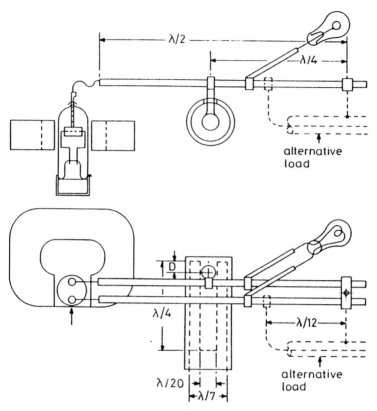

Figure 22.3   Magnetron circuit used for communications transmitter, 1937.
Source: *The GEC Journal of Research.*

situation that demanded a British re-armament policy. There was an imperative
need for the Services to be equipped with the most modern equipment that could
be engineered and manufactured in the short term.

Smith's committee recommended that the Committee of Imperial Defence
should ask the Board of Admiralty to sanction a substantial enlargement of
the Valve Section, at HM Signal School, and also that R&D, subject to Government control, should proceed in close co-operation with one or more industrial
organisations [24]. It was thought that the cost of such enlargement and the
subsequent programme of R&D would be divided between the Services.
Their views were to be borne in mind when the programme was drafted. These
deliberations caused an appreciable expansion of the Valve Section, in both
staff and facilities, and prompted the Admiralty to expand its contract with the
General Electric Company (GEC) to cover the R&D of special valves for Radar
Direction Finding.

Curiously, although de Forest had invented the audion valve in 1906, the
complete theory of the operation of the 3-electrode valve had not been

delineated until 1938. Triode valves are limited in their performance by three effects due to [25]:

1. the anode-grid inter-electrode capacitance;
2. the cathode lead inductance; and
3. the cathode-grid transit time of the thermionically emitted electrons.

The first of these constraints was described in 1919 by Miller and was well known in the 1930s: circuit techniques existed that could neutralise the adverse effect of the capacitance at high frequencies. Any valve, whether triode, tetrode or pentode, is limited in its utility by the damping effects produced in its grid-cathode circuit by the effects of the cathode lead inductance, and the finite transit times of the electrons travelling between the cathode and the grid. These limitations were only extensively investigated in 1936 and 1938 respectively. Prior to these years, valves had been developed for wireless telegraphy, wireless telephony, sound broadcasting and high definition television broadcasting, but in all these communication systems the valves had operated at frequencies below 50 MHz. Now, in the second half of the 1930–40 decade there was an immediate demand for valves that would function satisfactorily at metric and centimetric wavelengths.

Further thought on valves for radar was given at a meeting, held on 27th January 1939, which was attended by the Directors of Scientific Research of the three Services, and presided over by Sir Frank Smith. It was decided that the national interest necessitated still further developments, for communication purposes, of apparatus utilising ever decreasing wavelengths. The DSR, War Office, undertook to examine certain aspects of the problem, and as a result a contract was placed by the Ministry of Supply with Standard Telephones & Cables (STC) for the engineering of specific communication systems operating at very short wavelengths.

Contracts were also allocated to the British Thompson Houston Company and to Electric and Musical Industries Limited. The work was coordinated by the CVD (Communication: Valve Development Committee) which acted on behalf of the Army, the Royal Navy and the Royal Air Force. It operated through seven sub-committees, and shared out the work appropriately between them [26]. These contracts enabled new types of very high frequency valves to be developed and fabricated, and associated with them new circuit techniques (Fig. 22.4).

At GEC a major programme during 1939 was the development of a range of glass-to-metal and disc-seal valves and associated transmission line circuits for radar applications. The first of their 'micropup' (VT90) valves, for use in grounded anode oscillator circuits, was fabricated in 1939. Pairs of these valves could produce 200 kW of peak power at 600 MHz. Later disc seal valves could be operated at up to 3700 MHz and were a triumph of valve construction. The DET22, for example, used a planar cathode and grid, with the grid, wound with wire of 0.03 mm diameter, separated from the cathode by a spacing of just 0.07 mm [27].

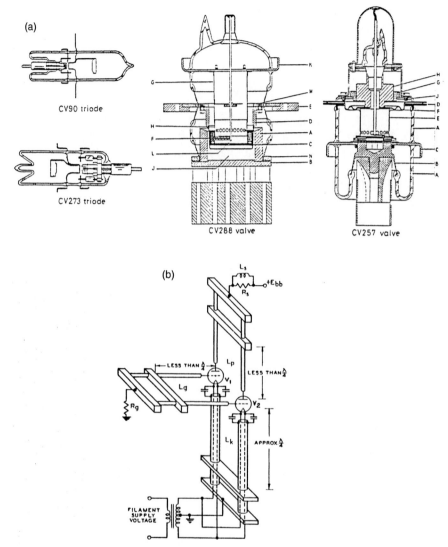

Figure 22.4   (a) Triodes for very short waves. (b) Use of transmission line circuit elements instead of lumped circuit elements.
Sources: (a) *Journal of The IEE*, 1946, **93**, Part IIIA, pp. 833–846. (b) Principles of radar (McGraw Hill, New York, 1946).

The needs of the Services for new valves and circuit techniques placed great demands on firms such as EMI, STC, and GEC. At GEC:

A thousand newcomers joined the Laboratory staff, mostly to work on valves and other devices, and in addition to their work on research and development, the Laboratories became a major pre-production source of valves for the Services, producing over 300,000 of 45 new types of valves during the war years. The type produced in the largest quantity was the CV58 (a 600 MHz diode) of which no less than 48,766 were made [28].

From this brief historical background it is apparent that DSR, Admiralty, who chaired the CVD, had considerable responsibility for the improvement of transmitting valves for RDF.

At the Bawdsey Research Station, Air Surface Vessel (ASV) radar work continued from July 1937 to May 1939 when activity commenced on Air Interception (AI) radar. The circuit techniques that had been progressed for ASV radar were applied to AI radar: they included the use of a squegging 200 MHz transmitter based on TY-150 valves, and a mixer which used a RCA 954 tube.

The AI antenna system comprised a radiating dipole and reflector, and four receiving dipoles, each having a reflector. The two pairs of dipoles were lobe switched to enable signals to be obtained from which the bearing and the elevation of a target could be determined. Experiments conducted during the period May to June 1939 showed that the AI Mark I radar had a maximum range of c. 3660 m and a minimum range of c. 460 m. Although this prototype was not entirely satisfactory the political situation in Europe led to hasty contracts being placed, in July 1939 with several well known industrial companies, for a Mark II version. Again, this was unsatisfactory and it was not until the Mark IV type was engineered – partly by Blumlein – that the Air Ministry had a useful AI radar. This metric radar when used with GCI (Ground Control Interception) radar, and the RAF's latest night fighter brought a cessation, in May 1941, to the nocturnal blitz attacks on Britain [29].

The 1.5 m (200 MHz) metric wave techniques employed for AI Marks I, II, III, IV, V and VI were much developed and by the summer of 1941 the technology had been applied to many different types of radar, including, among others, Air Surface Vessel (ASV) Marks I and II; Chain Home Low (CHL); Ground Control of Interception (GCI); Variable Elevation Beam (VEB); Type 286P; Searchlight Control (SLC), and AA No.3, Mark 3. Later Oboe operated at c. 200 MHz, and Monica worked at 225 MHz. These techniques would later be applied to the design of radio and television communication systems.

The radiating structures of some of these radars produced broad antenna beams that were not wholly suitable for the applications for which the radars were designed. The beams of the metric wavelength AI sets, for example, were such that the effective maximum range of the sets was equal to the height of the AI radar fitted aircraft above ground. Hence a night fighter flying at 25,000 ft (7620 m) could obtain a radar echo from a hostile target at ranges of up to c. 4.5 miles (c. 7.2 km), but when flying at just 5,000 ft (1520 m) the night fighter's target acquisition range was only c. 1.0 mile (c. 1.6 km). This limitation of AI metric radar follows from Fraunhofer diffraction theory which shows that the beam width, in radians (θ), at the half power points of a paraboloid antenna's polar response, is approximately λ/D, where λ is the wavelength and D is the diameter of the antenna; and that the gain of the antenna, relative to an isotropic radiator is $(\pi D/\lambda)^2$. Thus if λ is equal to 1.5 m (200 MHz), and D is 1.0 m, θ is c. 86° and the antenna gain is 4.4; but if λ is equal to 0.1 m (3,000 MHz) then, for the same antenna, θ is c. 5.8° and the antenna gain is 1,000.

These fundamental properties of a radiating structure confer several significant advantages on centimetric antennas and hence on centimetric radars (see Note 1 at end of chapter).

Some of these advantages were certainly known in 1939. On one of his visits to the Bawdsey Research Station, in connection with his responsibilities for transmitter valve development, C.S. Wright, DSR, Admiralty asked Dr E.G. Bowen what wavelength would be most desirable for airborne applications. Bowen's assessment was that to minimise ground return a beam of 10° was wanted and to obtain this from an antenna of 75 cm diameter – the largest that could be fitted into the nose of a fighter – a wavelength of 10 cm would be required. This agreed with Wright's own view of what was needed for naval applications [30].

Meantime, early in 1938, at about the time of the Austrian Anschluss, Sir Henry Tizard [31], the Chairman of the CSSAD, recognised that in the event of war with Germany there would be a demand for a large number of physicists. They would be needed not only to man the expanding early warning radar chain (CH), but also to devise new types of radar equipment. A possible source of these physicists would be the universities.

So, in the summer of 1939, because of the grim situation in Europe, it was agreed that about 80 university physicists should be introduced to RDF so that if war were declared their skills could be deployed to aid the war effort. Each of the physicists would spend a month at a CH station. Subsequently groups of nine to ten physicists, sworn to secrecy, from the Cavendish Laboratory, the Clarendon Laboratory, (University of Oxford), and the Universities of Birmingham, London, Manchester and Bristol were stationed, from 1st September 1939, at several CH radar sites, all of which had been placed on continuous alert from Good Friday.

One of the groups, from the University of Birmingham, comprising Professor M.L. Oliphant, Dr J.T. Randall, Mr H.A.H. Boot, Dr P.B. Moon, and Dr E.W. Titterton, and others was stationed at the CH station at Ventnor [32]. During their stay the party noted the limitations of the radar and the inability of the system to resolve individual aircraft in a group.

Now Oliphant, who had been informed at the close of 1938 by Sir Henry Tizard and Professor J.D. Cockcroft about radar progress and its problems, had visited Stanford University, among other places in the USA, in January 1939 and had acquired much information on the recent klystron work of the Varian brothers. From his Ventnor and Stanford visits Oliphant saw the possibility of improvements being made to the resolution problem if centimetric radio techniques could be applied to it. Gun laying, searchlight directing, airborne interception and air-to-surface vessel detection were all matters that would benefit from such an application – but only if sufficient transmitter power and receiver sensitivity could be obtained. Discussions between Oliphant and Admiralty staff followed his Ventnor experience. Subsequently in the autumn of 1939 the Communication: Valve Development Committee, assigned contracts to the universities of Birmingham, Oxford and Bristol for the development of transmitting and receiving valves able to function at 10 cm.

At Birmingham the Nuffield laboratory being constructed for the Department of Physics was renamed the Admiralty laboratory and it was there that secret valve work was undertaken.

The staff of the Laboratories comprised some members of the teaching staff who retained their appointments with reduced departmental duties and others who were appointed on the Admiralty's pay roll. Among the latter group were Dr J.T. Randall, Mr H.A.H. Boot, and Dr J. Sayers [33]. With the exception of Sayers, who was experienced in the radio techniques applicable to ionospheric research, the members of the team were all physicists, rather than electronic engineers, and had only a casual knowledge of radio. Oliphant felt that this ignorance was an advantage and not a handicap, for it meant that his staff would not be inhibited by prior learning and experience about what could and could not be done: they would be able to work from first principles.

Though the programme of research was officially sponsored by the Admiralty no detailed direction was issued by them. It was understood that the investigation of centimetric radio waves was the objective and to further this some allocation of staff to the various topics that had to be researched was necessary, as follows:

1. Oliphant, Sayers and Dawton (from the Royal Institution) were to inquire into the possibilities of the klystron – the only existing high power source of centimetric waves;
2. Moon and Nimmo were allocated the task of engaging in theoretical and experimental work on the klystron as an amplifier or as a detector;
3. Titterton was to construct a 50 cm oscillator using the GEC 'micropup' triode – which had been worked widely at 150 cm; and
4. Randall and Boot were to investigate the miniature receiving Barkhausen-Kurz receiving valve, which had been described by C.W. Rice [34] in the August 1936 issue of the *General Electric Review*, together with an attempt to obtain r.f. power from plasma oscillations in mercury discharge tubes. (These oscillations had been observed by Langmuir.)

In formulating this initial programme of tasks Oliphant was guided by developments that had been described in the technical literature, by discussions with staff at HM Signal School and by the recollections of conversations he had had with Dr Heil – one of the pioneers of velocity modulation methods. He had worked in the Cavendish Laboratory when Oliphant was there as a research worker.

Randall and Boot were unsuccessful and disenchanted with their project and in November 1939, 'at the risk of incurring some unpopularity from [their] fellow workers, [they] concentrated [their] thoughts on how [they] could combine the advantages of the klystron with what they believed to be the more favourable geometry of the magnetron' [35]. According to Shearman an element in their thinking may have been the seminal suggestion to Oliphant on a visit to the Admiralty establishment in Haslemere that the above combination might be a way forward to shorter wavelength operation. A further influencing factor was Rice's paper which gave an account of a concentric transmission line type of magnetron. In this the concentric line was the resonator.

The klystron was invented by the Varian brothers (with the aid of just a $100 grant from Stanford University) and was described by them in a paper [36] titled 'A high frequency oscillator and amplifier', published in May 1939. In this they stressed the fact that the conventional triode valve amplifier suffered from two 'fundamental difficulties' – with the resonant circuits and with the valve itself. They wrote:

As a resonant circuit the ordinary coil and condenser become unsatisfactory as the frequency is increased, not only because it becomes too small to be mechanically convenient but even more because the losses become too great and therefore the shunt impedance becomes too low. Another related and yet partially independent difficulty is the decline in frequency stability as the losses increase. These difficulties can be, and are, partially avoided by using such resonators as concentric lines; but because of limitations due to leads and other reasons the resonant circuit often remains a limiting factor. Improvement in resonant circuits is, therefore, of primary importance.

The second difficulty is associated with the finite transit time of the electron which leads to a loss of efficiency as the frequency of the operation increases. Although the transit time can be reduced by increasing the anode-cathode voltage and by reducing the physical dimensions of the valve, as in conventional commercial practice, the Varian brothers felt that there was a 'practical limit to useful results along this line [of attack]'. They preferred not to attempt the elimination of the effect of transit time, but to turn it to constructive use, as in the Barkhausen-Kurz valve and the magnetron. This objective and the application to the problem of a new type of resonator led then to the design, construction and testing of the klystron. One of their early valves worked at a wavelength of about 13 cm.

The history of resonators dates from 1894 when, in his 'Recent researches', J.J. Thomson dealt briefly with oscillation inside a cylinder. In the same year Sir Oliver Lodge demonstrated to members of the Royal Institution the radiation of electromagnetic waves from 'the inside of a hollow cylinder with sparks at the ends of a diameter; this being a feeble radiator, but a very persistent resonator' [37]. Three years later Lord Rayleigh [38] showed theoretically that, in perfectly conducting rectangular and circular cylinders, two classes of wave are possible. For each class there is a series of modes of excitation, with a cut-off frequency for each mode below which no propagation can occur.

In 1915 L. Silberstein reported work on the transmission of waves through hollow metal pipes, and the associated problem of the propagation of electromagnetic waves along a dielectric cylinder was studied theoretically by

D. Hondros and P. Debye and by O. Schriever [39,40,41]. Engineering interest in the properties of waveguides (including resonators) came in the early 1930s when work commenced independently at Bell Telephone Laboratories under the leadership of G.C. Southworth and at the Massachusetts Institute of Technology under the direction of W.L. Barrow. Both centres in 1936 published papers [42,43] which are now regarded as classics in their field. Southworth's paper was printed in the April issue of the *Bell System Technical Journal* and was complemented by a paper in the same volume on 'Hyper-frequency wave guides – mathematical basis', by three of Southworth's colleagues. They paid particular attention to the attenuation and impedance characteristics of hollow conducting cylinders.

Concurrently with this progress, W.W. Hansen, of Stanford University, was developing the theory of cavity resonators in connection with a scheme for producing high voltages. His 1937/38 paper [44] on 'A type of electrical resonator' dealt with the theory of the use of certain shapes – a cylinder, a sphere and a square prism – as electrical resonators and showed that in many respects they are equivalent to lumped circuits, with properly chosen circuit constants. He derived formulae for the lowest attainable frequency, the L, C, and Q values, and the series and shunt resistances of the above mentioned closed shapes. A particularly important contribution, for future (post 1937/38) centimetric valve development was his theory of coupling into these types of resonators. Subsequently, in 1939, Hansen wrote papers 'On resonators suitable for klystron oscillators' [45] (with R.D. Richtmyer) and 'On the resonant frequency of closed concentric lines' [46].

The papers of Hansen and of the Varian brothers were of course available in 1939 in Oliphant's department and, as Randall and Boot have noted, the 'intriguing new concepts were discussed at colloquia' in the department. The important features of resonators constructed in copper – low h.f. losses, wavelength stability, and potential capability of large heat dissipation – were known but also it was recognised that in designing high power klystrons it would be difficult to get sufficient power into a klystron's electron beam, which necessarily was of small cross-sectional area. Attempts to increase the power output by increasing the space current could be self-defeating, unless careful steps were taken to focus the electron, because such increases in current density could cause more electrons to spread outside the beam. These electrons would then strike one of the positive electrodes without contributing to the performance of the klystron. However the magnetron, in which the electrons do not form a beam, is free from this disadvantage.

With these known facts before them Randall and Boot had to consider how they could use the desirable features of cavity resonators in devising a new type of magnetron. Their solution [47,48], postulated in November 1939, is illustrated in Fig. 22.5a.

Six resonators, 1.2 cm in diameter, with slots 0.1 cm by 0.1 cm were employed, the length of the resonators being 4 cm. The type of resonator chosen was different to those considered by Hansen, because these could not be associated

*Figure 22.5   (a) Randall's and Boot's experimental magnetron. (b) GEC's production
E1189 cavity magnetron.*
Sources: (a) The Science Museum (b) Quineteq.

with the cylindrical anode and cathode of the magnetron. The only other forms
of short-wave resonator circuit familiar to Randall and Boot were Hertz's
original wire loop resonator and the short circuited λ/4 line.

An interesting anecdote about the design of the cavity resonator based on
Hertz's dipole was given by R. Millar in a 1975 issue of the *Sunday Times* [49].

Shortly before the outbreak of World War II, Randall lived in Aberystwyth,
and on Saturday afternoons it was his habit to delve among the technical books
in a bookshop just around the corner from his flat in Marine Parade. One
afternoon he found an old English translation of the papers of Heinrich Hertz
which described Hertz's experiments, including those which utilised small loops
of wire (with spark gaps) as detectors. And so, when Randall and Boot, one day
in November 1939, were giving some thought to the nature of the resonators

that could possibly be incorporated into the magnetron, Randall recalled the book he had found some months before. He noted that a 3-dimensional version of Hertz's wire loop resonator would give a cylinder with a slot down one side, and a 3-dimensional generation of the λ/4 short-circuited line would lead to a parallel-sided slot λ/4 deep. Either design was appropriate for an arrangement of a number of segments with cylindrical symmetry around a central cathode. Furthermore the designs were capable of being machined from a solid block of copper to allow of good heat transfer and, it was hoped, high power dissipation.

The resonator diameter of 12 mm was selected because H.M. McDonald in 1902 in his book *Electric waves* had calculated that the resonant wavelength of Hertz's wire loop resonator was 7.94 times its diameter and Randall and Boot were aiming to produce 10 cm radiation.

An oxide-coated cathode was thought to be too complex for a prototype and so the thickest tungsten wire, for which there was a filament transformer, was used, namely 0.75 mm diameter wire. Calculations indicated that for the expected frequency of oscillation an anode potential of 16 kV would be needed together with an axial magnetic flux density of 0.1 Wb/m^2.

The first anode block was machined by Boot, with the assistance of Mr M.P. Edwards, in December 1939, and was shown to Sir Edward Appleton and Sir Lawrence Bragg during a visit to the laboratory.

Immediately after Christmas 1939 work was transferred to the newly completed Nuffield Research Laboratory and the construction of the necessary power supply for testing the valve was commenced. Some delay was experienced in obtaining the high voltage rectifiers and it became necessary for Randall and Boot to make their own, which like the magnetron, were continually evacuated. This work was completed in February 1940.

On 21st February 1940, for the first time, the magnetron (Fig. 22.5a) continuously operated on the pump and with the glass-metal junctions closed with sealing wax, was excited. A c.w. power output of c. 400 W at a wavelength of 9.8 cm was obtained, via a coupling loop fitted into one of the resonators in a mid-section plane at right angles to the cylindrical axis.

Randall and Boot have written [32]:

The amount of power produced was, for the first time, capable of quite spectacular effects. It was uncomfortably hot to hold the hand near the output lead of the magnetron and a small arc sprayed off into the surrounding air. An attempt to estimate the power output was made by the very crude method of burning out successively larger 6 V filament lamps and it was found that the pressure neon lighting tubes, (about 1 m long by 37 mm diameter) could be lit to a brilliancy corresponding to a power consumption of 400 W, [6 A, 70 V at 50 Hz].

That the output had proved rather more than we had anticipated originally, is shown by the entries in laboratory notebooks of attempts to prove that the wavelength was not centimetric, but metric. The wavelength was measured however the next day by means of a pair of Lecher wires about 3 m long, and it was shown to be 9.8 cm.

The results were communicated, on 27th February 1940, to Sir Charles Wright and other government scientists. They arranged for Megaw of GEC, who had been concerned for some time with conventional magnetrons, to visit Birmingham in the hope that some assistance could be provided with the unfamiliar technological problems of constructing sealed-off cavity magnetrons of copper and glass with brazed joints. Shortly after Megaw's visit, in April 1940, Randall and Boot became acquainted at GEC with a number of skills connected with the making of seals, including, the method of joining copper to copper by means of a gold wire under pressure between the two surfaces at temperatures below 500 K.

Megaw soon designed an improved sealed-off version of Randall's and Boot's magnetron, resulting in a considerable reduction of the 22.7 kg magnet mass. An important factor in the development of the magnetron was the incorporation of an oxide-coated cathode. From the earliest days of magnetron research the diameter of the cathode had been kept small (about 3 per cent or less of the anode diameter), but in 1939, during discussions between Megaw and Gutton of SFR in Paris, it was concluded that a much larger cathode might be used in multi–segment valves without loss of efficiency. It was agreed that samples of the M16, Gutton's 8-segment valve, should be fitted with large cathodes and sent to the UK [50].

In May 1940 samples of the French M16 valve were taken to GEC, Wembley [12]. They had been fitted with large diameter oxide-coated cathodes and tests showed a pulse power of the order of 1 kW at 16 cm wavelength, with an efficiency of 15 per cent. Consequently it was decided to incorporate both thoriated tungsten, and oxide-coated, cathodes in the Wembley designs of the magnetron. Both types were completed together and showed on 29th June 1940 similar results – an output of 1 kW peak being obtained at 5–40 μs, 50 pps at 9.8 cm wavelength using a 2.3 kg permanent magnet. By 1st July 1940 the E1189 version of the valve (Fig. 22.5b) engineered by Megaw, was producing outputs of 4–5 kW at 9.8 cm; and on 17th July 1940, at a CVD meeting, an output of 12 kW was reported. Within a few months of the first test by Randall and Boot peak power outputs of up to 50 kW were obtained.

After the end of the Second World War, the resonant cavity magnetron was described as having 'had a more decisive influence on the war than any other single weapon'. The post-war Royal Commission on Awards to Inventors observed [33] that the inventions of the cavity magnetron and its associated strapping (by Sayers) 'were of outstanding brilliance and exceptional utility for offensive and defensive purposes'.

Though the history of electronics and communications post-1940 is outside the scope of this book, a few paragraphs on the war-time applications of some of the above mentioned devices will illustrate the great impact of science and engineering on modern warfare.

The invention of the magnetron together with the development of the klystron (Fig. 22.6) and the crystal detector, made centimetric radar and communication systems possible. Among the British centimetric radars subsequently engineered there were the following types: CHEL, AI Mark VIII, H₂S, AGLT, Type 271 and other naval radars, and GL Mark III.

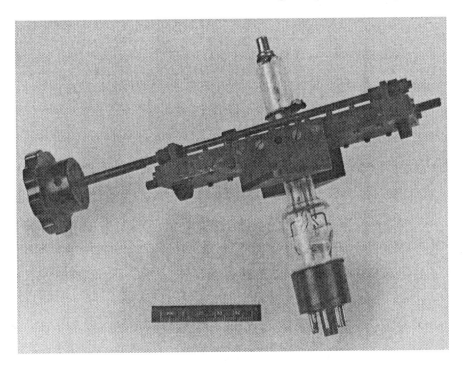

Figure 22.6   A Sutton tube.
Source: Quineteq.

The situation that faced Great Britain in June 1940 was dire. During the evacuation from Dunkirk (26th May–4th June 1940) the British Expeditionary Force had abandoned an immense amount of munitions, *viz*: 7,000 tons (7,110,000 kg) of ammunition, 90,000 rifles, 2,300 guns, 82,000 vehicles, 8,000 Bren guns and 400 anti-tank rifles [51]. Much had to be done to re-build the nation's defence capability. However, just over a month after the retreat from France the Battle of Britain commenced. It lasted from 10th July to 31st October 1940, a period in which the RAF lost 915 aircraft. All these losses had to be made good and the UK's factories and production centres were hard pressed. Additional resources were necessary to advance all the various inventions that were thought to be meritorious.

In August 1940 Sir Henry Tizard, the Chairman of the Committee for the Scientific Survey of Air Defence, and other members of a mission, known as the British Technical Mission [52], left the United Kingdom for the United States of America 'to enlist the help and powerful resources of the American radio manu-facturing industry'. It was appreciated that the mission would not 'get their help adequately without disclosing our technical information' [53], and so the dis-closures had to be given 'without conditions attaching to the offer'.

During the ensuing talks with US scientists all the UK's most precious and closely guarded secrets were handed over to the Americans [54,55]. The secrets included specifications, design plans, test reports, and where appropriate, actual hardware, relating to Asdic, RDX explosives, proximity fuses (of all types), rockets, gyro-gunsights, jet propulsion, radar (of all forms), the resonant cavity magnetron and micropup valves, inter alia. Among the secrets were Butement's proximity fuse circuit, which, in its basic configuration, remained unchanged throughout the US project on radio proximity fuses.

On the cavity magnetron, one US historian, P.J. Baxter [56] has said: '[It] was the most valuable cargo ever to reach these shores'. F.B. Llewellyn a former President of the Institute of Radio Engineers (USA) has reminisced about the test of the first British magnetron, given on 6th October 1940 at the Bell Telephone Laboratories, Whippany: 'It was a day to be remembered. The tube gave about 10 kW peak pulse output at a frequency in the vicinity of 3,000 MHz. This was a power about five times as great as was given by the triodes in the Mark I equipment and moreover was of a frequency over four times as high. Can you imagine our enthusiasm' [35].

The outcome of the test was an agreement to establish a radar research laboratory, funded by government money and staffed by the most capable physicists and engineers who could be found in the universities, to undertake the development of centimetric radars based on the resonant cavity magnetron. The laboratory became known as the Radiation Laboratory and was set up at the Massachusetts Institute of Technology in Cambridge, Massachusetts, USA.

An immense amount of R&D effort went into the design and manufacture of radar and related systems during World War II. After the end of hostilities, there was a view that the great body of information and new techniques in the electronics and high frequency fields should become available for peaceful uses. The Radiation Laboratory undertook the supervision of the task of presenting the collective accumulated wisdom of the many university, industrial and government laboratories in the USA, UK, and Canada and other Dominions in a written form. The result was the Radiation Laboratory Series of books totalling 28 volumes which were published in 1946–1947.

The Tizard Mission's disclosures also either initiated or spurred US efforts in the fields of radio proximity fuses and electrical gun laying predictors which were to have a very great impact against the V1s.

The United States proximity fuse (p.f.) programme was impressive by any standards and, in terms of the magnitude of the effort expended in developing and manufacturing a single entity, was probably second only to the atomic bomb project. The p.f. procurement contracts increased from a total value of $60 million in 1942 to $450 million in 1945, and eventually 2,000 interlocking suppliers and sub-suppliers were involved in producing material for the daily assembly of 70,000 fuses at five major plants. By V-J Day US firms were fabricating 101,000 special reserve p.f. batteries per day, and during the peak production period c. 1,000,000 miniature p.f. valves were being constructed every 2.5 (sic) days [57].

There were 22,000,000 radio proximity fuses made in the USA in the period 1940 to 1945, and of these 1,500,000 were fired at the enemy.

All these developments – centimetric radar, radio proximity fuses and electrical predictors – played a crucial role during the Vergeltungswaffe (V1) offensive against London and the south east of England in 1944. In particular the US SCR 584 radar, US M9 electrical predictor, and US T98 radio proximity fuse, when used with the British 93.98 mm Mk 2C power controlled AA gun proved highly effective in destroying the V1s. As an illustration, on 28th August 1944 94 V1s approached the coast of England: 23 were shot down by fighter aircraft, 65 were destroyed by the guns and 2 were brought down by the balloon barrage [58]. The best gun batteries accomplished a destruction rate of 40 shells fired to bring down one V1. This figure may be compared with the figure of 18,500 shells fired to destroy one enemy bomber during the 1940 night blitz attacks on London.

Also, the use of centimetric radar, (the British airborne ASV Mk III set designed, developed and engineered by EMI), was the crucial factor in the achievement of victory in the Battle of the Atlantic without which the war could not have been won.

In March 1943 – 43 months after the commencement of hostilities in September 1939 – the Axis powers' U-boats sank 627,000 tons of shipping. Two months later, in a single month, (May), following the introduction of the new radar into Coastal Command's planes, 41 U boats were sunk – more than the total number sunk during the whole of either 1940 or 1941. Shipping losses fell dramatically never to increase, for the remainder of the war, to the levels of the early years of the conflict.

By the end of the Second World War electronic techniques, both analogue and digital, had been greatly developed, and valves were available for all purposes for frequencies as high as 30 GHz. Valves existed that could produce more than 1 MW (peak power) output, and miniature, ruggedised, valves able to withstand immense linear and angular accelerations had been fabricated. Semi-conductor microwave detectors were effective. Specialised electrical predictors and digital computers had been designed and engineered, and systems utilising as many as 2,000 (sic) valves were in operation. The foundations of the principles of control theory and practice had been formulated.

All of these devices, systems and techniques enabled VHF FM stereophonic radio broadcasting, VHF and UHF television broadcasting, colour television, and microwave communications, to be implemented a few years after the cessation of hostilities. Subsequently, the development of transistor electronics and integrated circuit electronics led to the enormous growth of data communications, satellite communications, inter-planetary probe communications, and computing techniques. A history of these developments must be reserved for another volume on communications.

Figure 22.7   The No. 10 set.
Source: Quineteq.

Note 1

The advantages of centimetric antennas [59]

The advantages of centimetric antennas were listed in 1940 by R.A. Watson Watt (the radar pioneer) as [59]:

1. their ability to illuminate certain objects to the exclusion of others at relatively small angular separations; in particular to exclude the profound influence of ground-reflected rays;
2. their ability to obtain high discrimination in measurement of angle by the use of very steep-sided polar diagrams;
3. the economy of power concentration of most of the transferred energy within the pencil beam;
4. the reduction of unwanted signals, in particular naturally occurring radio noise and deliberate jamming signals by confining the reception substantially to a very small angular sector.

Subsidiary, but still important, advantages of such antennas are:

1. their ability to radiate and to receive efficiently signals modulated over a wide band of frequencies;
2. the intrinsically low level of atmospheric noise at centimetric wavelengths;
3. the exclusion of effects due to ionospheric bending;
4. the improved transportability of the equipment;
5. the increased effective mobility, for gun-fire control apparatus, conferred by an independence of site conditions.

Although Watson Watt had in mind radar applications, most of these advantages would be applicable to post war microwave communications, including the relaying of data, voice, music, and television signals. (A wartime microwave communications system, the No. 10 set, is shown in Fig. 22.7.)

References

1 WENSTROM, W.H.: 'Historical review of ultra-short wave progress', *Proc. I.R.E.*, 1932, **20**(1), pp. 95–112
2 YOUNG, P.: 'Power of speech' (George Allen, London, 1983), pp. 72–73
3 ANON.: ' "Feelers" for ships', *Wireless World*, 26th June 1936, pp. 623–624
4 HULL, A.W.: 'The effect of a uniform magnetic field on the motion of electrons between coaxial cylinders', *Phys. Rev.*, 1921, **18**, pp. 31–39
5 MONCRIEF, F.: 'Unproven sibling grows to be a giant', *Microwave Service News*, June 1978, pp. 36–39
6 ZACEK, A.: 'Ueber eine Methode zur erzeugung von sehr Kurzer elektro-magnetischen Wellen', *Zeit. F. Hochfrequenztechnik*, 1928, **32**, p. 172

7 YAGI, H.: 'Beam transmission of ultra short waves', *Proc. I.R.E.*, 1928, **16**, pp. 715–741

8 WHITE, W.C.: 'Some events in the early history of the oscillating magnetron', *Jour. of the Franklin Inst.*, 1952, **254**(3), pp. 197–204

9 HABANN, E.: 'Eine neue Generaturohre', *Zeit. F. Hochfrequenztechnik*, 1924, **24**(5), pp. 115–210

10 OKABE, K.: 'On the short-wave limit of magnetron oscillations', *Proc. I.R.E.*, 1929, **17**, pp. 652–659

11 POSTHUMUS, K.: 'Oscillations in a split anode magnetron', *Wireless Engineer and Experimental Wireless*, 1935, **12**, pp. 126–132

12 MEGAW, E.C.S.: 'The high-power magnetron, a review of early developments', *Jour. I.E.E.*, 1946, **93**, Part IIIA(5), pp. 977–984

13 BURNS, R.W.: 'Technology and air defence', *Transactions of the Newcomen Society*, 1995–96, **67**, pp. 161–181

14 NEALE, B.T.: 'CH – the first operational radar', chapter 8, pp. 132–150, in BURNS, R.W. (Ed.): 'Radar development to 1945' (IEE, London, 1986)

15 DOWDING, H.C.T.: 'Combined training exercise, 1937', 25th September 1937, AIR 16/67, PRO, Kew, UK

16 WOOD, D., and DEMPSTER, D.: 'The narrow margin' (Hutchinson, London, 1961), p. 7

17 BURNS, R.W.: 'Early history of the proximity fuse (1937–1940)', *IEE Proc-A*, May 1993, **140**(3), pp. 224–236

18 BURNS, R.W.: 'The life and times of A.D. Blumlein' (IEE, London, 2000), chapter 12, pp. 323–348

19 ANON.: 'Note on valve development and supply', 21st March 1940, AVIA 15/648, PRO, Kew, UK

20 BURNS, R.W.: 'Aspects of UK air defence from 1914 to 1935: some unpublished Admiralty contributions', *IEE Proc.*, 1989, **136**, Pt. A(6), pp. 267–278

21 WILLSHAW, W.E.: 'Microwave magnetrons: a brief history of research and development', *The GEC Journal of Research*, 1985, **3**(2), pp. 84–91

22 MEGAW, E.C.S.: 'The high power pulsed magnetron: notes on the contribution of the GEC Research Laboratories to the initial development', Report No. 8717, 30th August 1945, 9pp, Research Laboratories of the General Electric Company

23 BOWEN, E.G.: 'Radar days' (Adam Hilger, Bristol, 1987), pp. 51–52

24 CROWTHER, J.G., and WHIDDINGTON, R.: 'Science at war' (HMSO, London, 1947), p. 44

25 PARKER, P.: 'Electronics' (Edward Arnold, London, 1950), chapter 10, pp. 293–334

26 Ref. 24, p. 45

27 CLAYTON, R., and ALGAR, J.: 'The GEC Research Laboratories 1919–1984' (IEE, London, 1989), pp. 122–124

28 CLAYTON, R., and ALGAR, J. (Eds.): 'A scientist's war, the war diary of Sir Clifford Paterson 1939–45' (The IEE, London, 1991), p. 574

29 BURNS, R.W.: 'The development of methods of detecting hostile aircraft at night 1935–1941', *Transactions of the Newcomen Society*, 1997–1998, **69**(1), pp. 1–20

30 Ref. 23, p. 143

31 CLARK, R.W.: 'Tizard' (Methuen, London, 1965), p. 171

32 RANDALL, J.T., and BOOT, H.A.H.: 'Development of the multi-resonator magnetron in the University of Birmingham, (1939–1945)', The IEE Archives

33 Evidence presented to 'The Royal Commission on Awards to Inventors', 2nd, 3rd, and 4th May 1949, T166/10, PRO, Kew, UK

34 RICE, C.W.: 'Transmission and reception of centimetric radio waves', *General Electric Review*, August 1936, **39**(8), pp. 363–369

35 RANDALL, J.T., and BOOT, H.A.H.: 'Historical notes on the cavity magnetron', *IEEE Trans. on Electron Devices*, 1976, **ED-23**(7), pp. 724–729

36 VARIAN, R.H., and VARIAN, S.F.: 'A high frequency oscillator and amplifier', *Jour. of Applied Physics*, 1939, **10**, pp. 321–327

37 LODGE, O.: a report, *Proc. Roy. Inst.*, 1894, **14**, p. 321

38 RAYLEIGH, LORD.: 'On the passage of electric waves through tubes, or the vibrations of dielectric cylinders', *Phil. Mag*, 1897, **43**, pp. 125–132

39 HANDROS, D., and DEBYE, P.: 'Electromagnetic waves in dielectric wires', *Ann. Der Phys.*, 1910, **18**, pp. 465–476

40 ZAHN, H.: 'Detection of electromagnetic waves on dielectric wires', *Ann. der Phys*, 1916, **49**, pp. 907–933

41 SCHRIEVER, O.: 'Electromagnetic waves in dielectric conductors', *Ann. der Phys.*, 1920, **63**, pp. 645–673

42 SOUTHWORTH, G.C.: 'Hyper-frequency wave guides. General considerations and experimental results', *Bell System Technical Journal*, 1936, **15**, pp. 284–309

43 BARROW, W.L.: 'Transmission of electromagnetic waves in hollow tubes of metal', *Proc. I.R.E.*, 1936, **24**, pp. 1298–1328

44 HANSEN, W.W.: 'A type of electrical resonator', *Jour. of App. Phys*, 1938, **9**, pp. 654–663

45 HANSEN, W.W., and RICHTMYER, R.D.: 'On resonators suitable for klystron oscillators', *Jour. of App. Phys.*, 1939, **10**, pp. 189–194

46 HANSEN, W.W.: 'On the resonant frequency of closed concentric lines', *Jour. of App. Phys.*, 1939, **10**, pp. 38–43

47 RANDALL, J.T., and BOOT, H.A.H.: 'The cavity magnetron', *J.IEE*, 1946, **93**, Part IIIA(5), pp. 928–938

48 ANON.: The resonator magnetron. A new high power generator for microwaves', undated, The IEE Archives, (a four page hand written report, probably by Boot)

49 MILLER, R.: 'Secret weapon', *The Sunday Times Magazine*, 7th September 1975, pp. 8–10, 13, 15

50 MOLYNEUX-BERRY, R.B.: 'Dr Henri Gutton, French radar pioneer', chapter 4 in BURNS, R.W. (Ed.): 'Radar development to 1945' (IEE, London, 1988)

51 CHURCHILL, W.S.: 'Their finest hour' (Cassell, London, 1949), p. 320

52 Ref. 31, p. 249

53 Ref. 31, p. 251

54 British Technical Mission, 'Information given to the US Authorities', 28th October 1940, AVIA 53/629, PRO, Kew, UK

55 'Lists of secret technical information released to US Government and firms', AVIA 38/1009, PRO, Kew, UK
56 P.J. Baxter in 'Scientists against time' (Little, Brown, Boston, 1946)
57 BALDWIN, R.B.: 'The deadly fuse, secret weapon of World War II' (Jane's, London, 1980), p. 266
58 CHURCHILL. W.S.: 'Triumph and tragedy' (Cassell, London, 1954), pp. 47–60
59 WATSON-WATT, R.A.: 'Centimetric waves in RDF', a memorandum, March 1942, AVIA 10/54, PRO

Chapter 23

Epilogue

The history of communications exemplifies Polybius's statement that 'Opportunity is of great advantage in all things'. But though it is necessary for progress it is not sufficient in itself. Additional factors – inventiveness, conviction, commercial acumen, far-sightedness, fortitude and some luck – are required by those persons who seek to change society. Robert Hooke, in 1684, had a rudimentary notion for a signalling method, and he had the means, the telescope, to implement a practical system but he failed to advance his ideas. Chappe, showing much motivation, an awareness of the practical advantages of long-range signalling, and a great deal of resolution succeeded. Of course, Chappe had the good fortune that his developed method had an immediate application to military manoeuvres at a time when his country was being assailed by enemy forces.

Of the factors that have played an important part in the rise of electrical communications the following must be stressed: determination, patronage, the acquisition of patents, the formation of partnerships, personality, timeliness, and demand; and of these, determination is possibly the most important. This quality was demonstrated by both Marconi and Baird to a remarkable degree.

Their early investigations exemplified a statement of Benjamin Franklin: 'I have always thought that a man of tolerable abilities may work great changes, and accomplish great affairs among mankind, if he first forms a good plan, and cutting off all amusements or other employments that would divert his attention, makes the execution of that same plan his sole study and business.'

Neither Baird nor Marconi had any highly original suggestions to put forward at the outset of their investigations and both experimenters modelled their schemes on the ideas of others. Marconi's earliest transmitter was still essentially a coil and spark gap as used by Hertz (although the design of the spark-gap had been slightly changed to incorporate an improvement due to Righi), and a curved metal reflector situated behind it to direct the radiation to the receiver. Baird's earliest distant vision apparatuses were based on proposals that had been advanced by Nipkow and others in the 19th and early 20th centuries.

P.P. Eckersley, a former chief engineer of the BBC wrote in 1960:

Baird is to be honoured ... among those who see past immediate technical difficulties to an eventual achievement; Marconi did much the same with radio. Neither Baird nor Marconi were pre-eminently inventors or physicists; they had, however, that flair for picking about on the scrapheap of unrelated discoveries and assembling the bits and pieces to make something work and so revealing possibilities if not finality [1].

Nevertheless, notwithstanding their lack of profound intellect, and their inability to emulate Kelvin, Maxwell, Hertz and Lodge in original thought, Baird and Marconi succeeded where others had failed because they possessed qualities of patience, of concentration and of an overwhelming desire to succeed which enabled them to pursue their objectives with an indefatigable resolve.

Marconi set out on his life's work during the period that Maclaurin has characterised as 'the era of the entrepreneur inventor'. 'Success was dependent not only on the ability of the inventor to make a significant technical advance, but on his capacity to carry through a successful innovation – a rare combination of skills [2].' Entrepreneurial skill was needed to develop the radio industry in addition to inventive talent, scientific research and speculative capital. Marconi had the early advantage that progress in the new medium of wireless telegraphy was not determined by the existing large industrial organisations and research laboratories. But when Baird attempted to establish his low definition television system after 1926 there were several industrial giants, including the Westinghouse Electric and Manufacturing Company, General Electric, the Radio Corporation of America, Bell Telephone Laboratories, and Telefunken, that had well-endowed R&D laboratories and expert and well-experienced staffs, as well as the financial resources, to further television. Baird, unlike Marconi, faced formidable competition. Nevertheless companies were formed and the business strategies of these seemed to owe much to the formation and development of the Marconi companies.

There was another factor – timeliness – in Marconi's favour. When he commenced his work in 1895 a rudimentary demonstration of wireless telegraphy, that pointed the way forward, had already been given by Lodge in 1894. The development of wireless telegraphy, from the date of Hertz's 1886 experimental investigations of electromagnetic waves to 1894, was relatively short compared to the development of television. More than 50 years were to elapse from the date of Willoughby Smith's notification, in 1873, of the photo-conducting properties of selenium, before a crude demonstration of television could be given in 1925. During this period a profusion of patents and ideas from inventors, scientists and engineers from around the world had been published on 'seeing by electricity'. The way ahead was certainly not so obvious as in the case of wireless telegraphy. And even after Baird had given his January 1926 demonstration and indicated how television could be developed, the implementation of his system by others was not found to be particularly easy. The well-equipped British Admiralty Research Laboratory (ARL), at the National

Physical Laboratory, was not able to emulate Baird's achievement although Dr R.T. Beatty of ARL had been given a demonstration by Baird.

The possession of a master patent, such as A.G. Bell's telephone patent, is clearly of central importance in the commercial exploitation of an invention. Marconi, de Forest, Fessenden and Baird, among others, fully appreciated the need for a patent holding, unlike Hertz and (initially) Lodge. Though Hertz had shown that wireless waves have congruent properties to light waves, and Lodge had demonstrated their use for communications, both Hertz and Lodge lacked the far-sightedness necessary for the commercial applications of their ideas. They were academics who were interested, at first, only in extending the physics of electromagnetic wave propagation. This was in keeping with their professional appointments. Commercialisation, and the patenting of methods and apparatus, were not normally considered the province of university staff. Indeed, such processes would have been anathema to many scientists. Michael Faraday, the greatest experimental natural philosopher of the 19th century never applied for a single patent. Essentially, the dissemination of ideas, hypotheses, and experimental results, and the discussions that they provoked were the basis of uninhibited academic freedom and progress. Marconi, de Forest, Fessenden, Baird and others had no reservations about seeking to engage in business activities. They were entrepreneur inventors whose objectives were the commercial exploitation of their ideas. The telegraph, telephone, wireless and television pioneers all had the same policy with regard to the protection of their methods and apparatuses; they patented everything they could devise: for who can 'look into the seeds of time, and say which grain will grow and which will not [3]?' For them, knowledge for its own sake and academic success were of no interest, but knowledge that led to patents and financial success were the foundation of their being.

Branley's coherer patent provides an illustration of Marconi's modus operandi. Although Lodge was the first to conceive of using it as a wireless detector, he did not feel that he had made an important invention and consequently did not apply for a patent. Marconi made some improvements to the device and obtained a number of basic patents for its utilisation in wireless communications. Again his development work on the magnetic detector was based on a discovery by Lord Rutherford.

The principal patents in Marconi's name were on improvements relating to vertical antennas, coherers, magnetic detectors and methods of selective tuning. His British patent (no. 7777, filed in 1900) on selective resonance became one of the most famous in wireless history. The pioneer work on selective resonance had been undertaken by Oliver Lodge, and a patent had been issued in his name on his method of tuning, but when Marconi realised the importance of tuning in wireless communications he applied himself immediately to make Lodge's method more practical. Several suits were initiated by Marconi against his rivals, and were successfully concluded, but in 1911 the prior Lodge patent was upheld against Marconi.

The importance of patents in the field of communications cannot be overstated. Two further examples, involving Farnsworth and RCA, and Chevallier and RCA, will suffice to establish this point.

Philo Farnsworth, a 'lone inventor' of the 'Baird' type worked for many years from 1926 on an all-electronic system of television. His electron camera was based on an evacuated glass vessel, containing a photosensitive cathode and an anode, which he called an image dissector tube. R&D work on this and related cathode ray/photo-emissive devices resulted in 73 patents and 60 applications by 1938, of which Farnsworth contributed c. 75 per cent. Although his camera tube could not compete effectively with Zworykin's iconoscope tube nevertheless Farnsworth's patents covered a number of basic features of electron optics and were of some importance and worth. In particular Farnsworth's patent of 7th January 1927 and Zworykin's patent of 29th December 1923 led to an interference case being opened on 28th May 1932. Much legal wrangling between RCA and Farnsworth ensued but eventually RCA was persuaded that a settlement with Farnsworth was necessary. Negotiations led to a contract with Farnsworth Television dated 15th September 1939 according to which RCA was granted a licence at a cost of $1,000,000 over a period of ten years.

Again, RCA was obliged to enter into negotiations with P.E.L. Chevallier and F. Holweck in respect of their hard vacuum cathode ray tube. This had been seen by Zworykin when he had visited, late in 1928, the Laboratoire des Etablissements Edouard Belin, Paris. Zworykin immediately realised that their tube was superior to the type that he was using and bought a Holwek and Chevallier tube from the Laboratoire des Etablissements Edouard Belin. The result of Zworykin's journey was the purchase by RCA of Chevallier's patent of 1929 and the engineering of an electron gun, with electrostatic focusing, which came to be at the heart of all camera and display tubes built and sold by the RCA and some of its licensees. On this episode in television history, according to A. Abramson, Zworykin's biographer: 'Zworykin's trip to Paris changed the course of television history' [4].

Maclaurin has written: 'Marconi's contributions to the commercialisation of wireless made him more important as an innovator than as an inventor. But he succeeded in getting possession of many of the principal patents in the radio art, despite the fact that the most important wireless discoveries and inventions were not made by him or his associates' [5].

The directors of the various companies associated with Baird did not follow the patent policy of the directors of the Marconi companies and RCA and made an error of judgment in not seeking out and purchasing potentially worthwhile television patents. Too much reliance was placed on Baird's ability to further single-handedly the art of television. Hence, when mechanical systems of scanning gave way to electronic methods of scanning in the mid-1930s Baird Television Ltd was at a disadvantage and the patents of the company (which by the end of 1930 comprised the 88 patents of Baird and four from other members of the company) became almost valueless.

Practical/financial support and encouragement are necessary factors for an inventor's success. Apart from engaging in his life's work at an opportune time, the unknown 21-year old Marconi had the good fortune to receive patronage from W. Preece, the Engineer-in-Chief of the British Post Office, and the Post Office. And when Marconi's demonstrations before serving officers of the Army and the Royal Navy were successful his future prospects seemed assured. Marconi's rudimentary scheme of wireless telegraphy had an almost immediate application to signaling and to safety at sea. Baird was not so fortunate. He did not obtain any early support from the one body, the British Broadcasting Corporation, that could assist him and his television system had no direct relevance to safety.

The value of a partnership, comprising an outstanding inventor and a commercial entrepreneur, may be illustrated by several examples.

Cooke was inspired by a demonstration of a telegraph modelled on Schilling's efforts, but found he lacked the required knowledge to perfect his notions. When he approached Wheatstone, a noted experimentalist, success followed. The association of Cooke and Wheatstone led to a partnership, where each respected the other's skills to the advantage of their common interest. Following the initial meeting of Cooke and Wheatstone in 1837 the partnership flourished. Later Cooke wrote: 'The invention at once became a subject of public interest and I found that Mr Wheatstone was talking about it everywhere in the first person singular. . . . At length, in 1840, I required that our positions relatively to the invention and to each other should be ascertained by arbitration' [6].

The arbitrators were Sir Marc Brunel and Professor J.F. Daniel. In their judgment they said: 'it is to the united labours of the two gentlemen so well qualified for mutual assistance, that we must attribute the rapid progress which this important invention has made during the five years since they have been associated' [7].

Bain's failure to form a close relationship with a dynamic business associate was possibly the prime factor that led to his lack of success in later life. Schaffner's view that Bain 'was not a commercial man but his inventive powers were most wonderful' would seem to be correct [8].

Several examples of successful partnerships of the Cooke and Wheatstone type can be cited: for example Baird and Moseley, Farnsworth and Evanson, and Marconi and Isaacs. Baird said of Moseley (following their first meeting):

[He would] play a very prominent part in our future activities. Both Hutchinson's [O.G. Hutchinson was Baird's business partner] and my own knowledge of high finance were infantile and this was also true of our knowledge of journalism. We were both much upset because a certain wireless paper had seen fit to publish unfavourable criticisms of us which we considered utterly unfair and prejudiced.

We talked to Moseley about this and he said he would fix it and he did. The next issue of the paper contained an article by Moseley refuting all the previous attacks and hailing the invention as a great achievement. From that day on his position was established. He became one of the family and I welcomed him with open arms. What a relief to have a man about the house who took a real interest in our affairs [9].

Moseley made it his immediate mission in life to put Baird television on the map. 'Realising that television was struggling against odds, ignorance, scepticism and hostility', he has written, 'I decided to take up the cudgels on its behalf and determined to carry the fight through to the finish' [10]. He became Baird's self-appointed champion. He began to attend meetings of the Board, where he was a great help to Baird who was not at his best in the boardroom. Baird found the meetings of the various boards were long rigmaroles to be slept through. He had no patience with the reading of minutes and the propositions common to business, but Moseley was not the sort of man to fall asleep at board meetings or anywhere else where money was concerned.

With his knowledge of finance, journalism and publicity and his contacts in Fleet Street, Parliament and elsewhere, Moseley brought much valuable expertise and experience to bear on the side of the television pioneer.

Marconi was one of the first to foresee the commercial possibilities arising from the work of himself and the early radio pioneers. He worked energetically throughout his life to progress and extend wireless communications. His primary interest was in the technical development of the subject and his knowledge of business methods was such that, when he managed both the commercial and innovative aspects of his company, considerable difficulties arose. But when Godfrey Isaacs, a born businessman, was appointed managing director, thereby allowing Marconi more time to further his research and development work, the company prospered.

An agreeable personality and an ability to work with others is a desirable attribute for any successful partnership. Both Baird and Marconi had well-balanced personalities unlike, for example, Fessenden. After Marconi's death, Sir Ambrose Fleming, a friend of Marconi and Baird, wrote an appreciation of the great radio pioneer, which could equally well have applied to Baird [11].

In the first place, he was eminently utilitarian. His predominant interest was not in purely scientific knowledge per se, but in its practical application for useful purposes. He had a very keen appreciation of the subjects on which it was worthwhile to expend labour in the above respect.

He had enormous perseverance and powers of work. He was not discouraged by initial failures or adverse criticisms of his work. He had great powers of influencing others to assist him in the ends he had in view. He had remarkable gifts of invention and ready insight into the causes of failure and the means of remedy. He was also of equable temperament and never seemed to give way to impatience or anger. . . . He also owed a good deal to the loyal and efficient work of those who assisted him.

The recruitment of well-qualified staff, and the use of outside consultants, can greatly advance a new enterprise. Marconi fully appreciated the benefits that would accrue to the well-being of his companies from their expertise and experience. So far as is known none of the early Baird companies employed consultants and the staff of the companies lacked the formal academic research training characteristic of AT&T's, RCA's and EMI's R&D engineers. Possibly this difference stemmed from the different education backgrounds of the two inventors. Baird had followed a four year engineering course and engineering apprenticeship and had been awarded an Associateship of the Royal Technical

College, Glasgow. He was not an amateur but was probably as well qualified in engineering as the majority of the engineers who worked in industry in the 1920s. Marconi did not matriculate, did not follow an apprenticeship or a formal, structured course in engineering, and did not obtain a recognised qualification in this discipline; essentially he was an amateur inventor. He may have felt after his initial experiments that he lacked certain engineering skills and knowledge and that consequently staff possessing the requisite expertise and experience should be appointed. Marconi surrounded himself from an early stage with a group of very able engineers and technicians including Dr W.H. Eccles, Dr Erskine Murray, W.W. Bradfield, A. Gray, C.S. Franklin, H.J. Round and others By 1900 there were 17 professional engineers in the Marconi company in the UK. This was not the position, prior to c. 1932, in the Baird companies. Baird was the 'kingpin' of BTDC and BIT: the work that was undertaken in their laboratories was almost entirely based on his ideas and if anything happened to him that interrupted the flow of these ideas there was a possibility the companies would collapse. He had to be insured – for £150,000.

The need for far-sightedness or awareness of the likely development of a trend is another desirable quality for inventors. Ives, who was in charge of television development at Bell Telephone Laboratories, in 1929, could see no future for the use of cathode ray tubes in television receivers. 'This method of reception', wrote Ives, 'is old in the art and of very little promise. The images are quite small and faint and all the talk about this development promising display of television to large audiences is quite wild' [12].

Both Marconi and Baird initially had a 'blind spot'. Marconi considered the Morse code to be quite suitable for ship signalling and for transoceanic communications, and he saw no real need for wireless telephony. Like Baird, Marconi's approach to his work was pragmatic. He was not interested in investigating scientific fields which did not appear to have an immediate commercial viability. This blind spot was a temporary disadvantage for the Marconi company and the furtherance of communications. Luckily for Marconi some of the important early work on wirelesss telephony was undertaken by two of his rivals, de Forest and Fessenden, and neither inventor had access to financial or engineering resources or skills comparable to those of the Marconi companies. Baird's blind spot was his persistence – certainly until 1931 – in believing that low definition television, using the medium wave-band and mechanical methods of scanning, would suffice for popular television broadcasting. This persistence was against the advice he had been given in March 1929 by the Postmaster General that he should press on with experiments on a much lower band. Unfortunately for Baird both RCA and EMI, which had very well resourced R&D departments, were actively working towards an all-electronic, high definition solution to the television problem.

The evolution of high definition television by RCA and EMI demonstrates the excellence of engineering that results when vast resources are brought to bear on a problem.

Established in 1931, Electric and Musical Industries, with Sir Louis Sterling as the Managing Director, immediately began a private venture programme to further television. Shortly afterwards, Shoenberg became the Director of Research. His first task was to strengthen the research team. By June 1934 the research department comprised 114 persons, including 32 university graduates, nine of whom had PhDs, and ten of whom had been recruited from Oxford and Cambridge universities. In just over three years Shoenberg was able to offer to the Television Advisory Committee a 405-line, all-electronic, television system. It was an outstanding achievement – particularly during a period of depressed business conditions – and was made possible because 'the shrewd business man and financier, Sterling, completely trusted Shoenberg, an engineer, scientist or applied physicist with a large dash of the visionary' [13].

The achievement was remarkable from an engineering point of view. Prior to the commencement of 1930 radio communication/electronic techniques were primitive by modern standards. Bandwidths of equipment were of the order of several kHz, but, for high definition television, amplifiers, scanning circuits, pulse generators and pulse shaping circuits, modulators, power supplies, and radio transmitters had to be developed, ab initio, to handle signals having bandwidths of up to c. 3 MHz. In addition, the physics of phosphors, of secondary emission, and of photo-electric materials had to be researched; the electron optics of camera tubes and display tubes had to be formulated; and high vacuum and thin-film deposition techniques had to be evolved.

All of this engineering effort – which cost EMI c. £100,000 per year – was quite beyond the resources that a company set up to further the inventions of an individual innovator or inventor could bring to bear. And so, from the 1930s the role of the lone inventor, of the Marconi, de Forest, Fessenden, Farnsworth, and Baird types, began to decline. The large electrical manufacturing companies – GE, RCA, AT&T, Westinghouse, EMI, MWT, GEC, Telefunken, inter alia – with their large research departments, and the profits from the sales of their manufactured goods to sustain them, were the prime innovators in the communications field.

The evolution of high definition television in the early 1930s had an interesting sequel which highlighted the importance of 'technology transfer'. Following the appointment of Hitler as Chancellor of the Third Reich in January 1933 and the re-formation of the German Luftwaffe in March 1935, the need for an early warning radar system in the United Kingdom became urgent. Adventitiously, all the engineering advances that had led to high definition television could be utilised in the engineering of the UK's Chain Home surveillance radar: thus there was a technology transfer from television to radar. Soon, though, the requirement for radars operating at much higher frequencies became manifest and, in the UK, efforts were made to develop valves and techniques able to work at metric and centimetric wavelengths. When the resonant cavity magnetron valve was engineered, in 1940, it enabled microwave techniques to be devised, in the UK and the USA, which led to many centimetric radars. After the war these techniques were applied to communications engineering: the technology transfer was now from radar to UHF television and microwave communications generally, including satellite communications.

It seems salutary to end this history of communications with an illustration of the enormous advance that has taken place in radio communications. In 1901 Marconi endeavoured to transmit the Morse code signal for the letter 'S' from Poldhu, Cornwall to St Johns, Newfoundland, a distance of c. 2,500 miles (3,200 km). He used a transmitter having an output of c. 1800 W – Fleming estimated it as 10 to 12 kW – a sending antenna that consisted of 50 bare stranded copper wires suspended from a triatic stay, strained between two masts 160 ft (48.7 m) in height and 167 ft (61 m) apart, and a receiving kite antenna flown at 400 ft (122 m). Marconi recorded that he and Kemp heard the three dot signals of the Morse code signal, though the reception was greatly marred by atmospherics. Marconi's feat was not independently confirmed. Several decades later, but within the lifetime of a given individual, television viewers around the world were able to observe colour images of the outer planets radiated from an inter-planetary probe. The transmission distance was more than a million times longer than the Poldhu–St Johns propagation path. The power radiated was just 22 W.

References

1 MOSELEY, S.A.: 'John Baird' (Odhams, London, 1952), pp. 250–251
2 MACLAURIN, W.R.: 'Invention and innovation in the radio industry' (Macmillan, New York, 1949)
3 SHAKESPEARE, W.: 'Macbeth', 1, 3, 58
4 ABRAMSON, A.: 'Zworykin, pioneer of television' (University of Illinois, Chicago, 1995), p. 236
5 Ref. 2, p. 43
6 BOWERS, B.: 'Sir Charles Wheatstone FRS 1802–1875' (The IEE, London, 2001), p. 146
7 Ref. 6, p. 147
8 SCHAFFNER, T.P.: 'The telegraph manual: a complete history and description of the semaphoric, electric and magnetic telegraphs of Europe, Asia, Africa and America, ancient and modern' (Pudney and Russell, New York, 1859)
9 BURNS, R.W.: 'John Logie Baird, television pioneer' (The IEE, London, 2000)
10 MOSELEY, S.A.: 'Broadcasting in my time' (Rich and Cowan, 1935)
11 FLEMING, J.A.: Obituary, *Journal of the Royal Society of Arts*, 1837, **26**, pp. 57–62
12 IVES, H.E.: memorandum to H.P. Charlesworth, 16th December 1929, Case File 33089, 1, AT&T Archives
13 MCGEE, J.D.: '1971 Shoenberg Memorial Lecture', *The Royal Television Society Journal*, 1971, **13**, No. 9, May/June

Appendices

Appendix 1 (Ref.: BAGLEHOLE, K.C.: 'A century of service' (Cable & Wireless Ltd., 1969)

The formation of The Eastern and Associated Telegraph Companies

Company	Year	Action	Result	Year	
The Anglo-Mediterranean Tel. Co.	(1868)	Merged to form	The Eastern Telegraph Co.	1872	
The Falmouth, Gibraltar and Malta Tel. Co.	(1869)				
The British Indian Submarine Tel. Co.	(1869)				
The Marseilles, Algiers and Malta Tel. Co.	(1870)				
The British Indian Extension Co.	(1869)	Merged to form	The Eastern Extension, Australasia and China Tel. Co.	1873	
The China Submarine Tel. Co.	(1869)				
British Australian Tel. Co.	(1870)				
The Brazilian Submarine Tel. Co.	(1873)	The Brazilian Sub. Tel. Co. then	The Western Tel. Co.	1899	**The Eastern and Associated Tel. Cos.**
The Western & Brazilian Tel. Co.	(1873)				
Companhia Telegrafica Platino-Brazileira	(1872)	Taken over and became	The London-Platino Brazilian Tel. Co.	1878	
			The River Plate Tel. Co.	1865	
			The West Coast of America Tel. Co.	1877	
			The Eastern & South African Tel. Co.	1879	
			The African Direct Tel. Co.	1885	
			The West African Tel. Co.	1885	
			The Pacific & European Tel. Co.	1892	
			The Europe & Azores Tel. Co.	1893	
			Société Anonyme Belge de Cables Télégraphiques	1914	

Appendix 2 (Ref.: BAGLEHOLE, K.C.: 'A century of service' (Cable & Wireless Ltd., 1969)

The principal primary cables laid, for the period 1868–1928, by the companies that merged into Cable and Wireless Ltd

Date	Cable	Cable owner	Cable ship and owner
1868	Florida/Cuba	Cuba Submarine Tel. Co.	*Dacia*, India Rubber Gutta Percha & Telegraph Works Co.
	Malta/Alexandria Co.	Anglo-Med. Tel. Co.	*Scanderia*, TCM & *Chiltern*, TCM
1870	Suez/Aden/Bombay	Britsh Indian Sub. Tel. Co.	*Gr. Eastern*, Chartered by TCM, *Hibernia*, TCM & *Chiltern*, TCM
	PK/Carcavellos	Falmouth Gibraltar and Malta Tel. Co.	*Hibernia*, TCM
	Carcavellos/Gibraltar	Falmouth Gibraltar and Malta Tel. Co.	*Scanderia*, TCM
	Gibraltar/Malta	Falmouth Gibraltar and Malta Tel. Co.	*Scanderia*, TCM & *Edinburgh*, TCM
	Madras/Penang/Singapore	British Indian Extension Co.	*Scanderia*, TCM & *William Cory*, Chartered by TCM
	Singapore/Batavia	British Australian Tel. Co.	*Hibernia*, TCM
	Cuba/Jamaica & Puerto Rico/St. Thomas	West India and Panama Tel. Co.	*Dacia*, India Rubber Gutta Percha & Telegraph Works Co.
1871	St. Thomas/St. Kitts/Antigua/Martinique/St. Lucia/Barbados	West India and Panama Tel. Co.	*Dacia*, India Rubber Gutta Percha & Telegraph Works Co.
	St. Vincent/Grenada/Trinidad/Georgetown B.G.	West India and Panama Tel. Co.	*Dacia*, India Rubber Gutta Percha & Telegraph Works Co.
	Singapore/Saigon/Hong Kong	China Submarine Tel. Co.	*Agnes*, (?) & *Belgian*, (?) *Kangaroo*, TCM & *Minia*
	Java/Port Darwi	British Australia Tel. Co.	*Edinburgh*, TCM & *Hibernia*, TCM
1872	Puerto Rico/Jamaica	West India and Panama Tel. Co.	?
1873	Jamaica/Colon (Panama)	West India and Panama Tel. Co.	?
	Recife/Fortaleza/Sao Luiz/Belem	Western & Brazilian Co.	*Hooper*, (?)
	Recife/Bahia/Rio de Janeiro	Western & Brazilian Co.	*Hoope*, (?)
	Carcavellos/Madeira	Brazilian Submarine Co.	*Seine*, TCM

Appendix 2 — *continued*

Date	Cable	Cable owner	Cable ship and owner
1890	Zanzibar/Mombasa/Dar-es-Salaam	Eastern & South African Tel. Co.	*Recorder*, E. Ext. Australasia & China Tel. Co.
1891	Penang/Medan	E. Ext. Australasia & China Tel. Co.	*Recorde*, E. Ext. Australasia & China Tel. Co.
1893	Zanzibar/Seychelles/Mauritius	Eastern & South African Tel. Co.	*Scotia*, TCM
	Carcavellos/San Miguel/Fayal	Europe & Azores Tel. Co.	*Seine*, TCM
1894	Singapore/Labuan/Hong Kong	E. Ext. Australasia & China Tel. Co.	*Scotia*, TCM
1897	Jamaica/Turks Island/Bermuda	Direct West India Co.	*Scotia*, TCM
1899	Capetown/St. Helena	Eastern Tel. Co.	*Anglia*, TCM
	St. Helena/Ascension	Eastern Tel. Co.	*Seine*, TCM
1900	Belem/Recife/Rio de Janeiro/Montevideo	Western Tel. Co.	*Scotia*, TCM
	Ascension/St. Vincent	Eastern Tel. Co.	*Anglia*, TCM
1901	PK/Madeira/St. Vincent	Eastern Tel. Co.	*Anglia*, TCM & *Britannia*, TCM
	Mauritius/Rodrigues/Cocos/Freemantle	E. Ext. Australasia & China Tel. Co.	*Anglia*, TCM
	Durban/Mauritius	Eastern Tel. Co.	*Anglia*, TCM
	Ascension/Sierra Leone	Eastern Tel. Co.	*Anglia*, TCM
1902	Canada/Norfolk Island/Australia	Pacific Cable Board	*Colonia*, TCM
	Norfolk Island/New Zealand	Pacific Cable Board	*Anglia*, TCM
	Freemantle/Glenelg	Pacific Cable Board	*Scotia*, TCM
1906	St. Vincent/Fayal	Western Tel. Co.	*Colonia*, TCM
1910	Buenos Aires/Ascension	Western Tel. Co.	*Colonia*, TCM & *Cambria*, (?)
1913	Aden/Colombo/Penang	E. Ext. Australasia & China Tel. Co.	*Colonia*, TCM
1914	Singapore/Hong Kong	E. Ext. Australasia & China Tel. Co.	*Colonia*, TCM
1919	Rio de Janeiro/Ascension	Western Tel. Co.	*Colonia*, TCM
1920	Maranham/Barbados	Western Tel. Co.	*Stephan*, TCM
1922	Aden/Seychelles/Colombo	Eastern Tel. Co.	*Colonia*, TCM
1923	Suva/Auckland	Pacific Cable Board	*Stephan*, TCM

Appendix 2 – *continued*

Date	Cable	Cable owner	Cable ship and owner
1874	Madeira/St. Vincent/Recife Rio de Janeiro/Santos/	Brazilian Submarine Co.	*Hibernia*, TCM & *Edinburgh*, TCM
	Florianoplis/Rio Grande/Chuy	London Platino-Brazilian Cable Co.	*Ambassador*, (?)
1875	Rio Grande/Chuy/Maldonado/ Montevideo	London Platino-Brazilian Cable Co.	?
1875/ 1876	Lima/Mollendo/Arica/ Iquique/ Antofogasta/Serena/Valparaiso	West Coast of America Tel. Co.	*Dacia*, India Rubber Gutta Percha & Telegraph Works Co.
1876	Sydney/Nelson N.Z.	E. Ext. Australasia & China Tel. Co.	*Edinburgh*, TCM & *Hibernia*, TCM
1877	Rangoon/Penang	E. Ext. Australasia & China Tel. Co.	*Hibernia*, TCM & *Kangaroo*, TCM
1879	Penang/Malacca-Singapore/Java	E. Ext. Australasia & China Tel. Co.	*Scotia*, TCM & *Edinburgh*, TCM
	Durban/Delagoa Bay	Eastern & South African Tel. Co.	*Kangaroo*, TCM
	Delagoa/Mozambique	Eastern & South African Tel. Co.	*Seine*, TCM
	Mozambique/Zanzibar/Aden	Eastern & South African Tel. Co.	*Calabria*, TCM
1880	Hong Kong/Manila	E. Ext. Australasia & China Tel. Co.	*Calabria*, TCM
1883	Foojow/Shanghai	E. Ext. Australasia & China Tel. Co.	*Sherard Osborn* & *Scotia*, TCM
1884	Hong Kong/Foochow – Hong Kong/Macau	E. Ext. Australasia & China Tel. Co.	*Sherard Osborn*, E. Ext. Australasia & China Tel. Co.
1886	Bathurst/Sierra Leone/Accra/ Lagos/Bonny	African Direct Tel. Co.	*Britannia*, TCM
	Dakar/Bathurst/Conakry/Sierra Leone	West African Tel. Co.	*Silvertown, Dacia*, India Rubber Gutta Percha & Telegraph Works Co. & *Buccaneer*
	Grand Bassam/Accra/Sao Thome/Principe/Loanda	West African Tel. Co.	*Silvertown, Dacia*, India Rubber Gutta Percha & Telegraph Works Co. & *Buccaneer*
1889	Capetown/Port Nolloth/ Mossamedes	Eastern & South African Tel. Co.	*Scotia*, TCM
	Mossamedes/Benguela/Loanda	Eastern & South African Tel. Co.	*Scotia*, TCM
	Halifax/Bermuda	Halifax & Bermudas Tel. Co.	*Westmeath*, (?)
	Banjoewangie/Roebuck Bay	E. Ext. Australasia & China Tel. Co.	*Seine*, TCM

Appendix 3

Analysis of American and foreign television systems (D.G. FINK, 'Television standards and practice' (McGraw Hill, New York, 1943)

Number	Designation	Period of operation	When demon-strated	Scanning pattern			Synchronisation system (H = horizontal, V = vertical)	
				Lines frames fields	As-pect ratio	Type of motion	Per cent of carrier	Description of waveform
1	RMA standards	1939–1940	1939–1940	441/ 30–60	4:3	linear	20–25	H: single rectangular pulse; V: serrated with equalising pulses (see RMA standard M9–211
2	DuMont A (500 kc burst vertical synchronising pulse)	1939–1940	December, 1939 to date	variable (see note A)	4:3	linear	20–25	H: rectangular pulse; V: r–f burst during V blank with H superposed
3	DuMont B (Transmission of scanning waveforms)	1938–1939	1939	variable	4:3	any (note B)		synchronising inherent in transmission of scanning waveforms
4	Hazeltine (FM for synchronising)	1940	November, 1940	441/ 30–60	4:3	linear	100	any synchronising wave shape frequency modulated during H and V blank
5	RCA (507 lines, 495 lines later suggested)	1940	July, 1940	507/ 30–60 495/ 30–60	4:3	linear	20–25	waveform same as RMA
6	RCA (FM for sound)	1940	November, 1940	441/ 30–60	4:3	linear	20–25	RMA M9–211
7	RCA (FM sound quasi-FM for picture)	1940	October, 1940	441/ 30–60	4:3	linear		RMA M9–211
8	RCA (long integration synchronising pulse)	1940	October, 1940	441/ 30–60	4:3	linear	20–25	slots in vertical pulse at line frequency; V pulse = 9 H
9	Philco (525 lines)	1938–1940	1938–1940	525/ 30–60	4:3	linear	over 25	waveform same as RMA M9–211
10	Philco (605 lines)	1938–1940)	1938–1940	605/ 24–48	4:3	linear	over 25	waveform same as RMA M9–211
11	Philco (narrow vertical synchronising pulse)	1935–1938	up to 1938	441/ 30–60	4:3	linear	20–25	H: rectangular pulses: V: single narrow rectangular pulse, same level

	Transmitter characteristics						Other characteristics, apparatus employed, advantages claimed by sponsor, etc
Polarity of modulation	*Carrier attenuation*	*D.C. or a.c. transmission*	*Direction of polarisation*	*Total bandwidth*	*Picture sideband width*	*Carrier separation*	
negative	none	d.c.	horizontal	6 Mc	4–4.5 Mc	4.5 Mc	sound pre-emphasised sound and picture carriers of equal power; standards used sometime by RCA, GE. Farnsworth, Philco, CBS, Don Lee, General Television Corporation, Zenith and others
negative	none	d.c.	not specified	6 Mc	4–4.5 Mc	4.5 Mc	note A: designed for flexibility (continuous variability) in line and frame rates, including 15–30 as lower limit
negative (note B)	none (note B)	d.c. (note B)	not specified	6 Mc	4–4.5 Mc	4.5 Mc	note B: receiver automatically follows changes in scanning motion
negative	none	d.c.	not specified	6 Mc	4–4.5 Mc	4.5 Mc	FM modulator required for synchronising: high syncronising amplitude developed; improves picture modulation capability
negative	none	d.c.	horizontal	6 Mc	4–4.5 Mc	4.5 Mc	
negative	none	d.c.	horizontal	6 Mc	4–4.5 Mc	4.5 Mc	FM modulator for sound only; 75 kc maximum deviation
	none	d.c.	horizontal	6 Mc	4–4.5 Mc	4.5 Mc	7 kc maximum deviation for sound; 0.75 Mc maximum deviation for picture; improved transient response reported
negative	none	d.c.	horizontal	6 Mc	4–4.5 Mc	4.5 Mc	V pulse integrated in R–C circuit of long time constant: no equalising pulses used
negative	none	d.c.	vertical	6 Mc	4–4.5 Mc	4.5 Mc	sound carriers staggered
negative	none	d.c.	vertical	6 Mc	4–4.5 Mc	4.5 Mc	greater horizontal definition due to lower frame rate
negative	none	d.c.	vertical	6 Mc	4–4.5 Mc	4.5 Mc	

Appendix 3 – *continued*

Number	Designation	Period of operation	When demonstrated	Lines frames fields	Aspect ratio	Type of motion	Per cent of carrier	Description of waveform
				Scanning pattern			Synchronisation system (H = horizontal, V = vertical)	
12	Farnsworth (narrow vertical synchronising pulse)	1935–1938	up to 1938	441/ 30–60	4:3	linear	20–25	H: rectangular pulse: V: single narrow pulse of higher amplitude
13	GE (picture carrier 6db attenuated	1938	1938	441/ 30–60	4:3	linear	20–25	same as RMA M9–211
14	Kolorama (225 lines)	1935–1940	May, 1939	225/ 12–24	6:5	linear		by transmission pulse and power line
15	Sound during blanking single carrier system	prior 1940		note C	note C	note C	note C	note C: carrier frequency modulated during horizontal blanking interval with audio frequencies
16	CBS (3-colour system no. 3)	1940	August 1940	343/ 60–120	4:3	linear	note D	**American colour** adaptable to any synchronising signal that can control filter disks
16a	CBS combination colour and black and white	this system same as CBS no. 3, for colour transmissions: For black and white						
17	CBS (3-colour system no. 4)	1940		430/ 45–180	4:3	linear	note D	note D; quadruple interface
18	CBS (3-colour system no. 5)	1940		550/ 30–120	4:3	linear	note D	note D; quadruple interface
19	GE (2 colour system)	1940	November 1940	441/ 30–60	4:3	linear	20–25	RMA M9–211
20	British (BBC) standard	1936–1939	1936	405/ 25–50	5:4	linear	30	**Foreign** rectangular H pulses: serrated V pulse; no equalising pulses
21	Scophony (British) (in US also, in 1940)		1937	405/ 25–50	5:4	linear	30	BBC standard
22	early Baird (British)	1936–1937	1936	240/ 25	4:3	linear	40	8 per cent line synchronising pulses, single V pulse of 12 lines duration

Appendix 3 – *continued*

	Transmitter characteristics						Other characteristics, apparatus advantages claimed by sponsor, etc
Polarity of modulation	Carrier attenuation	D.C. or a.c. transmission	Direction of polarisation	Total bandwidth	Picture sideband width	Carrier separation	
negative	none	d.c.	not specified	6 Mc	4–4.5 Mc	4.5 Mc	
negative	6 db	d.c.	horizontal	6 Mc	4–4.5 Mc	4.5 Mc	narrower band at transmitter, wider band at receiver
positive	none	d.c.	vertical	100 kc/ 300 kc	300 kc		mechanical scanners. Transmitter on band at 2000 kc
negative	note C	d.c.	note C	6 Mc or less	4–4.5 Mc		note C: not specified audio frequencies substantially higher than one half line frequency suffer distortion; investigated by RCA, GE, Philico: suggested 1940 by Kaliman
television systems							
note D	note D	note D	note D	6 Mc	4–4.5 Mc	4.5 Mc	note D: not specified; mechanical filter disks or drums at transmitter and receiver
transmissions, no preference as to specifications is indicated							
note D	note D	note D	note D	6 Mc	4–4.5 Mc	4.5 Mc	line scan frequency approximately 19 400 p.p.s
note D	note D	note D	note D	6 Mc	4–4.5 Mc	4.5 Mc	line scan frequency 15 750 p.p.s; Colour coincidences 10 p.s
negative	none	d.c.	horizontal	6 Mc	4–4.5 Mc	4.5 Mc	dichromatic filter disks, synchronising on power line: odd lines always one colour, even lines always other colour
systems							
positive	none	d.c.	vertical	6 Mc	2.5 Mc	3.5 Mc	double sideband; 2cps/sec tolerance in line frequency for mechanical receivers
positive	none	d.c.	vertical	6 Mc	2.5 Mc	3.5 Mc	mechanical scanners, supersonic light valve; requires BBC tolerance on line synchronising
positive	none	d.c.	vertical	5 Mc	1.5 Mc	3.5 Mc	mechanical scanner at transmitter; live pick up by film only

Appendix 3 – *continued*

Number	Designation	Period of operation	When demon-strated	Scanning pattern			Synchronisation system (H = horizontal, V = vertical)	
				Lines frames fields	As-pect ratio	Type of motion	Per cent of carrier	Description of waveform
23	Baird 2-colour (British)		March 1938	120/ 16.6– 100	3:4	linear vertical	40	6 to 1 interlace by nonperiodic pulses
24	Baird 3-colour (British)		January 1938	120/ 16.6– 100	3:4	linear vertical	40	same as Baird 2-colour system
25	velocity modulation (Br.)		May 1934	60– 400/25	4:3		30	10 per cent line pulses; frame by ratchet circuit from line
26	French (PTT) standards		June 1937	440– 445/ 25–50	5:4	linear	30	
27	Barthelemy (French)		June 1937	450/ 25–50	5:4	linear	17 H 34 V	6 lines paired to form V pulse
28	German standards		1939–1940	441/ 25–50	5:4	linear	30	10 per cent line pulses; single 35 per cent V
29	German proposal		1937–1938	441/ 25–50	5:4	linear	30	burst of 1.1 Mc for 20 per cent line duration as V pulse; square H
30	Italian standards		1939	441/ 21–42	5:4	linear	30	same as German standards
31	Russian standards		1940	441/ 25–50	11.8	linear	20–25	essentially RMA M9–211

Appendix 3 – *continued*

	Transmitter characteristics						Other characteristics, apparatus employed, advantages claimed by sponsor, etc
Polarity of modulation	Carrier attenuation	D.C. or a.c. transmission	Direction of polarisation	Total bandwidth	Picture sideband width	Carrier separation	
positive	none	d.c.	vertical	5 Mc	2 Mc	3 Mc	flying spot pick up with rotating colour disks; projected large-screen image
positive	none	d.c.	vertical	5 Mc	2 Mc	3 Mc	same but three colours laboratory demonstration
positive	none	a.c.	vertical	3 Mc	1.5 Mc	no sound	c.r.t. light source on film; amplitude component added
positive		d.c.				4 Mc	standards embrace four systems
positive						4 Mc	Barthélemy interlace; rotating disk synchronising
positive	none	d.c.	vertical	5 Mc	2.0 Mc	2.8 Mc	picture carrier on high side of channel
positive	none	d.c.	vertical	5 Mc	2 Mc	2.8 Mc	
positive	none	d.c.	vertical	5 Mc	2 Mc	2.8 Mc	
negative	none	d.c.	horizontal	6 Mc	4–4.5 Mc	4.5 Mc	sound carrier lower in frequency than picture

Bibliography

ABRAMSON, A.: 'The history of television, 1880 to 1941' (McFarland, Jefferson, 1987)

ABRAMSON, A.: 'Zworykin, pioneer of television' (University of Illinois Press, Chicago, 1995)

AITKEN, H.G.J.: 'Syntony and spark – the origins of radio' (John Wiley, New York, 1976)

APPLEYARD, A.: 'The history of the Institution of Electrical Engineers' (IEE, London, 1939)

APPLEYARD, R.: 'Pioneers of electrical communication' (MacMillan, London, 1930)

ARCHER, G.L.: 'History of radio to 1926' (American Historical Society, New York, 1938)

ARCHER, G.L.: 'Big business and radio' (American Historical Society, New York, 1939)

BAGLEHOLE, K.C.: 'A century of service' (Anchor Press, Essex, 1969)

BAILEY, C.: 'The legacy of Rome' (Oxford University Press, 1924)

BAKER, T.T.: 'Wireless pictures and television' (Constable, London, 1926)

BAKER, T.T: 'The telegraphic transmission of photographs' (Constable, London, 1910)

BAKER, E.C.: 'Sir William Preece, FRS. Victorian engineer extraordinary' (Hutchinson, London, 1976)

BAKER, W.J.: 'A history of the Marconi Company' (Methuen, London, 1970)

BARNOUW, E.: 'A tower in Babel' (Oxford University Press, 1966)

BARNOUW, E.: 'The golden web' (Oxford University Press, 1968)

BARTY-KING, H.: 'Girdle around the earth' (Heinemann, London.1979)

BEAUCHAMP, K.G.: 'History of telegraphy' (IEE, London, 2001)

BELLOC, A.: 'La télégraphie historique' (Paris, 1894)

BLAKE, G.G.: 'History of radio telegraphy and telephony' (Radio Press, London, 1926)

BORDEAU, S.P.: 'Volts to Hertz, the rise of electricity' (Burgess Publishing, Minneapolis, 1982)

BOWERS, B.: 'Sir Charles Wheatstone FRS, 1802–1875' (IEE, London, 2001)

BOYER, J.M.J.: 'La transmission telegraphiques des images et des photographies' (Paris, 1864)

BOYNTON, L.: 'The Elizabethan Militia' (Routledge and Kegan Paul, London, 1967)

BRIGGS, A.: 'The golden age of wireless' (Oxford University Press, London, 1965)

BRIDGEWATER, T.H.: 'A A Campbell Swinton' (Royal Television Society, London, 1982)

BRIGHT, C.: 'Submarine telegraphs' (London, 1898)

BRUCE, R.V.: 'Bell' (Gallancz, London, 1973)

BRUCH, W.: 'Die Fernseh story' (Franckh'sche Verlagshandlung, Stuttgart, 1969)

BRYANT, J.H.: 'Heinrich Hertz' (IEEE Press, New York, 1988)

BURNS, R.W.: 'British television, the formative years' (IEE, London, 1986)

BURNS, R.W. (Ed.): 'Radar development to 1945' (IEE, London, 1988)

BURNS, R.W.: 'Television, an international history of the formative years' (IEE. London, 1998)

BURNS, R.W.: 'The life and times of A.D. Blumlein' (IEE, London, 2000)

BURNS, R.W.: 'John Logie Baird, television pioneer' (IEE, London, 2001)

CLAPHAM, J.H.: 'An economic history of modern Britain' (Cambridge University Press, 1963)

CLARK, R.W.: 'Tizard' (Methuen, London, 1965)

CLARKE, D.: 'The ingenious Mr Edgeworth' (Oldbourne, London, 1965)

CLAYTON, R., and ALGAR, J.: 'The GEC Research Laboratories 1919–1984' (IEE, London, 1989)

COOKE, W.F.: 'The electric telegraph: was it invented by Professor Wheatstone' (W.H. Smith, London, 1866)

CRAWLEY, W.T.: 'From telegraphy to television', (Warne, London, 1931)

DARMSTAEDTER, E.: 'Feuer – telegraphie im Alterturm' (Umschau, 1924)

DE FOREST, L.: 'Father of radio' (Chicago, 1950)

DE FOREST, L.: 'Television, today and tomorrow' (Dial Press, New York, 1942)

DOWSETT, H.M.: 'Wireless telephony and broadcasting' (Gresham Publishing Co., London, 1924)

DUNSHEATH, P.: 'A history of electrical engineering' (Faber and Faber, London, 1962)

EVERSON, G.: 'The story of television, the life of Philo T. Farnsworth' (Norton, New York, 1949)

FAGEN, M.D. (Ed.): 'A history of engineering and science in the Bell System' (Bell telephone Laboratories, 1975)

FAHIE, J.J.: 'A history of electric telegraphy to the year 1837' (F.N. Spon, London, 1884)

FAHIE, J.J.: 'A history of wireless telegraphy' (Blackwood, Edinburgh, 1899)

FLEMING, J.A.: 'The principles of electric wave telegraphy' (Longmans, Green, London, 1908)

FORBES, R.J.: 'Studies in ancient technology' (E.J. Brill, Leiden, 1955)

GAMBLE, J.: 'Essay on the different modes of communicating by signal' (London, 1797)

GILBERT, M.: 'The First World War: a complete history' (Henry Holt, New York, 1994)

GORHAM, M.: 'Broadcasting and television since 1900' (Dakers, London, 1952)

GUNTHER, R.T.: 'Early science at Oxford' (Dawsons, London, 1966)

HARLOW, A.F.: 'Old wires and new waves' (Arno Press, New York, 1971, reprint edition)

HARTCUP, G.: 'The war of invention' (Brassey's Defence Publishers, London, 1988)

HATSCHEK, P.: 'Electron optics' (American Photographic, Boston, 1948)

HERTZ, H.: 'Electric waves, being researches on the propagation of electric action with finite velocity through space' (MacMillan, New York, 1893)

HERTZ, J.: 'Heinrich Hertz, memoirs, letters and diaries', revised edition prepared by M. Hertz and C. Susskind (Eds.), (San Francisco Press, and Physik Verlag, GmbH, 1977)

HEZLET, A.: 'The electron and sea power' (Peter Davies, London, 1975)

HIGHTON, E.: 'The electric telegraph: its history and progress' (London, 1852)

HUBBARD, G.: 'Cooke and Wheatstone and the invention of the electric telegraph' (Routledge and Kegan Paul, London, 1965)

ISRAEL, P.: 'Edison. A life of invention' (John Wiley, New York, 1998)

JACOT, B.L., and COLLIER, D.M.B.: 'Marconi – Master of Space' (Hutchinson, London, undated)

JOLLY, W.P.: 'Sir Oliver Lodge. Psychical researcher and scientist' (Constable, London, 1974)

KIEVE, J.L.: 'The electric telegraph. A social and economic history' (David and Charles, Newton Abbot, 1973)

KINGSFORD, P.W.: 'Electrical engineers and workers' (Arnold, London, 1969)

KORN, A., and GLATZEL, B.: 'Handbuch der Phototelegraphie und Teleautographie' (Nemnich, Leipzig, 1911)

LARDNER, D.: 'Electric telegraph', revised and rewritten by E.B. Bright (London, 1867)

LODGE, O.: 'Advancing science' (Harcourt Brace, New York, 1931)

LODGE, O.J.: 'Past years – An Autobiography' (Hodder and Stroughton, London, 1931)

LODGE, O.J.: 'The work of Hertz and his successors' (Electrician Printing and Publishing Co., London, 1894)

MacLAURIN, W.R.: 'Invention and innovation in the radio industry' (MacMillan, New York, 1949)

MARTIN, M.J.: 'Wireless transmission of photographs' (Wireless Press, London, 1916)

MARTIN, M.J.: 'The electrical transmission of photographs' (Pitman, London, 1921)

MAXWELL, J.C.: 'Treatise on electricity and magnetism' (Clarendon Press, Oxford, 1873)

MILLAR, A.H.: 'James Bowman Lindsay and other pioneers of inventions' (MacLeod, Dundee, 1925)

MOTTELAY, P.F.: 'The bibliographical history of electricity and magnetism' (Griffin, London, 1922)

NORMAN, B.: 'Here's looking at you' (BBC and RTS, London, 1984)

OBERLIESEN, R.: 'Information, Daten und Signale' (Deutsches Museum, Rowohlt, 1982)

PARSONS, W.B.: 'Engineers and engineering in the Renaissance' (MIT Press, 1939)

PAWLEY, E.: 'BBC engineering 1922–1972' (BBC Publications, 1972)

PHILLIPS, V.J.: 'Early radio wave detectors' (IEE, London, 1980)

POCOCK, R.: 'The early British radio industry' (Manchester University Press, 1988)

POCOCK, R., and GARRATT, G.R.M.: 'The origin of marine radio' (HMSO, London, 1972)

POPHAM, H.: 'Telegraphic signal' (London, 1800)

PRESCOTT, G.B.: 'History, theory and present practice of the electric telegraph' (Trubner, London, 1860)

PRIESTLY, J.: 'History and present state of electricity' (London, 1767)

RAWLINSON, G.: 'The history of Herodotus' (London, 1889)

REID, J.D.: 'The telegraph in America, its founders, promoters and noted men' (Derby Bros., New York, 1879)

ROBIDA, A.: 'The XXth century, the conquest of the regions of the air' (Paris, 1884)

RONALDS, F.: 'Description of an electric telegraph and some other electrical apparatus' (London, 1823)

ROSSI, L.: 'Trajan's column and the Trajan wars' (Cornell University press, 1971)

ROUTLEDGE, R.: 'Discoveries and inventions of the nineteenth century' (George Routledge, London, 1903)

ROWLANDS, P., and WILSON, P.J. (Eds.).: 'Oliver Lodge and the invention of radio' (PD Publications, Liverpool, 1994)

RUHMER, E.: 'Wireless telephony' trans. from the German by J. Erskine Murray, with an appendix on R.A. Fessenden by the translator (Crosby Lockwood, London, 1904)

RYDER, J.D., and FINK, D.G.: 'Engineers and electrons; a century of electrical progress' (IEEE Press, New York, 1984)

SABINE, R.: 'The history and progress of the electric telegraph' (London, 1867)

SARTON, G.: 'A history of science' (Harvard University Press, 1959)

SHIERS, G.: 'Early television. A bibliographic guide to 1940' (Garland Publishing, New York, 1997)

SOLYMAR, L.: 'Getting the message' (Oxford University Press, 1999)

STANDAGE, T.: 'The Victorian internet' (Weidenfeld and Nicolson, London, 1998)

STURMEY, S.G.: 'The economic development of radio' (Duckworth, London, 1952)

SUN TZU.: 'Ping Fa', English translation and introduction by R.T. Ames, (Ballantine Books, New York, 1993)

THOMPSON, R.L.: 'Wiring a continent: history of the telegraph industry in the United States, 1832–1866' (Princeton University Press, 1947)

THOMPSON, S.P.: 'The life of Lord Kelvin' (Macmillan, London, 1910)

THROWER, K.R.: 'History of the British radio valve to 1940' (MMA International, England, 1992)

TOLSTOY, J.: 'James Clerk Maxwell' (Canongate, Edinburgh, 1981)

TURNBULL, L.: 'The electromagnetic telegraph' (A. Hart, Philadelphia, 1853)

TYNE, G.F.J.: 'Saga of the vacuum tube' (H.W. Sams, Indianopolis, 1977)

URBANITZKY, A.R.von.: 'Electricity in the service of man' (Cassell, London, 1886)

VERRALL, A.W.: 'The Agamemnon of Aeschylus' (MacMillan, London, 1904)

VYVYAN, R.N.: 'Wireless over thirty years' (George Routledge, London, 1933)

WATSON-WATT, R.A.: 'Three steps to victory' (Odhams Press, London, 1957)

WILKINS, J.: 'Mercury, or the secret and swift messenger, showing how a man, with privacy and speed, may communicate his thoughts to a friend at any distance' (London, 1641)

WILSON, G.: 'The old telegraphs' (Phillimore, London, 1976)

WORCESTER, Marquess of.: 'Century of the names and scantlings of such inventions as at present I can call to mind to have tried and perfected' (London, 1666)

WYLDE, J. (Ed.).: 'The industries of the world' (London Printing and Publishing Company, c. 1886)

YOUNG, P.: 'Power of speech' (George Allen, London, 1983)

Index

Printed in the United Kingdom
by Lightning Source UK Ltd.
105217UKS00001B/1-21